MECHANICAL ENGINEERING

A Series of Textbooks and Reference Books

Editor

L. L. Faulkner

Columbus Division, Battelle Memorial Institute
and Department of Mechanical Engineering
The Ohio State University
Columbus, Ohio

1. *Spring Designer's Handbook*, Harold Carlson
2. *Computer-Aided Graphics and Design*, Daniel L. Ryan
3. *Lubrication Fundamentals*, J. George Wills
4. *Solar Engineering for Domestic Buildings*, William A. Himmelman
5. *Applied Engineering Mechanics: Statics and Dynamics*, G. Boothroyd and C. Poli
6. *Centrifugal Pump Clinic*, Igor J. Karassik
7. *Computer-Aided Kinetics for Machine Design*, Daniel L. Ryan
8. *Plastics Products Design Handbook, Part A: Materials and Components; Part B: Processes and Design for Processes*, edited by Edward Miller
9. *Turbomachinery: Basic Theory and Applications*, Earl Logan, Jr.
10. *Vibrations of Shells and Plates*, Werner Soedel
11. *Flat and Corrugated Diaphragm Design Handbook*, Mario Di Giovanni
12. *Practical Stress Analysis in Engineering Design*, Alexander Blake
13. *An Introduction to the Design and Behavior of Bolted Joints*, John H. Bickford
14. *Optimal Engineering Design: Principles and Applications*, James N. Siddall
15. *Spring Manufacturing Handbook*, Harold Carlson
16. *Industrial Noise Control: Fundamentals and Applications*, edited by Lewis H. Bell
17. *Gears and Their Vibration: A Basic Approach to Understanding Gear Noise*, J. Derek Smith
18. *Chains for Power Transmission and Material Handling: Design and Applications Handbook*, American Chain Association
19. *Corrosion and Corrosion Protection Handbook*, edited by Philip A. Schweitzer
20. *Gear Drive Systems: Design and Application*, Peter Lynwander
21. *Controlling In-Plant Airborne Contaminants: Systems Design and Calculations*, John D. Constance
22. *CAD/CAM Systems Planning and Implementation*, Charles S. Knox
23. *Probabilistic Engineering Design: Principles and Applications*, James N. Siddall
24. *Traction Drives: Selection and Application*, Frederick W. Heilich III and Eugene E. Shube
25. *Finite Element Methods: An Introduction*, Ronald L. Huston and Chris E. Passerello

KINEMATICS AND DYNAMICS OF MACHINERY

26. *Mechanical Fastening of Plastics: An Engineering Handbook*, Brayton Lincoln, Kenneth J. Gomes, and James F. Braden
27. *Lubrication in Practice: Second Edition*, edited by W. S. Robertson
28. *Principles of Automated Drafting*, Daniel L. Ryan
29. *Practical Seal Design*, edited by Leonard J. Martini
30. *Engineering Documentation for CAD/CAM Applications*, Charles S. Knox
31. *Design Dimensioning with Computer Graphics Applications*, Jerome C. Lange
32. *Mechanism Analysis: Simplified Graphical and Analytical Techniques*, Lyndon O. Barton
33. *CAD/CAM Systems: Justification, Implementation, Productivity Measurement*, Edward J. Preston, George W. Crawford, and Mark E. Coticchia
34. *Steam Plant Calculations Manual*, V. Ganapathy
35. *Design Assurance for Engineers and Managers*, John A. Burgess
36. *Heat Transfer Fluids and Systems for Process and Energy Applications*, Jasbir Singh
37. *Potential Flows: Computer Graphic Solutions*, Robert H. Kirchhoff
38. *Computer-Aided Graphics and Design: Second Edition*, Daniel L. Ryan
39. *Electronically Controlled Proportional Valves: Selection and Application*, Michael J. Tonyan, edited by Tobi Goldoftas
40. *Pressure Gauge Handbook*, AMETEK, U.S. Gauge Division, edited by Philip W. Harland
41. *Fabric Filtration for Combustion Sources: Fundamentals and Basic Technology*, R. P. Donovan
42. *Design of Mechanical Joints*, Alexander Blake
43. *CAD/CAM Dictionary*, Edward J. Preston, George W. Crawford, and Mark E. Coticchia
44. *Machinery Adhesives for Locking, Retaining, and Sealing*, Girard S. Haviland
45. *Couplings and Joints: Design, Selection, and Application*, Jon R. Mancuso
46. *Shaft Alignment Handbook*, John Piotrowski
47. *BASIC Programs for Steam Plant Engineers: Boilers, Combustion, Fluid Flow, and Heat Transfer*, V. Ganapathy
48. *Solving Mechanical Design Problems with Computer Graphics*, Jerome C. Lange
49. *Plastics Gearing: Selection and Application*, Clifford E. Adams
50. *Clutches and Brakes: Design and Selection*, William C. Orthwein
51. *Transducers in Mechanical and Electronic Design*, Harry L. Trietley
52. *Metallurgical Applications of Shock-Wave and High-Strain-Rate Phenomena*, edited by Lawrence E. Murr, Karl P. Staudhammer, and Marc A. Meyers
53. *Magnesium Products Design*, Robert S. Busk
54. *How to Integrate CAD/CAM Systems: Management and Technology*, William D. Engelke
55. *Cam Design and Manufacture: Second Edition*; with cam design software for the IBM PC and compatibles, disk included, Preben W. Jensen
56. *Solid-State AC Motor Controls: Selection and Application*, Sylvester Campbell
57. *Fundamentals of Robotics*, David D. Ardayfio

58. *Belt Selection and Application for Engineers*, edited by Wallace D. Erickson
59. *Developing Three-Dimensional CAD Software with the IBM PC*, C. Stan Wei
60. *Organizing Data for CIM Applications*, Charles S. Knox, with contributions by Thomas C. Boos, Ross S. Culverhouse, and Paul F. Muchnicki
61. *Computer-Aided Simulation in Railway Dynamics*, by Rao V. Dukkipati and Joseph R. Amyot
62. *Fiber-Reinforced Composites: Materials, Manufacturing, and Design*, P. K. Mallick
63. *Photoelectric Sensors and Controls Selection and Application*, Scott M. Juds
64. *Finite Element Analysis with Personal Computers*, Edward R. Champion, Jr., and J. Michael Ensminger
65. *Ultrasonics: Fundamentals, Technology, Applications: Second Edition, Revised and Expanded*, Dale Ensminger
66. *Applied Finite Element Modeling: Practical Problem Solving for Engineers*, Jeffrey M. Steele
67. *Measurement and Instrumentation in Engineering: Principles and Basic Laboratory Experiments*, Francis S. Tse and Ivan E. Morse
68. *Centrifugal Pump Clinic: Second Edition, Revised and Expanded*, Igor J. Karassik
69. *Practical Stress Analysis in Engineering Design: Second Edition, Revised and Expanded*, Alexander Blake
70. *An Introduction to the Design and Behavior of Bolted Joints: Second Edition, Revised and Expanded*, John H. Bickford
71. *High Vacuum Technology: A Practical Guide*, Marsbed H. Hablanian
72. *Pressure Sensors: Selection and Application*, Duane Tandeske
73. *Zinc Handbook: Properties, Processing, and Use in Design*, Frank Porter
74. *Thermal Fatigue of Metals*, Andrzej Weronski and Tadeusz Hejwowski
75. *Classical and Modern Mechanisms for Engineers and Inventors*, Preben W. Jensen
76. *Handbook of Electronic Package Design*, edited by Michael Pecht
77. *Shock-Wave and High-Strain-Rate Phenomena in Materials*, edited by Marc A. Meyers, Lawrence E. Murr, and Karl P. Staudhammer
78. *Industrial Refrigeration: Principles, Design and Applications*, P. C. Koelet
79. *Applied Combustion*, Eugene L. Keating
80. *Engine Oils and Automotive Lubrication*, edited by Wilfried J. Bartz
81. *Mechanism Analysis: Simplified and Graphical Techniques, Second Edition, Revised and Expanded*, Lyndon O. Barton
82. *Fundamental Fluid Mechanics for the Practicing Engineer*, James W. Murdock
83. *Fiber-Reinforced Composites: Materials, Manufacturing, and Design, Second Edition, Revised and Expanded*, P. K. Mallick
84. *Numerical Methods for Engineering Applications*, Edward R. Champion, Jr.
85. *Turbomachinery: Basic Theory and Applications, Second Edition, Revised and Expanded*, Earl Logan, Jr.
86. *Vibrations of Shells and Plates: Second Edition, Revised and Expanded*, Werner Soedel

87. *Steam Plant Calculations Manual: Second Edition, Revised and Expanded*, V. Ganapathy
88. *Industrial Noise Control: Fundamentals and Applications, Second Edition, Revised and Expanded*, Lewis H. Bell and Douglas H. Bell
89. *Finite Elements: Their Design and Performance*, Richard H. MacNeal
90. *Mechanical Properties of Polymers and Composites: Second Edition, Revised and Expanded*, Lawrence E. Nielsen and Robert F. Landel
91. *Mechanical Wear Prediction and Prevention*, Raymond G. Bayer
92. *Mechanical Power Transmission Components*, edited by David W. South and Jon R. Mancuso
93. *Handbook of Turbomachinery*, edited by Earl Logan, Jr.
94. *Engineering Documentation Control Practices and Procedures*, Ray E. Monahan
95. *Refractory Linings Thermomechanical Design and Applications*, Charles A. Schacht
96. *Geometric Dimensioning and Tolerancing: Applications and Techniques for Use in Design, Manufacturing, and Inspection*, James D. Meadows
97. *An Introduction to the Design and Behavior of Bolted Joints: Third Edition, Revised and Expanded*, John H. Bickford
98. *Shaft Alignment Handbook: Second Edition, Revised and Expanded*, John Piotrowski
99. *Computer-Aided Design of Polymer-Matrix Composite Structures*, edited by Suong Van Hoa
100. *Friction Science and Technology*, Peter J. Blau
101. *Introduction to Plastics and Composites: Mechanical Properties and Engineering Applications*, Edward Miller
102. *Practical Fracture Mechanics in Design*, Alexander Blake
103. *Pump Characteristics and Applications*, Michael W. Volk
104. *Optical Principles and Technology for Engineers*, James E. Stewart
105. *Optimizing the Shape of Mechanical Elements and Structures*, A. A. Seireg and Jorge Rodriguez
106. *Kinematics and Dynamics of Machinery*, Vladimír Stejskal and Michael Valášek
107. *Shaft Seals for Dynamic Applications*, Les Horve

Additional Volumes in Preparation

Heat Exchanger Design Handbook, T. Kuppan

Mechanical Engineering Software

Spring Design with an IBM PC, Al Dietrich

Mechanical Design Failure Analysis: With Failure Analysis System Software for the IBM PC, David G. Ullman

To our sons

Vladimír and Pavel

Jan and Michael

KINEMATICS AND DYNAMICS OF MACHINERY

VLADIMÍR STEJSKAL
MICHAEL VALÁŠEK
Czech Technical University
Prague, Czech Republic

Marcel Dekker, Inc. New York • Basel • Hong Kong

Library of Congress Cataloging-in-Publication Data

Stejskal, Vladimír.
Kinematics and dynamics if machinery / Vladimír Stejskal, Michael Valášek ; translated by Edvard Nichtburger.
p. cm. — (Mechanical engineering ; 106)
Translated from Czech.
Includes bibliographical references and index.
ISBN 0-8247-9731-0 (hardcover : alk. paper)
1. Machinery, Kinematics of. 2. Machinery, Dynamics of. I. Valášek, Michael. II. Title. III. Series: Mechanical engineering (Marcel Dekker, Inc.) ; 106.
TJ175.S77 1996 96-13969
621.8'11—dc20 CIP

All statements, opinions, recommendations, technical data and information contained herein are based on tests and experience that the authors believe to be reliable, but their accuracy or freedom from error or omission is not guaranteed. Marcel Dekker, Inc., and the authors make no warranty with regard to the MASS software, its accuracy, or its suitability for any purpose other than that specified in the text. This software is licensed solely on an "as is" basis. The only warranty made with respect to the MASS software diskette is that the diskette medium on which the software is recorded is free of defects. Marcel Dekker, Inc., will replace a diskette found to be defective if such defect is not attributable to misuse by the purchaser or his agent. The defective diskette must be returned within 10 days to:

Customer Service
Marcel Dekker, Inc.
Cimarron Rd., P.O. Box 5005
Monticello, NY 12701

Comments or suggestions, or bug reports concerning the MASS software are welcome and should be sent to:

Professor Vladimír Stejskal or Professor Michael Valášek
stejskal@fsik.cvut.cz valasek@fsik.cvut.cz

Department of Mechanics
Faculty of Mechanical Engineering
Czech Technical University in Prague
Karlovo nam. 13, 121 35 Praha 2
Czech Republic

The publisher offers discounts on this book when ordered in bulk quantities. For more information, write to Special Sales/Professional Marketing at the address below.

This book is printed on acid-free paper.

MARCEL DEKKER, INC.
270 Madison Avenue, New York, New York 10016

Current printing (last digit):
10 9 8 7 6 5 4 3 2 1

PRINTED IN THE UNITED STATES OF AMERICA

Preface

Mechanisms have been and continue to be the basis of all machines. The previously strictly mechanical materialization of the whole machine function is at present substituted to a varying extent by hydraulic, electric and electronics subsystems controlled by computer, i.e. now so-called mechatronical systems. Nevertheless, in order to act with forces or to handle objects, mechanisms are used. For the design and control of these machines the knowledge of kinematic and dynamic properties of these machines is crucial. The use of computers in this technical field has made it possible to solve complex real problems which had been unsolvable using traditional methods. The book summarizes the state of the art of so-called multibody system kinematics and dynamics, which enable the solution or problems pertaining to kinematics and dynamics of machinery. The book describes methods forming the basis of one part of Computer Aided Engineering (CAE) or Computational Mechanics, namely automated analysis and synthesis of multibody systems (MBS).

This book is unique in the description of all present relevant approaches to multibody systems and in their comparative evaluation. There are several books dealing with multibody systems. However, each of them describes only the particular author's method, without comparison to other possible approaches. The unique feature of this book and its main distinction from other books is that it presents not only the authors' method of solution but also a survey and comparison of the most important methods. From the authors' point of view this is the main advantage of this book. The reader can study and compare all the available methods from a unified point of view and in just one book, without having to study many other books.

The book will assist all those interested in the design of mechanisms such as robots, manipulators, building machines, textile machines, vehicles, airplanes, satellites, ships, biomechanical systems (vehicle crash simulators, barrier tests, human motion studies, etc.), controlled mechanical systems, mechatronical devices and many others. Readers will get an overwiev of suitable approaches to kinematic and dynamic problems of multibody systems, methods of solution, and their software implementation. This will enable them to design complex spatial mechanisms using computers. This book may serve as a textbook to provide the necessary mechanical foundation to robotics, vehicle dynamics, mechatronics, biomechanics or general machinery.

The general approach is characterized by a systematic explanation of basic concepts of kinematics and dynamics of multibody systems and, based on them, a comparative explanation of all advanced techniques using computational mechanics. From this point of view all different concepts of mechanics can ultimately be formulated in matrices.

The treatment of kinematic methods begins with the traditional vector approach and proceeds to one that stresses matrix methods as the most suitable and the most computationally efficient. The other unique feature of the approach used is the systematic attention to the computational complexity of different methods of solution. CPU time remains an important consideration for solving problems. Many other topics such as the possibility of body orientation description, including singular cases, are treated in detail.

The general approach to the dynamics is based on the equivalence between Newton–Euler and Lagrange equations of motion. All other approaches are explained as special cases of this equivalence because of particular choice of coordinates, introduction of substitutions, different notations, etc. All described methods are compared from the point of computational efficiency, and from this point of view special attention is devoted to nontraditional approaches such as recursive or residual ones. Not only are methods of analysis treated but methods of synthesis and optimization as well. The book concludes with a brief chapter on further topics for research.

This book is appropriate as a text for senior or postgraduate courses in mechanical engineering dealing with subjects of analysis and synthesis of mechanisms, fundamentals of robotics, vehicle dynamics, mechatronics and multibody modelling of different systems. It is realistically possible to cover all essential contents of the book in two semesters, one devoted to kinematics and the other to dynamics. A basic knowledge of mechanics and calculus is assumed.

The book may also attract the attention of practical engineers and researchers in the field of machine design and dynamics, and also in the field of biomechanics and mechatronics. They will find methods and algorithms for the solution of their everyday problems and explanation of algorithms on which commercial software packages are based. An understanding of these multibody system methods is important in order to use the software packages correctly and to avoid the possibility of failure due to peculiar and special cases in general theory.

Acknowledgments

We would like to pay our great respects to the memory of our late teacher Dr. Vladimír Brát, who further developed the matrix method, thereby making possible the routine solution of spatial mechanisms. The results of his work are summarized in his book (Brát, 1981). We continue with his work and further develop his results.

We gratefully acknowledge the support and inspiration to write this book given to us by Professor Michael J. Pratt and Professor Karel Juliš.

Finally, the authors should like to thank to their wives and parents for their support, encouragement and patience, without which this book could never have appeared.

Vladimír Stejskal and Michael Valášek

Contents

Preface .... *iii*

Chapter 1 Introduction 1

Chapter 2 Mechanism Design and Computer Aided Design (CAD) 5

2.1 The Nature of Engineering Design 5
2.2 The Design of Mechanisms and Machines 10
2.3 What May Be Left to a Computer 12
2.4 Principles of CAD 14
2.4.1 The Method of Successive Approximations 14
2.4.2 Hierarchical Structure of Models 14
2.4.3 Transformation of Models 16
2.4.4 Man-Machine Dialogue — A Palette of Tools 16
2.4.5 Knowledge Engineering 17
2.4.6 Structure of CAD Software 17
2.5 The Problem of Automated Analysis and Synthesis in CAD of Spatial Mechanisms 18

Chapter 3 Structure Analysis of Mechanical Systems 21

3.1 Kinematic Pairs 21
3.2 Kinematic Chains and System Groups 22
3.3 Mechanisms 29
3.4 Compartments 30
3.5 Structural Description of an MBS 31

3.5.1 MBS Formula 31
3.5.2 Application of Graph Theory 32
3.5.3 Graph Manipulation Language — GRM 42

Chapter 4 Description of Point and Body Motion 45

4.1 Vector Description 46
4.1.1 Point and Body Motion 46
4.1.2 Simultaneous Motion 51
4.1.3 Vector Quantities, Expressions and Transformations 56
4.1.4 Homogeneous Coordinates 60
4.2 Matrix Notation 60
4.2.1 Point and Body Motion 61
4.2.1.1 Position 61
4.2.1.2 Velocity 63
4.2.1.3 Acceleration 66
4.2.2 Inverse Motion 68
4.2.3 Simultaneous Motion 70
4.2.3.1 Position 70
4.2.3.2 Velocity 72
4.2.3.3 Acceleration 73
4.2.3.4 Notes 75
4.2.4 Other Transformations 75
4.3 Quaternion Notation 77
4.3.1 Introduction of Quaternions 78
4.3.2 Dual Quaternions 82
4.3.3 Shortened Dual Quaternions 84
4.4 Screw Notation 88
4.5 Basic Motions 94
4.5.1 Basic Matrices 94
4.5.2 Permutation Matrices 101
4.5.3 Basic Quaternions 103
4.5.4 Basic Screws 104
4.6 Computational Complexity of Different Notations 105

4.7 Special Cases ... 107
4.7.1 Spherical Motion ... 107
4.7.2 Planar Motion ... 107

Chapter 5 Kinematic Analysis of MBSs with Lower Kinematic Pairs ... 109

5.1 Lower Kinematic Pairs ... 109
5.1.1 Revolute Pair ... 111
5.1.2 Prismatic Pair ... 112
5.1.3 Cylindrical Pair ... 113
5.1.4 Helical Pair ... 113
5.1.5 Spherical Pair ... 114
5.1.5.1 Euler Angles ... 115
5.1.5.2 Cardan Angles ... 120
5.1.5.3 Roll-Pitch-Yaw ... 123
5.1.5.4 Euler Parameters ... 124
5.1.6 Flat Pair ... 128
5.2 Simple Open Kinematic Chains ... 129
5.3 Applications to Kinematic Chains ... 130
5.4 Simple Mechanisms ... 133
5.4.1 Matrix Methods of Solving the Position Problem 134
5.4.1.1 The Closed Loop Method ... 134
5.4.1.2 The Disconnected Loop Method ... 135
5.4.2 The Matrix Method of Determining Velocities and Accelerations ... 136
5.4.3 The Solution of General Point and Body Motion 137
5.5 Applications to Simple Mechanisms ... 138
5.5.1 A Universal Joint ... 138
5.5.2 RRSRR Mechanism ... 144
5.6 Composed Mechanisms ... 147
5.7 Applications to Composed Mechanisms ... 147
5.7.1 A Wheel Suspension Mechanism ... 147
5.7.2 A Robot Mechanism ... 149
5.8 The Basic Matrix Method ... 152
5.9 Planar and Spherical Mechanisms ... 155

5.10 The Transmission Matrix (Velocity Transformation Matrix) ... 155
5.11 Other Approaches ... 158
5.11.1 Denavit-Hartenberg's Method ... 158
5.11.2 The Generalized Vector Method ... 161
5.11.3 The Generalized Method of Disconnecting a Loop ... 164
5.11.4 The Natural Coordinates Method ... 167
5.11.5 The Compartment Method in Kinematics ... 170
5.11.6 The Geometric Diagram and the Parametric Method in Kinematics ... 172
5.12 Numerical Solution ... 177
5.13 Analytical Solution ... 182

Chapter 6 MBSs with Higher Kinematic Pairs ... 197

6.1 Mechanisms with a General (Smooth) Pair ... 198
6.2 Applications ... 204
6.2.1 A Spatial Parabolic Cam ... 204
6.2.2 A Tripod Driving Device ... 206
6.3 Mechanisms with a General Roll Pair ... 208
6.4 Applications to MBSs with Roll Pairs 209
6.4.1 A Planar Tyre ... 209
6.4.2 A Mobile Robot ... 211

Chapter 7 Overconstrained and Singular Cases of MBSs ... 213

7.1 Mathematical Background ... 214
7.2 Physical Interpretation in Kinematics ... 219
7.3 Applications ... 223

Chapter 8 Kinetostatic Analysis of MBSs ... 231

8.1 General Solution of the Problem ... 231
8.2 The Parametric Method in Statics ... 240

8.3 MBS Static Analysis: The Equilibrium Position 242
8.4 Applications 248
8.4.1 Universal Joint Kinetostatics 248
8.4.2 The Equilibrium Position of a System of Bars .253
8.4.3 The Equilibrium Position of a Wheel Suspension Mechanism 254
8.4.4 RRSRR Mechanism Kinetostatics by the Parametric Method 255

Chapter 9 MBS Dynamic Analysis 257

9.1 Multibody System Description 258
9.2 Basic Approaches to the Dynamics of MBSs 260
9.2.1 Approaches Based on Vector Mechanics 261
9.2.1.1 Newton-Euler Approach 261
9.2.1.2 D'Alembert's Principle 265
9.2.2 Approaches Based on Analytical Mechanics 267
9.2.2.1 The Principle of Virtual Work 268
9.2.2.2 Jourdain's Principle 268
9.2.2.3 Gauss's Principle 269
9.2.2.4 Hamilton's Principle 269
9.2.2.5 Lagrange's Equations 270
9.2.2.6 Hamilton's Canonical Equations 271
9.3 Lagrange's Equations of Mixed Type 272
9.3.1 Lagrange's Equations Written in Quasi-coordinates 277
9.4 Equivalence between Vector and Analytical Approaches 281
9.5 Various Multibody Formalisms 289
9.5.1 Non-recursive Multibody Formalisms 290
9.5.1.1 NEWEUL Approach 290
9.5.1.2 MESA VERDE Approach 294
9.5.1.3 Kane's Approach 298
9.5.1.4 ADAMS Approach 301
9.5.1.5 DADS Approach 303
9.5.1.6 Natural Coordinates Approach 304

9.5.1.7 Composite Rigid-Body Approach 309
9.5.1.8 Inverse Multibody Formalism 318
9.5.1.9 Residual Algorithm 318
9.5.2 Recursive Multibody Formalisms 319
9.5.2.1 Open-Loop Systems 320
9.5.2.2 Closed-Loop Systems 323
9.5.3 Computational Complexity of Multibody Formalisms 332
9.6 Physical Interpretation of Lagrange Multipliers 335
9.7 The Compartment Method in Dynamics 346
9.7.1 Compartment Model of an Arbitrary Physical System 347
9.7.2 General Transformation of a Compartment into a Macromodel Satisfying Kirchhoff's Laws 350
9.8 Numerical Solution 353
9.8.1 Solution in Physical Coordinates 356
9.8.1.1 Baumgarte Stabilization 356
9.8.1.2 Different Variational Formulation 358
9.8.1.3 DAE Solvers 360
9.8.2 Solution in Independent Coordinates 361
9.8.2.1 Projection Methods of Determining **R** 363
9.8.2.2 Orthogonalizing Methods of Determining **R** 366
9.9 Change of MBS Structure in the Course of Motion .. 368
9.10 MBS with Flexible Elements 375
9.11 MBS with Feedback Control 383
9.12 Applications 383
9.12.1 Car Front Wheel Trapezoidal Suspension ... 383
9.12.2 Car Axle on Passing a Surface Unevenness... 386
9.12.3 New Car Wheel Mechanism 387

Chapter 10 General Computer Systems for MBS Analysis Automation 389

10.1 System KIN 392
10.2 System OKAM2 392
10.3 System DYN 395

10.4 System ODAM 399

Chapter 11 Accuracy and Sensitivity of MBSs ... 405

11.1 Kinematic Accuracy 406
11.1.1 Kinematic Accuracy of a Simple Open Chain 406
11.1.2 Kinematic Accuracy of a Simple Closed Chain 412
11.1.3 Kinematic Accuracy of Mixed Chains 415
11.1.4 Evaluation of Kinematic Accuracy and Tolerance Spaces 416
11.1.5 Calibration for Kinematic Errors 420
11.2 Dynamic Accuracy 422
11.3 Applications 428
11.3.1 Kinematic Accuracy of a Robot 428
11.3.2 Feeding Machine Dynamic Accuracy 432

Chapter 12 MBS Synthesis and Optimization 435

12.1 Synthesis of Transmission Mechanisms 439
12.2 Synthesis of Guiding Mechanisms 444
12.3 Applications 444
12.3.1 A Planar Guiding Mechanism 444
12.3.2 Car Front Wheel Suspension 452
12.4 Dynamic Synthesis 452

Chapter 13 Actual Research Problems and Further Developments 463

References 469

Index 489

Chapter 1

Introduction

Mechanisms have been and continue to be the basis of all machines. The previously strictly mechanical realization of the whole machine function is at present to a varying extent substituted by hydraulic, electric and electronic subsystems. Nevertheless, in order to act with force or to handle objects, mechanisms are used; these may be controlled by microprocessors, as in the case of industrial robots, cars with semiactive suspension and antiblocking devices.

For the design and control of these machines, methods originating in the era of classical mechanics and especially in the 19th century served their purpose until the end of 1950's. The methods of traditional mechanics might have been considered a closed subject at that time. The use of computers in this technical field has made it possible to solve complex real problems which had been unsolvable using traditional, often graphical, methods. It has become necessary to develop new methods of problem formulation and solution, as the classical procedure required individual construction of a mathematical model by a human being, which for more complex problems reaches the limits of human capability. Since the 1960's a rigorous development of methods and formalisms of mechanics and a reevaluation of classical approaches has taken place.

Computers have had a similar effect in other engineering disciplines. The integration of such computer usage in engineering has led to the formation of integrated systems of Computer Aided Design (CAD), Computer Aided Manufacturing (CAM), Computer Aided Process Planning (CAPP), etc., or, speaking in general terms, of Computer Aided Engineering (CAE). This book describes methods forming the basis of one CAE system, namely the system of automated analysis and synthesis of mechanical systems, so-called multibody systems (MBS).

The principal aim of this book is to present the reader with a survey of the most important methods of mechanics of spatial mechanisms from the aspect of their suitability for computer processing. From this viewpoint the book is going to assist all those interested in the design of mechanisms, e.g. robots, cars, building machines, biomechanics. Chapter 2 presents the interconnection between methods of mechanics dealt with in subsequent chapters and general principles of CAD systems. Chapter 3 treats the structural properties of mechanisms. Chapter 4 deals with various approaches to the description of the state of motion of mechanisms used both in kinematics and dynamics. The basic procedures of kinematic solution of mechanisms with lower and higher kinematic pairs are introduced in Chapters 5 and 6. These procedures form the basis of mechanism design. The topic of Chapter 7 is the analysis of so-called singular cases. In Chapter 8 the static analysis of mechanisms is solved. Chapter 9 is, after Chapter 5, the second most important chapter of the book. It contains the description of solution procedures of mechanism dynamics. Examples from actually existing CAE program systems based on methods presented in this book are given in Chapter 10. Chapters 11 and 12 deal with applications of these methods in mechanism synthesis. The closing Chapter 13 presents some topics for further development in the field of MBS mechanics.

In this book we use many different quantities and variables – scalars, vectors, matrices, tensors, quaternions, screws, dual quantities. We use a different designation for these quantities, but at the end we represent all of them by matrices.

A geometric vector is printed with an arrow, e.g. $\vec{r}$; its elements are written in parentheses, e.g. $\vec{r} = (r_1, r_2, ..., r_n)$. An algebraic vector is written using bolface letters, e.g. $\mathbf{r}$. Its elements are written in column vector matrix. A quaternion is designated by the "roof" above the letter, e.g. $\hat{\mathbf{r}}$. Other quantities are directly represented by matrices.

Matrices are designated by bold letters. Their elements are written in square brackets [ ] and are divided by commas only if it is necessary for the sake of clarity. Column matrices are written by the symbol ⌈ ⌉ or transposition T

$$\mathbf{r} = \begin{bmatrix} r_1 \\ r_2 \\ \vdots \\ r_n \end{bmatrix} = \lceil r_1, r_2, \ldots, r_n \rceil = [r_1, r_2, ..., r_n]^T$$

A diagonal matrix is written as a row matrix by means of the symbol ⌈ ⌋, e.g. a diagonal matrix $\mathbf{D}$ with elements $d_1, d_2, ..., d_n$ on its diagonal is written

$$\mathbf{D} = \begin{bmatrix} d_1 & & & \\ & d_2 & & \\ & & \ddots & \\ & & & d_n \end{bmatrix} = \lceil d_1, d_2, ..., d_n \rfloor$$

The transposition of a matrix is designated by the superscript T, the inversion by the superscript -1 and the determinant by the symbol det, e.g. $\mathbf{A}^T, \mathbf{A}^{-1}, \det \mathbf{A}$. The trace of matrix $\mathbf{A}$ is designated by $\mathrm{tr}\,\mathbf{A}$.

For the frequently used trigonometric functions sine and cosine we use the abbreviations s and c, e.g.

$$\begin{aligned} \mathrm{s}\varphi &= \sin\varphi \\ \mathrm{c}\varphi &= \cos\varphi \end{aligned}$$

and also

$$\begin{aligned} \mathrm{s}_1 &= \mathrm{s}\varphi_1 = \sin\varphi_1 \\ \mathrm{c}_1 &= \mathrm{c}\varphi_1 = \cos\varphi_1 \end{aligned}$$

Chapter 2

Mechanism Design and Computer Aided Design (CAD)

Mechanisms are the basis of most machines. Their analysis and synthesis is essential in mechanical engineering. In the following we shall deal with the methods of computer automation of these activities which are an integral part of machine design. The starting point for the development of methods of analysis and synthesis of mechanisms is therefore the general methodology of engineering design from the point of view of the creator of a CAD system.

In this Chapter we shall deal with both the classical and the computer aided general principles of CAD system activities and of the design of machines and mechanisms.

2.1 The Nature of Engineering Design

The nature of construction design is at present the subject of thorough study. In spite of the results obtained (Altshuller, 1984; Yoshikawa, 1983; Hubka, 1984) an integral theory of constructing does not yet exist. Let us therefore consider the evident and conscious steps in the work of an engineer. The construction design activities can be classified from various points of view (Valášek, 1988). Our approach is to interpret the constructing of a technological object as the creating of a system. It is a multilevel, hierarchical, iterative process. The respective hierarchical levels of this pro-

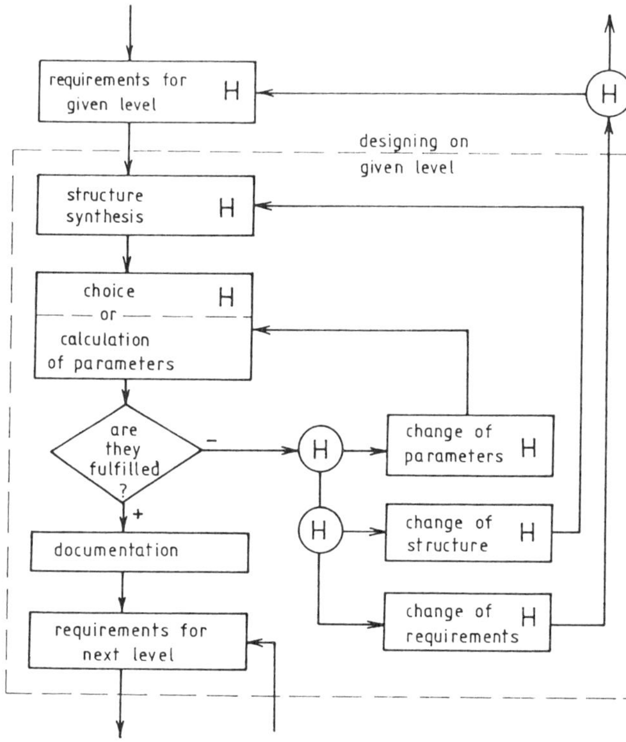

Figure 2.1. General model of design process.

cess correspond with system levels of the description of the created object. Because of its complexity, the whole process is decomposed into a series of partial processes, which follow each other and also influence each other through feedback. Each stage of the process has the same structure of fundamental steps, as shown in Figure 2.1.

In the first step the construction designer receives the specification of his task at the given level. At the beginning he has at his disposal his own knowledge of constructions already developed, whether from his own experience or from other sources, e.g. literature. He creates a classified and more precise version of the specification, checks its completeness and proves its consistency. The designer searches his own knowledge for similar principles and constructions relating to the specification. He either finds what is to be created or finds a similar construction that may be changed to suit the given specification. The change may be insignificant (e.g. change of size) or more essential (e.g. change of structure). Should he be unable to find a similar construction, he must use his knowledge for the structure synthesis of the created object. The structure synthesis of the created object is the second design step, even if it can be done through choice of a known structure of a previous construction. Structure represents the quality attributes of objects, which in most cases do not have the character of continuous variables; they are usually discrete. The structure synthesis of a created object includes in the case of a spatial mechanism: the choice of the kind of mechanism (linkage, cam), the number and kind of kinematic pairs, the number of elements, the choice of driving and driven elements, etc. The structure synthesis of created objects is exactly the kind of creative activity of human beings that has not yet been fully explored.

In the following steps the designer uses the empirical and theoretical knowledge of the appropriate sciences in relation to the created object on the criteria with which the object must comply to achieve its real existence, e.g. mobility, force balance, rigidity, strength, energy balance. In general, we speak of quality criteria with which the created object must comply. Not only the quality criteria, but also the algorithms of their evaluation on the objects, relate to the knowledge associated with the respective sciences. These algorithms make it possible to formulate most quality criteria mathematically as equalities or inequalities for functions of parameters of created objects of the given structure.

Parameters are defined as quantitatively expressible properties of the created object, e.g. size, mass, material properties, as well as dynamical and functional object characteristics, e.g. velocity, acceleration, deformation, stress, force, efficiency. It is, however, necessary to distinguish the parameters of the model which describes the created object from the parameters of the computational model of the created object. In the former case all quantitative properties are parameters; in the latter case those parameters consist of proper parameters, the values of which we know or choose, and of variables, the values of which are to be found or determined.

Not all quality criteria can be expressed in mathematical terms. We often have quality criteria which are evaluated in some other (non-mathematical, non-analytical) way, e.g. empirical quality criteria for car surfaces or blade surfaces, ergonomic or aesthetic criteria. The number of non-analytical criteria decreases steadily with the development of science, but many of them remain and their fundamental reducibility to mathematical relations (in the present sence of the term) has not yet been proved.

If the set of conditions for the analytical quality criteria uniquely determines the unknown parameters of the object, it is possible to invert the problem and, using the required quality criteria together with mathematical relations, determine the parameters of the created object. The designer usually treats partial cases in exactly the same way. With the development of mathematics and electronic data processing attempts came to develop the analytical approach to constructing and to creating methods of fully automated design of the created object from given requirements. Even if for a moment we leave aside the complicated question of solvability of the respective mathematical relations, we are still confronted with the fundamental problem of the relation between the number of criteria and parameters. In most cases the number of parameters is much higher than the number of mathematically expressible criteria. It is therefore impossible to carry out the automatic analytical construction because of lack of uniqueness of the solution of mathematical relations. However this lack of uniqueness allows the optimization of the object according to a certain viewpoint. Optimizing, i.e. selecting an extreme out of many variants, removes or reduces the lack of uniqueness. Thus using additional optimizing criteria we increase the original number of quality criteria for the object. We must, however, be aware of the difficulties and problems connected with expressing the criteria for optimization. We always have many and various demands on the created object and these, when singled out, can be expressed as optimization conditions with a clear mathematical formulation. This is not always the case when they are combined. To express mathematically the resulting combined criteria may be extremely difficult or impossible, and this would need to be solved by using an incomplete quasiheuristic criterion, the analytical formulation of which is difficult.

Another problem arises with the non–analytical quality criteria which cannot as yet be expressed mathematically. The situation is usually even more complicated with regard to the solvability of the given mathematical relations with the given computational methods on the given computer technology in an acceptable time.

To conclude, we may say that the designer determines the parameters of the created object either analytically (when it is really possible to determine inversely the necessary parameters from the quality criteria) or by a simple choice based on experience (or other criteria that are difficult to formulate); neither of the two methods may be left out. On the object so created, with chosen parameters, he then evaluates the quality criteria. If they do not hold, the designer changes the object parameters or even the structure of the whole specification.

It is difficult to describe exactly the circumstances in which the designer decides after a negative quality criterion evaluation to use qualitatively different approaches to change the created object. These are the approaches which are at present almost exclusively concerned with human factors. Several factors are significant — the degree of noncompliance with the quality criteria and the significance of affected criteria, the degree of exhaustion of possible ranges of changeable parameters, the number of previous changes and their success rate, experience with the given structure or with other, alternative structures, experience in realization of similar specification, etc. All these and other factors lead eventually to a decision to suggest the nature of a qualitative change, its range and contents. For the proposal to change parameters the designer uses his own experience or the calculated quality criteria sensitivities for each parameter. We thus get further steps in the designer's activity — the choice or calculation of object parameters, the evaluation of object quality criteria, or the proposal to change the parameter, structure or specification of the given created object.

The designer must be able to communicate his results to other people. The results of his work at the given construction level are the starting point for the next stage. We thus come to the last step of designer activities at the given construction level: documentation of the given level and specification for the next level. The first thing the specification for the next level must contain is the decomposition of the problem into elements of the lower level. The steps in the work of the designer with an indisputable human part for which we have no formal theory are designated in Figure 2.1 with H for "human".

The created object is always a complicated hierarchical system. The construction process always forms several system levels given by the relation: elements $\leftrightarrow$ system. An element is a part of a system, but on the other hand an element itself is a system composed of several elements and the system again is a part — an element — of a larger system. Connected with these are the levels given by the material properties of the substance from which the object is created and the levels given by the interaction of the created object with its environment. The construction process is carried out on each system level of the created object.

Considering construction as a system approach by object creation leads — just as any other system approach — to the problem of solution of the so-called system paradox (Sadovskij, 1974). The solution to the problem of creating a given system or understanding a given system is possible only if the problem of its creation or understanding as an element of a larger system is solved. This problem, however, can be solved only if the problem of its creation or understanding as a system has been solved. This double relation between the understanding of elements of a system and of a system of its elements appears at all levels of system description; it causes every system cognition to be, as a matter of principle, relative. Let us consider as an example "design a machine". The size of its parts will be derived from their loads. These in turn are affected by the mass of the machine parts which in turn is affected by the size of the parts. Yet these sizes are, to start with, not known.

The way out of this paradoxical situation lies in stepwise approximations using knowingly limited and incomplete ideas. From this starting point at one system level we derive our ideas of the next system level, and from here again backwards the ideas of the first system level. We shall be successful if the differences between the knowledge obtained in each step decrease. In an ideal case we shall get two successive steps with identical, and hence true, knowledge. This is at present the only known solution to the system approach paradox. We describe it as the method of successive approximations, whereby by repeating the process with the help of stepwise approximations, the originally imprecise knowledge tends to be more precise as it approaches its final form.

2.2 The Design of Mechanisms and Machines

The process of designing mechanisms and machines is identical with the general description of an engineering design above. Problem specification is given by the purpose and function of the machine together with the fact that it is a part of a larger system. Let us take, for example, a manipulator. Its purpose is to move a certain object. Its function as a machine can in most cases be expressed as the requirement to act with a certain force along a certain path in space, with a given precision in each case. A machine is nowadays no more a purely mechanical device. It can contain other elements — electric, hydraulic, pneumatic, electronic, etc. A frequent requirement is to secure the correct function of the machine even by variable and previously unknown outside influences. This requirement is usually met by a control system, e.g. using microprocessors. Up-to-date control systems of machines are also capable of substituting whole mechanical subsystems of a machine, which is very much of an advantage when there are sophisticated requirements on the machine function, e.g. on the movement of a machine work element. This approach may be of advantage also with regard to other viewpoints, e.g. machine weight, function flexibility, manufacturing complexity, etc. Such approach is called mechatronics from the words *mecha*nism or *mecha*nics and elec*tronics*.

According to the specification, the designer carries out the decomposition of functions on individual machine subsystems. For the mechanical subsystem he designs according to its function, appropriate kinematical structure and the essential sizes of the mechanism, in a way which is capable of realizing the required movement of the work element. From the given load and chosen kinematical structure he determines, for the time being, the forces acting on the mechanism. These are the basis for the provisional determination of mechanism element dimensions and for the choice of their material. On this basis the designer can start to carry out the successive dynamic and detailed strength computations of the machine.

During the whole process of designing the machine, situations may be encountered where the design for various reasons is unsuitable and it is necessary to repeat a variable number of previous design steps. The designer must comply in his design with a number of criteria, e.g. the realization of the proper function — transfer of forces, transformation of movement, accuracy, rigidity, physical realizability of the machine, strength, manufacturability and other criteria such as reliability, etc. The designer cannot describe in his calculations the processes in the real machine in their full generality. To be able to carry out the computations he must use simplifying assumptions, on the basis of which he creates the computational model of the designed machine. In order to be adequate this model must reflect substantial properties of real phenomena. Therefore the designer must check out the validity of the simplifying assumptions used during the design (e.g. rigidity, linear yield, plasticity, disregarding the velocity of elastic wave propagation velocity etc.). Assumptions must also be made concerning the interaction of the machine with its environment. The most important matter in this respect is the tuning of frequency properties of the machine environment system, e.g. resonance prevention, isolation of vibrations, etc.

There are many parameters in machine design, the values of which are not uniquely determined by the above criteria. On the other hand, we have several requirements for the machine design which do not follow a strict rule, e.g. minimum mass, frequency, energy consumption, price. The designer uses these criteria and the appearance of some free parameters to optimize the machine design.

In spite of all the development of science and technology, there are several criteria in the design process which either do not measure up to the complexity of reality or do not allow a strict mathematical description. For these and other reasons the role of the designer and his experience is indispensable.

The designer is capable of substituting a series of design steps and loops on the basis of his experience. In the process of designing he can also take into account other global criteria in a heuristic way, thus enabling a better embedding of the machine in the system above the machine and so better fulfilling its purpose. A good example is the approach of Wittenbauer (1923) with the criterion of smallest possible machine work irregularities and the studies of how the single parts affect the behaviour and the function of the machine as a whole. This fruitful approach was later developed by Benedict and Tesar (1978).

2.3 What May Be Left to a Computer

On the basis of the above analysis it is possible to decide which parts of the engineering design process are more or less suitable for computer realization or support.

On the one hand the human being — the designer — is indispensable in the process of creating an object in several design steps (in Figure 2.1 designated by H). This is especially the case with regard to the object structure synthesis, original choice of parameters or choice of parameters that cannot be determined from the quality criteria, choice of method (change of parameters, structure or specification), noncompliance with quality criteria and the problem of decomposition in partial levels (specification).

On the other hand there are many steps of the engineering design process which may be left to a computer. This is the case especially with regard to the computation of object parameters and evaluation of (analytical) quantity criteria of the created object. In the process of object creation a large quantity of structure and parameter values are formed and these must be efficiently stored, retrieved and amended. Thus the storing and editing of created object data is a computer realizable design step.

This activity (the documentation step in Figure 2.1) is spread out over the whole design process. The quantity of this data increases with the number of created object levels and construction stages. Whereas the capability to carry out fast and precisely voluminous calculations is essential for computations, the key point is the capability of precise and fast storage and precise and fast output of graphical documentation on graphic display units or plotters. A further advantage is the storage of constructions carried out previously, standards and other data necessary for the construction process.

A significant part of the construction documentation and its storage is the graphical documentation in the form of various types of charts. It is not necessary to have these drawings physically on paper; it is sufficient to store and transfer them via computer. Nevertheless, drawings are still an indispensable means of communication among designers, between designers and technologists, etc. Instead of charts and drawings we have files with defined protocols.

There are many steps in the engineering design process that are more or less capable of being expressed as algorithms. However, the form of their computer realization differs completely from classical computational procedures. This is the case with presenting human knowledge in its language or logical form. Also the way it is used is different. Instead of a strict determinative sentence we have a loosely-binding piece of language-formulated advice. We may thus classify the engineering procedures into three groups:

– The first group contains the procedures that may be completely described by algorithms and efficiently solved using present computer technology. These procedures may be completely given over to computers. These are the classical CAD algorithms.

– The second group contains procedures which give the results in a way which is, to a certain extent, similar to a human procedure. Such procedures can be realized on computers only by means of artificial intelligence (AI) and the user gets the result in the form of alternative recommendations. The choice is then left to him. These are, for example, the so-called expert systems.

– The third group contains procedures whose execution and result lie outside the range of what we can do using present computer technology.

The first group may be characterised as doing the work of a reckoner, librarian and draftsman; the second group represents an advisor.

A discussion of the impossibility of automatic object construction design by computers without human participation sometimes arises in connection with the above analysis. From the above considerations it is possible to give the essential reasons for the near impossibility of automatic construction (at least now and in the near future):

a) The necessity of using the method of successive approximations leads to construction procedures which are iterative in principle and which require the distinction at different system levels between the essential and the superficial; all of this lies for the time being in the human domain.

b) The non-existence of a calculus for the computation of object structure. We are able to calculate only object parameters, not object structure. We determine the object structure merely on the basis of previous experience. It is the mutual interdependence of structure and parameters on different system levels which causes engineering work to be an iterative process requiring the indispensable input of man and his experience.

c) Automatic construction would be possible only on the basis of determining construction parameters from the required quality criteria for the construction. This is not generally possible as we usually have to determine more parameters than the number of given conditions, and the problem does not have a unique solution. Furthermore, the determination of parameters would require the inverse solution of mathematical models. The inverse solution of partial and ordinary differential equations represents a so-called incorrect problem. The solution of incorrect problems is problematic (Tichonov and Arsenin, 1986). The inversion is generally possible only for mathematical models in the form of algebraic transcendental equations.

2.4 Principles of CAD

2.4.1 The Method of Successive Approximations

The method of successive approximations is the fundamental principle of designing and of CAD. Its importance is immense. The fundamental repetitive and iterative nature of designing and of CAD follows from this method, as well as the indispensable role of humans in the designing process because of their capability of recognizing in reality and knowledge the essential and the superficial. This is necessary for processing incomplete and limited ideas and is for the present AI techniques possible only with great difficulty.

2.4.2 Hierarchical Structure of Models

From the principle of successive approximations, it follows that many different ideas of an object have to be designed before the object design is complete. The most efficient way of manipulating such ideas is the use of models. Modelling is therefore very typical for any engineering activity. A further fundamental principle of designing and CAD is the systematic use of models at all stages of object creating.

We may distinguish (Valášek, 1988) two kinds of models: descriptive models and operational models. The former contain the information about the state of object development and about the knowledge of it. Examples are diagrams, drafts, drawings, charts, descriptions of technological procedures, manuals, etc.

The latter simulate the function of the created object or produce further data about the object. Examples are computational models (equations, simulation programs), numerical control (NC) programs, formal deduction by the inference engine of an expert system, etc.

The difference between the two kinds of models is clearly defined although at the same time subtle. The distinction is that descriptive models only keep data, but operational models generate new data about the object. For example, the record of an equation is a descriptive model, but the calculation according to it is an operational model.

The descriptive and operational models correspond to each other as a structure and a process of a system. Every operational model originates from some descriptive model and every descriptive model is created by some operational model. For example, an engineer describes the structure and parameters of a designed object, and from them the equations and simulations of object function are concluded. From simulations, further parameters or changes of object structure are concluded. Besides this mutual correspondence between descriptive and operational models, we can distin-

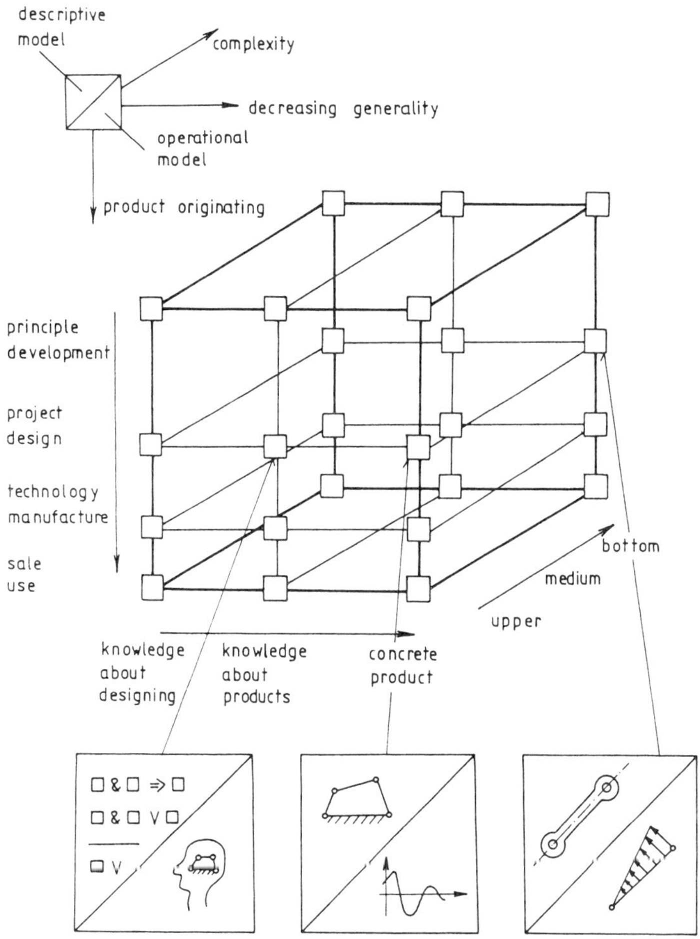

Figure 2.2. Hierarchical structure of models in CAD.

guish the relations of hierarchy given by different views — the complexity of particular models, the stages of object development and the generality of knowledge about the object being designed (Figure 2.2).

Among descriptive models we can distinguish topological models, which obtain only the structure of system element connections, and metric models, which obtain the complete spatial geometric shapes (i.e. solid models).

By the hierarchy of complexity among operational (computational) models we can distinguish between computational models of bottom level (partial differential equations), of medium level (ordinary differential equations) and of upper level (transcendental equations). Mutual correspondence of these models of a created object as well as different properties of direct and inverse problem solutions of these models can be described.

2.4.3 Transformation of Models

A further principle of CAD is the systematic use of possible mutual transformation of models aided by the use of a computer.

We can distinguish the transformation of a descriptive model to the corresponding operational model and the transformation between descriptive models. The transformation of a descriptive model to an operational model may be realized almost automatically (from the theoretical viewpoint completely automatically (Valášek, 1987c, 1988)). A highly efficient way of using CAD is to let a man create the descriptive model and to carry out automatically the simulation of its function by its operational model. The design process is accelerated because the attention of the designer is concentrated only on the succesive steps of object idea approximations and is not disturbed by the preparation of object function investigation. The methods of these transformations are the main topic of this book. The transformation between descriptive models on different hierarchical levels is, however, not automatically realizable, as one descriptive model contains more information than the other. The information lacking for the transformation is usually supplied interactively by humans, e.g. using different descriptive model editors.

2.4.4 Man-Machine Dialogue — A Palette of Tools

From the method of successive approximations the need for human participation in CAD is obvious. We therefore get the principle of man-machine dialogue. The problem is to create a user friendly CAD system that is flexible enough and can easily be extended. But how do we build man-machine dialogues without sticking too much to present ways of designing?

A possible solution is to organize the CAD and the man-machine dialogue as a palette of tools. The user chooses from the palette the required tools, and tools may be added or modified. The intermediate results of a particular tool operation are stores in particular descriptive models which represent the initial data for further tools. The tools in the palette can be grouped into a complex tool which automates a part of the design. Alternatively the complex tool can be split into partial tools when used for a new or unusual object.

CAD according to these requirements can be realized by the methods of object-oriented programming. It is possible to realize it by a special command language. The command language is like a traditional programming language, but it is interpretative and interactive, and the semantics of its instructions are interactively exchangeable according to the palette tools.

This principle is to achieve CAD by programming the engineering methods of problem solving rather then programming the designing procedures for particular parts and elements of the designed structure or machine (Valášek, 1988). The tools of the palette are methods of solution and manipulation of particular engineering abstract (ideal) objects (elements) such as rigid body, elastic body, ideal spring, ideal damper, ideal capacitor.

2.4.5 Knowledge Engineering

Only a part of a designer-engineer's knowledge can be expressed algorithmically, by equations, as the results of generally valid formal theory, etc. If we need to realize as much as possible in CAD by computer, we have to use methods of AI, particularly expert systems. Thus, a further part of human knowledge about designing can be computer aided; only the purely intuitive and creative part remains in the human domain.

What can expert systems contribute? Expert systems as a formalized human experience can be especially useful in designing steps in Figure 2.1 which need human participation (indicated by H) (Valášek, 1988). Expert systems can formalize human knowledge in cases where general formal theory does not exist. The design process includes many iterative cycles controlled by the method of successive approximations. Human experience enables us to join directly the inputs and outputs of these cycles without iterations. Hence it is possible to determine directly the patterns of solution from input demands or the properties of design from its patterns. For this purpose we use expert systems of a diagnostic kind which can obviously accelerate the finding of a solution.

In many cases human experience enables us to determine a sequence of design stages leading to the solution of a particular design problem. Each stage can be solved by a particular CAD (with the use of a particular special expert system). Each stage starts from a known object structure and includes known sequences of decisions from a known class of possible solutions. For this purpose we use expert systems of a generative kind. They can save a lot of tedious work in generating objects.

2.4.6 Structure of CAD Software

From the fundamental principles of CAD operation we can conclude fundamental parts of CAD structure and its software.

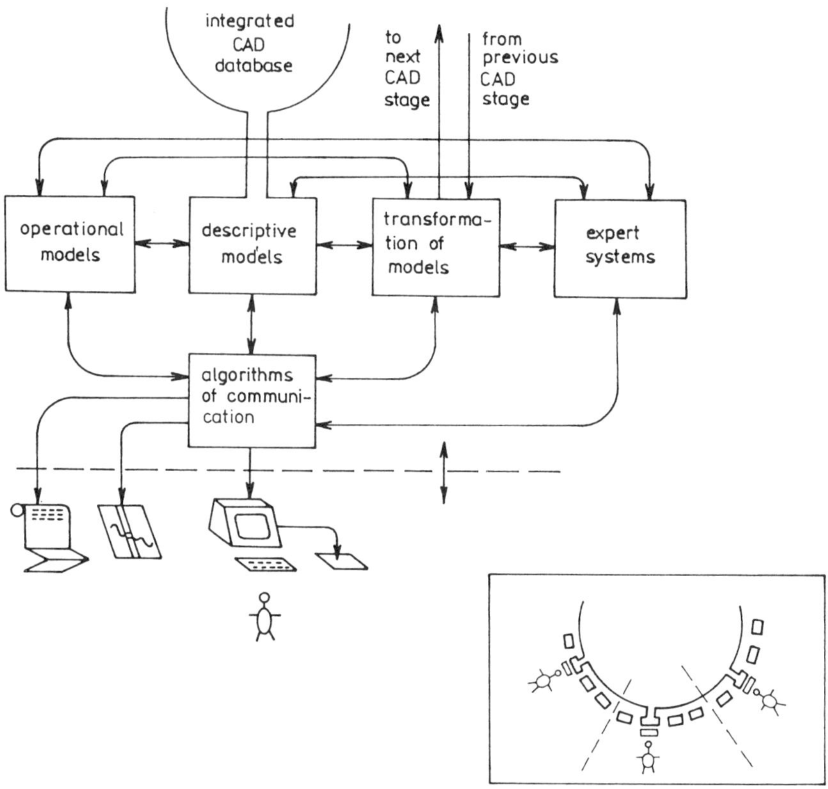

Figure 2.3. General structure of a CAD software.

In Figure 2.3 these fundamental parts are described in a general form which can be specialised and detailed for any particular CAD. A great proportion of the corresponding software modules can be realized independently of the CAD of the particular object (machine, product).

2.5 The Problem of Automated Analysis and Synthesis in CAD of Spatial Mechanisms

Spatial mechanisms are models of real machines or their parts consisting usually of several members, the spatial motions of which are bounded by kinematic constraints. The simplest model usually contains rigid members

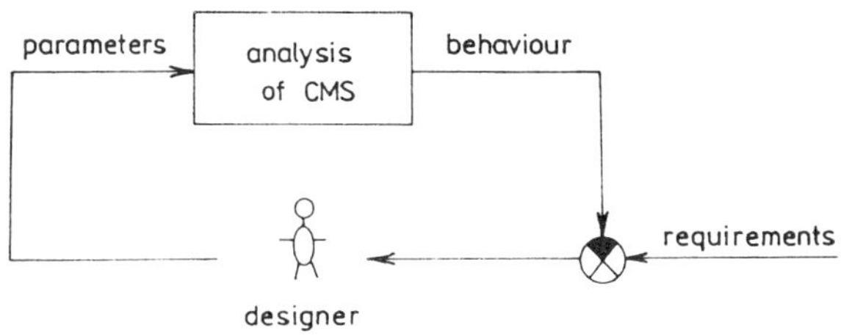

Figure 2.4. Feedback synthesis realized by designer.

bounded by ideal kinematic pairs, ideal springs and dampers. In some cases the model must be more realistic. For this reason the real kinematic pairs (with friction and backlashes) or elastic members are taken into consideration. The elastic properties of member are usually taken into account by finite element modelling.

The computation for the quality criteria evaluation or for the determination (synthesis) of parameters is the basic component of a spatial mechanism CAD system.

This is the topic dealt with in this book. We shall describe the ways of composing a kinematic, static and dynamic model of a spatial mechanism. We shall deal with a numerical solution of the respective mathematical (operational) model using a computer. Further we shall show from the viewpoint of a CAD system the ways of automating the use of such models for a certain spatial mechanism. As in Section 2.4.3 we shall use the case of realization of automated transformation of a descriptive model into an operational model. From the point of view of classical classification we distinguish between the analysis and synthesis of mechanisms. The task of analysis is the determination of the time function of the investigated action on the designed machine and on its model. The solution of the analysis problem forms the basis for evaluation of quality criteria. The task of synthesis is the determination of structure and parameters of the machine model under consideration. There is no general formal theory for the structure computation (synthesis) at our disposal. The designer must use his intuition. The parameter synthesis may, however, be done in a number of ways:

- the use of the designer's own intuition,
- the inverse solution of a (simplified) operational model,
- the inverse solution of an operational model using designer-realized feedback (Figure 2.4).

The required quality criteria (e.g. "mass centre motion of a certain body should follow a horizontal straight line") form the starting point for the parameter synthesis. Their mathematical formulation together with the mathematical model for the analysis of the mechanism represent the conditions that are to be fulfilled. In many cases these may be considered to be equations for unknown parameters. This may be understood as the inverse solution of a mathematical (operational) model for the mechanism analysis. The problem of inverse solution may, however, be incorrect (Tichonov and Arsenin, 1986), e.g. in the case of differential equations we may be confronted with incorrect problems, the solution of which is difficult or impossible. The original mathematical model would therefore be simplified for the purpose of parameter synthesis. To give an example, the solution of this problem is simpler for algebraic transcendental equations.

Should the direct solution of the problem of parameter synthesis be difficult or impossible, and the required criteria do not have to be fulfilled exactly, e.g. "the mass centre of a certain body should follow only approximately a horizontal straight line", an inverse solution using feedback is possible (Figure 2.4). The basis of this procedure is the analysis for chosen parameters. The results of the analysis are compared with the requirements; based on this comparison, the designer then chooses another parameter change. The synthesis is completed when the required quality criteria are subjectively fulfilled. In this case it is not necessary to limit the complexity of the mathematical model used for the analysis.

In this book we shall deal with the formulation of mathematical models for the analysis and synthesis of spatial mechanism parameters and with the automation of this process for certain mechanisms. It follows from the above that these are the key problems in creating a spatial mechanism CAD system. It follows further that the starting point for any task is always the composition of a mathematical model of the analysis. Similarly the starting point for the solution of mechanical problems is always the description of motion, i.e. kinematics. Therefore we shall concentrate most of our attention on kinematics and analysis of spatial mechanisms.

Chapter 3

Structure Analysis of Mechanical Systems

Every mechanical system is, first of all, a system of bodies. It can, however, also contain other elements, e.g. elastic and damping elements. The bodies are in most cases rigid, and the elastic and damping elements are considered to have no mass. It is therefore a discrete system. Although there are mechanical systems formed by freely movable bodies, in practice there is usually a system of bodies whose mutual movement is either completely or partially bounded by constraints. That is why, historically, such a system has been described as a Constrained Mechanical System (CMS), but now it is usually called a Multibody System (MBS).

The essential characteristic of an MBS is its structure. In describing it, we consider the number of bodies forming the system and their mutual constraints without regard to dimensions from the mere topological viewpoint. For this purpose, expressions such as kinematic pair, kinematic chain, system group, mobility, degree of freedom, etc. are used.

The general methods of describing and creating systems of bodies are dealt with in the theory of system composition (also called Structure Analysis and Synthesis). A summary follows of some of the results of this theory that are relevant to the purpose of this book.

3.1 Kinematic Pairs

Two bodies whose relative motion is bound by a constraint form a kinematic pair. We usually use the term "kinematic pair" with the appropriate adjective for the type of constraint, e.g. revolute kinematic pair, spherical kinematic pair, etc.

We distinguish between lower and higher kinematic pairs. In a lower kinematic pair the two bodies contact each other in a surface. The revolute, prismatic, helical, cylindrical, spherical and flat kinematic pairs belong to this class. Elements which make contact in a point or curve (line, abscissa) form a higher kinematic pair. There is no unified terminology for particular higher kinematic pairs. These are, for example, point contact of a sphere and a plane, point contact of bodies of general shape, line contact of a cylinder or cone with a plane, etc.

For the number of independent coordinates uniquely determining the relative position of two constrained bodies, we use the term "degree of freedom" of a given kinematic pair. Alternatively we may introduce the term "kinematic pair class". A kinematic pair is of the j-th class if it diminishes the relative motion of linked bodies by j degrees of freedom, i.e. j scalar constraint conditions correspond to the given kinematic pair. It follows that such a pair has $(6 - j)$ independent coordinates. The degree of motion is the fundamental characteristic quantity of kinematic pairs.

One of the bodies of a system is usually considered to be the reference element, and the position of other bodies is determined in relation to this reference body. If the reference body is stationary, the term "frame" is used.

The coordinates in the definition of freedom of motion can be linear or angular, absolute (measured with regard to the frame) or relative.

A summary of lower kinematic pairs, their properties and chart symbols is shown in Table 3.1. The higher kinematic pairs are described in detail in Chapter 6.

3.2 Kinematic Chains and System Groups

The individual bodies of a multibody system are its elements. An element connected to its neighbouring elements by s kinematic pairs is an element of degree s. An element of degree 1 is also called unary (Figure 3.1a), of degree 2 binary (Figure 3.1b), of degree 3 ternary (Figure 3.1c), etc. A number of bodies linked by kinematic pairs form a kinematic chain. The simplest description of its structure is by a chart, where an abscissa represents a body and a circle represents an arbitrary kinematic pair (Figure 3.2). The configuration described by a polygon, as for example in Figure 3.2, represents a loop. On the basis of the presence of loops in a structure we distinguish:

– closed kinematic chains, if they contain one or more loops so that each element and each kinematic pair is contained in at least one of them (Figure 3.3a),

TABLE 3.1. Kinematic pairs

Pair name	Pair chart symbol	Pair symbol	Degree of freedom	Class
revolute	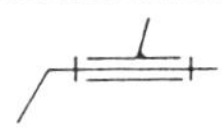	R	1	5
prismatic	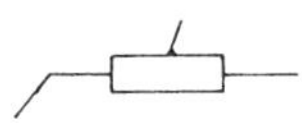	P	1	5
helical		H	1	5
cylindrical		C	2	4
spherical	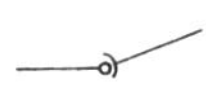	S	3	3
flat		F	3	3
general	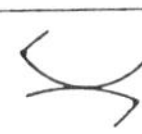	G	5	1
arbitrary	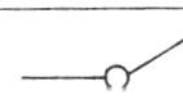	A	i	6 - i

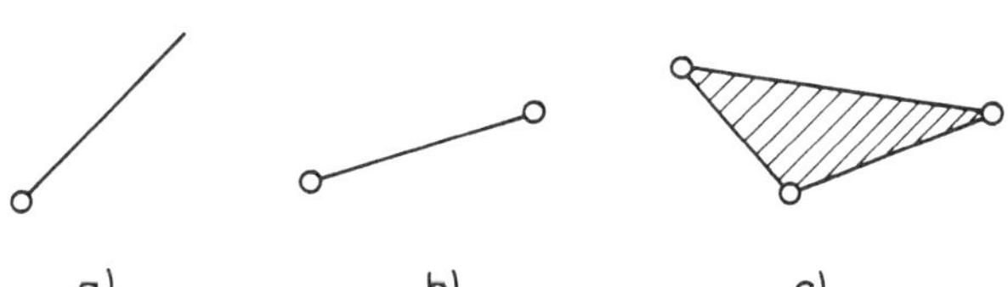

Figure 3.1. Unary, binary and ternary elements.

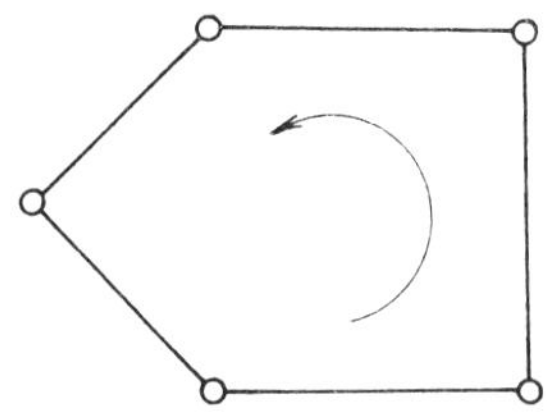

Figure 3.2. Kinematic loop.

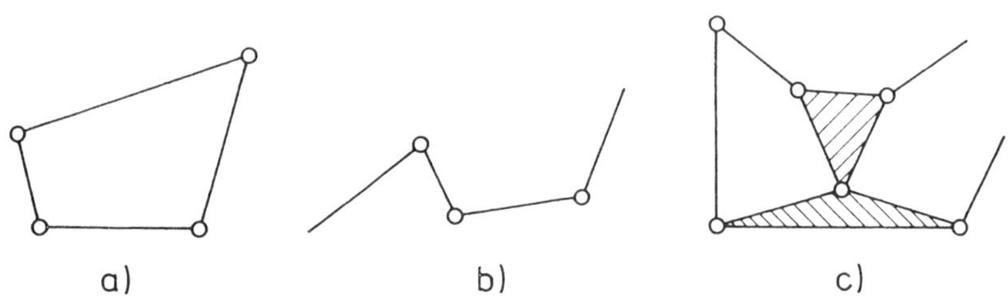

Figure 3.3. Kinematic chains.

– open kinematic chains, if they contain no loop (Figure 3.3b),

– mixed kinematic chains, if they are a combination of the above chains (Figure 3.3c).

On the basis of the kind of curves obtained as trajectories of points by the relative motion of any two bodies of the chain, we may classify the chains as planar, spherical and spatial.

Another classification may distinguish:

– simple chains — containing only binary elements (Figure 3.3a),

– complex chains — containing at least one element of degree 3 or higher (Figure 3.3c).

These classifications do not exclude each other. We can therefore speak of a simple spatial closed chain (Figure 3.3a), a complex planar mixed chain (Figure 3.3c), etc.

The configuration, i.e. the mutual position of all chain elements, is described by the coordinates of a kinematic chain. These are the independent coordinates of kinematic pairs belonging to the chain. The coordinates of all kinematic pairs of open chains are independent. The basic characteristic property of a kinematic pair is its freedom of motion expressed by the number i of independent coordinates that are necessary to determine the position of all its elements uniquely. This may be obtained as follows.

Every free body in space has six degrees of freedom. For a system consisting of n elements, one of which is considered fixed (basic frame), the $n-1$ movable elements have $6(n-1)$ degrees of freedom. Each kinematic pair of class j diminishes the freedom of motion of the system by j degrees of freedom. Designating the number of kinematic pairs of class j as d_j, it follows that the freedom of motion of the particular system

$$i = 6(n-1) - \sum_{j=1}^{5} j d_j \tag{3.1}$$

This is referred to in literature as the Grübler or Kutzbach formula (Assur, 1952; Angeles, 1982). This formula can be used as a formula of orientation for any MBS with the exception of singular cases (see Chapter 7).

Using formula (3.1) for a simple open chain, we can understand in terms of the number i the degree of freedom of the last n-th chain element. This means for $i > 6$ that a certain position of the n-th element may be reached by $i - 6$ different chain configurations.

For the special case of planar MBS the formula (3.1) has the form

$$i = 3(n-1) - \sum_{j=1}^{2} j d_j \tag{3.2}$$

This is valid only for regular cases.

There is a special significance to kinematic chains which do not change their mobility after being connected to an arbitrary system. Kinematic chains defined in this way are called system groups. Connecting them to or disconnecting them from a given system enables given systems to be modified or structurally new systems to be created while maintaining the original freedom of motion. The term "system group" has been introduced for the classification of planar mechanisms used by Assur (1952) and further investigated by Artobolevskij (1975). We shall now summarize the findings of those investigations.

Limiting ourselves for the sake of simplicity to planar systems containing only kinematic pairs of class 2, and designating by m the number of elements of a system group, we obtain from (3.2) the condition

$$3m = 2d_2 \tag{3.3}$$

according to which the number of system group elements is always even. The simplest of all is therefore a binary group (Figure 3.4a). More complex system groups can be created by joining two bodies and three kinematic pairs to the original group. In this way we get ternary, biternary, triternary,

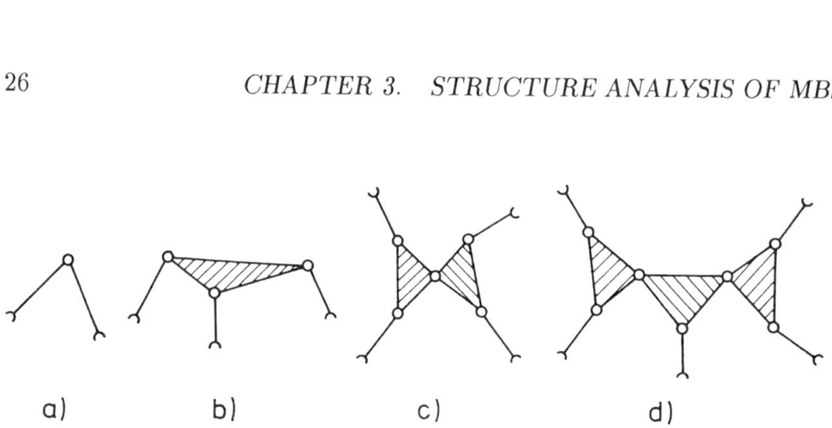

Figure 3.4. Binary, ternary, biternary and triternary groups.

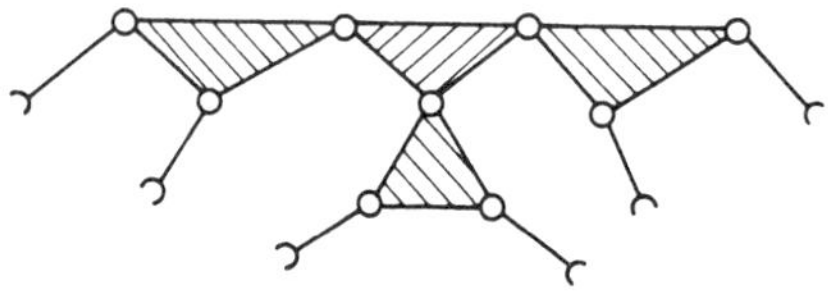

Figure 3.5. System group of class 2.

etc. groups. They are shown in Figure 3.4, where the semicircles designate the kinematic pairs through which the group is to be joined to a system. All system groups created in the above way belong to class 1 of Assur's classification.

Class 2 consists of system groups created by modifying groups of class 1 in such a way that the binary element of the internal ternary element is substituted by another ternary element with two binary ones. For example, modifying the group according to Figure 3.4d we obtain a system group according to Figure 3.5.

System groups of class 3 are characterized by a loop, and system groups of class 4 by two or more loops. Examples of such groups are shown in Figures 3.6 and 3.7. A system group can be defined by the requirement that its joining to a (basic) system would not change the mobility of the

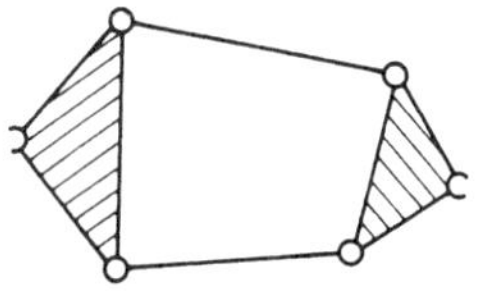

Figure 3.6. System group of class 3.

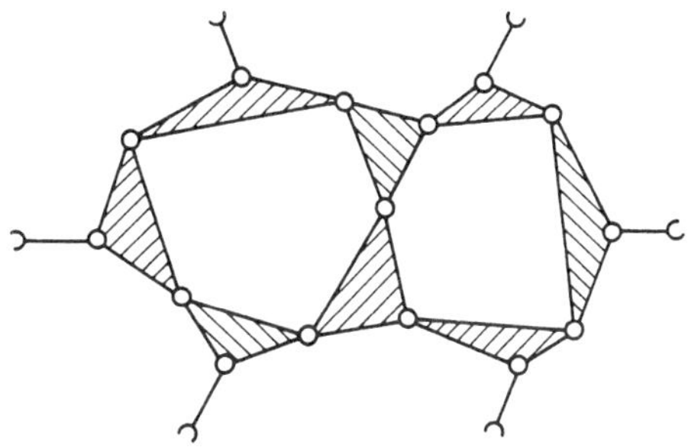

Figure 3.7. System group of class 4.

whole; this also holds in the case of spatial systems. As neither of the system group bodies is a frame, the relation

$$6n = \sum_{j=1}^{5} j d_j \tag{3.4}$$

between the number n elements of a system group and number d_j of kinematic pairs of class j in the group must hold according to (3.1). Formula (3.4) allows more variability than for planar systems. Its validity is, however, not universal. It fails especially in so-called singular cases, although it may happen in regular cases that a combination of bodies and kinematic pairs for which (3.4) holds is not realizable (e.g. a unary element with two spherical kinematic pairs); in other cases, the type of relative motion, i.e. the type of kinematic pair in the given class, is also significant.

The essential spatial system groups are given in Figure 3.8. Their composition is described in Tab. 3.2.

The idea of creating systems with the help of system groups can be used in two ways.

The first is the classification of systems. The advantage of this approach lies in its transparency and simplicity; this is only true, however, for systems where the number of elements is small. With more complex systems, the number of combinations by this method is too confusing and the problem of isomorphisms arises. This is the reason for attempts to obtain an analytical basis for the classification of mechanisms (Uicker and Raicu, 1975; Agrawal and Rao, 1987).

The second lies in the possibility of independently describing the kinematic solutions of system groups using the knowledge of motion at the points where they are joined in the main system. The solution of the whole system can then be obtained by composing partial solutions in principle

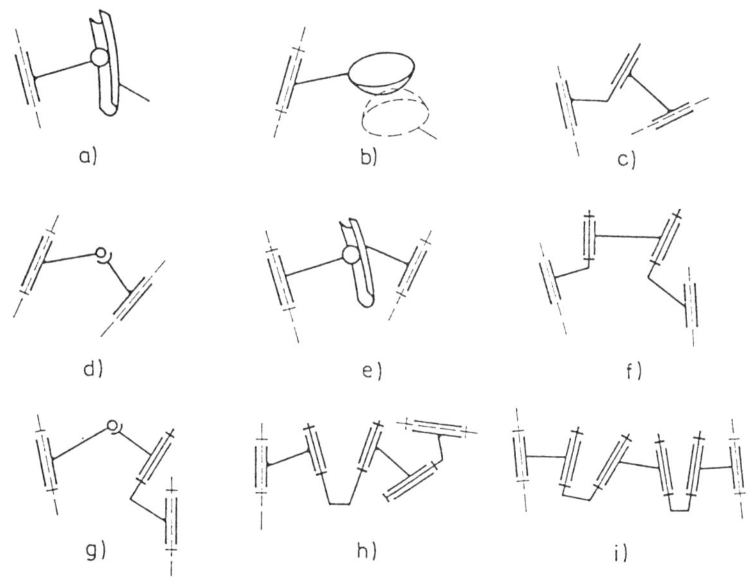

Figure 3.8. Spatial system groups.

in the same way as by composing system structures. This approach has already been followed by Assur and its significance has increased with the possibility of using computers (Brát and Lederer, 1973; Novotný, 1983). It has also been developed by Tesar (Benedict and Tesar, 1978; Freeman and Tesar, 1982) in the field of dynamics. An advantage of this approach is that it is much easier to follow the influence of individual element groups on the behaviour of the whole.

TABLE 3.2. Essential spatial system groups

Fig. 3.8	n	d_1	d_2	d_3	d_4	d_5
a	1		1		1	
b	1	1				1
c	2				3	
d	2			1	1	1
e	2		1			2
f	3				2	2
g	3			1		3
h	4				1	4
i	5					6

3.3 Mechanisms

Turning an element of a closed kinematic chain into a frame gives rise to a mechanism. A mechanism can also be defined as a mechanical device to transform motion and transfer forces or to cause points and bodies to follow paths of given properties. In a narrower sense of the word a system of this type with one degree of freedom is considered to be a mechanism.

Elements whose motion is given are the driving elements; the rest are called driven elements. The same designation can be used for the respective kinematic pairs. The driving elements are also called input elements. The element of the mechanism which carries out the motion for which the mechanism has been designed is called an output element. The rest are transfer elements. If the input element motion is described by coordinate $\varphi(t)$ and the output element motion by $s(t), s(\varphi)$ is the so-called input–output relation. According to the type of this relation we distinguish between mechanisms with constant transfer ratio (the input–output relation is linear) and non-constant ratio (nonlinear input–output relation).

Regarding mechanisms from the synthesis viewpoint leads to the classification into transmission mechanisms and guidance mechanisms. Transmission mechanisms serve to transfer motion or forces between input and output elements. Guidance mechanisms secure the prescribed motion of an element or its point. Mechanisms are divided into planar, spherical and spatial mechanisms, in the same way as kinematic chains.

According to other criteria they can be divided into independent groups such as linkages (containing only lower kinematic pairs), cam mechanisms, gear mechanisms, straight line mechanisms, hydraulic mechanisms and pneumatic mechanisms.

From the viewpoint of solving a mechanism we are especially interested in the number of coordinates characterising the position of all mechanism elements and their division into dependent and independent coordinates. The number p of all coordinates of all kinematic pairs is given by the relation

$$p = \sum_{j=1}^{5}(6-j)d_j$$

A mechanism with i degrees of freedom must have i independent coordinates. The number of dependent coordinates is then given by the difference

$$\begin{aligned} p - i &= \sum_{j=1}^{5}(6-j)d_j - [6(n-1) - \sum_{j=1}^{5} jd_j] = \\ &= 6\sum_{j=1}^{5} d_j - 6n + 6 = 6 \end{aligned}$$

taking into account the fact that for simple mechanisms the number of kinematic pairs equals the number of elements, i.e.

$$\sum_{j=1}^{5} d_j = n$$

This leads to the following theorem.

Theorem 3.1. The kinematic pairs of a simple mechanism with i degrees of freedom have i independent and 6 dependent coordinates.

This theorem holds for all mechanisms regardless of the type of kinematic pairs used in the mechanism.

3.4 Compartments

Approaches developed in system theory can be successfully applied to the study of more complex MBS's. One of the approaches is the diakoptic method based on the fact that even very complex systems are formed by joining relatively simple parts. The behaviour of the whole system is the result of interactions between its parts. These interactions typically occur in strictly limited and usually spatially small areas which we shall call nodes. These simple parts with variously concentrated and divided properties of the original real object are called compartments. We shall describe the constraints to their environment with the help of so-called node variables, by means of which we may express all the interactions being considered between compartments. We shall describe the behaviour of each compartment by means of independent equations in node variables. The model of a whole complex system would then be obtained by connecting the models of particular compartments into a whole and by the equations relating their node variables arising from the connection structure of compartments into a whole.

In the case of an MBS, the element, or a group of elements such as a system group, may be a compartment. Nodes, which are the places of interaction, are usually kinematic pairs. The forces and velocities in these may usually be taken as node variables. The behaviour of compartments may be described using the laws of mechanics, the constraints between them expressing the equality of velocities and forces in kinematic pairs.

The application of the above principles will be described in more detail in Chapters 5 and 9.

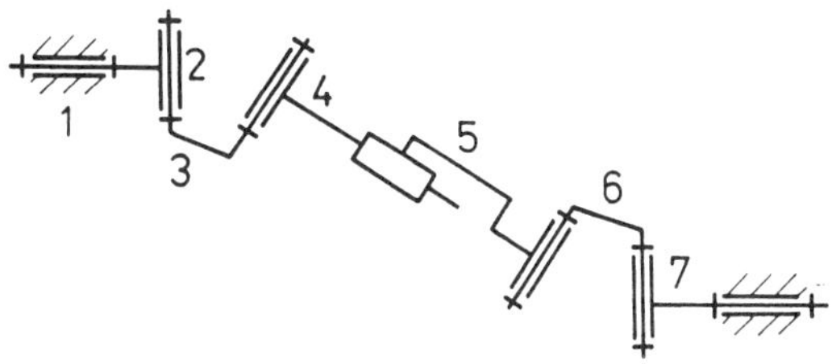

Figure 3.9. Mechanism with one degree of freedom.

3.5 Structural Description of an MBS

For the structural description of an MBS the topological data alone are essential. The metrical data (dimensions) are not. The essentials are usually considered to be: the number and type of elements, the number and type of kinematic pairs and their location on bodies. From these we can deduce further data on structure, such as the number of basic loops, the number and type of kinematic chains and system groups.

The simplest, and for human beings the most understandable, expression of the structure of an MBS is its kinematic chart. To draw such charts we use chart symbols for kinematic pairs according to Table 3.1.

3.5.1 MBS Formula

Using less space than is needed for a kinematic chart, we can describe the structure of an MBS by a formula, where hyphens represent bodies and letters represent the type of kinematic pair (see Table 3.1). For example, the structure of a mechanism consisting of seven elements and serving to transform rotational motion between passing shafts drawn in Figure 3.9 can be described by the formula

$$\underline{\overline{R}} - R - R - P - R - R - \underline{\overline{\overline{R}}}$$

In this formula, symbols of kinematic pairs connected with the frame are underlined, symbols with one line above represent an input kinematic pair, and those with two lines above an output kinematic pair. In the case of simple mechanisms we can use, without causing any misunderstanding, the more concise form

$$RRRPRRR$$

In a similar way, the mechanism of front wheel suspension of a car as in

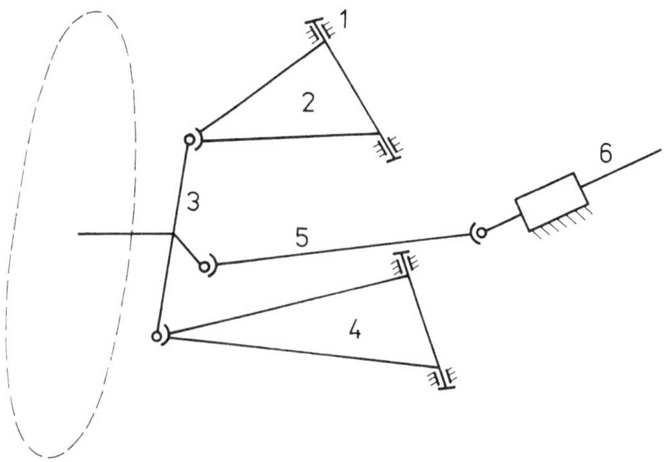

Figure 3.10. Mechanism of front wheel suspension.

Figure 3.10 can be described by formula

$$\begin{array}{ccc} \underline{R}-S & - & S-\underline{R} \\ & | & \\ & S & -\ S-\underline{P} \end{array}$$

It is obvious from this formula that the mechanism has five elements with two independent loops. One element is of degree three, and the rest of degree two. The mechanism contains two revolute, four spherical and one prismatic kinematic pairs.

3.5.2 Application of Graph Theory

MBS analysis and synthesis automation cannot be done without using computers. One of the problems that must be solved for this purpose is the creation of a description of the structure of the solved MBS that is simple for the user and understandable for the computer. With regard to CAD theory according to Chapter 2, we need a topological descriptive model of the created object.

A possible approach is to apply the results of graph theory, where relatively simple answers can be found even to such questions as the search for independent loops and determination of the smallest number of transformations between two MBS elements. The main definitions and some basic

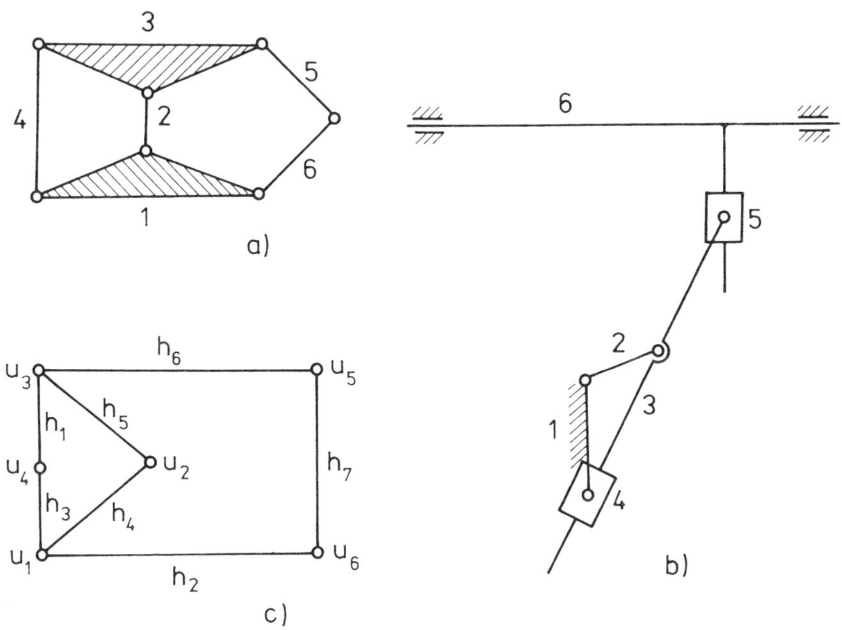

Figure 3.11. Mechanism of a shaping machine.

results of graph theory are given without proof, to the extent necessary for the following chapters. For details see (Koenig et al., 1967; Chen Wai-Kai, 1971; Mayeda, 1972).

The usual definition of an abstract graph is as follows. Let $\mathcal{U}$ be a non-empty set, and $\overline{\mathcal{H}}$ the set of all subsets of $\mathcal{U}$ with exactly two elements. Let us choose an arbitrary subset $\mathcal{H} \subset \overline{\mathcal{H}}$. The ordered pair $(\mathcal{U}, \mathcal{H})$ is called a graph (or an abstract graph) and designated $\mathbf{G}(\mathcal{U}, \mathcal{H})$. The elements of set $\mathcal{U}$ are called nodes, and the two–element subsets of set $\mathcal{U}$ are called edges.

The definition of graphs shows an analogy with kinematic charts. These also have two types of objects — bodies and kinematic pairs. We can choose a correspondence:

$$\text{MBS body} \leftrightarrow \text{node of graph } \mathbf{G}$$

$$\text{MBS kinematic pair} \leftrightarrow \text{edge of graph } \mathbf{G}$$

Graph $\mathbf{G}(\mathcal{U}, \mathcal{H})$ is then called the associated graph of a mechanical system.

A correspondence body $\leftrightarrow$ edge, pair $\leftrightarrow$ node is certainly also possible. It leads, however, to problems with bodies of third and higher degree which, when represented by edges, would not comply with the definition of an edge as a two-element set of nodes.

For the sake of simplicity we shall often use the terms body/node and edge/kinematic pair interchangeably. The symbols $\mathcal{U}, \mathcal{H}$ will have a double meaning, either as the set of nodes or edges or as the number of elements of those sets, i.e. the number of nodes or edges.

As an example of the above terms, Figure 3.11a shows the kinematic chart of a Stephenson chain; according to the usual principles of drawing kinematic charts, bodies are represented by abscissas and geometrical forms, and kinematic pairs of class 2 in the planar classification are represented by circles. Figure 3.11b represents an application of a Stephenson chain, where element 1 has become the frame, as a shaping machine mechanism. The individual kinematic pairs are either translation or revolute. Figure 3.11c shows the associated graph of the system described, where nodes (bodies) are represented as circles, and edges (kinematic pairs) as abscissas. The node indices have been chosen according to the numbers of bodies; edge indices have been chosen arbitrarily. For the sake of simplicity, the chosen example is a planar mechanism. The extension to spatial MBS's is easy.

Charts shown in Figure 3.11 are for engineers a traditional, condensed and very suitable means of exchange of information. Computers which are not equipped with recognition of graphical information require a different input description of graph structure, e.g. a list of all kinematic pairs characterised by a number of linked bodies. For the graph in Figure 3.11c we obtain: (1, 2), (2, 3), (1, 4), (4, 3), (3, 5), (5, 6), (6, 1). Kinematic pairs described in this way are given in arbitrary order. The orientation, if need be, can be expressed by the order of the nodes. The storage requirement for k kinematic pairs is $2k$, as each kinematic pair is represented by a pair of numbers.

This method can be modified by storing the body numbers in two fields P, Q. The former field stores the numbers of the origin bodies, the latter the numbers of end bodies of the respective kinematic pairs. For the graph in Figure 3.11c we get

$$P[1,\ 2,\ 1,\ 4,\ 3,\ 5,\ 6]$$
$$Q[2,\ 3,\ 4,\ 3,\ 5,\ 6,\ 1]$$

Another storage method is a list of all bodies which contains for each body a list of connected bodies. In our case this is written 1 : 2, 4, 6; 2 : 1, 3; 3 : 2, 4, 5; 4 : 1, 3; 5 : 3, 6; 6 : 1, 5. This method is advantageous for algorithms searching for a path from one body to another.

For the purpose of further discussion it is advantageous to characterize the MBS structure with the help of a complete node–edge incidence matrix

$\mathbf{A}_c$, where its elements are defined as follows:

$$a_{ij} = \begin{cases} 1 & i\text{-th body is incident with } j\text{-th kinematic pair} \\ 0 & \text{otherwise} \end{cases}$$

For the graph in Figure 3.11c we have

$$\begin{array}{ccrc} & & \text{kinematic} & \\ & & \text{pairs}: & \begin{array}{ccccccc} 1 & 2 & 3 & 4 & 5 & 6 & 7 \end{array} \\ \mathbf{A}_c & = & \begin{array}{r} \text{bodies}: 1 \\ 2 \\ 3 \\ 4 \\ 5 \\ 6 \end{array} & \left[\begin{array}{ccccccc} & 1 & 1 & 1 & & & \\ & & & 1 & 1 & & \\ 1 & & & & 1 & 1 & \\ 1 & & 1 & & & & \\ & & & & & 1 & 1 \\ & 1 & & & & & 1 \end{array} \right] \end{array}$$

The matrix $\mathbf{A}_c$ contains all the information on the given graph and has the following properties:

a) There are exactly two 1's in each column.

b) The number of 1's in each line equals the degree of body (e.g. body 3 is a ternary body).

c) Renumeration of kinematic pairs or bodies is represented by column or line permutation.

d) The order of matrix $\mathbf{A}_c$ is $U - 1$, where U is the number of all bodies.

e) The number of non-zero matrix elements equals twice the number of kinematic pairs.

Due to property a) one line (e.g. the last one) may be left out, giving the reduced inicidence matrix $\mathbf{A}$. From property d) it follows that the matrix $\mathbf{A}$ contains at least one square submatrix of order $U - 1$. The algorithm for the construction of matrix $\mathbf{A}$ on the basis of the above methods of graph description is simple. It is also possible to specify it directly. Most methods of MBS analysis require the knowledge of loops. The theory of graphs makes it possible to find the loops using a pure algebraic method. Before we show this method, it is necessary to introduce new terms.

In the definition of MBS the use of massless elastic damping elements joining some bodies of the system has been accepted. An example of such a system is shown in Figure 3.12a; the respective associated graph is shown in Figure 3.12b. This is a more general concept of graph. There are two kinds of edges shown in this graph. The full-line edges again represent kinematic pairs; the broken-line edges represent force action between the respective bodies (e.g. through springs, dampers, or even electromagnetic or gravitational fields). Sometimes even the force link can be described as a kinematic pair with 6 degrees of freedom (Wittenburg, 1977). More force

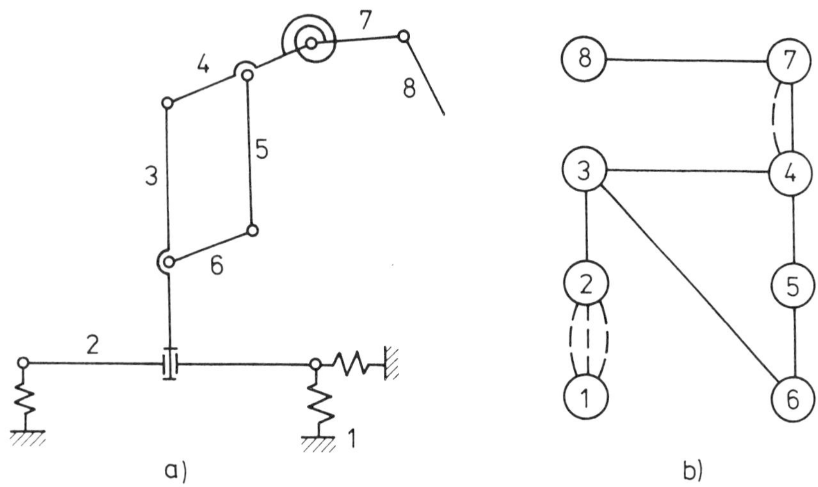

Figure 3.12. Mechanical system containing massless elastic damping elements.

constraints between the same bodies are represented by parallel broken-line edges. We shall use this concept of generalized associated graphs for the solution of MBS static or dynamic problems. For the purpose of kinematics, where the term "force constraint" does not appear, it is possible to leave out the broken-line edges in Figure 3.12b, resulting in a disconnected graph with two components formed by the following nodes (2, 3, 4, 5, 6, 7, 8), (1); these components may be kinematically described as separate entities. As we are dealing at the moment only with the kinematics of MBS, we shall limit our discussion to connected finite graphs, i.e. graphs where each node can be reached using the graph edges from any other node, and where the number of nodes and edges is finite.

A connection between two nodes described by a sequence of nodes and edges is called a sequence. A sequence where each node appears not more than once is a path. The number of edges of a path is the length of a path. A path where the start node and end node are the same node is a loop (circle). A connected graph containing no circle is a tree. A graph $\mathbf{G}$ is called a subgraph of $\mathbf{G}'$ if the sets of nodes and edges of the graph $\mathbf{G}$ are subsets of the sets of nodes and edges of the graph $\mathbf{G}'$. A subgraph of a graph which contains all nodes of the graph and which is a tree is called the graph skelet. Skelet edges are branches; all other edges are chords. A graph can have several skelets.

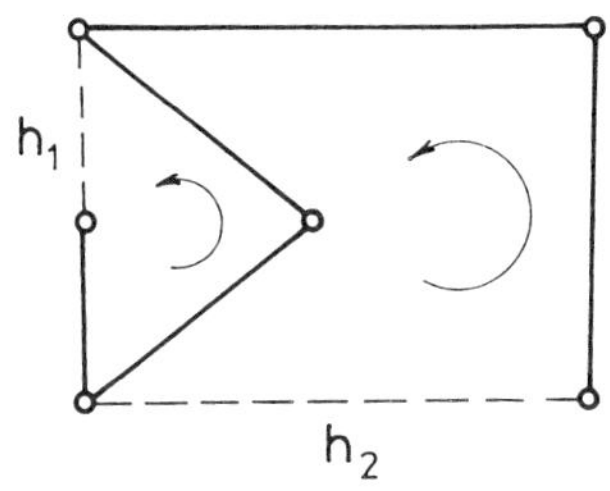

Figure 3.13. Skelet and basic loops of a mechanism.

A theorem in graph theory states that a skelet of each connected graph with H edges and U nodes has $U-1$ branches. Hence obviously the $H-U+1$ other edges are chords. Circles created by using chords of a given skelet are basic circles. The determination of a skelet is not unique, and the set of basic circles is not uniquely determined. The number of elements of such a set is, however, invariant. This is expressed in terms of mechanics by

Theorem 3.2: The number l of independent loops of a complex mechanism with n elements and d kinematic pairs is given by the relation

$$l = d - n + 1 \tag{3.5}$$

Because a square submatrix of order $U-1$ of an incidence matrix is regular if and only if its columns correspond to a certain skelet in a graph, it is possible to find a certain set of independent loops in an algebraic way. It is possible to find $7-6+1=2$ basic circles in the graph of Figure 3.11c. Basic circles shown in Figure 3.13 correspond to the skelet drawn as a thick line with chords h_1, h_2. A complete matrix $\mathbf{B}_c$ of circles defined by elements

$$b_{ij} = \begin{cases} 1 & i\text{-th circle contains the } j\text{-th edge} \\ 0 & \text{otherwise} \end{cases}$$

is a convenient description of circles in a graph.

The matrix $\mathbf{B}_c$ has dimensions (S, H) where S is the number of loops and H is the number of edges. Its properties are as follows:

a) The number of 1's in a line is the number of edges in the circle, i.e. its length.

b) Renumeration of edges or circles corresponds to column or line permutation.

c) The rank of matrix $\mathbf{B}_c$ is $H-U+1$.

Due to property c) we do not lose any information in $\mathbf{B}_c$ by keeping only linearly independent lines. We thus get the matrix $\mathbf{B}$ of basic circles and dimensions $(H - U + 1, H)$. Through column permutation it is always possible to obtain the form

$$\mathbf{B} = [\mathbf{E}, \mathbf{B}_v]$$

where matrix $\mathbf{B}_v$ of the type $(H - U + 1, U - 1)$ corresponds to branches of a certain skelet and the unit matrix $\mathbf{E}$ corresponds to chords. In our case

$$\mathbf{B} = \left[\begin{array}{cc|ccccc} 1 & 0 & 1 & 1 & 1 & 0 & 0 \\ 0 & 1 & 0 & 1 & 1 & 1 & 1 \end{array}\right]$$

We shall use the above results to obtain two pieces of knowledge in the theory of system composition relating to the number of coordinates and mobility:

a) The number of all kinematic pairs of an MBS with l independent loops and n elements is, according to (3.5), $d = l + n - 1$. As $6 - j$ coordinates correspond to each kinematic pair of class j, the number of coordinates in the system being studied is

$$p = \sum_{j=1}^{5} (6 - j) d_j$$

The number of dependent coordinates is given by the difference

$$\begin{aligned} p - i &= \sum_{j=1}^{5} (6 - j) d_j - [6(n - 1) - \sum_{j=1}^{5} j d_j] = \\ &= 6 \sum_{j=1}^{5} d_j - 6(n - 1) = 6(n + l - 1) - 6(n - 1) = 6l \end{aligned}$$

This result is expressed by

Theorem 3.3: A complex MBS with i degrees of freedom and l independent loops has i independent and $6l$ dependent coordinates.

b) The skelet of an associated MBS with n bodies, d kinematic pairs and l independent loops is formed by $d - l$ branches and l chords. As skelet edges correspond to an open simple kinematic chain, with independent coordinates of kinematic pairs, the degrees of freedom of kinematic pairs belonging to the skelet are

$$i_k = \sum_{m=1}^{d-l} f_m$$

where f_m are the degrees of freedom of the m-th kinematic pair. The actual mobility of the MBS is, of course, smaller by the degrees of freedom

i_t of kinematic pairs forming the chords. It is

$$i_t = \sum_{j=1}^{5} j d_{t_j}$$

where d_{t_j} is the number of kinematic pairs of class j relating to chords. The degrees of freedom i of the MBS are

$$i = i_k - i_t = \sum_{m=1}^{d-l} f_m - \sum_{j=1}^{5} j d_{t_j} \tag{3.6}$$

For a planar MBS, eq. (3.6) reduces to

$$i = \sum_{m=1}^{d-l} f_m - 2d_{t_2} - d_{t_1} \tag{3.7}$$

For a planar linkage MBS where all kinematic pairs allow only the value 1 for the degrees of freedom, eq. (3.7) reduces to the very simple expression

$$i = d - 3l \tag{3.8}$$

It is interesting to note that the formula does not contain the number of bodies; it follows for instance that a planar joint mechanism ($i = 1$) can have only 4, 7, 10, 13, ... kinematic pairs.

The logical complement to the circles matrix is the complete matrix $\mathbf{C}_c$ of cutsets with elements

$$c_{ij} = \begin{cases} 1 & \text{if the } i\text{-th section contains the } j\text{-th edge} \\ 0 & \text{otherwise} \end{cases}$$

whereby a cutset is the set $\mathbf{C}$ of edges with following properties:

a) by removing a set $\mathbf{C}$ of edges the graph splits into two parts (an isolated node is also a part).

b) no subset of $\mathbf{C}$ has property a).

From the sklet properties, it follows that a cutset always contains at least one branch. The set of cutsets corresponding to skelet branches is the set of basic cutsets. The complete matrix $\mathbf{C}_c$ of cutsets is of rank $U - 1$. Hence the lines of matrix $\mathbf{C}_c$ are not linearly independent. Keeping only independent lines corresponding to a certain skelet, we get the matrix $\mathbf{C}$ of basic cutsets. This may always be changed to the form

$$\mathbf{C} = [\mathbf{C}_t, \mathbf{E}]$$

where matrix $\mathbf{C}_t$ of the type $(U-1, H-U+1)$ corresponds to chords of a certain skelet and the unit matrix $\mathbf{E}$ corresponds to its branches. In our case

$$\mathbf{C} = \left[\begin{array}{cc|ccccc} 1 & 0 & 1 & 0 & 0 & 0 & 0 \\ 1 & 1 & 0 & 1 & 0 & 0 & 0 \\ 1 & 1 & 0 & 0 & 1 & 0 & 0 \\ 0 & 1 & 0 & 0 & 0 & 1 & 0 \\ 0 & 1 & 0 & 0 & 0 & 0 & 1 \end{array}\right]$$

A number of useful relations hold for the matrices which have been introduced. For the sake of completeness, they are listed as follows. Matrices $\mathbf{A}, \mathbf{B}$ and $\mathbf{B}, \mathbf{C}$ are, using for all three the same numeration of edges, orthogonal; thus

$$\mathbf{A}\mathbf{B}^T = \mathbf{0} \tag{3.9a}$$
$$\mathbf{B}\mathbf{C}^T = \mathbf{0} \tag{3.9b}$$

Let us choose a certain skelet in graph $\mathbf{G}(\mathcal{U}, \mathcal{H})$. The incidence matrix may always be written in the form

$$\mathbf{A} = [\mathbf{A}_t, \mathbf{A}_v] \tag{3.10}$$

where the submatrix $\mathbf{A}_t$ of type $(U-1, H-U+1)$ corresponds to chords and a submatrix $\mathbf{A}_v$ of type $(U-1, U-1)$ corresponds to branches of the selected skelet. It holds further that

$$\mathbf{B}_v = \mathbf{C}_t^T \quad \text{and} \quad \mathbf{C}^T = \mathbf{B}_v^T \tag{3.11a}$$

respectively

$$\mathbf{B}_v^T = \mathbf{C}_t = \mathbf{A}_v^{-1}\mathbf{A}_t \tag{3.11b}$$

These relationships assume the same order of edges in matrices $\mathbf{B}, \mathbf{C}$ and they enable us to calculate the matrices $\mathbf{B}, \mathbf{C}$ from matrix $\mathbf{A}$ or, having one of the matrices $\mathbf{B}, \mathbf{C}$, to calculate the other. It is in general not possible to calculate $\mathbf{A}$ from $\mathbf{B}, \mathbf{C}$.

All operations for non-oriented graphs must be carried out in modulo 2 algebra, where addition and multiplication are defined by the following tables

addition	0	1
0	0	1
1	1	0

multiplication	0	1
0	0	0
1	0	1

The commutative, associative and distributive laws hold.

This summary does not exhaust the possibilities for the application of graph theory to the description and analysis of MBS. In many cases it is convenient to use oriented graphs, where every edge (kinematic pair) is given an orientation (direction). This is significant especially in those cases where the position of i-th and $(i+1)$-th bodies are described in relative coordinates. We must then distinguish whether the position of body $(i+1)$ is expressed in the space of body i or vice versa. In the former case we use the given transformation; in the latter case we must use the inverse transformation. We give several examples of such an approach in Chapters 4 and 5.

It is also possible to define topological matrices for oriented graphs. A complete node–edge incidence matrix $\vec{\mathbf{A}}_c$ of an oriented graph is defined as follows

$$a_{ij} = \begin{cases} +1 & \text{if edge } j \text{ begins in node } i \text{ (positive} \\ & \text{incidence of node } i \text{ with pair } j) \\ -1 & \text{if edge } j \text{ ends in node } i \\ & \text{(negative incidence)} \\ 0 & \text{no incidence of edge } j \text{ with node } i \end{cases}$$

For the graph in Figure 3.14 we obtain:

$$\begin{array}{lll} & \text{kinematic} & \\ & \text{pairs :} & \begin{matrix} \;1 & \;\;2 & \;\;3 & \;\;4 & \;\;5 & \;\;6 & \;\;7 \end{matrix} \\ & \text{bodies : } \begin{matrix} 1 \\ 2 \\ 3 \\ 4 \\ 5 \\ 6 \end{matrix} & \\ \vec{\mathbf{A}}_c = & & \begin{bmatrix} & 1 & -1 & 1 & & & \\ & & & -1 & 1 & & \\ 1 & & & & -1 & -1 & \\ -1 & & 1 & & & & \\ & & & & & 1 & -1 \\ & -1 & & & & & 1 \end{bmatrix} \end{array}$$

In a similar manner we may introduce the reduced incidence matrix $\vec{\mathbf{A}}$ of an oriented graph. This may be used as an easy method for the determination of the number of graph skelets, which is

$$k = \det(\vec{\mathbf{A}}\vec{\mathbf{A}}^T)$$

Analogously we can define the matrix $\vec{\mathbf{B}}$ of circles and matrix $\vec{\mathbf{C}}$ of cutsets of an oriented graph, etc. (Stejskal, 1977a). Relations which hold are similar to those for topological matrices of non-oriented graphs

$$\begin{array}{lll} \vec{\mathbf{A}}\vec{\mathbf{B}}^T = \mathbf{0}, & \vec{\mathbf{A}} = [\vec{\mathbf{A}}_t, \vec{\mathbf{A}}_v], & \vec{\mathbf{B}} = [\mathbf{E}, \vec{\mathbf{B}}_v] \\ \vec{\mathbf{B}}\vec{\mathbf{C}}^T = \mathbf{0}, & \vec{\mathbf{B}}_v = -\vec{\mathbf{C}}_t^T, & \vec{\mathbf{B}}_v^T = -\vec{\mathbf{C}}_t \\ \vec{\mathbf{C}} = [\vec{\mathbf{C}}_t, \mathbf{E}], & \vec{\mathbf{C}}_t = -\vec{\mathbf{B}}_v^T, & \vec{\mathbf{B}}_v^T = -\vec{\mathbf{A}}_v^{-1}\vec{\mathbf{A}}_t \end{array} \tag{3.12}$$

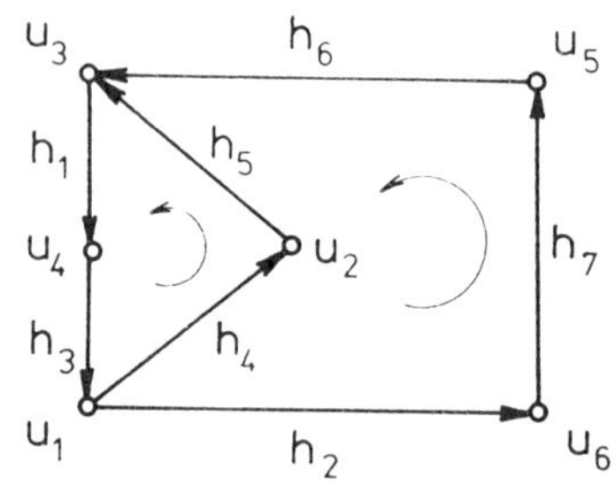

Figure 3.14. Oriented graph of a mechanism.

All operations in these relations are carried out using standard algebra.

The next step in generalization is to assign values to edges and nodes. We thus get an edge-valued or node-valued graph. The edge values may express, for example, the kinematic pair type expressed by symbols R, P, C, S or by an arbitrary numeric code — degree of freedom, coordinates and their values, etc. In the case of force constraints, these may be parameters of springs, dampers, etc. Node values may express dimensions, mass, momentum and coordinates of elements.

An example of a software tool to express and manipulate all given graph types is described in the following section.

3.5.3 Graph Manipulation Language — GRM

A graph manipulation language (GRM) has been created for the purpose of easy graph specification and manipulation (Drátovský, 1978). It serves to specify and modify input data (graph structures) for further processing. GRM is an example of a general tool for computer processing of topological descriptive models. GRM is a conversational tool, enabling the user at any time to display and inspect the actual status of the created structure. A program written in GRM is a sequence of commands, all of which have the same syntax. The individual GRM commands are completely independent of each other, except that they all process the same created structure. The commands are processed sequentially in their turn, each command being translated and immediately interpreted. GRM allows easy definition and implementation of new commands and is therefore flexible for new requirements related to the purpose for which it was originally designed.

As an example of the use of GRM, let us describe the graph associated with the shaping mechanism of Figure 3.11 and let us find the shortest path between nodes FRAME and SLEEVE2. We shall use the designation

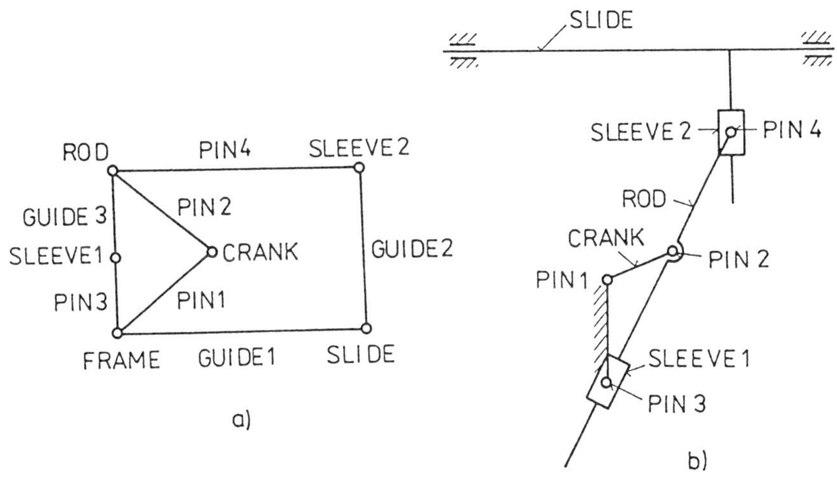

Figure 3.15. Shaping machine mechanism and its associated graph.

according to Figure 3.15.

DIsorientate	
GRaph	((FRAME CRANK ROD SLEEVE1 SLEEVE2 SLIDE) (PIN1 FRAME CRANK) (PIN2 CRANK ROD) (PIN 3 FRAME SLEEVE1) (PIN4 ROD SLEEVE2) (GUIDE1 FRAME SLIDE) (GUIDE2 SLIDE SLEEVE2))
Attributes	((FRAME (FRAME ( ))) (PAIR (PIN1 (ROT)) (PIN2 (ROT)) (PIN3 (ROT)) (PIN4 (ROT)) (GUIDE1 (PRISM))(GUIDE2 (PRISM)))
List	
WAy	(FRAME SLEEVE2)
ENd	

(Note eg. "DIsorientate" means that one has to type in only "DI"; the system will recognise the rest of the command automatically.) The first command sets the non-oriented graph mode. The second command defines the graph with the help of a node list and an expression list of the type (edge, node, node). The third command assigns a node or edge p property i with value d with the help of a list of expressions of type $(i(p(d)))$. The List command causes the printout of information on the given graph. The WAy command searches for the shortest path between given nodes. The last command ENd terminates the program run.

Chapter 4

Description of Point and Body Motion

The basis for all methods of MBS analysis and synthesis is the kinematics of point and body motion. The motion of an isolated point, i.e. of a body modelled by a point, is, however, quite rare in practice. By point motion we understand the motion of a significant point of a certain body. The motion of an isolated body is similarly quite rare. We usually study the systems of bodies constrained by kinematic pairs and force elements – multibody systems. However, the description of the motion of the whole multibody system is based on the description of motion of individual bodies and significant points.

The kinematics of a point of a body can be solved from various viewpoints and using various formalisms. We shall use the analytical approach as our starting point in this book. For the description and solution of problems we shall use the matrix formalism, the main advantage of which is its conciseness and the easy programming of resultant relations. This advantage is so significant that we use it even if the computational method is based on some other formalism, e.g. quaternion formalism (Wehage, 1984). The drawback of matrix formalism is its lack of lucidity because of the algebraic description of vector or tensor quantities and relationships between them.

For an easier understanding of the meaning of matrix symbols in kinematics, we shall first introduce the main results in traditional vector formalism and show their matrix description. We shall then use just the matrix formalism.

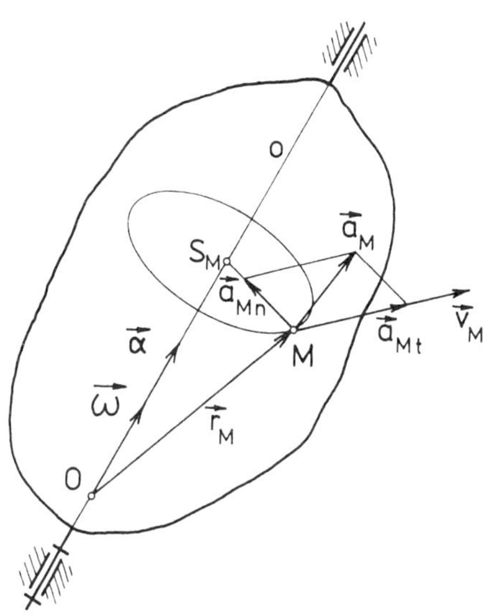

Figure 4.1. Revolute motion of a body.

4.1 Vector Description

4.1.1 Point and Body Motion

The classification of spatial motion of bodies usually distinguishes between simple (translational, rotational, spherical) and complex (screw, general) motion.

The simplest spatial motion is translational motion. A line connecting any two points of a body carrying out translational motion maintains the same direction in space. Paths of all points of the body are congruent curves distinguished only by their displacement; the velocities and accelerations of all points are the same. Translational motion of a body is given by the motion of an arbitrary point of the body. The state of motion of a body is described by vectors of velocity $\vec{v}$ and acceleration $\vec{a}$. These are free vectors as they can be assigned to any point.

Revolute motion is carried out by a body containing one line that is permanently stationary. This line is the axis of revolution, Figure 4.1. On this line lie the angular velocity vector $\vec{\omega}$ and the angular acceleration vector $\vec{\alpha}$. These two vectors specify the state of motion of the body. Both vectors are bound to the axis o. The velocity $\vec{v}_M$ and acceleration $\vec{a}_M$ of an arbitrary point M, whose position relative to the origin on the axis of revolution is given by the radius vector $\vec{r}_M$, are given by expressions

$$\vec{v}_M = \vec{\omega} \times \vec{r}_M \tag{4.1}$$
$$\vec{a}_M = \vec{\alpha} \times \vec{r}_M + \vec{\omega} \times (\vec{\omega} \times \vec{r}_M) = \vec{a}_{Mt} + \vec{a}_{Mn} \tag{4.2}$$

where $\vec{a}_{Mt}$ is the tangential and $\vec{a}_{Mn}$ the normal component of acceleration; $\times$ is the symbol of vector product.

To derive the above expressions we would use the relationship

$$\left[\frac{\mathrm{d}\vec{c}}{\mathrm{d}t}\right]_a = \left[\frac{\mathrm{d}\vec{c}}{\mathrm{d}t}\right]_b + \vec{\omega}_{ab} \times \vec{c} \tag{4.3}$$

for the derivative of an arbitrary vector $\vec{c}$ defined in space b which is moving relative to space a with angular velocity $\vec{\omega}_{ab}$ (Bottema and Roth, 1979). The bracket subscripts state the space in which the differentiation is carried out.

A body for which one point is permanently stationary carries out a spherical motion. This point is the centre O of spherical motion. For such motion the *Euler Theorem* (1776) holds: Any displacement of a body carrying out a spherical motion can be expressed as revolution about a certain line passing through the centre of spherical motion. This line is called the orientational axis of rotation (Nikravesh, 1988). The state of motion of a spherically moving body is described by vectors $\vec{\omega}$ of angular velocity and $\vec{\alpha}$ of angular acceleration. The two vectors are in general not collinear; they are bounded to the centre of spherical motion, Figure 4.2. The velocity $\vec{v}_M$ and acceleration $\vec{a}_M$ of an arbitrary point can be computed using relations that are formally the same as (4.1), (4.2). The physical reality is, however, not the same. During revolution the vectors $\vec{\omega}, \vec{\alpha}$ lie in a fixed axis of revolution; in spherical motion the two vectors lie in different carrying lines, the position of which changes with time.

General spatial motion is the most general type of motion. It is a complex motion and for its solution it is convenient to decompose it into simple motions. Two methods of decomposition are currently in use.

One of these is based on *Chasles Theorem* (1831) (Bottema and Roth, 1979) stating that: a general displacement in three-dimensional space can be represented by a unique screw displacement, i.e. a general spatial instantaneous motion of a body can be decomposed to the translation in the direction of a certain axis and to the rotation about the same axis.

We shall call the second method the basic decomposition. According to this, we can imagine the general spatial motion as composed of two simultaneous motions: the carrying (translational) motion given by the motion of a selected reference point R, and the spherical motion relative to the translational motion with the reference point R as its centre. The state

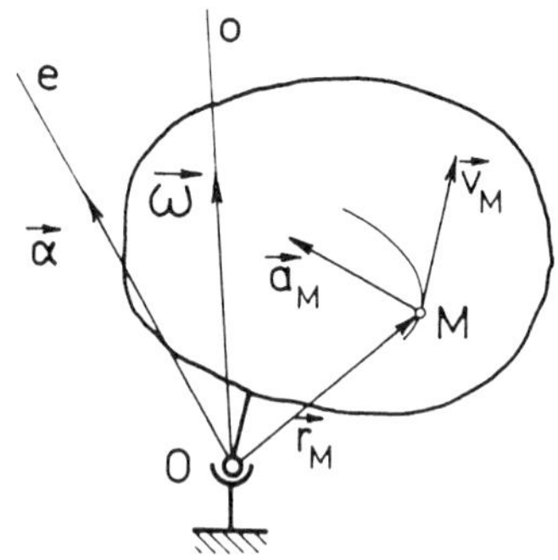

Figure 4.2. Spherical motion of a body.

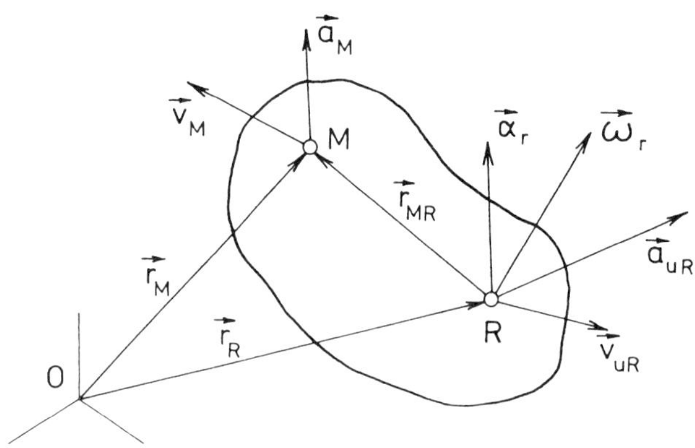

Figure 4.3. General spatial motion of a body.

of motion of a body is determined by five vectors: radius vector $\vec{r}_R$ of the reference point R, translational velocity $\vec{v}_{uR}$, translational acceleration $\vec{a}_{uR}$, relative angular velocity $\vec{\omega}_r$ and relative angular acceleration $\vec{\alpha}_r$, Figure 4.3. Each vector is of a different character. The radius vector $\vec{r}_R$ is bound to a fixed origin, vectors $\vec{v}_{uR}, \vec{a}_{uR}$ are free vectors, while vectors $\vec{\omega}_r, \alpha_r$ are bound to a (moving) reference point. All of them are functions of time. The position of an arbitrary point M in basic space (Figure 4.3) is given by the equation

$$\vec{r}_M = \vec{r}_{MR} + \vec{r}_R \tag{4.4}$$

Its velocity $\vec{v}_M$ and acceleration $\vec{a}_M$ are given by the following relations

$$\vec{v}_M = \frac{\mathrm{d}\vec{r}_M}{\mathrm{d}t} = \vec{v}_r + \vec{v}_u = \vec{\omega}_r \times \vec{r}_{MR} + \vec{v}_{uR} \tag{4.5}$$

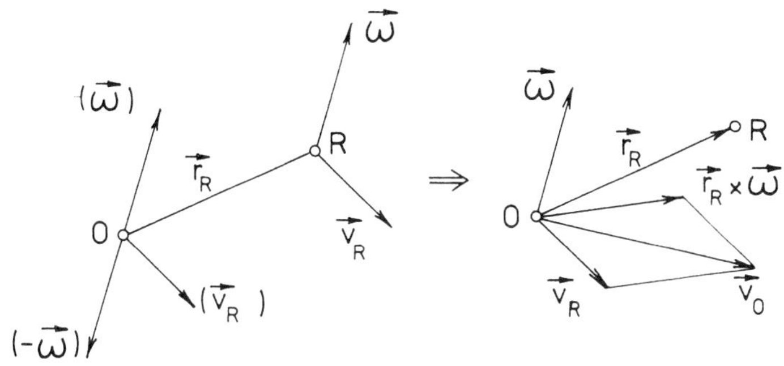

Figure 4.4. Velocities in different reference points.

$$\begin{aligned} \vec{a}_M &= \frac{\mathrm{d}\vec{v}_M}{\mathrm{d}t} = \vec{a}_r + \vec{a}_u = \vec{\alpha}_r \times \vec{r}_{MR} + \vec{\omega}_r \times (\vec{\omega}_r \times \vec{r}_{MR}) + \\ &\quad + \vec{a}_{uR} = \vec{a}_{rt} + \vec{a}_{rn} + \vec{a}_u \end{aligned} \tag{4.6}$$

where $\vec{a}_{rt}$ is the tangential, and $\vec{a}_{rn}$ the normal component of the relative acceleration and $\vec{r}_{MR}$ is the radius vector giving the position of point M relative to point R.

As the motion is described relative to a chosen point R, the question arises as to how this description changes when another reference point, e.g. O, is chosen instead of R. It can be proved (Bottema and Roth, 1979) that the angular quantities do not depend on the choice of reference point. It is therefore not necessary to use subscripts for them. However, the kinematic quantities characterising translational motion do change. These can be computed using equations (4.5) and (4.6) for the point (leaving out the subscript u).

$$\vec{v}_O = \vec{\omega} \times (-\vec{r}_R) + \vec{v}_R = \vec{r}_R \times \vec{\omega} + \vec{v}_R \tag{4.7}$$

$$\begin{aligned} \vec{a}_O &= \vec{\alpha} \times (-\vec{r}_R) + \vec{\omega} \times (\vec{\omega} \times (-\vec{r}_R)) + \vec{a}_R = \\ &= \vec{r}_R \times \vec{\alpha} + \vec{\omega} \times (\vec{r}_R \times \vec{\omega}) + \vec{a}_R \end{aligned} \tag{4.8}$$

The above relations can be interpreted as follows.

Let us introduce in point O two vectors $\vec{\omega}, -\vec{\omega}$ with the same absolute values and the same directions with opposite orientation, Figure 4.4. This can be interpreted as moving vector $\vec{\omega}$ by parallel translation to point O and creating a pair of vectors $\vec{\omega}, -\vec{\omega}$. This pair can be expressed (Sommerfeld, 1952) by the free vector $\vec{r}_R \times \vec{\omega}$. Adding it to the vector $\vec{v}_R$ we get, according to (4.7), the velocity $\vec{v}_O$ of the translational motion of the given reference point O. A free acceleration vector $\vec{\omega} \times (\vec{r}_R \times \vec{\omega})$ dependent on velocities

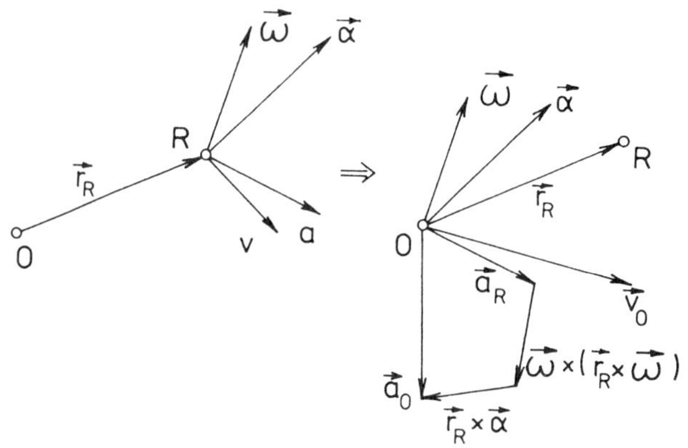

Figure 4.5. Accelerations in different reference points.

also corresponds to the translation of vector $\vec{\omega}$. Manipulating vector $\vec{\alpha}$ in a similar way to vector $\vec{\omega}$ gives the translational acceleration $\vec{a}_O$ at reference point O in accordance with (4.8), Figure 4.5.

Let us return once more to the problem of position and let us describe it using scalars, Figure 4.6. It is the problem of motion of point M of a body performing a general spatial motion. We have chosen a Cartesian system of coordinates $O_b x_b y_b z_b$ in the space of the body and moving along with the body. The basic system $O_a x_a y_a z_a$ is stationary. System $O_b x'_a y'_a z'_a$ is a translationally moving space; O_b is the point of reference.

Designating the coordinates of point M in space a by symbols x_{aM}, y_{aM}, z_{aM}, those of the same point M in space b by symbols x_{bM}, y_{bM}, z_{bM} and those of point O_b in system a by symbols a_1, a_2, a_3, we can, using analytical geometry, substitute the vector equation

$$\vec{r}_{aM} = \vec{r}_{bM} + \vec{r}_{ab}$$

by three scalar equations

$$\begin{aligned} x_{aM} &= a_{11}x_{bM} + a_{12}y_{bM} + a_{13}z_{bM} + a_1 \\ y_{aM} &= a_{21}x_{bM} + a_{22}y_{bM} + a_{23}z_{bM} + a_2 \\ z_{aM} &= a_{31}x_{bM} + a_{32}y_{bM} + a_{33}z_{bM} + a_3 \end{aligned} \tag{4.9}$$

where a_{ij} is the direction cosine of the angle between i-th axis of system a and j-th axis of system b.

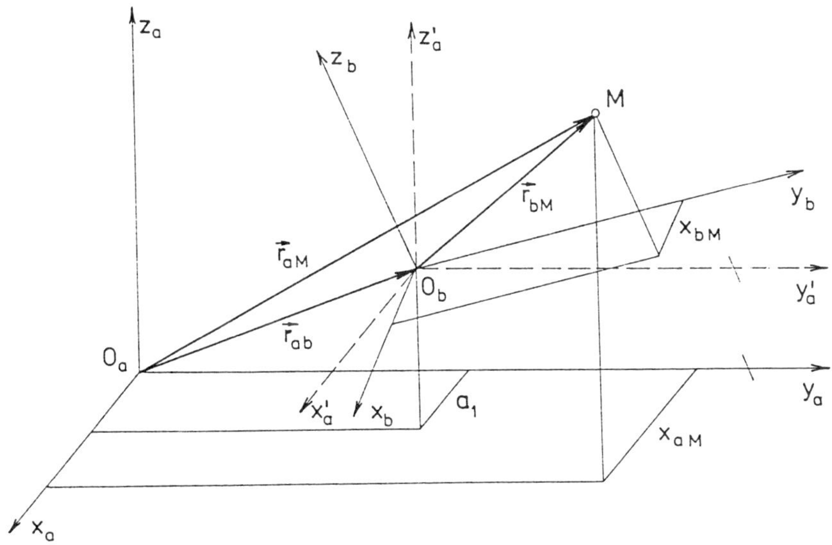

Figure 4.6. Scalar description of point M motion.

4.1.2 Simultaneous Motion

The kinematics of point and body discussed in the previous sections also appears to be sufficient for the study of kinematic relations in a system of bodies. It is usually convenient to consider the motion of bodies as motion composed of partial simultaneous motions, as this is in any case the manner in which the motion of bodies is realized in practice. We shall give in this section general relations for the description of the state of motion of a body with two simultaneous motions both of which can be general spatial motions. The constructional realization of partial motions is usually simpler and would be contained in general relations as special cases. The more complicated cases of three or more simultaneous motions are briefly discussed at the end of this Section.

Let us use the designation in Figure 4.7. We shall examine the motion of body 3 determined by the motion of a system of coordinates $O_3x_3y_3z_3$ rigidly attached to body 3, with regard to a fixed system $O_1x_1y_1z_1$ of the basic frame (motion 13). We shall use the fact that body 3 also carries out a motion relative to body 2, whose motion is determined by the motion of a system which is firmly connected with it $O_2x_2y_2z_2$; this, in its turn, carries out a motion relative to the basic frame 1. All these motions are general

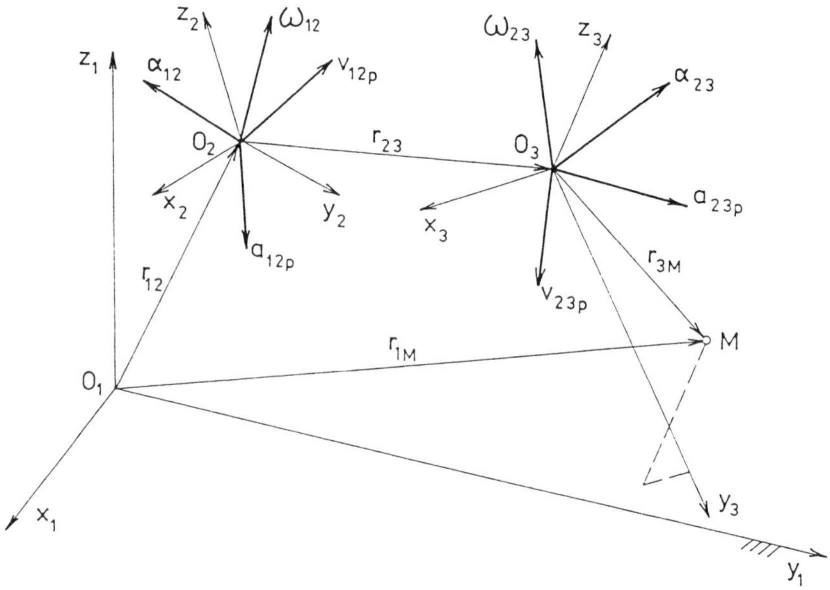

Figure 4.7. Simultaneous motion.

spatial motions and we shall express them using basic decomposition with origins O_2, O_3 as reference points, Figure 4.7.

Motion 12 is determined by vectors

$$\vec{r}_{12},\ \vec{v}_{12p} = \left[\frac{\mathrm{d}\vec{r}_{12}}{\mathrm{d}t}\right]_1,\ \vec{a}_{12p} = \left[\frac{\mathrm{d}\vec{v}_{12p}}{\mathrm{d}t}\right]_1,\ \vec{\omega}_{12},\ \vec{\alpha}_{12} = \left[\frac{\mathrm{d}\vec{\omega}_{12}}{\mathrm{d}t}\right]_1,$$

Motion 23 is determined by vectors

$$\vec{r}_{23},\ \vec{v}_{23p} = \left[\frac{\mathrm{d}\vec{r}_{23}}{\mathrm{d}t}\right]_2,\ \vec{a}_{23p} = \left[\frac{\mathrm{d}\vec{v}_{23p}}{\mathrm{d}t}\right]_2,\ \vec{\omega}_{23},\ \vec{\alpha}_{23} = \left[\frac{\mathrm{d}\vec{\omega}_{23}}{\mathrm{d}t}\right]_2,$$

where the subscript p designates carrying (translational) motion of respective basic decompositions, and bracket subscripts designate the space in which the derivation is carried out. To describe the fact that the resultant motion $3:1$ is composed of translational motion $2:1$ and relative motion $3:2$ we shall use the following symbolic notation

$$13 = 12 + 23 \tag{4.10}$$

Let us first determine the angular quantities of the resultant motion.

The time derivative of any vector quantity in space 1 is, according to (4.3)

$$\begin{aligned}\left[\frac{d\vec{c}}{dt}\right]_1 &= \left[\frac{d\vec{c}}{dt}\right]_2 + \vec{\omega}_{12} \times \vec{c} \\ \left[\frac{d\vec{c}}{dt}\right]_1 &= \left[\frac{d\vec{c}}{dt}\right]_3 + \vec{\omega}_{13} \times \vec{c}\end{aligned}$$

Subtracting the two equations, we obtain

$$\left[\frac{d\vec{c}}{dt}\right]_2 = \left[\frac{d\vec{c}}{dt}\right]_3 + (\vec{\omega}_{13} - \vec{\omega}_{12}) \times \vec{c}$$

and according to (4.3) also

$$\left[\frac{d\vec{c}}{dt}\right]_2 = \left[\frac{d\vec{c}}{dt}\right]_3 + \vec{\omega}_{23} \times \vec{c}$$

Comparing the two relations gives

$$\vec{\omega}_{13} = \vec{\omega}_{12} + \vec{\omega}_{23} \tag{4.11}$$

Hence in simultaneous motion we obtain the angular velocity as the vector sum of velocities of partial motions.

The time derivative of (4.11) in space 1 yields

$$\begin{aligned}\left[\frac{d\vec{\omega}_{13}}{dt}\right]_1 &= \left[\frac{d\vec{\omega}_{12}}{dt}\right]_1 + \left[\frac{d\vec{\omega}_{23}}{dt}\right]_1 \\ \vec{\alpha}_{13} &= \vec{\alpha}_{12} + \left[\frac{d\vec{\omega}_{23}}{dt}\right]_2 + \vec{\omega}_{12} \times \vec{\omega}_{23} \\ \vec{\alpha}_{13} &= \vec{\alpha}_{12} + \vec{\alpha}_{23} + \vec{\alpha}_R \end{aligned} \tag{4.12}$$

where $\vec{\alpha}_R$ is Resal's acceleration. In this case,

$$\vec{\alpha}_R = \vec{\omega}_{12} \times \vec{\omega}_{23} \tag{4.13}$$

The position of point M (fixed in space 3) in basic space is given by the radius vector

$$\vec{r}_{1M} = \vec{r}_{12} + \vec{r}_{23} + \vec{r}_{3M} \tag{4.14}$$

For the velocity we obtain the relation

$$\vec{v}_{1M} = \left[\frac{d\vec{r}_{1M}}{dt}\right]_1 = \left[\frac{d\vec{r}_{12}}{dt}\right]_1 + \left[\frac{d\vec{r}_{23}}{dt}\right]_1 + \left[\frac{d\vec{r}_{3M}}{dt}\right]_1$$

where

$$\begin{aligned}
\left[\frac{\mathrm{d}\vec{r}_{12}}{\mathrm{d}t}\right]_1 &= \vec{v}_{12p} \\
\left[\frac{\mathrm{d}\vec{r}_{23}}{\mathrm{d}t}\right]_1 &= \left[\frac{\mathrm{d}\vec{r}_{23}}{\mathrm{d}t}\right]_2 + \vec{\omega}_{12} \times \vec{r}_{23} = \vec{v}_{23p} + \vec{\omega}_{12} \times \vec{r}_{23} \\
\left[\frac{\mathrm{d}\vec{r}_{3M}}{\mathrm{d}t}\right]_1 &= \left[\frac{\mathrm{d}\vec{r}_{3M}}{\mathrm{d}t}\right]_3 + \vec{\omega}_{13} \times \vec{r}_{3M} = \vec{0} + (\vec{\omega}_{12} + \vec{\omega}_{23}) \times \vec{r}_{3M}
\end{aligned}$$

It follows that

$$\begin{aligned}
\vec{v}_{1M} &= \vec{v}_{12p} + \vec{\omega}_{12} \times (\vec{r}_{23} + \vec{r}_{3M}) + \vec{v}_{23p} + \vec{\omega}_{23} \times \vec{r}_{3M} \\
\vec{v}_{1M} &= \vec{v}_{12p} + \vec{v}_{12s} + \vec{v}_{23p} + \vec{v}_{23s} \\
\vec{v}_{1M} &= \vec{v}_{12M} + \vec{v}_{23M} \qquad (4.15)
\end{aligned}$$

where subscript s designates (relative) spherical motion of respective basic decomposition. The resultant (absolute) velocity of point M is the vector sum of translational and relative velocities. The translational velocity $\vec{v}_{12M}$ of point M corresponds to the assumption that point M is, for motion 12, a point on body 2. Similarly the relative velocity $\vec{v}_{23M}$ is the velocity of point M in the independently considered relative motion 23.

The solution for acceleration is more complicated. Let us start using the definition

$$\vec{a}_{1M} = \left[\frac{\mathrm{d}\vec{v}_{1M}}{\mathrm{d}t}\right]_1 = \left[\frac{\mathrm{d}\vec{v}_{12M}}{\mathrm{d}t}\right]_1 + \left[\frac{\mathrm{d}\vec{v}_{23M}}{\mathrm{d}t}\right]_1$$

where

$$\begin{aligned}
\left[\frac{\mathrm{d}\vec{v}_{12M}}{\mathrm{d}t}\right]_1 &= \left[\frac{\mathrm{d}\vec{v}_{12p}}{\mathrm{d}t}\right]_1 + \left[\frac{\mathrm{d}(\vec{\omega}_{12} \times (\vec{r}_{23} \times \vec{r}_{3M}))}{\mathrm{d}t}\right]_1 = \\
&= \vec{a}_{12p} + \vec{\alpha}_{12} \times \vec{r}_{2M} + \vec{\omega}_{12} \times \left(\left[\frac{\mathrm{d}\vec{r}_{23}}{\mathrm{d}t}\right]_1 + \left[\frac{\mathrm{d}\vec{r}_{3M}}{\mathrm{d}t}\right]_1\right) = \\
&= \vec{a}_{12p} + \vec{\alpha}_{12} \times \vec{r}_{2M} + \vec{\omega}_{12} \times (\vec{v}_{23p} + \vec{\omega}_{12} \times \vec{r}_{23} + \\
&\quad + \vec{\omega}_{13} \times \vec{r}_{3M}) = \vec{a}_{12p} + \vec{\alpha}_{12} \times \vec{r}_{2M} + \vec{\omega}_{12} \times \vec{v}_{23M} + \\
&\quad + \vec{\omega}_{12} \times (\vec{\omega}_{12} \times \vec{r}_{2M}) = \\
&= \vec{a}_{12p} + \vec{a}_{12Mt} + \vec{\omega}_{12} \times \vec{v}_{23M} + \vec{a}_{12Mn} \\
\left[\frac{\mathrm{d}\vec{v}_{23M}}{\mathrm{d}t}\right]_1 &= \left[\frac{\mathrm{d}\vec{v}_{23p}}{\mathrm{d}t}\right]_1 + \left[\frac{\mathrm{d}(\vec{\omega}_{23} \times \vec{r}_{3M})}{\mathrm{d}t}\right]_1 = \\
&= \vec{a}_{23p} + \vec{\omega}_{12} \times \vec{v}_{23p} + \left[\frac{\mathrm{d}(\vec{\omega}_{23} \times \vec{r}_{3M})}{\mathrm{d}t}\right]_2 + \vec{\omega}_{12} \times
\end{aligned}$$

$$
\begin{aligned}
& \times(\vec{\omega}_{23} \times \vec{r}_{3M}) = \vec{a}_{23p} + \vec{\omega}_{12} \times \vec{v}_{23p} + \vec{\alpha}_{23} \times \vec{r}_{3M} + \\
& + \vec{\omega}_{23} \times (\vec{\omega}_{23} \times \vec{r}_{3M}) + \vec{\omega}_{12} \times (\vec{\omega}_{23} \times \vec{r}_{3M}) = \vec{a}_{23p} + \\
& + \vec{\alpha}_{23} \times \vec{r}_{3M} + \vec{\omega}_{23} \times (\vec{\omega}_{23} \times \vec{r}_{3M}) + \vec{\omega}_{12} \times \vec{v}_{23M} = \\
= \;& \vec{a}_{23p} + \vec{a}_{23Mt} + \vec{a}_{23Mn} + \vec{\omega}_{12} \times \vec{v}_{23M}
\end{aligned}
$$

and finally

$$
\begin{aligned}
\vec{a}_{1M} &= \vec{a}_{1p} + \vec{a}_{12Mt} + \vec{a}_{12Mn} + \vec{a}_{23p} + \vec{a}_{23Mt} + \vec{a}_{23Mn} + 2\vec{\omega}_{12} \times \vec{v}_{23M} \\
\vec{a}_{1M} &= \vec{a}_{12} + \vec{a}_{23} + \vec{a}_{Cor} \qquad (4.16)
\end{aligned}
$$

The resultant (absolute) acceleration of point M is the vector sum of translational, relative and Coriolis accelerations, where the latter is defined by the relation

$$\vec{a}_{Cor} = 2\vec{\omega}_{12} \times \vec{v}_{23M} \qquad (4.17)$$

Subscript t in the above relations designates the tangential component of acceleration corresponding to the axis of angular acceleration of relative spherical motion. Subscript n designates the normal component relative to the actual axis of revolution of the relative spherical motion.

The relations for the calculation of kinematic quantities of motion of body n composed of $n-1$ simultaneous motions can be obtained by generalising the above procedure. They are as follows

$$1n = 12 + 23 + 34 + \cdots + (n-1)n \qquad (4.18)$$

$$\vec{\omega}_{1n} = \sum_{j=1}^{n-1} \vec{\omega}_{j,j+1} \qquad (4.19)$$

$$\vec{\alpha}_{1n} = \sum_{j=1}^{n-1} \vec{\alpha}_{j,j+1} + \sum_{j=1}^{n-1} \left(\vec{\omega}_{j,j+1} \times \sum_{k=j+1}^{n-2} \vec{\omega}_{k,k+1} \right) \qquad (4.20)$$

$$\vec{v}_{1M} = \sum_{j=1}^{n-1} \vec{v}_{j,j+1,M} \qquad (4.21)$$

$$\vec{a}_{1M} = \sum_{j=1}^{n-1} \vec{a}_{j,j+1,M} + 2\sum_{j=1}^{n-1} \left(\vec{\omega}_{j,j+1} \times \sum_{k=j+1}^{n-2} \vec{v}_{k,k+1,M} \right) \qquad (4.22)$$

For example for three simultaneous motions

$$
\begin{aligned}
14 &= 12 + 23 + 34 \\
\vec{\omega}_{14} &= \vec{\omega}_{12} + \vec{\omega}_{23} + \vec{\omega}_{34} \\
\vec{\alpha}_{14} &= \vec{\alpha}_{12} + \vec{\alpha}_{23} + \vec{\alpha}_{34} + \vec{\omega}_{12} \times (\vec{\omega}_{23} + \vec{\omega}_{34}) + \vec{\omega}_{23} \times \vec{\omega}_{34} \\
\vec{v}_{1M} &= \vec{v}_{12M} + \vec{v}_{23M} + \vec{v}_{34M} \\
\vec{a}_{1M} &= \vec{a}_{12M} + \vec{a}_{23M} + \vec{a}_{34M} + 2(\vec{\omega}_{12} \times (\vec{v}_{23M} + \vec{v}_{34M}) + \vec{\omega}_{23} \times \vec{v}_{34M})
\end{aligned}
$$

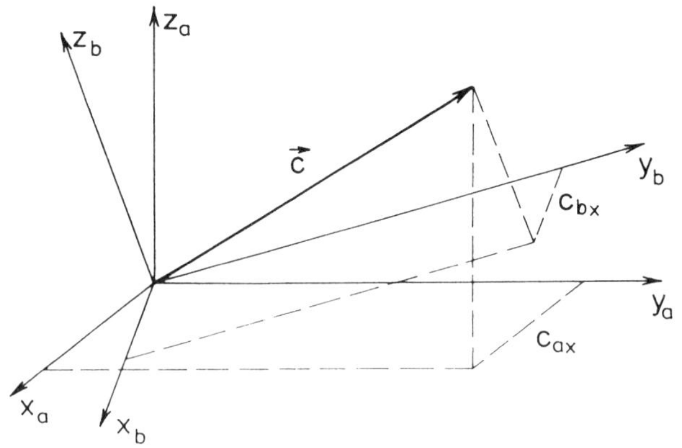

Figure 4.8. Vector coordinates in different spaces.

4.1.3 Vector Quantities, Expressions and Transformations

A vector quantity (or, more concisely, a vector) in a three-dimensional space can be described as an ordered triplet of numbers which signify the oriented length of vector projections on to the axes of coordinates. These numbers are the vector coordinates.

It is evident that the same vector can be described by coordinates in various systems of coordinates. For example, vector $\vec{c}$ from Figure 4.8. can be written alternatively using column matrices as follows

$$\begin{aligned}
\vec{c}_a &\equiv \begin{bmatrix} c_x \\ c_y \\ c_z \end{bmatrix}_a \equiv [c_x, c_y, c_z]_a^T \equiv \lceil c_x, c_y, c_z \rceil_a \\
\vec{c}_b &\equiv \begin{bmatrix} c_x \\ c_y \\ c_z \end{bmatrix}_b \equiv [c_x, c_y, c_z]_b^T \equiv \lceil c_x, c_y, c_z \rceil_b \qquad (4.23)
\end{aligned}$$

where the subscript designates the space in which the coordinates are expressed.

The vector coordinates as introduced above are scalar quantities. It is, however, also possible to express vectors in a right-handed Cartesian system as a linear combination of perpendicular unit vectors $\vec{i}, \vec{j}, \vec{k}$ (the so-called base of the given space). The coefficients of the linear combination are vector coordinates. For example

$$\vec{c}_a = c_x \begin{bmatrix} 1 \\ 0 \\ 0 \end{bmatrix}_a + c_y \begin{bmatrix} 0 \\ 1 \\ 0 \end{bmatrix}_a + c_z \begin{bmatrix} 0 \\ 0 \\ 1 \end{bmatrix}_a = c_x\vec{i}_a + c_y\vec{j}_a + c_z\vec{k}_a \qquad (4.24)$$

Expressions $c_x\vec{i}_a, c_y\vec{j}_a, c_z\vec{k}_a$ are the so-called components of vector $\vec{c}_a$. They are also vectors.

We shall now introduce for vectors in (4.4) and (4.9), respectively, column matrices $\mathbf{u}_{aM} = \lceil x_{aM}, y_{aM}, z_{aM} \rceil$, $\mathbf{u}_{bM} = \lceil x_{bM}, y_{bM}, z_{bM} \rceil$, $\mathbf{u}_{ab} = \lceil a_1, a_2, a_3 \rceil$. Using the above notation changes equation (4.9) into the following form

$$\mathbf{u}_{aM} = \mathbf{S}_{ab}\mathbf{u}_{bM} + \mathbf{u}_{ab}, \qquad (4.25)$$

where $\mathbf{S}_{ab} = [a_{ij}]$, $i, j = 1, 2, 3$ is the matrix of directional cosines. Equation (4.25) is equivalent to the vector equation (4.4).

Only vectors expressed in the same system can be compared. For the equation (4.4) we implicitly assumed all vectors to be expressed in the same basic space a. The equivalence of equations (4.4) and (4.25) requires the vector components contained in (4.25) also to be expressed in system a. This is not evident in the notation used. Should it be necessary to state precisely that, for example, vector $\vec{u}_{bM}$ is expressed by its components in space a, we shall use the symbol of space a as the first subscript, i.e. $\mathbf{u}_{a,bM}$. The equation (4.4) thus becomes

$$\mathbf{u}_{a,aM} = \mathbf{u}_{a,bM} + \mathbf{u}_{a,ab} \qquad (4.26)$$

Comparing (4.26) and (4.25) we obtain (Figure 4.6)

$$\mathbf{u}_{a,bM} = \mathbf{u}_{a',bM} = \mathbf{S}_{ab}\mathbf{u}_{b,bM} \qquad (4.27)$$

as systems a, a' are only translationally displaced. The transformational relation quoted holds generally for any vector quantity $\mathbf{c}$, Figure 4.8, in the form

$$\mathbf{c}_a = \mathbf{S}_{ab}\mathbf{c}_b \qquad (4.28)$$

This property is so fundamental that it enables a definition to be given of vectors as quantities complying with the given transformation.

The matrix which expresses the rotational displacement of the two systems is

$$\mathbf{S}_{ab} = [\mathbf{i}_{a,b}, \mathbf{j}_{a,b}, \mathbf{k}_{a,b}] = \begin{bmatrix} c x_a x_b & c x_a y_b & c x_a z_b \\ c y_a x_b & c y_a y_b & c y_a z_b \\ c z_a x_b & c z_a y_b & c z_a z_b \end{bmatrix} \qquad (4.29)$$

where $cy_a x_b$ is the cosine of the angle of axes y_a, x_b. The matrix is orthogonal and

$$\mathbf{S}_{ab}^{-1} = \mathbf{S}_{ab}^T = \mathbf{S}_{ba} \tag{4.30}$$

For right-handed systems

$$\det\mathbf{S}_{ab} = +1 \tag{4.31}$$

Thus transformation (4.28) is orthogonal.

An important consequence follows from the problem of eigenvalues of matrix $\mathbf{S}_{ab}$ defined by the relation

$$\mathbf{S}_{ab}\mathbf{p}_{b\nu} = \lambda_\nu \mathbf{p}_{b\nu}, \quad \nu = 1, 2, 3 \tag{4.32}$$

where λ_ν is the ν-th eigenvalue and $\mathbf{p}_{b\nu}$ the ν-th eigenvector. It can be proved that $\lambda_1 = 1$ and $\lambda_{2,3}$ are complex associated for any matrix $\mathbf{S}_{ab}$ (Goldstein, 1950). Substituting $\lambda_1 = 1$ into (4.32) and comparing with (4.27) we obtain

$$\mathbf{u}_{a,bM} = \mathbf{u}_{b,bM} \tag{4.33}$$

Hence, for any matrix $\mathbf{S}_{ab}$, there is a real vector with the same coordinates in spaces a and b. The result quoted enables a more precise formulation to be given of the *Euler Theorem* in Section 4.1: Any displacement of a body with one fixed point can be carried out as a rotation of a certain angle around an axis which coincides with the fixed point and whose position is determined by the eigenvector $\mathbf{p}_{b1}$ corresponding to eigenvalue $\lambda_1 = 1$ from (4.32).

The relations describing MBS kinematics contain vectors and their products. Further we shall use equivalently vectors $\vec{c}$ and their column-wise matrix representations $\mathbf{c}$ and we shall not distinguish between vectors and column-wise matrices. We shall use the following vector or matrix notation for the scalar product of vectors $\mathbf{p}$ and $\mathbf{q}$

$$\mathbf{p} \cdot \mathbf{q} = \mathbf{p}^T\mathbf{q} = \mathbf{q}^T\mathbf{p} = \mathrm{tr}(\mathbf{p}\mathbf{q}^T) = \mathrm{tr}(\mathbf{q}\mathbf{p}^T) \tag{4.34}$$

The scalar product is an important means of calculating the angle φ of two vectors. In this case

$$\mathbf{p} \cdot \mathbf{q} = |\mathbf{p}||\mathbf{q}| \cos\varphi \tag{4.35}$$

The result of a scalar product is a scalar quantity (tensor of order zero) and is therefore independent of the choice of system of coordinates. The scalar product is thus invariant in orthogonal transformation. This is proved by the following. Let

$$\mathbf{p}_a = \mathbf{S}_{ab}\mathbf{p}_b, \quad \mathbf{q}_a = \mathbf{S}_{ab}\mathbf{q}_b \tag{4.36}$$

Hence

$$\mathbf{p}_a^T\mathbf{q}_a = (\mathbf{S}_{ab}\mathbf{p}_b)^T(\mathbf{S}_{ab}\mathbf{q}_b) = \mathbf{p}_b^T\mathbf{S}_{ab}^T\mathbf{S}_{ab}\mathbf{q}_b = \mathbf{p}_b^T\mathbf{q}_b \tag{4.37}$$

Let us consider instead of matrix $\mathbf{S}_{ab}$ a so far unknown transformation matrix. Computing the square of length of vector in system a and b yields, from an equation analogous to (4.37), for the fulfillment of the evident requirement of same vector length in both system, the necessary orthogonality of the transformation matrix. Similarly, it can be shown that an orthogonal transformation preserves the magnitude of the vector angle.

The result of a vector product of vectors $\mathbf{p}, \mathbf{q}$ is the vector

$$\mathbf{s} = \mathbf{p} \times \mathbf{q} = -\mathbf{q} \times \mathbf{p} \tag{4.38}$$

the coordinates of which can be obtained by operations

$$\mathbf{s} = \mathbf{P}\mathbf{q} = -\mathbf{Q}\mathbf{p} = \mathbf{Q}^T\mathbf{p} = \frac{1}{2}(\mathbf{P}\mathbf{q} - \mathbf{Q}\mathbf{p}) \tag{4.39}$$

where

$$\mathbf{P} = \begin{bmatrix} 0 & -p_z & p_y \\ p_z & 0 & -p_x \\ -p_y & p_x & 0 \end{bmatrix}, \quad \mathbf{Q} = \begin{bmatrix} 0 & -q_z & q_y \\ q_z & 0 & -q_x \\ -q_y & q_x & 0 \end{bmatrix} \tag{4.40}$$

are skew-symmetric operators of the vector product.

Vector product transformation from space b to space a is more complicated. Let us at first approach the problem without using vector product operators.

Let (4.36) hold again. Hence using (4.38)

$$\mathbf{s}_a = \mathbf{S}_{ab}\mathbf{s}_b = \mathbf{S}_{ab}(\mathbf{p}_b \times \mathbf{q}_b)$$

and also

$$\mathbf{s}_a = \mathbf{p}_a \times \mathbf{q}_a = \mathbf{S}_{ab}\mathbf{p}_b \times \mathbf{S}_{ab}\mathbf{q}_b$$

Hence by comparing

$$\mathbf{S}_{ab}(\mathbf{p}_b \times \mathbf{q}_b) = \mathbf{S}_{ab}\mathbf{p}_b \times \mathbf{S}_{ab}\mathbf{q}_b \tag{4.41}$$

Another approach to the problem of vector product transformation is possible by transforming the operator and the vector on which the operation is to be carried out. Using (4.40)

$$\begin{aligned} \mathbf{s}_a &= \mathbf{S}_{ab}\mathbf{s}_b = \mathbf{S}_{ab}(\mathbf{P}_b\mathbf{q}_b) = \mathbf{S}_{ab}\mathbf{P}_b\mathbf{q}_b = \mathbf{S}_{ab}\mathbf{P}_b\mathbf{S}_{ab}^T\mathbf{S}_{ab}\mathbf{q}_b = \\ &= (\mathbf{S}_{ab}\mathbf{P}_b\mathbf{S}_{ab}^T)(\mathbf{S}_{ab}\mathbf{q}_b) = \mathbf{P}_a\mathbf{q}_a \end{aligned} \tag{4.42}$$

Hence the relation for the transformation of vector product operator

$$\mathbf{P}_a = \mathbf{S}_{ab}\mathbf{P}_b\mathbf{S}_{ab}^T = \mathbf{S}_{ab}\mathbf{P}_b\mathbf{S}_{ba} \tag{4.43}$$

For other formulae, the procedure is analogous. We present only the fundamental ones using operators (4.40).

$$\begin{aligned} \mathbf{a} \times (\mathbf{b} \times \mathbf{c}) &= (\mathbf{a} \cdot \mathbf{c})\mathbf{b} - (\mathbf{a} \cdot \mathbf{b})\mathbf{c} = \mathbf{ABc} = (\mathbf{a}^T\mathbf{c})\mathbf{b} - (\mathbf{a}^T\mathbf{b})\mathbf{c} = \\ &= -\mathbf{ACb} = \mathbf{AC}^T\mathbf{b} \qquad (4.44) \\ (\mathbf{a} \times \mathbf{b}) \times \mathbf{c} &= (\mathbf{a} \cdot \mathbf{c})\mathbf{b} - (\mathbf{b} \cdot \mathbf{c})\mathbf{a} \qquad (4.45) \end{aligned}$$

4.1.4 Homogeneous Coordinates

In addition to the common expression of vectors in three-dimensional space through three coordinates, we can use so-called homogeneous coordinates. They represent in general an object in an n-dimensional space as entity in an $(n+1)$-dimensional space, where the corresponding perspective projection forms the original object in n-dimensional space. This can be expressed by adding a coordinate — scale factor — to each vector. We shall call such a vector an extended vector. For example, the radius vector $\mathbf{u}_{aM}$ of point M can be written in Cartesian coordinates $\mathbf{u}_{aM} = \lceil x_{aM}, y_{aM}, z_{aM} \rceil$. The same vector in homogeneous coordinates is the extended radius vector

$$\mathbf{r}_{aM} = \lceil a_1, a_2, a_3, a_4 \rceil \qquad (4.46)$$

where

$$x_{aM} = \frac{a_1}{a_4}, \quad y_{aM} = \frac{a_2}{a_4}, \quad z_{aM} = \frac{a_3}{a_4}$$

The same extended vector can be expressed by coordinates of different size which we obtain by multiplying the original coordinates including the scale factor by a certain non-zero constant. For example, vector $\lceil 2, 3, 4 \rceil$ can be expressed as an extended vector $\lceil 2, 3, 4, 1 \rceil$ or $\lceil 20, 30, 40, 10 \rceil$ or $\lceil -4, -6, -8, -2 \rceil$ etc.

The null vector corresponding to the origin is represented as $\lceil 0, 0, 0, w \rceil$ where w is an arbitrary non-zero scale factor. This is how bound vectors are expressed. Free vectors representing directions are written as extended vectors in the form $\lceil x, y, z, 0 \rceil$. Extended vector $\lceil 0, 0, 0, 0 \rceil$ is not defined.

4.2 Matrix Notation

The matrix notation of methods of MBS kinematic and dynamic analysis and synthesis is the basic tool used in this book. Its advantages are, above all, its compactness, the concise computational complexity and the notational richness of the contents, where one symbol describes geometrical quantities of a body as well as velocities and accelerations. This notational conciseness gains in significance especially when solving large systems.

4.2.1 Point and Body Motion

4.2.1.1 Position

Let us look again at equation (4.25), expressing the position of point M in system a. Its disadvantage is the sum in the equation. This follows especially from the fact that vector transformation from space b into space a is in general not a linear transformation. This is so only in the special case of spherical motion, where $\mathbf{u}_{ab} = \mathbf{0}$. One of the possibilities for making the description more concise is the use of homogeneous coordinates. This makes it possible in general to write a nonlinear vector transformation between two spaces in general spatial motion at least formally as a linear transformation of their homogeneous description. The importance of this is immense because of its conciseness, compactness and the ability to survey the description. We shall use this method in the following, choosing without loss of generality the value 1 for the scale factor of homogeneous coordinates.

We shall express the radius vectors $\mathbf{u}_{aM}, \mathbf{u}_{bM}$ of point M in space a and b using their extended radius vectors

$$\mathbf{r}_{aM} = \begin{bmatrix} \mathbf{u}_{aM} \\ 1 \end{bmatrix}, \quad \mathbf{r}_{bM} = \begin{bmatrix} \mathbf{u}_{bM} \\ 1 \end{bmatrix} \tag{4.47}$$

We shall further rewrite equation (4.25) expressing the decomposition of general spatial motion in the form

$$\begin{bmatrix} \mathbf{u}_{aM} \\ 1 \end{bmatrix} = \begin{bmatrix} \mathbf{S}_{ab} & \mathbf{u}_{ab} \\ \mathbf{0} & 1 \end{bmatrix} \begin{bmatrix} \mathbf{u}_{bM} \\ 1 \end{bmatrix} \tag{4.48}$$

which can be understood even completely formally as an addition of the identity $1 = 1$ to equation (4.25). In the detailed form of (4.48) we get

$$\left[\begin{array}{c} x_{aM} \\ y_{aM} \\ z_{aM} \\ \hline 1 \end{array}\right] = \left[\begin{array}{ccc|c} a_{11} & a_{12} & a_{13} & a_1 \\ a_{21} & a_{22} & a_{23} & a_2 \\ a_{31} & a_{32} & a_{33} & a_3 \\ \hline 0 & 0 & 0 & 1 \end{array}\right] \left[\begin{array}{c} x_{bM} \\ y_{bM} \\ z_{bM} \\ \hline 1 \end{array}\right] \tag{4.49}$$

which can be written as a symbolic matrix equation

$$\mathbf{r}_{aM} = \mathbf{T}_{ab}\mathbf{r}_{bM} \tag{4.50}$$

expressing a homogeneous transformation of extended vectors where the transformation matrix (transformation operator) is

$$\mathbf{T}_{ab} = \begin{bmatrix} \mathbf{S}_{ab} & \mathbf{u}_{ab} \\ \mathbf{0} & 1 \end{bmatrix} \tag{4.51}$$

Our understanding of the transformation (4.50) has been so far more or less static. Should body b be in motion relative to body a (the position of both bodies being given by the position of the respective coordinate systems with the same designation), we can interpret relation (4.50) as the expression of motion $b:a$. Then $\mathbf{r}_{aM}, \mathbf{T}_{ab}$ are functions of time, $\mathbf{r}_{bM}$ is a constant and the equation

$$\mathbf{r}_{aM}(t) = \mathbf{T}_{ab}(t)\mathbf{r}_{bM} \tag{4.52}$$

expresses the motion of an arbitrary point M of body b in space a. Its form

$$\mathbf{u}_{aM}(t) = \mathbf{S}_{ab}(t)\mathbf{u}_{bM} + \mathbf{u}_{ab}(t) \tag{4.53}$$

is the matrix formulation of the basic decomposition (4.4) of general spatial motion. Vector $\mathbf{u}_{ab}(t)$ describes the translational motion; matrix $\mathbf{S}_{ab}(t)$ expresses relative spherical motion.

Should point M also be in motion relative to space b, the equation (4.50) takes the form

$$\mathbf{r}_{aM}(t) = \mathbf{T}_{ab}(t)\mathbf{r}_{bM}(t) \tag{4.54}$$

In detail

$$\mathbf{u}_{aM}(t) = \mathbf{S}_{ab}(t)\mathbf{u}_{bM}(t) + \mathbf{u}_{ab}(t) \tag{4.55}$$

Both these latter equations express the kinematics of simultaneous motion of point and body.

In the following we shall need the matrix $\mathbf{T}_{ab}^{-1}$ expressing the inverse transformation to (4.50). This can be computed from its definition

$$\mathbf{T}_{ab}\mathbf{T}_{ab}^{-1} = \mathbf{T}_{ab}^{-1}\mathbf{T}_{ab} = \mathbf{E}_4, \tag{4.56}$$

where $\mathbf{E}_4$ is the unit matrix of dimension 4. Using divided matrices

$$\mathbf{T}_{ab}^{-1} = \begin{bmatrix} \mathbf{A} & \mathbf{B} \\ \mathbf{C} & D \end{bmatrix}, \quad \mathbf{E}_4 = \begin{bmatrix} \mathbf{E}_3 & \mathbf{0} \\ \mathbf{0}^T & 1 \end{bmatrix}$$

according to (4.56) we have

$$\begin{bmatrix} \mathbf{S}_{ab} & \mathbf{u}_{ab} \\ \mathbf{0}^T & 1 \end{bmatrix} \begin{bmatrix} \mathbf{A} & \mathbf{B} \\ \mathbf{C} & D \end{bmatrix} = \begin{bmatrix} \mathbf{E}_{33} & \mathbf{0} \\ \mathbf{0}^T & 1 \end{bmatrix}$$

Hence in detail

$$\begin{aligned} \mathbf{S}_{ab}\mathbf{A} + \mathbf{u}_{ab}\mathbf{C} &= \mathbf{E}_3 \\ \mathbf{S}_{ab}\mathbf{B} + \mathbf{u}_{ab}D &= \mathbf{0} \\ \mathbf{0}^T.\mathbf{A} + 1.\mathbf{C} &= \mathbf{0}^T \\ \mathbf{0}^T.\mathbf{B} + 1.D &= 1 \end{aligned}$$

and further

$$\mathbf{A} = \mathbf{S}_{ab}^{-1}, \quad \mathbf{B} = -\mathbf{S}_{ab}^{-1}\mathbf{u}_{ab}, \quad \mathbf{C} = \mathbf{0}^T, \quad D = 1$$

Using (4.30) also

$$\mathbf{T}_{ab}^{-1} = \mathbf{T}_{ba} = \begin{bmatrix} \mathbf{S}_{ab}^T & -\mathbf{S}_{ab}\mathbf{u}_{ab} \\ \mathbf{0} & 1 \end{bmatrix} \tag{4.57}$$

matrices $\mathbf{T}_{ab}, \mathbf{T}_{ab}^{-1}$ are, unlike $\mathbf{S}_{ab}, \mathbf{S}_{ab}^{-1}$, not orthogonal; in the case of determinants, however, the following relation holds

$$\det\mathbf{T}_{ab} = \det\mathbf{T}_{ab}^{-1} = \det\mathbf{S}_{ab} = \det\mathbf{S}_{ab}^{-1} = +1 \tag{4.58}$$

4.2.1.2 Velocity

As (4.50) is a linear transformation, we can introduce analogously the computation of velocities. The time derivative of equation (4.52), assuming $\mathbf{r}_{bM}$ = const, gives

$$\mathbf{v}_{aM} = \dot{\mathbf{T}}_{ab}\mathbf{r}_{bM} \tag{4.59}$$

where

$$\mathbf{v}_{aM} = \dot{\mathbf{r}}_{aM} = \lceil \dot{\mathbf{u}}_{aM}, 0 \rceil = \lceil \dot{x}_{aM}, \dot{y}_{aM}, \dot{z}_{aM}, 0 \rceil \tag{4.60}$$

is the extended velocity of point M in space a.

For solving MBS velocities it is practical to introduce a separate symbol for the velocity of a body. We therefore introduce matrix $\mathbf{V}_{b,ab}$ for the velocity of motion $b : a$, which is by definition

$$\dot{\mathbf{T}}_{ab} = \mathbf{T}_{ab}\mathbf{V}_{b,ab} \tag{4.61}$$

so that

$$\mathbf{V}_{b,ab} = \mathbf{T}_{ab}^{-1}\dot{\mathbf{T}}_{ab} = \begin{bmatrix} \mathbf{S}_{ab}^T\dot{\mathbf{S}}_{ab} & \mathbf{S}_{ab}^T\dot{\mathbf{u}}_{a,ab} \\ \mathbf{0} & 0 \end{bmatrix} \tag{4.62}$$

In the same way we introduce the matrix $\boldsymbol{\Omega}_{b,ab}$ for the angular velocity of $b : a$ motion by the relation

$$\dot{\mathbf{S}}_{ab} = \mathbf{S}_{ab}\boldsymbol{\Omega}_{b,ab} \tag{4.63}$$

so that

$$\boldsymbol{\Omega}_{b,ab} = \mathbf{S}_{ab}^{-1}\dot{\mathbf{S}}_{ab} = \mathbf{S}_{ab}^T\dot{\mathbf{S}}_{ab} \tag{4.64}$$

$\boldsymbol{\Omega}_{b,ab}$ is a skew-symmetric singular matrix. The above statement follows from the orthogonality of matrix $\mathbf{S}_{ab}$. Differentiating

$$\mathbf{S}_{ab}^T\mathbf{S}_{ab} = \mathbf{E}_3 \tag{4.65}$$

with respect to time, we get

$$\dot{\mathbf{S}}_{ab}^T \mathbf{S}_{ab} + \mathbf{S}_{ab}^T \dot{\mathbf{S}}_{ab} = \mathbf{0} \tag{4.66}$$

Hence and from (4.64)

$$\boldsymbol{\Omega}_{b,ab}^T = (\mathbf{S}_{ab}^T \dot{\mathbf{S}}_{ab})^T = \dot{\mathbf{S}}_{ab}^T \mathbf{S}_{ab} = -\mathbf{S}_{ab}^T \dot{\mathbf{S}}_{ab} = -\boldsymbol{\Omega}_{b,ab} \tag{4.67}$$

proving skew-symmetry. Singularity follows from the fact that the diagonal elements of a skew-symmetric matrix (3,3) are zeros.

A skew-symmetric singular matrix represents a vector. In this case it is the vector $\boldsymbol{\omega}_{ab}$ of angular velocity of relative spherical motion $b : a$ (Figure 4.6), which can be expressed with the help of its components

$$\boldsymbol{\omega}_{ab} = \lceil \omega_x, \omega_y, \omega_z \rceil_{ab} \tag{4.68}$$

or in form (4.40)

$$\boldsymbol{\Omega}_{b,ab} = \begin{bmatrix} 0 & -\omega_z & \omega_y \\ \omega_z & 0 & -\omega_x \\ -\omega_y & \omega_x & 0 \end{bmatrix}_{b,ab} \tag{4.69}$$

Equation (4.64) also confirms that $\boldsymbol{\omega}_{ab}$ is independent of the choice of point of reference.

Substituting (4.64) into (4.62) we get the velocity matrix in the form

$$\mathbf{V}_{b,ab} = \begin{bmatrix} \boldsymbol{\Omega}_{b,ab} & \mathbf{S}_{ab}^T \dot{\mathbf{u}}_{a,ab} \\ \mathbf{0} & 0 \end{bmatrix} \tag{4.70}$$

from which the physical meaning of each submatrix is evident. The left upper submatrix $\boldsymbol{\Omega}_{b,ab}$ expresses the angular velocity of relative spherical motion, and the right upper submatrix $\mathbf{S}_{ab}^T \dot{\mathbf{u}}_{a,ab}$ expresses the velocity of translational motion. Both velocities are represented by components in space b and hold for basic decomposition of $b : a$ motion with point of reference O_b. Matrix $\mathbf{V}_{b,ab}$ can be more precisely designated as the velocity matrix of general spatial motion $b : a$ in basic decomposition with point of reference O_b.

For the extended absolute velocity of a general point M of body b we obtain, after substituting (4.61) into (4.59), the relation

$$\mathbf{v}_{aM} = \mathbf{T}_{ab} \mathbf{V}_{b,ab} \mathbf{r}_{bM} \tag{4.71}$$

in which the expression $\mathbf{V}_{b,ab} \mathbf{r}_{bM}$ represents the absolute velocity of point M expressed by its components in space b. Using (4.51) and (4.70) we get the matrix equation

$$\dot{\mathbf{u}}_{a,aM} = \mathbf{S}_{ab} \boldsymbol{\Omega}_{b,ab} \mathbf{u}_{b,bM} + \dot{\mathbf{u}}_{a,ab} \tag{4.72}$$

which corresponds to vector equation (4.5) and where $\dot{\mathbf{u}}_{a,ab} = \lceil \dot{a}_1, \dot{a}_2, \dot{a}_3 \rceil$ is the translational velocity of point O_b expressed by components in space a. The expression $\mathbf{S}_{ab}\mathbf{\Omega}_{b,ab}\mathbf{u}_{b,bM}$ is the velocity of point M in relative spherical motion with centre O_b expressed by components in space a and $\dot{\mathbf{u}}_{a,aM} = \lceil \dot{x}_{aM}, \dot{y}_{aM}, \dot{z}_{aM} \rceil$ is the absolute (resultant) velocity of point M in general spatial motion of body b expressed by components in space a.

The comparison of element $\mathbf{S}_{ab}\mathbf{\Omega}_{b,ab}\mathbf{u}_{b,bM}$ from (4.72) and element $\boldsymbol{\omega}_r \times \mathbf{r}_{MR}$ from (4.5) yields the justification for expressing the vector of angular velocity as a skew-symmetric matrix.

Should the point M also be in motion relative to space b, equation (4.54) would be used and we would obtain

$$\mathbf{v}_{aM} = \dot{\mathbf{T}}_{ab}\mathbf{r}_{bM} + \mathbf{T}_{ab}\mathbf{v}_{bM}, \tag{4.73}$$

where

$$\mathbf{v}_{bM} = \dot{\mathbf{r}}_{bM} = \lceil \dot{\mathbf{u}}_{bM}, 0 \rceil = \lceil \dot{x}_{bM}, \dot{y}_{bM}, \dot{z}_{bM}, 0 \rceil \tag{4.74}$$

is the extended velocity of point M relative to space b. Substituting (4.61) into (4.73) we obtain

$$\mathbf{v}_{aM} = \mathbf{T}_{ab}\mathbf{V}_{b,ab}\mathbf{r}_{bM} + \mathbf{T}_{ab}\mathbf{v}_{bM} \tag{4.75}$$

and hence for $\dot{\mathbf{u}}_{a,aM}$

$$\dot{\mathbf{u}}_{a,aM} = \mathbf{S}_{ab}\mathbf{\Omega}_{b,ab}\mathbf{u}_{b,bM} + \dot{\mathbf{u}}_{a,ab} + \mathbf{S}_{ab}\dot{\mathbf{u}}_{b,bM} \tag{4.76}$$

where $\dot{\mathbf{u}}_{b,bM} = \lceil \dot{x}_{bM}, \dot{y}_{bM}, \dot{z}_{bM} \rceil$ is the velocity of point M relative to space b expressed by components in space b.

Instead of equation (4.59), (4.71), (4.72) we would have (4.73), (4.75), (4.76).

As alternatives to definitions (4.61), (4.62) we can introduce

$$\dot{\mathbf{T}}_{ab} = \mathbf{V}_{a,ab}\mathbf{T}_{ab}, \quad \mathbf{V}_{a,ab} = \dot{\mathbf{T}}_{ab}\mathbf{T}_{ab}^{-1} = \mathbf{T}_{ab}\mathbf{V}_{b,ab}\mathbf{T}_{ab}^{-1} \tag{4.77}$$

$$\dot{\mathbf{S}}_{ab} = \mathbf{\Omega}_{a,ab}\mathbf{S}_{ab}, \quad \mathbf{\Omega}_{a,ab} = \dot{\mathbf{S}}_{ab}\mathbf{S}_{ab}^T \tag{4.78}$$

Matrix

$$\mathbf{V}_{a,ab} = \begin{bmatrix} \mathbf{\Omega}_{a,ab}, & -\mathbf{\Omega}_{a,ab}\mathbf{u}_{a,ab} + \dot{\mathbf{u}}_{a,ab} \\ \mathbf{0} & 0 \end{bmatrix} \tag{4.79}$$

is then the velocity matrix of general spatial motion $b : a$ by basic decomposition with point of reference O_a. The angular velocity of relative spherical motion represented by the upper left submatrix $\mathbf{\Omega}_{a,ab}$ has not changed; however it is expressed by components in space a. The right upper submatrix expressing velocity of translational motion is different from (4.70) and corresponds to the vector equation (4.7) and Figure 4.4.

For the extended absolute velocity of a general point M of body b we get, by substituting (4.77) into (4.59), the expression

$$\mathbf{v}_{aM} = \mathbf{V}_{a,ab}\mathbf{T}_{ab}\mathbf{r}_{bM} \tag{4.80}$$

which yields

$$\dot{\mathbf{u}}_{a,aM} = \boldsymbol{\Omega}_{a,ab}\mathbf{S}_{ab}\mathbf{u}_{b,bM} + \boldsymbol{\Omega}_{a,ab}\mathbf{u}_{a,ab} - \boldsymbol{\Omega}_{a,ab}\mathbf{u}_{a,ab} + \dot{\mathbf{u}}_{a,ab}$$

Hence using transformation

$$\boldsymbol{\Omega}_{a,ab} = \mathbf{S}_{ab}\boldsymbol{\Omega}_{b,ab}\mathbf{S}_{ab}^{T} \tag{4.81}$$

according to (4.43), and after some modification, we get an equation analogous to (4.72).

From the two given possible definitions of velocity matrix we shall use the former, i.e.

$$\mathbf{V}_{ab} = \mathbf{V}_{b,ab}$$

as this is more natural and leads to simpler relations. We shall leave out the subscript b whenever there is no danger of misunderstanding.

4.2.1.3 Acceleration

Differentiating (4.59) with respect to time, we obtain for acceleration the relation

$$\mathbf{a}_{aM} = \ddot{\mathbf{T}}_{ab}\mathbf{r}_{bM}, \tag{4.82}$$

where

$$\mathbf{a}_{aM} = \dot{\mathbf{v}}_{aM} = \ddot{\mathbf{r}}_{aM} = [\ddot{\mathbf{u}}_{aM}, 0] = [\ddot{x}_{aM}, \ddot{y}_{aM}, \ddot{z}_{aM}, 0] \tag{4.83}$$

is the extended absolute acceleration of point M expressed by components in space a.

Matrix $\ddot{\mathbf{T}}_{ab}$ can be computed with the help of previously introduced matrices from the relation (4.61)

$$\ddot{\mathbf{T}}_{ab} = \dot{\mathbf{T}}_{ab}\mathbf{V}_{b,ab} + \mathbf{T}_{ab}\dot{\mathbf{V}}_{b,ab} \tag{4.84}$$

We shall call the matrix

$$\mathbf{A}_{b,ab} = \dot{\mathbf{V}}_{b,ab} \tag{4.85}$$

the acceleration matrix of $b : a$ motion by basic decomposition with point of reference O_b. Substituting (4.61) and (4.85) into (4.84) we obtain

$$\ddot{\mathbf{T}}_{ab} = \mathbf{T}_{ab}(\mathbf{A}_{b,ab} + \mathbf{V}_{b,ab}^{2}) = \mathbf{T}_{ab}\mathbf{B}_{b,ab} \tag{4.86}$$

where we have introduced the relation

$$\mathbf{B}_{b,ab} = \mathbf{A}_{b,ab} + \mathbf{V}^2_{b,ab} \tag{4.87}$$

to describe the complete acceleration matrix of $b : a$ motion by basic decomposition with point of reference O_b.

Differentiating in the same way (4.63) with respect to time, we can compute that

$$\ddot{\mathbf{S}}_{ab} = \dot{\mathbf{S}}_{ab}\boldsymbol{\Omega}_{b,ab} + \mathbf{S}_{ab}\dot{\boldsymbol{\Omega}}_{b,ab} = \mathbf{S}_{ab}(\boldsymbol{\Omega}^2_{b,ab} + \dot{\boldsymbol{\Omega}}_{b,ab}) \tag{4.88}$$

where the matrix

$$\mathcal{A}_{b,ab} = \dot{\boldsymbol{\Omega}}_{b,ab} = \begin{bmatrix} 0 & -\alpha_z & \alpha_y \\ \alpha_z & 0 & -\alpha_x \\ -\alpha_y & \alpha_x & 0 \end{bmatrix}_{b,ab} \tag{4.89}$$

representing vector

$$\boldsymbol{\alpha}_{ab} = \lceil \alpha_x, \alpha_y, \alpha_z \rceil_{ab} \tag{4.90}$$

is called the angular acceleration matrix of $b : a$ motion. Substituting the previously derived relations and modifying (4.87) yields the matrix

$$\mathbf{B}_{b,ab} = \begin{bmatrix} \mathcal{A}_{b,ab} + \boldsymbol{\Omega}^2_{b,ab}, & \mathbf{S}^T_{ab}\ddot{\mathbf{u}}_{a,ab} \\ \mathbf{0} & 0 \end{bmatrix} \tag{4.91}$$

This characterises the acceleration of body b by general spatial $b : a$ motion expressed by basic decomposition with point of reference O_b. The left upper submatrix in (4.91) represents the acceleration in relative spherical motion; the right upper submatrix represents the acceleration of point of reference O_b. All quantities are expressed by components in space b.

Using (4.86) we get from (4.82) a simple relation for the computation of acceleration

$$\mathbf{a}_{aM} = \mathbf{T}_{ab}\mathbf{B}_{b,ab}\mathbf{r}_{bM} \tag{4.92}$$

Hence

$$\ddot{\mathbf{u}}_{a,aM} = \mathbf{S}_{ab}\mathcal{A}_{b,ab}\mathbf{u}_{b,bM} + \mathbf{S}_{ab}\boldsymbol{\Omega}^2_{b,ab}\mathbf{u}_{b,bM} + \ddot{\mathbf{u}}_{a,ab} \tag{4.93}$$

corresponding to vector equation (4.6). The term $\mathcal{A}_{b,ab}\mathbf{u}_{b,bM}$ represents the tangential and term $\boldsymbol{\Omega}^2_{b,ab}\mathbf{u}_{b,bM}$ the normal component of the acceleration of point M in relative spherical motion; both components are expressed in space b. Multiplying those terms from the left by matrix $\mathbf{S}_{ab}$ is equivalent to transformation of the components into space a. The term $\ddot{\mathbf{u}}_{a,ab}$ is the translational acceleration, and $\ddot{\mathbf{u}}_{a,aM}$ the absolute acceleration of point M expressed in space a.

Should point M also be in motion in the space b, we would start by using relation (4.73) and would finally obtain

$$\mathbf{a}_{aM} = \ddot{\mathbf{T}}_{ab}\mathbf{r}_{bM} + 2\dot{\mathbf{T}}_{ab}\mathbf{v}_{bM} + \mathbf{T}_{ab}\mathbf{a}_{bM} \tag{4.94}$$

where

$$\mathbf{a}_{bM} = \dot{\mathbf{v}}_{bM} = \ddot{\mathbf{r}}_{bM} = [\ddot{\mathbf{u}}_{b,bM}, 0] = [\ddot{x}_{bM}, \ddot{y}_{bM}, \ddot{z}_{bM}, 0] \tag{4.95}$$

is the extended acceleration of point M in space b. We would further obtain

$$\mathbf{a}_{aM} = \mathbf{T}_{ab}\mathbf{B}_{b,ab}\mathbf{r}_{bM} + 2\mathbf{T}_{ab}\mathbf{V}_{b,ab}\mathbf{v}_{bM} + \mathbf{T}_{ab}\mathbf{a}_{bM} \tag{4.96}$$

We would substitute equations (4.94), (4.96) for equations (4.82), (4.92).

Alternatively, relations for acceleration can be obtained using equations (4.77), (4.78). The corresponding relations are not given here since we shall not make further use of this definition.

4.2.2 Inverse Motion

Starting from equation (4.54), and using the complete notation, let us rewrite it in the form

$$\mathbf{r}_{a,aM}(t) = \mathbf{T}_{ab}(t)\mathbf{r}_{b,bM}(t) \tag{4.97}$$

This equation describes the $b : a$ motion, i.e. the observer is stationary in space a and observes the motion of body b and its point M. We shall call this motion the basic motion. Sometimes, however, we wish to observe the inverse (reciprocal) $a : b$ motion, where the observer is stationary in space b and observes the motion of body a and its point M. For the inverse motion, we get from (4.97), interchanging indices a, b the equation

$$\mathbf{r}_{b,bM}(t) = \mathbf{T}_{ba}(t)\mathbf{r}_{a,aM}(t) \tag{4.98}$$

Formally modifying this from (4.97), we get

$$\mathbf{r}_{b,bM}(t) = \mathbf{T}_{ab}^{-1}(t)\mathbf{r}_{a,aM}(t) \tag{4.99}$$

Comparing (4.99), (4.98) yields

$$\mathbf{T}_{ba} = \mathbf{T}_{ab}^{-1} \tag{4.100}$$

The transformation matrix of inverse motion equals the inverse transformation matrix of basic motion.

To obtain relations for velocities and accelerations of inverse motion, we start with the equation

$$\mathbf{T}_{ab}\mathbf{T}_{ba} = \mathbf{E}_4 \tag{4.101}$$

Differentiating with respect to time gives

$$\dot{\mathbf{T}}_{ab}\mathbf{T}_{ba} + \mathbf{T}_{ab}\dot{\mathbf{T}}_{ba} = \mathbf{0} \tag{4.102}$$

Before proceeding, we must decide how to describe general spatial motions $b : a$ and $a : b$. Formally simple relations can be obtained using basic decomposition in both cases. The reference point for motion $b : a$ is again the point O_b, and for motion $a : b$ the point O_a. For modifying equation (4.102), matrix $\dot{\mathbf{T}}_{ab}$ from equation (4.61) is used and $\dot{\mathbf{T}}_{ba}$ is then obtained by exchanging indices in (4.61). As

$$\dot{\mathbf{T}}_{ba} = \mathbf{T}_{ba}\mathbf{V}_{a,ba} \tag{4.103}$$

it is possible instead of (4.102) to write

$$\mathbf{T}_{ab}\mathbf{V}_{b,ab}\mathbf{T}_{ab}^{-1} + \mathbf{V}_{a,ab} = \mathbf{0} \tag{4.104a}$$

or, after multiplying the latter equation by matrix $\mathbf{T}_{ab}^{-1}$ from the left and by matrix $\mathbf{T}_{ab}$ from the right,

$$\mathbf{V}_{b,ab} + \mathbf{T}_{ba}\mathbf{V}_{a,ab}\mathbf{T}_{ba}^{-1} = \mathbf{0} \tag{4.104b}$$

The velocity matrix $\mathbf{V}_{b,ab}$ of basic motion and $\mathbf{V}_{a,ab}$ of inverse motion, by basic decomposition and reference point O_b and O_a, respectively, are linked by equations (4.104). The physical meaning of their submatrices has been explained above.

The relations between acceleration matrices of both motions are to be obtained by differentiating equation (4.102) with respect to time. Hence, modifying

$$\dot{\mathbf{T}}_{ab}\mathbf{V}_{b,ab}\mathbf{T}_{ba} + \mathbf{T}_{ab}\dot{\mathbf{V}}_{b,ab}\mathbf{T}_{ba} + \mathbf{T}_{ab}\mathbf{V}_{b,ba}\dot{\mathbf{T}}_{ba} + \dot{\mathbf{V}}_{a,ab} = \mathbf{0} \tag{4.105}$$

after substitution

$$\dot{\mathbf{V}}_{b,ab} = \mathbf{A}_{b,ab}, \quad \dot{\mathbf{V}}_{a,ab} = \mathbf{A}_{a,ab} \tag{4.106}$$

yields

$$\mathbf{T}_{ab}\mathbf{V}_{b,ab}^2\mathbf{T}_{ba} + \mathbf{T}_{ab}\mathbf{A}_{b,ab}\mathbf{T}_{ba} + \mathbf{T}_{ab}\mathbf{V}_{b,ab}\mathbf{T}_{ba}\mathbf{V}_{a,ab} + \mathbf{A}_{a,ba} = \mathbf{0} \tag{4.107}$$

and finally, after substituting from (4.104a),

$$\mathbf{T}_{ab}\mathbf{A}_{b,ab}\mathbf{T}_{ab}^{-1} + \mathbf{A}_{a,ba} = \mathbf{0} \tag{4.108a}$$

and

$$\mathbf{A}_{b,ab} = \mathbf{T}_{ba}\mathbf{A}_{a,ba}\mathbf{T}_{ba}^{-1} \tag{4.108b}$$

respectively. The acceleration matrices of both motions are linked by equation (4.108).

In a similar way, relations for angular velocity matrices and angular acceleration matrices can be found to be

$$\begin{aligned}
\mathbf{S}_{ab}\boldsymbol{\Omega}_{b,ab}\mathbf{S}_{ab}^{T} + \boldsymbol{\Omega}_{a,ba} &= \mathbf{0} &\quad (4.109a)\\
\boldsymbol{\Omega}_{b,ab} + \mathbf{S}_{ba}\boldsymbol{\Omega}_{a,ba}\mathbf{S}_{ba}^{T} &= \mathbf{0} &\quad (4.109b)\\
\mathbf{S}_{ab}\mathcal{A}_{b,ab}\mathbf{S}_{ab}^{T} + \mathcal{A}_{a,ba} &= \mathbf{0} &\quad (4.110a)\\
\mathcal{A}_{b,ab} + \mathbf{S}_{ba}\mathcal{A}_{a,ba}\mathbf{S}_{ba}^{T} &= \mathbf{0} &\quad (4.110b)
\end{aligned}$$

The physical meaning of the symbols used has been explained above.

4.2.3 Simultaneous Motion

In this section, we shall derive relations in matrix notation for computing kinematic quantities in simultaneous motion.

4.2.3.1 Position

Let us assume a body a, which we shall designate also as body 1. Let there be a body 2 in motion relative to 1, a body 3 in motion relative to 2, etc., ..., a body $n \equiv b$ in motion relative to $n-1$, Figure 4.9. The resultant motion $b : a$, which is to be examined, is composed of a series of simultaneous motions, each of which can be a general spatial motion. Schematically this can be described by a symbolic equation

$$ab \equiv 1n = 12 + 23 + \cdots + (n-2)(n-1) + (n-1)n \tag{4.111}$$

see also (4.10). The motion of a general point M of body b is described for individual motions according to (4.50) by relations

$$\mathbf{r}_{1M} = \mathbf{T}_{12}\mathbf{r}_{2M}, \quad \mathbf{r}_{2M} = \mathbf{T}_{23}\mathbf{r}_{3M}, \quad \ldots \quad, \mathbf{r}_{n-1,M} = \mathbf{T}_{n-1,n}\mathbf{r}_{nM} \tag{4.112}$$

Substituting consecutively radius vectors $\mathbf{r}_{2M}, ..., \mathbf{r}_{n-1,M}$ into the first one equation of (4.112) yields a matrix equation

$$\mathbf{r}_{1M} = \mathbf{T}_{12}\mathbf{T}_{23}\ldots\mathbf{T}_{n-1,n}\mathbf{r}_{nM} \tag{4.113}$$

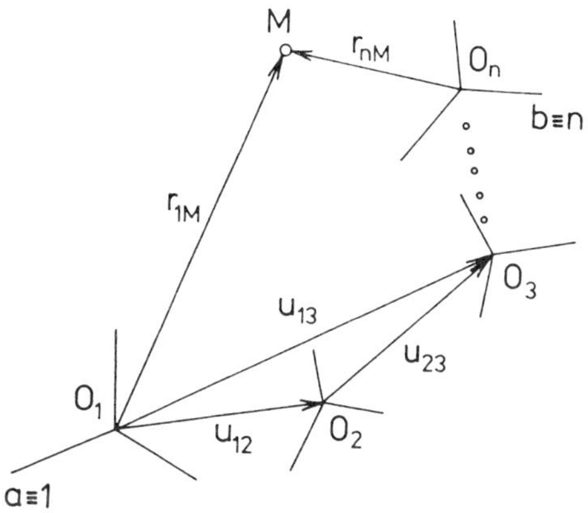

Figure 4.9. Simultaneous motion of n bodies.

expressing the motion of point M in space 1, i.e. the resultant motion. This can also be described by

$$\mathbf{r}_{1M} = \mathbf{T}_{1n}\mathbf{r}_{nM} \tag{4.114}$$

Comparing (4.113) and (4.114) it follows that the transformation matrix of the resultant motion is given by the product of transformation matrices of partial motions, i.e.

$$\mathbf{T}_{1n} = \mathbf{T}_{12}\mathbf{T}_{23}\ldots\mathbf{T}_{n-1,n} \tag{4.115}$$

The transformation matrix of composed motion is to be obtained by multiplying the transformation matrix of carrying motion from the right hand side consecutively by each relative motion matrix.

Another interpretation of equation (4.113) is transforming consecutively the motion of point M in space $b \equiv n$ into space $(n-1), (n-2), \ldots, 2, 1$. By transformation composition the transformation matrix is to be used for left hand side multiplication. Obtaining the transformation matrix of inverse motion is simple. According to (4.100)

$$\begin{aligned}\mathbf{T}_{n1} &= (\mathbf{T}_{12}\mathbf{T}_{23}\ldots\mathbf{T}_{n-1,n})^{-1} = \mathbf{T}_{n-1,n}^{-1}\cdots\mathbf{T}_{23}^{-1}\mathbf{T}_{12}^{-1} = \\ &= \mathbf{T}_{n,n-1}\ldots\mathbf{T}_{32}\mathbf{T}_{21}\end{aligned} \tag{4.116}$$

The product of transformation matrices preserves their characteristic submatrix structure according to (4.51) and also the submatrix significance. Let us show this in an example of two simultaneous motions composed as follows

$$13 = 12 + 23 \tag{4.117}$$

According to (4.115), Figure 4.9

$$\begin{aligned}\mathbf{T}_{13} &= \mathbf{T}_{12}\mathbf{T}_{23} = \begin{bmatrix} \mathbf{S}_{12} & \mathbf{u}_{12} \\ \mathbf{0} & 1 \end{bmatrix} \begin{bmatrix} \mathbf{S}_{23} & \mathbf{u}_{23} \\ \mathbf{0} & 1 \end{bmatrix} = \\ &= \begin{bmatrix} \mathbf{S}_{12}\mathbf{S}_{23}, & \mathbf{S}_{12}\mathbf{u}_{23} + \mathbf{u}_{12} \\ \mathbf{0}, & 1 \end{bmatrix} = \begin{bmatrix} \mathbf{S}_{13} & \mathbf{u}_{13} \\ \mathbf{0} & 1 \end{bmatrix}\end{aligned} \tag{4.118}$$

The direction cosines matrix

$$\mathbf{S}_{13} = \mathbf{S}_{12}\mathbf{S}_{23} \tag{4.119}$$

is a product of orthogonal matrices and is therefore itself orthogonal.

4.2.3.2 Velocity

For the resultant velocity, the relation according to (4.59) is

$$\mathbf{v}_{1M} = \dot{\mathbf{T}}_{1n}\mathbf{r}_{nM} \tag{4.120}$$

After substituting from (4.115) and differentiating with respect to time we obtain

$$\begin{aligned}\mathbf{v}_{1M} &= (\dot{\mathbf{T}}_{12}\mathbf{T}_{23}\dots\mathbf{T}_{n-1,n} + \mathbf{T}_{12}\dot{\mathbf{T}}_{23}\dots\mathbf{T}_{n-1,n} + \cdots + \\ &\quad + \mathbf{T}_{12}\mathbf{T}_{23}\dots\dot{\mathbf{T}}_{n-1,n})\mathbf{r}_{nM}\end{aligned} \tag{4.121}$$

and using (4.61)

$$\begin{aligned}\mathbf{v}_{1M} &= (\mathbf{T}_{12}\mathbf{V}_{12}\mathbf{T}_{23}\dots\mathbf{T}_{n-1,n} + \mathbf{T}_{12}\mathbf{T}_{23}\mathbf{V}_{23}\dots\mathbf{T}_{n-1,n} + \cdots + \\ &\quad + \mathbf{T}_{12}\mathbf{T}_{23}\dots\mathbf{T}_{n-1,n}\mathbf{V}_{n-1,n})\mathbf{r}_{nM} = \\ &= (\mathbf{T}_{12}\mathbf{V}_{12}\mathbf{T}_{2n} + \mathbf{T}_{13}\mathbf{V}_{23}\mathbf{T}_{3n} + \cdots + \\ &\quad + \mathbf{T}_{1k}\mathbf{V}_{k-1,k}\mathbf{T}_{kn} + \cdots + \mathbf{T}_{1n}\mathbf{V}_{n-1,n})\mathbf{r}_{nM}\end{aligned} \tag{4.122}$$

Using (4.71) we derive the relation

$$\mathbf{v}_{1M} = \mathbf{T}_{1n}\mathbf{V}_{1n}\mathbf{r}_{nM} \tag{4.123}$$

Comparing (4.123) and (4.122) and modifying yields for the velocity matrix of resultant motion

$$\begin{aligned}\mathbf{V}_{1n} &= \mathbf{T}_{n-1,n}^{-1}\mathbf{T}_{n-2,n-1}^{-1}\cdots\mathbf{T}_{23}^{-1}\mathbf{V}_{12}\mathbf{T}_{23}\dots\mathbf{T}_{n-1,n} + \\ &\quad + \mathbf{T}_{n-1,n}^{-1}\mathbf{T}_{n-2,n-1}^{-1}\cdots\mathbf{T}_{34}^{-1}\mathbf{V}_{23}\mathbf{T}_{34}\dots\mathbf{T}_{n-1,n} + \\ &\quad \vdots \\ &\quad + \mathbf{V}_{n-1,n} = \\ &= \mathbf{T}_{2n}^{-1}\mathbf{V}_{12}\mathbf{T}_{2n} + \mathbf{T}_{3n}^{-1}\mathbf{V}_{23}\mathbf{T}_{3n} + \cdots + \\ &\quad + \mathbf{T}_{kn}^{-1}\mathbf{V}_{k-1,k}\mathbf{T}_{kn} + \cdots + \mathbf{V}_{n-1,n}\end{aligned} \tag{4.124}$$

Using (4.64) we obtain in a similar way for the angular velocity of resultant motion

$$\begin{aligned}\boldsymbol{\Omega}_{1n} &= \mathbf{S}_{n-1,n}^T\mathbf{S}_{n-2,n-1}^T\ldots\mathbf{S}_{23}^T\boldsymbol{\Omega}_{12}\mathbf{S}_{23}\ldots\mathbf{S}_{n-1,n}+\\ &\quad+\mathbf{S}_{n-1,n}^T\mathbf{S}_{n-2,n-1}^T\ldots\mathbf{S}_{34}^T\boldsymbol{\Omega}_{23}\mathbf{S}_{34}\ldots\mathbf{S}_{n-1,n}+\\ &\quad\vdots\\ &\quad+\boldsymbol{\Omega}_{n-1,n}=\\ &= \mathbf{S}_{2n}^T\boldsymbol{\Omega}_{12}\mathbf{S}_{2n}+\mathbf{S}_{3n}^T\boldsymbol{\Omega}_{23}\mathbf{S}_{3n}+\cdots+\\ &\quad+\mathbf{S}_{kn}^T\boldsymbol{\Omega}_{k-1,k}\mathbf{S}_{kn}+\cdots+\boldsymbol{\Omega}_{n-1,n}\end{aligned} \tag{4.125}$$

The form of relation (4.124) and (4.125) shows that we are dealing with a transformation of partial velocity and angular velocity matrices into space n in which the matrices $\mathbf{V}_{1n}, \boldsymbol{\Omega}_{1n}$ are expressed. Precisely designated: $\mathbf{V}_{1n} = \mathbf{V}_{n,1n}, \boldsymbol{\Omega}_{1n} = \boldsymbol{\Omega}_{n,1n}$.

For better understanding, let us give the relations for velocities in the simple case of two simultaneous motions in (4.117) with complete indexing

$$\mathbf{V}_{3,13} = \mathbf{T}_{23}^{-1}\mathbf{V}_{2,12}\mathbf{T}_{23} + \mathbf{V}_{3,23} \tag{4.126}$$

$$\boldsymbol{\Omega}_{3,13} = \mathbf{S}_{23}^T\boldsymbol{\Omega}_{2,12}\mathbf{S}_{23} + \boldsymbol{\Omega}_{3,23} \tag{4.127}$$

and hence

$$\begin{aligned}\mathbf{v}_{1M} &= \mathbf{T}_{13}\mathbf{V}_{3,13}\mathbf{r}_{3M} = \mathbf{T}_{13}(\mathbf{T}_{23}^{-1}\mathbf{V}_{2,12}\mathbf{T}_{23}+\mathbf{V}_{3,23})\mathbf{r}_{3M} =\\ &= \mathbf{T}_{12}\mathbf{V}_{2,12}\mathbf{r}_{2M} + \mathbf{T}_{13}\mathbf{V}_{3,23}\mathbf{r}_{3M}\end{aligned} \tag{4.128}$$

Substituting the required matrices into (4.128) yields a matrix equation

$$\dot{\mathbf{u}}_{1,1M} = \dot{\mathbf{u}}_{1,12} + \mathbf{S}_{12}\boldsymbol{\Omega}_{2,12}\mathbf{u}_{2M} + \mathbf{S}_{12}\dot{\mathbf{u}}_{2,23} + \mathbf{S}_{13}\boldsymbol{\Omega}_{3,23}\mathbf{u}_{3M} \tag{4.129}$$

corresponding to the vector equation (4.15), where

$$\begin{aligned}\mathbf{v}_{12p} &= \dot{\mathbf{u}}_{1,12}, &\quad \mathbf{v}_{12s} &= \mathbf{S}_{12}\boldsymbol{\Omega}_{2,12}\mathbf{u}_{2M}\\ \mathbf{v}_{23p} &= \mathbf{S}_{12}\dot{\mathbf{u}}_{2,23}, &\quad \mathbf{v}_{23s} &= \mathbf{S}_{13}\boldsymbol{\Omega}_{3,23}\mathbf{u}_{3M}\end{aligned} \tag{4.130}$$

The meaning of the symbols used was given in Section 4.1.2.

4.2.3.3 Acceleration

The resultant acceleration of point M, according to (4.82), is given by the expression

$$\mathbf{a}_{1M} = \ddot{\mathbf{T}}_{1n}\mathbf{r}_{nM} \tag{4.131}$$

and according to (4.92)

$$\mathbf{a}_{1M} = \mathbf{T}_{1n}(\mathbf{A}_{1n} + \mathbf{V}_{1n}^2)\mathbf{r}_{nM} = \mathbf{T}_{1n}\mathbf{B}_{1n}\mathbf{r}_{nM} \tag{4.132}$$

where, because of (4.85) and (4.89)

$$\mathbf{A}_{1n} = \dot{\mathbf{V}}_{1n}, \qquad \mathcal{A}_{1n} = \dot{\boldsymbol{\Omega}}_{1n} \tag{4.133}$$

Writing the given relations using partial motion matrices leads to complicated expressions. We shall therefore neither use nor show them here. We shall give them in another form in Section 4.5. An example for the simple case of two simultaneous motions from (4.117) is:

$$\begin{aligned} \mathbf{a}_{1M} &= (\ddot{\mathbf{T}}_{12}\mathbf{T}_{23} + 2\dot{\mathbf{T}}_{12}\dot{\mathbf{T}}_{23} + \mathbf{T}_{12}\ddot{\mathbf{T}}_{23})\mathbf{r}_{3M} = \\ &= (\mathbf{T}_{12}(\mathbf{A}_{12} + \mathbf{V}_{12}^2)\mathbf{T}_{23} + 2\mathbf{T}_{12}\mathbf{V}_{12}\mathbf{T}_{23}\mathbf{V}_{23} + \\ &\quad + \mathbf{T}_{12}\mathbf{T}_{23}(\mathbf{A}_{23} + \mathbf{V}_{23}^2))\mathbf{r}_{3M} = \\ &= (\mathbf{T}_{12}\mathbf{B}_{12}\mathbf{T}_{23} + 2\mathbf{T}_{12}\mathbf{V}_{12}\mathbf{T}_{23}\mathbf{V}_{23} + \mathbf{T}_{13}\mathbf{B}_{23})\mathbf{r}_{3M} \qquad (4.134) \\ \mathbf{A}_{3,13} &= \dot{\mathbf{V}}_{3,13} = \mathbf{T}_{32}\mathbf{V}_{2,32}\mathbf{V}_{2,12}\mathbf{T}_{23} + \mathbf{T}_{23}^{-1}\mathbf{A}_{2,12}\mathbf{T}_{23} + \\ &\quad + \mathbf{T}_{23}^{-1}\mathbf{V}_{2,12}\mathbf{T}_{23}\mathbf{V}_{3,23} + \mathbf{A}_{3,23} \qquad (4.135) \\ \mathcal{A}_{3,13} &= \dot{\boldsymbol{\Omega}}_{3,13} = \mathbf{S}_{32}\boldsymbol{\Omega}_{2,32}\boldsymbol{\Omega}_{2,12}\mathbf{S}_{23} + \mathbf{S}_{23}^T\mathcal{A}_{2,12}\mathbf{S}_{23} + \\ &\quad + \mathbf{S}_{23}^T\boldsymbol{\Omega}_{2,12}\mathbf{S}_{23}\boldsymbol{\Omega}_{3,23} + \mathcal{A}_{3,23} \qquad (4.136) \end{aligned}$$

Substituting the required matrices and the detailed form of (4.134) yields a matrix equation corresponding to the vector equation (4.16). Finally we obtain

$$\begin{aligned} \mathbf{a}_{1M} &= \mathbf{T}_{12}\mathbf{B}_{2,12}\mathbf{T}_{23}\mathbf{r}_{3M} + 2\mathbf{T}_{12}\mathbf{V}_{2,12}\mathbf{T}_{23}\mathbf{V}_{3,23}\mathbf{r}_{3M} + \mathbf{T}_{13}\mathbf{B}_{3,23}\mathbf{r}_{3M} = \\ &= \mathbf{T}_{12}\mathbf{B}_{2,12}\mathbf{r}_{2M} + 2\mathbf{T}_{12}\mathbf{V}_{2,12}\mathbf{v}_{2M} + \mathbf{T}_{13}\mathbf{B}_{3,23}\mathbf{r}_{3M} \qquad (4.137) \end{aligned}$$

Rewriting using submatrices yields

$$\begin{aligned} \mathbf{T}_{12}\mathbf{B}_{2,12}\mathbf{r}_{2M} &\rightarrow \mathbf{S}_{12}\mathcal{A}_{2,12}\mathbf{u}_{2M} + \mathbf{S}_{12}\boldsymbol{\Omega}_{2,12}^2\mathbf{u}_{2M} + \ddot{\mathbf{u}}_{1,12} = \\ &= \mathbf{a}_{12Mt} + \mathbf{a}_{12Mn} + \mathbf{a}_{1p} \qquad (4.138) \\ 2\mathbf{T}_{12}\mathbf{V}_{2,12}\mathbf{v}_{2M} &\rightarrow 2\mathbf{S}_{12}\boldsymbol{\Omega}_{2,12}\dot{\mathbf{u}}_{2M} = \mathbf{a}_{Cor} \qquad (4.139) \\ \mathbf{T}_{13}\mathbf{B}_{3,23}\mathbf{r}_{3M} &\rightarrow \mathbf{S}_{13}\mathcal{A}_{3,23}\mathbf{u}_{3M} + \mathbf{S}_{13}\boldsymbol{\Omega}_{3,23}^2\mathbf{u}_{3M} + \mathbf{S}_{12}\ddot{\mathbf{u}}_{2,23} = \\ &= \mathbf{a}_{23Mt} + \mathbf{a}_{23Mn} + \mathbf{a}_{23p} \qquad (4.140) \end{aligned}$$

Symbols introduced in 4.1.2 were used in the last two equations.

4.2.3.4 Notes

For the sake of uniqueness, vectors $\mathbf{r}, \mathbf{v}$ and $\mathbf{a}$ have been called extended radius vector, extended velocity vector and extended acceleration vector. In the following, we shall leave out the adjective "extended" when there is no danger of confusing the extended vectors for similar vectors $\mathbf{u}, \dot{\mathbf{u}}, \ddot{\mathbf{u}}$. We shall also leave out the subscripts a and b designating the space in which the quantity is expressed. They were used in previous paragraphs for easier understanding of the physical meaning of derived relations.

4.2.4 Other Transformations

Homogeneous transformation (4.50) of extended vectors leads to the description of other transformations, namely those of perspective projection (Paul, 1981). These transformations are of advantage first in the non-traditional use of the described matrix notation outside mechanics. This is the case, for example, in computer graphics, where the potential for concise description of spatial motions using these transformations is very valuable. Secondly, the transformations described give a complete interpretation of all elements of the transformation matrix $\mathbf{T}$.

The first case we shall deal with is deformation of object size. Transformation

$$\mathbf{T} = \begin{bmatrix} a & 0 & 0 & 0 \\ 0 & b & 0 & 0 \\ 0 & 0 & c & 0 \\ 0 & 0 & 0 & 1 \end{bmatrix} \tag{4.141}$$

expresses uniform object stretch with factor a along axis x, factor b along axis y and factor c along z. Let us, for example, transform the radius vector $\mathbf{u}_M = \lceil x_M, y_M, z_M \rceil$ of point M

$$\begin{bmatrix} ax_M \\ by_M \\ cz_M \\ 1 \end{bmatrix} = \begin{bmatrix} a & 0 & 0 & 0 \\ 0 & b & 0 & 0 \\ 0 & 0 & c & 0 \\ 0 & 0 & 0 & 1 \end{bmatrix} \begin{bmatrix} x_M \\ y_M \\ z_M \\ 1 \end{bmatrix} \tag{4.142}$$

A cube transforms by this means into a right parallelepiped. Transformation

$$\mathbf{T} = \begin{bmatrix} s & 0 & 0 & 0 \\ 0 & s & 0 & 0 \\ 0 & 0 & s & 0 \\ 0 & 0 & 0 & 1 \end{bmatrix} \tag{4.143}$$

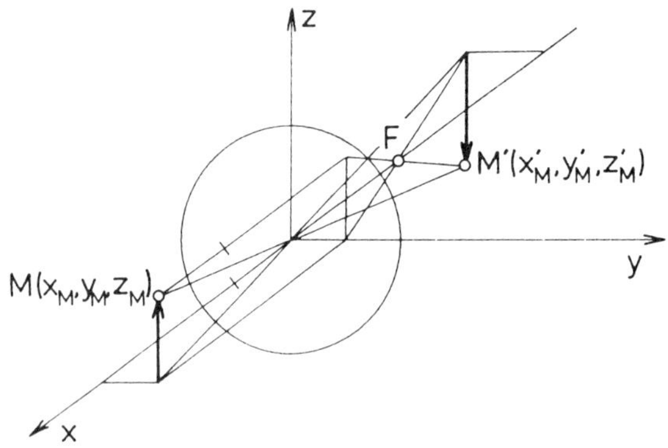

Figure 4.10. Transformation of vector by simple lens.

expresses a homothety with factor s relative to the origin of the system of coordinates. This transformation is, because of the definition of homogeneous coordinates, equivalent to the transformation

$$\mathbf{T} = \begin{bmatrix} 1 & 0 & 0 & 0 \\ 0 & 1 & 0 & 0 \\ 0 & 0 & 1 & 0 \\ 0 & 0 & 0 & 1/s \end{bmatrix} \tag{4.144}$$

Another case to be treated is that of perspective transformations. Let us assume a mapping determined by a simple lens in the origin and with its axis along the x-axis (Figure 4.10) with positive focal length f. Let this transformation map point M on to point M'. Due to the fact that rays pass through the centre of the lens unchanged

$$\frac{y_M}{x_M} = \frac{y_{M'}}{x_{M'}}, \qquad \frac{z_M}{x_M} = \frac{z_{M'}}{x_{M'}} \tag{4.145}$$

Further, due to the fact that rays parallel to the lens axis pass through the focal point

$$\frac{y_M}{f} = \frac{y_{M'}}{x_{M'} + f} \quad , \quad \frac{z_M}{f} = \frac{z_{M'}}{x_{M'} + f} \tag{4.146}$$

$x_{M'}, y_{M'}, z_{M'}$ are negative, and f positive. Modifying these equations yields

$$x_{M'} = \frac{x_M}{1 - \frac{x_M}{f}} \quad , \quad y_{M'} = \frac{y_M}{1 - \frac{x_M}{f}} \quad , \quad z_{M'} = \frac{z_M}{1 - \frac{x_M}{f}} \tag{4.147}$$

This transformation can be expressed by a homogeneous transformation matrix

$$\mathbf{T} = \begin{bmatrix} 1 & 0 & 0 & 0 \\ 0 & 1 & 0 & 0 \\ 0 & 0 & 1 & 0 \\ -1/f & 0 & 0 & 1 \end{bmatrix} \tag{4.148}$$

as dividing by a scale factor does not change the description of the object

$$\begin{bmatrix} x_{M'} \\ y_{M'} \\ z_{M'} \\ 1 \end{bmatrix} = \begin{bmatrix} x_M \\ y_M \\ z_M \\ 1-(x_M/f) \end{bmatrix} = \begin{bmatrix} 1 & 0 & 0 & 0 \\ 0 & 1 & 0 & 0 \\ 0 & 0 & 1 & 0 \\ -1/f & 0 & 0 & 1 \end{bmatrix} \begin{bmatrix} x_M \\ y_M \\ z_M \\ 1 \end{bmatrix} \tag{4.149}$$

Similarly, analogous transformations along the y-axis and the z-axis have $-1/f$ in the place of element (4,2) or (4,3).

Another useful case is planar reflection. For example, reflection on the xy-plane is expressed by the homogeneous transformation matrix

$$\mathbf{T} = \begin{bmatrix} 1 & 0 & 0 & 0 \\ 0 & 1 & 0 & 0 \\ 0 & 0 & -1 & 0 \\ 0 & 0 & 0 & 0 \end{bmatrix} \tag{4.150}$$

Reflections on other coordinate planes can be expressed in a similar way. Reflections on a general plane are to be obtained by composing reflection and motion.

Using the transformations described above, the required spatial image of the examined object is to be constructed. For the projection into, for example, the xy-plane, the transformation matrix

$$\mathbf{T} = \begin{bmatrix} 1 & 0 & 0 & 0 \\ 0 & 1 & 0 & 0 \\ 0 & 0 & 0 & 0 \\ 0 & 0 & 0 & 1 \end{bmatrix} \tag{4.151}$$

can be used. The situation is analogous for other projections.

4.3 Quaternion Notation

There are more formalisms available for describing the point and body spatial motion kinematics. They can be classified into synthetic formalisms (vector method, trigonometric method) and analytic formalism (matrix method, quaternions, screw description, special unitary matrix method,

Pauli spin matrix method) (Rooney, 1977). These formalisms are mutually equivalent due to the isomorphism with the Klein spatial motion group (Karger and Novák, 1978). These isomorphisms are not simple and straightforward, they are not trivial, and knowledge of them yields valuable information. We shall therefore describe the use of quaternions for the description of motion in a modification analogous to the transformation matrix method.

4.3.1 Introduction of Quaternions

A quaternion is a particular mathematical object. It is an ordered arrangement $(u_0, \vec{u})$, where u_0 is a real number and $\vec{u}$ a real three-dimensional vector. A quaternion is written as the sum of a scalar and vector quantity

$$\hat{\mathbf{u}} = u_0 + \vec{u} = u_0 + u_1\vec{i} + u_2\vec{j} + u_3\vec{k} \tag{4.152}$$

where $\vec{i}, \vec{j}, \vec{k}$ designate orthogonal unit vectors. Vector $\vec{u}$ can be expressed as a column matrix

$$\mathbf{u} = \lceil u_1, u_2, u_3 \rceil \tag{4.153}$$

and similarly the whole quaternion

$$\underline{\mathbf{u}} = \lceil u_0, u_1, u_2, u_3 \rceil \tag{4.154}$$

Equality and addition of quaternions is defined as equality and addition of its components. The neutral element of addition $\lceil 0, 0, 0, 0 \rceil$ is called the null quaternion.

Multiplication of quaternions has a more complicated definition

$$\hat{\mathbf{w}} = \hat{\mathbf{u}}\hat{\mathbf{v}} = u_0 v_0 - \vec{u}.\vec{v} + u_0\vec{v} + v_0\vec{u} + \vec{u} \times \vec{v} \tag{4.155}$$

where $\vec{u}.\vec{v}$ is the scalar product and $\vec{u} \times \vec{v}$ the vector product of vectors $\vec{u}, \vec{v}$, whereas $u_0\vec{v}$ and $v_0\vec{u}$ are the products of scalar quantities u_0 and v_0 and vector quantities $\vec{v}$ and $\vec{u}$ respectively. Multiplication of quaternions is not commutative. The product rule (4.155) is the particular quaternion property giving them their particular usage.

The conjugated quaternion $\hat{\mathbf{u}}^*$ is defined as

$$\hat{\mathbf{u}}^* = u_0 - \vec{u} \tag{4.156}$$

For the product of conjugated quaternions holds

$$(\hat{\mathbf{u}}\hat{\mathbf{v}})^* = \hat{\mathbf{v}}^*\hat{\mathbf{u}}^* \tag{4.157}$$

The quaternion norm is defined as

$$|\hat{\mathbf{u}}|^2 = \hat{\mathbf{u}}^*\hat{\mathbf{u}} = \hat{\mathbf{u}}\hat{\mathbf{u}}^* = u_0^2 + |\mathbf{u}|^2 \tag{4.158}$$

The inverse quaternion to quaternion $\hat{\mathbf{u}}$, when $\hat{\mathbf{u}}\hat{\mathbf{u}}^* \neq 0$, is

$$\hat{\mathbf{u}}^{-1} = \frac{\hat{\mathbf{u}}^*}{|\hat{\mathbf{u}}|^2} = \frac{\hat{\mathbf{u}}^*}{\hat{\mathbf{u}}\hat{\mathbf{u}}^*} \tag{4.159}$$

Multiplication of quaternions can with advantage be expressed in matrix form (Wehage, 1984)

$$\begin{aligned} \underline{\mathbf{w}} &= \begin{bmatrix} w_0 \\ \mathbf{w} \end{bmatrix} = \begin{bmatrix} u_0 v_0 - \mathbf{u}^T\mathbf{v} \\ u_0\mathbf{v} + v_0\mathbf{u} + \tilde{\mathbf{u}}\mathbf{v} \end{bmatrix} = \begin{bmatrix} v_0 u_0 - \mathbf{v}^T\mathbf{u} \\ v_0\mathbf{u} + u_0\mathbf{v} - \tilde{\mathbf{v}}\mathbf{u} \end{bmatrix} = \\ &= \begin{bmatrix} u_0 & -u_1 & -u_2 & -u_3 \\ u_1 & u_0 & -u_3 & u_2 \\ u_2 & u_3 & u_0 & -u_1 \\ u_3 & -u_2 & u_1 & u_0 \end{bmatrix} \begin{bmatrix} v_0 \\ v_1 \\ v_2 \\ v_3 \end{bmatrix} = \\ &= \begin{bmatrix} v_0 & -v_1 & -v_2 & -v_3 \\ v_1 & v_0 & v_3 & -v_2 \\ v_2 & -v_3 & v_0 & v_1 \\ v_3 & v_2 & -v_1 & v_0 \end{bmatrix} \begin{bmatrix} u_0 \\ u_1 \\ u_2 \\ u_3 \end{bmatrix} \end{aligned} \tag{4.160}$$

The scalar multiplication by vector "$\vec{u}$." is here expressed by multiplying by the matrix $\mathbf{u}^T$, and vector multiplication "$\vec{u}\times$" is expressed by multiplying with skew-symmetric matrix $\tilde{\mathbf{u}} = \mathbf{U}$ (see Section 4.1.3).

The quaternion product can then be written in a compact form using matrices

$$\begin{aligned} \mathbf{u}^+ &= \begin{bmatrix} 0 & -\mathbf{u}^T \\ \mathbf{u} & \tilde{\mathbf{u}} \end{bmatrix}, \quad \overset{+}{\underline{\mathbf{u}}} = \begin{bmatrix} u_0 & -\mathbf{u}^T \\ \mathbf{u} & u_0\mathbf{E}_3 + \tilde{\mathbf{u}} \end{bmatrix} = u_0\mathbf{E}_4 + \mathbf{u}^+ \\ \mathbf{v}^- &= \begin{bmatrix} 0 & -\mathbf{v}^T \\ \mathbf{v} & -\tilde{\mathbf{v}} \end{bmatrix}, \quad \overset{-}{\underline{\mathbf{v}}} = v_0\mathbf{E}_4 + \mathbf{v}^- \end{aligned} \tag{4.161}$$

By expanding the products it can be proved that (4.160) can be written as

$$\underline{\mathbf{w}} = \overset{+}{\underline{\mathbf{u}}}\,\underline{\mathbf{v}} = \overset{-}{\underline{\mathbf{v}}}\,\underline{\mathbf{u}} \tag{4.162}$$

The unit quaternion is the quaternion with a unit norm. It can be used to express orthonormal transformation of vectors and quaternions from one space into another, where one of those spaces has rotated with regard to the other through an angle Φ around an axis passing a fixed point in the direction of unit vector $\vec{d}$. This rotation can be expressed by a unit quaternion

$$\hat{\mathbf{e}} = e_0 + \vec{e} = \cos\frac{\Phi}{2} + \sin\frac{\Phi}{2}\vec{d} \tag{4.163}$$

Its components are the so-called *Euler parameters* (Wittenburg, 1977).

The orthogonal transformation of quaternions $\hat{\mathbf{u}}_b$ or vector $\vec{u}_b$ from space b in motion to stationary space a is as follows

$$\hat{\mathbf{u}}_a = \hat{\mathbf{e}}_{ab}\hat{\mathbf{u}}_b\hat{\mathbf{e}}^*_{ab}, \qquad \vec{u}_a = \hat{\mathbf{e}}_{ab}\vec{u}_b\hat{\mathbf{e}}^*_{ab} \tag{4.164}$$

The same transformation can be expressed by two quaternions differing in sign of e_0. For the sake of uniqueness we assume that e_0 is always positive.

The same transformation of vectors is expressed by direction cosines $\mathbf{S}_{ab}$ by equation (4.28).

From (4.164) it follows that the composition of orthogonal transformation can be described by multiplying quaternions

$$\hat{\mathbf{u}}_a = \hat{\mathbf{e}}_{ab}\hat{\mathbf{e}}_{bc}\hat{\mathbf{u}}_c\hat{\mathbf{e}}^*_{bc}\hat{\mathbf{e}}^*_{ab} \tag{4.165}$$

$$\hat{\mathbf{e}}_{ac} = \hat{\mathbf{e}}_{ab}\hat{\mathbf{e}}_{bc} \tag{4.166}$$

To understand the connections between transformation matrices and quaternions, it is important to be able to express Euler parameters $\hat{\mathbf{e}}_{ab}$ from elements of direction cosine matrix $\mathbf{S}_{ab}$ and vice versa. Let us describe the algorithm (Wehage, 1984).

Using (4.155) the equation (4.164) can be written as

$$\begin{aligned}\hat{\mathbf{u}}_a = {} & (e^2_{0,ab} + \vec{e}_{ab}.\vec{e}_{ab})u_{0,b} + (e^2_{0,ab} - \vec{e}_{ab}.\vec{e}_{ab})\vec{u}_b + 2((\vec{u}_b.\vec{e}_{ab})\vec{e}_{ab} + \\ & + e_{0,ab}(\vec{e}_{ab} \times \vec{u}_b))\end{aligned} \tag{4.167}$$

Further $\hat{\mathbf{e}}_{ab}$ is a unit quaternion

$$e^2_{0,ab} + \vec{e}_{ab}.\vec{e}_{ab} = 1, \qquad e^2_{0,ab} - \vec{e}_{ab}.\vec{e}_{ab} = 2e^2_{0,ab} - 1 \tag{4.168}$$

Hence for quaternions in (4.164)

$$\underline{\mathbf{u}}_a = \mathbf{S}'_{ab}\underline{\mathbf{u}}_b \tag{4.169}$$

where

$$\mathbf{S}'_{ab} = \left[\begin{array}{c|c} 1 & \mathbf{0} \\ \hline \mathbf{0} & (2e^2_{0,ab} - 1)\mathbf{E}_3 + 2(\mathbf{e}_{ab}\mathbf{e}^T_{ab} + e_{0,ab}\tilde{\mathbf{e}}_{ab}) \end{array}\right] \tag{4.170}$$

and analogously for vectors in (4.29)

$$\mathbf{S}_{ab} = 2(e_{0,ab}^2 - 1)\mathbf{E}_3 + 2(\mathbf{e}_{ab}\mathbf{e}_{ab}^T + e_{0,ab}\tilde{\mathbf{e}}_{ab}) = \begin{bmatrix} s_{11} & s_{12} & s_{13} \\ s_{21} & s_{22} & s_{23} \\ s_{31} & s_{32} & s_{33} \end{bmatrix}$$

$$\mathbf{S}_{ab} = 2\begin{bmatrix} e_0^2 + e_1^2 - \frac{1}{2} & e_1e_2 - e_0e_3 & e_1e_3 + e_0e_2 \\ e_1e_2 + e_0e_3 & e_0^2 + e_2^2 - \frac{1}{2} & e_2e_3 - e_0e_1 \\ e_1e_3 - e_0e_3 & e_2e_3 + e_0e_1 & e_0^2 + e_3^2 - \frac{1}{2} \end{bmatrix} \tag{4.171}$$

Due to skew-symmetry of the matrix

$$(e_{0,ab}\tilde{\mathbf{e}}_{ab})^T = -e_{0,ab}\tilde{\mathbf{e}}_{ab} \tag{4.172}$$

it follows from (4.171) that

$$\begin{aligned} \mathbf{S}_{ab} + \mathbf{S}_{ab}^T &= 2((2e_{0,ab}^2 - 1)\mathbf{E}_3 + 2\mathbf{e}_{ab}\mathbf{e}_{ab}^T) \\ \mathbf{S}_{ab} - \mathbf{S}_{ab}^T &= 4e_{0,ab}\tilde{\mathbf{e}}_{ab} \end{aligned} \tag{4.173}$$

Let us construct matrix $\mathbf{B}$

$$\mathbf{B} = \left[\begin{array}{c|ccc} \text{tr } \mathbf{S}_{ab} & -s_{23} & -s_{13} & -s_{12} \\ \hline s_{32} & s_{11} & s_{12} & s_{13} \\ s_{13} & s_{21} & s_{22} & s_{23} \\ s_{21} & s_{31} & s_{32} & s_{33} \end{array}\right] \tag{4.174}$$

From (4.168) and (4.169)

$$\text{tr } \mathbf{S}_{ab} = s_{11} + s_{22} + s_{23} = 3(2e_{0,ab}^2 - 1) + 2\mathbf{e}_{ab}^T\mathbf{e}_{ab} = 4e_{0,ab}^2 - 1 \tag{4.175}$$

We can now write matrix $\mathbf{B}'$ representing relations (4.173), (4.175) in compact form

$$\mathbf{B}' = \mathbf{B} + \mathbf{B}^T + (1 - \text{tr } \mathbf{S}_{ab})\mathbf{E}_4 = 4\tilde{\mathbf{e}}_{ab}\tilde{\mathbf{e}}_{ab}^T = 4\begin{bmatrix} e_0e_0 & e_0e_1 & e_0e_2 & e_0e_3 \\ e_1e_0 & e_1e_1 & e_1e_2 & e_1e_3 \\ e_2e_0 & e_2e_1 & e_2e_2 & e_2e_3 \\ e_3e_0 & e_3e_1 & e_3e_2 & e_3e_3 \end{bmatrix} \tag{4.176}$$

A simple algorithm for the calculation of quaternion $\hat{\mathbf{e}}_{ab}$ from matrix $\mathbf{S}_{ab}$ follows from equation (4.176):

1. Let k equal the index of the highest absolute value of elements on the matrix $\mathbf{B}'$ diagonal, $k = 0, 1, 2, 3$.

2. Let us compute

$$e'_{m,ab} = \frac{b'_{km}}{2\sqrt{b'_{kk}}}, \quad m = 0, 1, 2, 3 \tag{4.177}$$

3. Let

$$\begin{aligned} e_{m,ab} &= e'_{m,ab} \operatorname{sgn} e'_{0,ab}, \quad m = 0,1,2,3 \\ \operatorname{sgn} e'_{0,ab} &= \begin{cases} -1 & \text{for} \quad e'_{0,ab} < 0 \\ +1 & \text{for} \quad e'_{0,ab} \geq 0 \end{cases} \end{aligned} \tag{4.178}$$

The first step improves the numerical computational precision, but it is not indispensable. The last step ensures that $e_{0,ab} \geq 0$.

4.3.2 Dual Quaternions

Unit quaternions made possible the description of orthogonal transformation of a coordinate system and of body motion given by rotation around an axis passing the fixed origin. For the description of general spatial motion, dual quaternions must be used.

A dual quaternion is defined as an ordered pair of two quaternions $[\hat{\mathbf{u}}_1; \hat{\mathbf{u}}_2]$. Identity and addition of dual quaternions are defined by identity and addition of components respectively. The null dual quaternion is the neutral element $[\hat{\mathbf{0}}; \hat{\mathbf{0}}]$ of addition.

The dual quaternion can be written as

$$\begin{aligned} \hat{\mathbf{U}} &= [\hat{\mathbf{u}}_1; \hat{\mathbf{u}}_2] = \hat{\mathbf{u}}_1[\hat{\mathbf{1}}; \hat{\mathbf{0}}] + \hat{\mathbf{u}}_2[\hat{\mathbf{0}}; \hat{\mathbf{1}}] = \hat{\mathbf{u}}_1 + \hat{\mathbf{u}}_2\varepsilon = \\ &= [(u_{10}, \vec{u}_1); (u_{20}; \vec{u}_2)] \end{aligned} \tag{4.179}$$

where ε is a dual number introduced by W. Clifford (Dimentberg, 1978) with the property

$$\varepsilon^2 = \varepsilon\varepsilon = [\hat{\mathbf{0}}; \hat{\mathbf{1}}][\hat{\mathbf{0}}; \hat{\mathbf{1}}] = [\hat{\mathbf{0}}; \hat{\mathbf{0}}] \tag{4.180}$$

Multiplication of dual quaternions is defined as

$$\hat{\mathbf{W}} = \hat{\mathbf{U}}\hat{\mathbf{V}} = [\hat{\mathbf{u}}_1; \hat{\mathbf{u}}_2][\hat{\mathbf{v}}_1; \hat{\mathbf{v}}_2] = [\hat{\mathbf{u}}_1\hat{\mathbf{v}}_1; \hat{\mathbf{u}}_1\hat{\mathbf{v}}_2 + \hat{\mathbf{u}}_2\hat{\mathbf{v}}_1] = [\hat{\mathbf{w}}_1; \hat{\mathbf{w}}_2] \tag{4.181}$$

It is not commutative and has zero divisors $[\hat{\mathbf{0}}; \hat{\mathbf{u}}]$.

Both real numbers and vectors are special cases of quaternions, where both dual quaternion components are real numbers or vectors. The operations on dual quaternions introduced above therefore define operations on dual numbers and dual vectors. It is often advantageous to write dual quaternions in the form $\hat{\mathbf{U}} = \hat{\mathbf{u}}_1 + \varepsilon\hat{\mathbf{u}}_2$ and then treat them algebraically using the convention $\varepsilon^2 = 0$.

The conjugated dual quaternion belonging to dual quaternion (4.179) is

$$\hat{\mathbf{U}}^* = \hat{\mathbf{u}}_1^* + \hat{\mathbf{u}}_2^*\varepsilon = [\hat{\mathbf{u}}_1^*; \hat{\mathbf{u}}_2^*] \tag{4.182}$$

The dual conjugated dual quaternion belonging to dual quaternion (4.179) is

$$\hat{\mathbf{U}}_c^* = \hat{\mathbf{u}}_1^* - \hat{\mathbf{u}}_2^* \varepsilon = [\hat{\mathbf{u}}_1^*; -\hat{\mathbf{u}}_2^*] \tag{4.183}$$

For multiplication of dual quaternions the following hold

$$\begin{aligned}(\hat{\mathbf{U}}\hat{\mathbf{V}})^* &= \hat{\mathbf{V}}^*\hat{\mathbf{U}}^* \\ (\hat{\mathbf{U}}\hat{\mathbf{V}})_c &= \hat{\mathbf{U}}_c\hat{\mathbf{V}}_c \\ (\hat{\mathbf{U}}\hat{\mathbf{V}})_c^* &= \hat{\mathbf{V}}_c^*\hat{\mathbf{U}}_c^* \end{aligned} \tag{4.184}$$

The norm of a dual quaternion is a dual number (Yang and Freudenstein, 1964)

$$|\hat{\mathbf{U}}|^2 = \hat{\mathbf{U}}^*\hat{\mathbf{U}} = \hat{\mathbf{U}}\hat{\mathbf{U}}^* = [|\hat{\mathbf{u}}_1|^2; 2(u_{10}u_{20} + \vec{u}_1.\vec{u}_2)] \tag{4.185}$$

The dual quaternion inversion

$$\hat{\mathbf{U}}^{-1} = [\hat{\mathbf{u}}_1^{-1}; -\hat{\mathbf{u}}_1^{-1}\hat{\mathbf{u}}_2\hat{\mathbf{u}}_1^{-1}] \tag{4.186}$$

is defined for $|\hat{\mathbf{u}}_1| \neq 0$. A unit dual quaternion is a dual quaternion whose norm is a unit quaternion

$$|\hat{\mathbf{U}}| = |\hat{\mathbf{U}}|^2 = [1; 0] \tag{4.187}$$

A unit dual quaternion describes general spatial motion and transformations (Bottema and Roth, 1979). The position of a point of a given radius vector $\vec{u} = (x, y, z)$ is described by a dual quaternion

$$\hat{\mathbf{U}} = [\hat{\mathbf{1}}; \hat{\mathbf{u}}] = \hat{\mathbf{1}} + (0 + \vec{u})\varepsilon = [1, 0, 0, 0; 0, x, y, z] \tag{4.188}$$

The transformation of a dual quaternion $\hat{\mathbf{U}}$ describing a point on to a dual quaternion $\hat{\mathbf{U}}'$ given by motion expressed by dual quaternion $\hat{\mathbf{A}}$ is given by (Stehno, 1973)

$$\hat{\mathbf{U}}' = \hat{\mathbf{A}}\hat{\mathbf{U}}\hat{\mathbf{A}}_c^* \tag{4.189}$$

The composition of motion is described by a dual quaternion product. This follows from the equation

$$\hat{\mathbf{U}}_1 = \hat{\mathbf{A}}_{12}\hat{\mathbf{A}}_{23}\hat{\mathbf{U}}_3\hat{\mathbf{A}}_{c,23}^*\hat{\mathbf{A}}_{c,12}^* \tag{4.190}$$

Hence, comparing with equation (4.189),

$$\hat{\mathbf{A}}_{13} = \hat{\mathbf{A}}_{12}\hat{\mathbf{A}}_{23} \tag{4.191}$$

Assuming the motion is a translation determined by vector $\vec{t}$

$$\hat{\mathbf{A}} = [\ \hat{\mathbf{1}}; \frac{1}{2}\vec{t}\ ] \tag{4.192}$$

If the motion is revolution around an axis passing a fixed origin described by Euler parameters $\hat{\mathbf{e}}$, then

$$\hat{\mathbf{A}} = [\hat{\mathbf{e}}; \hat{\mathbf{0}}] \tag{4.193}$$

If the motion is a revolution described by Euler parameters $\hat{\mathbf{e}}$ around an axis passing a point whose radius vector is $\vec{t}$, the corresponding dual quaternion $\hat{\mathbf{A}}$ is to be obtained using the above special cases (4.192), (4.193) according to relation

$$\hat{\mathbf{A}} = [\ \hat{\mathbf{1}}; \frac{1}{2}\vec{t}\ ][\ \hat{\mathbf{e}}; \hat{\mathbf{0}}\ ][\ \hat{\mathbf{1}}; -\frac{1}{2}\vec{t}\ ] = [\ \hat{\mathbf{e}}; \vec{t} \times \vec{e}\ ] \tag{4.194}$$

This relation expresses the composition of resultant motion from the following motions: translation of the origin to a point on the axis of revolution, revolution around an axis passing the origin, and translation of the origin to the original position. The validity of relations (4.192)–(4.194) can be verified by an elementary calculation following the above rules. The application of these rules for kinematical description is however laborious and computationally complicated. We shall therefore introduce a simpler concept in the following sections.

4.3.3 Shortened Dual Quaternions

Let a spatial motion of a body with a coordinate system b fixed to it be described in a coordinate system a as a decomposition into translational motion with a reference point in the origin of b determined by vector $\vec{t} = (t_x, t_y, t_z)$ and into relative spherical motion around origin O_b described by quaternion $\hat{\mathbf{e}}$, whose elements e_0, e_1, e_2, e_3 are Euler parameters. This motion may be described by a shortened dual quaternion (SDQ)

$$\hat{\mathbf{K}}_{ab} = [\ \hat{\mathbf{e}}; \vec{t}\ ] = [\ (e_0, \vec{e}); \vec{t}\ ] = [e_0, e_1, e_2, e_3; t_x, t_y, t_z] \tag{4.195}$$

SDQ describes the position of body b in coordinate system a in such a way that the first quaternion $\hat{\mathbf{e}}$ determines the angular orientation of the body, and the second is a mere vector, specifying the position of the system b origin in coordinate system a.

For the sake of simplicity in composing motions and transformations described by SDQ, we define the SDQ product as follows:

$$\hat{\mathbf{W}} = \hat{\mathbf{U}}\hat{\mathbf{V}} = [\hat{\mathbf{u}}; \vec{u}][\hat{\mathbf{v}}; \vec{v}] = [\hat{\mathbf{u}}\hat{\mathbf{v}}; \hat{\mathbf{u}}\vec{v}\hat{\mathbf{u}}^* + \vec{u}] = [\hat{\mathbf{w}}; \vec{w}] \tag{4.196}$$

This definition of multiplying SDQ's enables the structural equivalence of matrix and quaternion notations of point and body motion to be maintained.

The motion of point M whose position in a coordinate system b is determined by vector $\vec{u}_{bM}$ can be described using SDQ's as follows:

$$[\hat{0}; \vec{u}_{aM}] = [\hat{\mathbf{e}}_{ab}; \vec{t}_{ab}][\hat{0}; \vec{u}_{bM}] = [\hat{0}; \hat{\mathbf{e}}_{ab}\vec{u}_{bM}\hat{\mathbf{e}}^*_{ab} + \vec{t}_{ab}] \tag{4.197}$$

or

$$\hat{\mathbf{U}}_{aM} = \hat{\mathbf{K}}_{ab}\hat{\mathbf{U}}_{bM} \tag{4.198}$$

where

$$\hat{\mathbf{U}}_{aM} = [\hat{0}; \vec{u}_{aM}], \qquad \hat{\mathbf{U}}_{bM} = [\hat{0}; \vec{u}_{bM}], \qquad \hat{\mathbf{K}}_{ab} = [\hat{\mathbf{e}}_{ab}; \vec{t}_{ab}] \tag{4.199}$$

Relations (4.197)–(4.199) are analogies of vector relations (4.4) and matrix relations (4.25), (4.50).

For the inverse motion $a : b$ we obtain, in analogy to relation (4.57),

$$\hat{\mathbf{K}}^{-1}_{ab} = [\hat{\mathbf{e}}^*_{ab}; -\hat{\mathbf{e}}^*_{ab}\vec{t}_{ab}\hat{\mathbf{e}}_{ab}] \tag{4.200}$$

$$\hat{\mathbf{K}}_{ba} = \hat{\mathbf{K}}^{-1}_{ab} \tag{4.201}$$

For the composition of two simultaneous motions according to (4.117) we obtain in analogy to (4.118)

$$\begin{aligned} \hat{\mathbf{K}}_{13} &= \hat{\mathbf{K}}_{12}\hat{\mathbf{K}}_{23} = [\hat{\mathbf{e}}_{12}; \vec{t}_{12}][\hat{\mathbf{e}}_{23}; \vec{t}_{23}] = \\ &= [\hat{\mathbf{e}}_{12}\hat{\mathbf{e}}_{23}; \hat{\mathbf{e}}_{12}\vec{t}_{23}\hat{\mathbf{e}}^*_{12} + \vec{t}_{12}] \end{aligned} \tag{4.202}$$

In this way, the position is described by SDQ's in analogy with the matrix description.

The definition of multiplication (4.196) gives no advantage in terms of computational complexity. The following definition is equivalent and computationally more convenient.

$$\hat{\mathbf{W}} = \hat{\mathbf{U}}\hat{\mathbf{V}} = [\hat{\mathbf{u}}; \vec{u}][\hat{\mathbf{v}}; \vec{v}] = [\hat{\mathbf{u}}\hat{\mathbf{v}}; \mathbf{S}(\hat{\mathbf{u}})\vec{v} + \vec{u}] = [\hat{\mathbf{w}}; \vec{w}] \tag{4.203}$$

where $\mathbf{S}(\hat{\mathbf{u}})$ designates the operation of computation of the direction cosines matrix of a transformation described by quaternion $\hat{\mathbf{u}}$. This operation is given by relation (4.171).

The computation of $\hat{\mathbf{u}}\vec{v}\hat{\mathbf{u}}^*$ in (4.202) requires $2 \times$ (16 multiplications + 12 additions) or $2 \times$ (12 multiplications + 16 additions) for 2 quaternion products. The computation of $\mathbf{S}(\hat{\mathbf{u}})\vec{v}$ requires 10 multiplications + 10 additions for the computation of $\frac{1}{2}\mathbf{S}$ according to (4.171), a further 9 multiplications + 6 additions for the vector product and 3 additions for doubling the halved elements of the resultant vector. In total, this gives 19 multiplications and 19 additions, which is much better.

In addition, definition (4.203) allows the maintenance of structural equivalence of expressing velocities and accelerations for both quaternion and matrix descriptions. The structural equivalence fails by using (4.196), because the time derivative of $\hat{\mathbf{u}}\vec{v}\hat{\mathbf{u}}^*$

$$\frac{\mathrm{d}}{\mathrm{d}t}(\hat{\mathbf{u}}\vec{v}\hat{\mathbf{u}}^*) = \left(\frac{\mathrm{d}}{\mathrm{d}t}\hat{\mathbf{u}}\right)\vec{v}\hat{\mathbf{u}}^* + \hat{\mathbf{u}}\vec{v}\left(\frac{\mathrm{d}}{\mathrm{d}t}\hat{\mathbf{u}}^*\right) \tag{4.204}$$

destroys the possibility of analogy with (4.59) and further linked rules.

To avoid this obstacle we introduce other ways of SDQ multiplication (Valášek and Stejskal, 1988).

$$\hat{\mathbf{U}}\hat{\mathbf{V}}^L = [\hat{\mathbf{u}}_1;\vec{u}_2][\hat{\mathbf{v}}_1;\vec{v}_2] = [\hat{\mathbf{u}}_1\hat{\mathbf{v}}_1;\vec{v}_2] \tag{4.205}$$

$$\hat{\mathbf{U}}^R\hat{\mathbf{V}} = [\hat{\mathbf{u}}_1;\vec{u}_2]^R[\hat{\mathbf{v}}_1;\vec{v}_2] = [\hat{\mathbf{u}}_1\hat{\mathbf{v}}_1;2\vec{u}_1\times\vec{v}_2+\vec{u}_2] \tag{4.206}$$

together with the following rules: $[\ldots]^L$ quaternion in (4.205) is not left associative in multiplication. $[\ldots]^R$ quaternion in (4.206) is not right associative in multiplication. These rules mean for the calculation of the product of those quaternions

$$\hat{\mathbf{U}}\hat{\mathbf{V}}\ldots\hat{\mathbf{W}}[\ldots]^L = (\hat{\mathbf{U}}\hat{\mathbf{V}}\ldots\hat{\mathbf{W}})[\ldots]^L \tag{4.207}$$

$$[\ldots]^R\hat{\mathbf{U}}\hat{\mathbf{V}}\ldots\hat{\mathbf{W}} = [\ldots]^R(\hat{\mathbf{U}}\hat{\mathbf{V}}\ldots\hat{\mathbf{W}}) \tag{4.208}$$

The following relations do not hold:

$$\hat{\mathbf{U}}\hat{\mathbf{V}}\ldots\hat{\mathbf{W}}[\ldots]^L = \hat{\mathbf{U}}\hat{\mathbf{V}}\ldots(\hat{\mathbf{W}}[\ldots]^L) \tag{4.209}$$

$$[\ldots]^R\hat{\mathbf{U}}\hat{\mathbf{V}}\hat{\mathbf{W}} = ([\ldots]^R\hat{\mathbf{U}})\hat{\mathbf{V}}\ldots\hat{\mathbf{W}} \tag{4.210}$$

Let us further introduce another dual quaternions conjugation operation $[\ldots]_\varepsilon$. Let

$$\hat{\mathbf{K}}_{ab} = [\hat{\mathbf{e}}_{ab};\vec{u}_{ab}] \tag{4.211}$$

then

$$\hat{\mathbf{K}}_{ab,\varepsilon} = [\hat{\mathbf{e}}_{ab};\vec{0}] \tag{4.212}$$

Further

$$\hat{\mathbf{K}}^*_{ab,\varepsilon} = [\hat{\mathbf{e}}^*_{ab};\vec{0}] \tag{4.213}$$

We can now introduce a direct analogy to the computation of the time derivative of transformation according to (4.61), as follows:

$$\frac{\mathrm{d}}{\mathrm{d}t}\hat{\mathbf{K}}_{ab} = \hat{\mathbf{K}}^L_{ab,\varepsilon}\hat{\mathbf{W}}^R_{b,ab} \tag{4.214}$$

where $\hat{\mathbf{W}}^R_{b,ab}$ is the velocity quaternion defined as

$$\hat{\mathbf{W}}^R_{b,ab} = \hat{\mathbf{K}}^*_{ab,\varepsilon}\frac{\mathrm{d}}{\mathrm{dt}}\hat{\mathbf{K}}_{ab} = \hat{\mathbf{K}}^*_{ab,\varepsilon}[\dot{\hat{\mathbf{e}}}_{ab};\dot{\vec{u}}_{ab}] \tag{4.215}$$

Also

$$\hat{\mathbf{W}}^R_{b,ab} = [\hat{\mathbf{e}}^*_{ab}\dot{\hat{\mathbf{e}}}_{ab}; \hat{\mathbf{e}}^*_{ab}\vec{u}_{ab}\hat{\mathbf{e}}_{ab}]^R = [\hat{\mathbf{e}}^*_{ab}\dot{\hat{\mathbf{e}}}_{ab}; \mathbf{S}(\hat{\mathbf{e}}^*_{ab})\vec{u}_{ab}]^R \tag{4.216}$$

Both quaternions in (4.216) have a simple physical meaning. To interpret the former, let us start with the identity

$$\hat{\mathbf{e}}^*\hat{\mathbf{e}} = \hat{\mathbf{e}}\hat{\mathbf{e}}^* = 1 \tag{4.217}$$

which derivative by time yields

$$\dot{\hat{\mathbf{e}}}^*\hat{\mathbf{e}} + \hat{\mathbf{e}}^*\dot{\hat{\mathbf{e}}} = 0 \tag{4.218}$$

Modifying (4.218) yields

$$\dot{\hat{\mathbf{e}}}^*\hat{\mathbf{e}} = -\hat{\mathbf{e}}^*\dot{\hat{\mathbf{e}}} = -(\dot{\hat{\mathbf{e}}}^*\hat{\mathbf{e}})^* \tag{4.219}$$

corfirming that we are dealing with a vector. Let us designate

$$\vec{c} = \hat{\mathbf{e}}^*\dot{\hat{\mathbf{e}}} \tag{4.220}$$

By left multiplying of (4.220) by quaternion $\hat{\mathbf{e}}$ we obtain the result

$$\dot{\hat{\mathbf{e}}} = \hat{\mathbf{e}}\vec{c} \tag{4.221}$$

In the following we shall differentiate the basic transformation relation

$$\hat{\mathbf{a}}_a = \hat{\mathbf{e}}_{ab}\hat{\mathbf{a}}_b\hat{\mathbf{e}}^*_{ab} \tag{4.222}$$

with respect to time. We obtain

$$\begin{aligned}
\dot{\hat{\mathbf{a}}}_a &= \dot{\hat{\mathbf{e}}}_{ab}\hat{\mathbf{a}}_b\hat{\mathbf{e}}^*_{ab} + \hat{\mathbf{e}}_{ab}\mathbf{a}_b\hat{\mathbf{e}}^*_{ab} + \hat{\mathbf{e}}_{ab}\dot{\hat{\mathbf{a}}}_b\dot{\hat{\mathbf{e}}}^*_{ab} = \\
&= \hat{\mathbf{e}}_{ab}(\vec{c}\hat{\mathbf{a}}_b + \dot{\hat{\mathbf{a}}}_b - \hat{\mathbf{a}}_b\vec{c})\hat{\mathbf{e}}^*_{ab} = \\
&= \hat{\mathbf{e}}_{ab}(\dot{\hat{\mathbf{a}}}_b + 2\vec{c}\times\vec{a}_b)\hat{\mathbf{e}}^*_{ab}
\end{aligned} \tag{4.223}$$

Comparing the result with matrix equation (4.72) or vector equation (4.5) shows that we are dealing with velocity transformation and

$$2\vec{c}\times\vec{a}_b = \boldsymbol{\omega}_{b,ab}\times\vec{a}_b \tag{4.224}$$

As the last equation must be fulfilled for any vector $\vec{a}_b$ it is also true that

$$\boldsymbol{\omega}_{b,ab} = 2\vec{c} = 2\hat{\mathbf{e}}^*\dot{\hat{\mathbf{e}}} \tag{4.225}$$

The former of the two quaternions in (4.216) therefore has the meaning of halved angular velocity of motion $b : a$ expressed in space b.

The latter of the two quaternions can be written as

$$\mathbf{S}(\hat{\mathbf{e}}^*_{ab})\vec{\dot{u}}_{ab} = \mathbf{S}^T_{ab}\vec{\dot{u}}_{ab} = \vec{\dot{u}}_{b,ab} = \vec{v}_{ab} \tag{4.226}$$

and has the meaning of velocity of system b origin relative to system a expressed in space b. This explains the term "shortened dual velocity quaternion $\hat{\mathbf{W}}^R_{b,ab}$" and proves the analogy.

As in (4.214) we introduce the direct analogy of the computation of the second time derivation of transformation according to (4.84), (4.86), (4.91) as follows:

$$\frac{\mathrm{d}^2}{\mathrm{d}t^2}\hat{\mathbf{K}}_{ab} = \hat{\mathbf{K}}^L_{ab,\varepsilon}\hat{\mathbf{Z}}^R_{b,ab} = \hat{\mathbf{K}}^L_{ab,\varepsilon}(\hat{\mathbf{Y}}^R_{b,ab} + \hat{\mathbf{W}}^R_{b,ab,\varepsilon}\hat{\mathbf{W}}^R_{b,ab}) \tag{4.227}$$

The following holds

$$\hat{\mathbf{Y}}^R_{b,ab} = \frac{\mathrm{d}}{\mathrm{d}t}\hat{\mathbf{W}}^R_{b,ab} \tag{4.228}$$

$$\hat{\mathbf{Y}}^R_{b,ab} = [\hat{\mathbf{e}}^*_{b,ab}\ddot{\hat{\mathbf{e}}}_{b,ab} + \dot{\hat{\mathbf{e}}}^*_{b,ab}\dot{\hat{\mathbf{e}}}_{b,ab}; \mathbf{S}(\hat{\mathbf{e}}^*_{ab})\vec{\ddot{u}}_{ab} + \dot{\mathbf{S}}(\hat{\mathbf{e}}^*_{ab})\vec{\dot{u}}_{ab}]^R \tag{4.229}$$

$$\hat{\mathbf{W}}^R_{b,ab,\varepsilon} = [\hat{\mathbf{e}}^*_{ab}\dot{\hat{\mathbf{e}}}_{ab}; \mathbf{0}]^R \tag{4.230}$$

We shall call the quaternion $\hat{\mathbf{Y}}^R$ the acceleration quaternion and quaternion $\hat{\mathbf{Z}}^R$ the complete acceleration quaternion. If, when using these relations, two desired SDQ's appear in the product, it is necessary to carry out the computation according to the following:

$$\hat{\mathbf{K}}_1\hat{\mathbf{U}}^L_\varepsilon\hat{\mathbf{W}}^R_U\hat{\mathbf{K}}_2\hat{\mathbf{V}}^L_\varepsilon\hat{\mathbf{W}}^R_V\hat{\mathbf{K}}_3 = ((((\hat{\mathbf{K}}_1)\hat{\mathbf{U}}^L_\varepsilon)(\hat{\mathbf{W}}^R_U(\hat{\mathbf{K}}_2)))\hat{\mathbf{V}}^L_\varepsilon)(\hat{\mathbf{W}}^R_V(\hat{\mathbf{K}}_3)) \tag{4.231}$$

4.4 Screw Notation

One of a series of very attractive formalisms for the description and solution of problems in mechanics is the screw theory (Yang and Freudenstein, 1964; Dimentberg, 1971, 1978). This theory has its starting point in the common vector notation. However, it distinguishes free vectors, dealt with by the common vector notation, and bound vectors, which are described by screw theory. In many problems of rigid body mechanics, it is possible to substitute vectors representing various quantities, e.g. force or displacement, by equivalent vectors which maintain the conditions of the original problems. The equivalence of these vectors is determined by the equality of their moments relative to an arbitrary point in space, and is preserved by their displacement along the line with which they coincide. They are therefore called line vectors. Examples of these are forces and angular velocities.

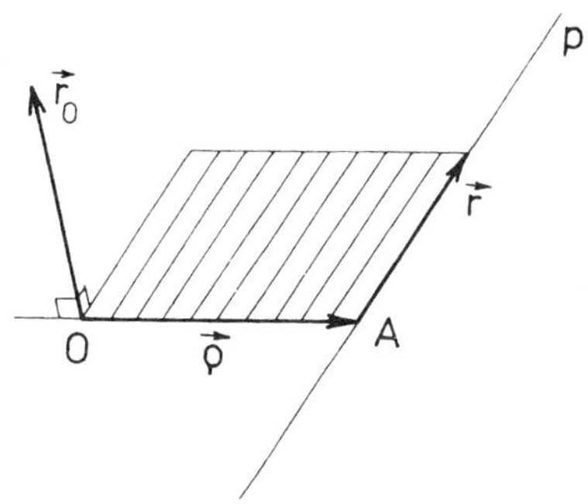

Figure 4.11. Line vector and its moment.

A line vector determined by vector $\vec{r}$ and point A or line p respectively where it acts, can be determined from the viewpoint of equivalence by vector $\vec{r}$ and moment $\vec{r}^{\,0} = \vec{\varrho} \times \vec{r}$, $\vec{\varrho} = \overrightarrow{OA}$ relative to an arbitrary point O, only (Figure 4.11).

A group of line vectors can be substituted by the main vector of the system, which is the sum of all vectors, and the main moment of the system, which is the sum of the moments of all the vectors relative to a given point O. This point is called the line vector system substitution point. Changing the point of substitution to O' preserves the main vector $\vec{r}$ of the system. For the main moment $(\vec{r}^{\,0})'$ it holds

$$(\vec{r}^{\,0})' = \vec{r}^{\,0} + \vec{\varrho} \times \vec{r}, \quad \vec{\varrho} - \overrightarrow{O'O} \tag{4.232}$$

Points of substitution do exist, for which the main moment $\vec{r}^{\,0}$ is collinear with the main vector $\vec{r}$. Let us take a point of substitution O for which $\vec{r}$ and $\vec{r}^{\,0}$ are not collinear. Let us then consider the line perpendicular to $\vec{r}$ and $\vec{r}^{\,0}$ passing through O and let us find the point S of this line for which

$$\overrightarrow{SO} = \frac{\vec{r}^{\,0} \times \vec{r}}{|\vec{r}|^2} \tag{4.233}$$

Let S be the new point of substitution; using (4.232) we have

$$(\vec{r}^{\,0})_S = \vec{r}^{\,0} + \left(\frac{\vec{r}^{\,0} \times \vec{r}}{|\vec{r}|^2} \right) \times \vec{r} = \frac{\vec{r}^{\,0} . \vec{r}}{|\vec{r}|^2} \vec{r} = p\vec{r} \tag{4.234}$$

This expresses the collinearity of the main moment and the main vector with p as a constant. This is a property of all points of a line passing through S parallel to the main vector, and no other line has this property. This line is called the central axis of the system.

An equivalent line vector system substitution for an arbitrary spatial point given by pair $(\vec{r}, \vec{r}^{\,0})$ is called *motor* (*mo*ment vec*tor*). Motor $(\vec{r}, \vec{r}^{\,0})$, whose moment $\vec{r}^{\,0}$ is collinear with vector $\vec{r}$ (with p as a constant), is called screw (with parameter p). The line with which $\vec{r}$ and $\vec{r}^{\,0}$ coincide is called the screw axis. It is a central axis of the system.

Motor and screw are line vector system alternatives. Screw is a special case of a motor. Screw $(\vec{r}, p\vec{r})$ completely determines motor $(\vec{r}, \vec{r}^{\,0})$ for an arbitrary space point and this motor, vice versa, fully determines this screw. Let us therefore choose an arbitrary fixed point, in following the coordinate system origin, and let us substitute each considered screw in this point by the corresponding motor. As the point is fixed, every screw is completely described by the pair $(\vec{r}, \vec{r}^{\,0})$ of substitution in origin. The Cartesian coordinates of vectors $\vec{r}, \vec{r}^{\,0}$, i.e. 6 quantities, are called the Plücker screw coordinates.

The most important matter for the analytical description of screws is the introduction of dual quantities and the *principle of transference*. Dual quantities have been introduced in Section 4.3.2. Using them we can represent the motor $(\vec{r}, \vec{r}^{\,0})$ by the dual vector $\vec{r} + \varepsilon\vec{r}^{\,0}$. We shall develop below the screw description so that the principle of transference is valid. This enables the complete analogy of vector calculus and screw calculus to be preserved in the notation of dual vector calculus.

We have introduced the basic arithmetic operations on dual quantities. Let us consider dual numbers and a function $\mathbf{F}$ of a dual variable $\mathbf{X} = x + x^0\varepsilon$, which can in general be expressed as

$$\mathbf{F}(\mathbf{X}) = \mathbf{F}(x + x^0\varepsilon) = f(x, x^0) + \varepsilon g(x, x^0) \tag{4.235}$$

For the compliance with the transference principle,

$$\mathbf{F}(\mathbf{X}) = f(x) + \varepsilon x^0 f'(x) \tag{4.236}$$

can be obtained. Hence for differentiable real variable functions f

$$e^{\mathbf{X}} = e^x + \varepsilon x^0 (e^x)' = e^x + \varepsilon x^0 e^x = e^x(1 + \varepsilon x^0)$$

However

$$e^{\varepsilon x^0} = e^{0+\varepsilon x^0} = 1 + \varepsilon x^0$$

Therefore

$$\begin{aligned} e^{\mathbf{X}} &= e^{x+\varepsilon x^0} = e^x e^{\varepsilon x^0} && (4.237\text{a}) \\ \sin \mathbf{X} &= \sin x + \varepsilon x^0 \cos x && (4.237\text{b}) \\ \cos \mathbf{X} &= \cos x - \varepsilon x^0 \sin x && (4.237\text{c}) \\ \tan \mathbf{X} &= \tan x + \varepsilon x^0 / \cos^2 x && (4.237\text{d}) \end{aligned}$$

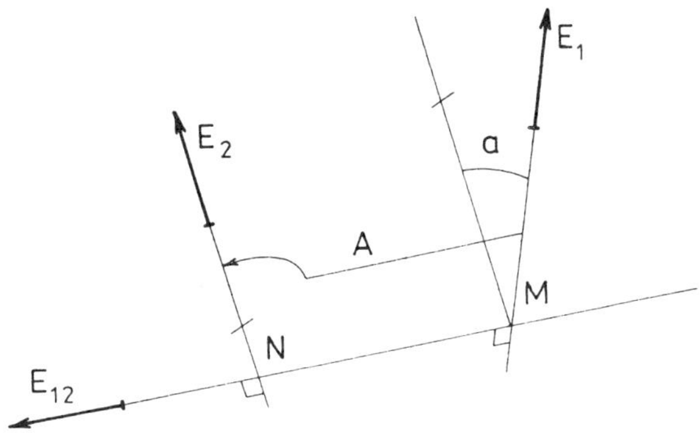

Figure 4.12. Unit screw and dual number.

The screw whose vector is a unit vector and whose parameter is zero is called the unit screw.

A screw and dual number product is a special case of a product of dual quaternions, where one dual quaternion represents a dual vector representing a screw and the other dual quaternion is a dual number. Every screw $\vec{\mathbf{R}} = \vec{r} + \varepsilon\vec{r}^{\,0}$ can be expressed as a product $\vec{\mathbf{E}}\vec{\mathbf{A}}$ of unit screw $\vec{\mathbf{E}}$ with the same axis and a dual number

$$\mathbf{A} = |\vec{r}|e^{\varepsilon p} \tag{4.238}$$

where p is the screw parameter. This dual number is called the modulus of the screw. Let us consider two axes with unit screws $\vec{\mathbf{E}}_1$ and $\vec{\mathbf{E}}_2$ (Figure 4.12). The dual number $\mathbf{A}$, where

$$\mathbf{A} = a + \varepsilon a^0 \tag{4.239}$$

which expresses the motion required to identify $\vec{\mathbf{E}}_1$ with $\vec{\mathbf{E}}_2$, is called the dual angle $\mathbf{A}$ of the given axes. It is a screw motion with angle a and distance a^0 along a unit screw $\vec{\mathbf{E}}_{12}$ determined by the skew lines transversal of axes of screws $\vec{\mathbf{E}}_1, \vec{\mathbf{E}}_2$. In this case,

$$\vec{\mathbf{E}}_2 = (\cos\mathbf{A} + \vec{\mathbf{E}}_{12}\sin\mathbf{A})\vec{\mathbf{E}}_1 \tag{4.240}$$

Let us consider screws $\vec{\mathbf{R}}_1, \vec{\mathbf{R}}_2$ with axes $\vec{\mathbf{E}}_1, \vec{\mathbf{E}}_2$, as in Figure 4.12. As scalar product of screws $\vec{\mathbf{R}}_1, \vec{\mathbf{R}}_2$, we designate the dual number equal to the scalar product of their motors in origin. Then

$$\vec{\mathbf{R}}_1.\vec{\mathbf{R}}_2 = |\vec{\mathbf{R}}_1||\vec{\mathbf{R}}_2|\cos\mathbf{A} \tag{4.241}$$

A screw whose motor in origin equals the vector product of their motors in origin is the screw product of screws $\vec{\mathbf{R}}_1, \vec{\mathbf{R}}_2$, so

$$\vec{\mathbf{R}}_1 \times \vec{\mathbf{R}}_2 = \vec{\mathbf{E}}_{12}|\vec{\mathbf{R}}_1||\vec{\mathbf{R}}_2| \sin \mathbf{A} \tag{4.242}$$

As the projection of screw $\vec{\mathbf{R}}$ on to unit screw $\vec{\mathbf{E}}$, we designate the dual number equal to their scalar product $\vec{\mathbf{R}}.\vec{\mathbf{E}}$.

Unit vectors $\vec{i}, \vec{j}, \vec{k}$ of coordinate axes also represent unit screws $\vec{i}+\varepsilon\vec{0}, \vec{j}+\varepsilon\vec{0}, \vec{k}+\varepsilon\vec{0}$. Let the dual angles of screw $\vec{\mathbf{R}}$ to those axes be

$$\mathbf{A} = a + \varepsilon a^0, \quad \mathbf{B} = b + \varepsilon b^0, \quad \mathbf{C} = c + \varepsilon c^0 \tag{4.243}$$

Let us construct the projections of screw $\vec{\mathbf{R}}$ on to coordinate axes.

$$\begin{aligned} \mathbf{X} &= x + \varepsilon x^0 = \vec{\mathbf{R}} \cos \mathbf{A} \\ \mathbf{Y} &= y + \varepsilon y^0 = \vec{\mathbf{R}} \cos \mathbf{B} \\ \mathbf{Z} &= z + \varepsilon z^0 = \vec{\mathbf{R}} \cos \mathbf{C} \end{aligned} \tag{4.244}$$

Screw $\vec{\mathbf{R}}$ equals the sum of its orthogonal components on the axes

$$\begin{aligned} \vec{\mathbf{R}} &= \vec{i}\mathbf{X} + \vec{j}\mathbf{Y} + \vec{k}\mathbf{Z} = \vec{i}(x + \varepsilon x^0) + \vec{j}(y + \varepsilon y^0) + \\ &+ \vec{k}(z + \varepsilon z^0) \end{aligned} \tag{4.245}$$

which are called the dual orthogonal screw coordinates. For screws $\vec{\mathbf{R}}_1, \vec{\mathbf{R}}_2$, let

$$\vec{\mathbf{R}}_l = \vec{i}\mathbf{X}_l + \vec{j}\mathbf{Y}_l + \vec{k}\mathbf{Z}_l, \quad l = 1, 2$$

Hence

$$\begin{aligned} \vec{\mathbf{R}}_1.\vec{\mathbf{R}}_2 &= \mathbf{X}_1\mathbf{X}_2 + \mathbf{Y}_1\mathbf{Y}_2 + \mathbf{Z}_1\mathbf{Z}_2 \\ \vec{\mathbf{R}}_1 \times \vec{\mathbf{R}}_2 &= \begin{vmatrix} \vec{i} & \vec{j} & \vec{k} \\ \mathbf{X}_1 & \mathbf{Y}_1 & \mathbf{Z}_1 \\ \mathbf{X}_2 & \mathbf{Y}_2 & \mathbf{Z}_2 \end{vmatrix} \\ \vec{\mathbf{R}}_1 + \vec{\mathbf{R}}_2 &= \vec{i}(\mathbf{X}_1 + \mathbf{X}_2) + \vec{j}(\mathbf{Y}_1 + \mathbf{Y}_2) + \vec{k}(\mathbf{Z}_1 + \mathbf{Z}_2) \end{aligned} \tag{4.246}$$

Relations (4.242)–(4.246) demonstrate the complete analogy to vector calculus.

The screw notation enables an easy description to be made of the general spatial motion of line vectors and therefore also of bodies. There are three different ways of expressing line vector motion.

The motion can be expressed as a transformation of coordinates of the screw representing the line vector. This transformation from one coordinate system to the other is expressed by a screw affinor $\mathbf{A}_{21}$. Then

$$\vec{\mathbf{R}}_2 = \mathbf{A}_{21}\vec{\mathbf{R}}_1 \tag{4.247a}$$

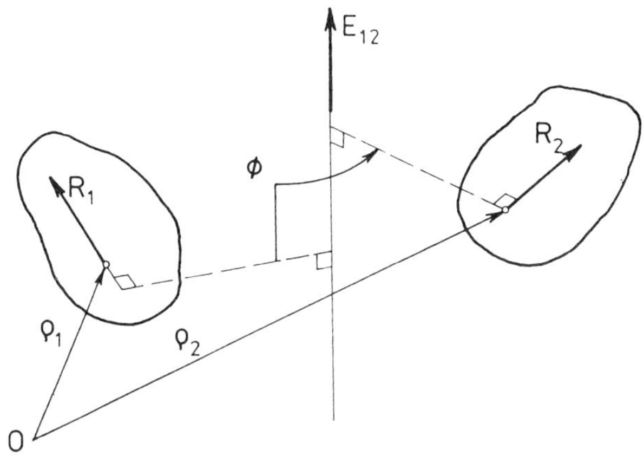

Figure 4.13. Screw motion of a body.

and

$$\begin{bmatrix} \mathbf{X}_2 \\ \mathbf{Y}_2 \\ \mathbf{Z}_2 \end{bmatrix} = \begin{bmatrix} \cos \mathbf{A}_{11} & \cos \mathbf{A}_{12} & \cos \mathbf{A}_{13} \\ \cos \mathbf{A}_{21} & \cos \mathbf{A}_{22} & \cos \mathbf{A}_{23} \\ \cos \mathbf{A}_{31} & \cos \mathbf{A}_{32} & \cos \mathbf{A}_{33} \end{bmatrix} \begin{bmatrix} \mathbf{X}_1 \\ \mathbf{Y}_1 \\ \mathbf{Z}_1 \end{bmatrix} \tag{4.247b}$$

where $\mathbf{A}_{mn}$ are dual angles of the respective axes of both coordinate systems. Composition of motion is expressed by multiplying screw affinor matrices.

Another possibility is to use the dual analogy of the Rodrigues formula. (Bottema and Roth, 1979) Let a body carry out a screw motion (Figure 4.13) given by the dual angle $\mathbf{\Phi} = \varphi + \varepsilon\varphi^0$ along a unit screw $\vec{\mathbf{E}}_{12}$. Each screw $\vec{\mathbf{R}}_1$ (representing a line vector) firmly fixed with the body moves to screw $\vec{\mathbf{R}}_2$.

$$\vec{\mathbf{R}}_2 = \vec{\mathbf{R}}_1 + \frac{2\vec{\mathbf{\Theta}}}{1+\vec{\mathbf{\Theta}}^2} \times (\vec{\mathbf{R}}_1 + \vec{\mathbf{\Theta}} \times \vec{\mathbf{R}}_1) \tag{4.248}$$

where the dual vector of motion is

$$\vec{\mathbf{\Theta}} = \vec{\mathbf{E}}_{12} \tan \frac{\mathbf{\Phi}}{\mathbf{2}} \tag{4.249}$$

For composing two screw motions with dual vectors $\vec{\mathbf{\Theta}}_1$ and $\vec{\mathbf{\Theta}}_2$ into a resultant motion given by dual vector $\vec{\mathbf{\Theta}}$,

$$\vec{\mathbf{\Theta}} = \frac{\vec{\mathbf{\Theta}}_1 + \vec{\mathbf{\Theta}}_2 - \vec{\mathbf{\Theta}}_1 \times \vec{\mathbf{\Theta}}_2}{1 - \vec{\mathbf{\Theta}}_1 . \vec{\mathbf{\Theta}}_2} \tag{4.250}$$

The third possible description of the motion is again the use of dual quaternions, but this time as in Section 4.3.2. In the case of body motion as in Figure 4.13, the screw $\vec{\mathbf{R}}_1$ moves over to screw $\vec{\mathbf{R}}_2$ according to the relation (Hamilton theorem) (Bottema and Roth, 1979)

$$\vec{\mathbf{R}}_2 = \left(\cos\frac{\mathbf{\Phi}}{2} + \vec{\mathbf{E}}_{12}\sin\frac{\mathbf{\Phi}}{2}\right)\vec{\mathbf{R}}_1\left(\cos\frac{\mathbf{\Phi}}{2} - \vec{\mathbf{E}}_{12}\sin\frac{\mathbf{\Phi}}{2}\right) \tag{4.251}$$

Relation (4.240) is a special and important case of (4.251). Unlike relation (4.188), in this case $\vec{\mathbf{R}}_i = \vec{r}_i + \varepsilon(\vec{\varrho}_i \times \vec{r}_i) = [\vec{0} + \vec{r}_i; \vec{0} + \vec{\varrho}_i \times \vec{r}_i]$. Motion composition is expressed as the product of respective dual quaternions.

The screw notation is attractive for its consistent construction of analogy: spherical motion $\longrightarrow$ general spatial motion. However, the resulting formalism has the disadvantage of computational complexity (see Section 4.6).

4.5 Basic Motions

Let coordinate systems of bodies a and b be identical at the beginning of the motion. Let body b carry out the kind of motion (revolution or translation) where, for some pair of axes with the same designation, the axes of both systems coincide. This kind of motion is called basic motion. In this case, the transformation matrix, velocity and acceleration matrices or analogously transformation SDQ, velocity and acceleration quaternions or corresponding screw entities have a particularly simple form. We speak therefore of basic matrices, basic quaternions and basic screws. We shall designate translations in direction x, y, and z respectively by indices Z1, Z2, Z3, or in the case of revolution around those axes by indices Z4, Z5, Z6. The basic matrices, quaternions or screws are functions of one coordinate. We shall write it in parentheses after the matrix, quaternion or screw symbol.

The significance of basic motions lies in the fact that every spatial motion of a body can be composed of basic motions. The respective basic matrices, quaternions or screws are very simple and every transformation matrix, SDQ or screw can be expressed as a product of transformation matrices, SDQ or screw of basic motions. This is convenient for both manual and computer solution of the description of point or body motion.

4.5.1 Basic Matrices

Transformation, velocity and acceleration matrices of basic motions are to be obtained from (4.51), (4.62), (4.91).

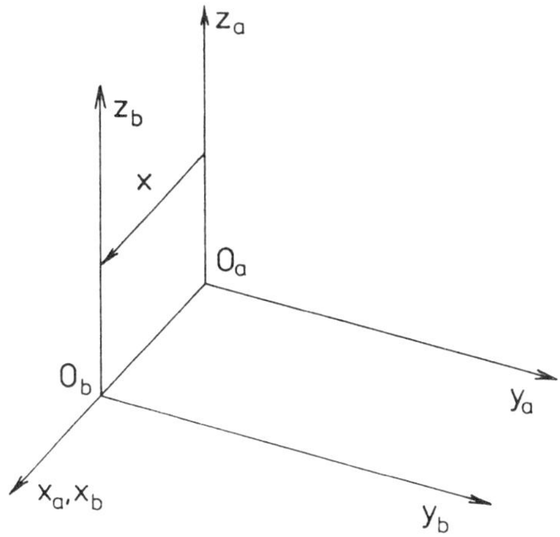

Figure 4.14. Basic motion Z1.

1. Translational motion in the direction of the axis $o \equiv x_a \equiv x_b$ (Figure 4.14). Let us designate the displacement x. It then follows that

$$\mathbf{T}_{Z1}(x) = \begin{bmatrix} 1 & 0 & 0 & x \\ 0 & 1 & 0 & 0 \\ 0 & 0 & 1 & 0 \\ 0 & 0 & 0 & 1 \end{bmatrix} \tag{4.252a}$$

For the computation of velocity and acceleration, let us introduce a matrix differential operator $\mathbf{D}_{Z1}(\dot{x})$ as follows:

$$\mathbf{D}_{Z1}(\dot{x}) = \begin{bmatrix} 0 & 0 & 0 & \dot{x} \\ 0 & 0 & 0 & 0 \\ 0 & 0 & 0 & 0 \\ 0 & 0 & 0 & 0 \end{bmatrix} \tag{4.252b}$$

Then according to (4.62), (4.85), (4.87)

$$\mathbf{V}_{Z1} = \mathbf{D}_{Z1}(\dot{x}) = \mathbf{D}_{Z1}(1)\mathbf{E}(\dot{x}) \tag{4.252c}$$

$$\mathbf{A}_{Z1} = \mathbf{D}_{Z1}(\ddot{x}) = \mathbf{D}_{Z1}(1)\mathbf{E}(\ddot{x}) \tag{4.252d}$$

$$\mathbf{B}_{Z1} = \mathbf{A}_{Z1} \tag{4.252e}$$

where, for the sake of completeness, we use the unit matrix operator $\mathbf{E}(p)$

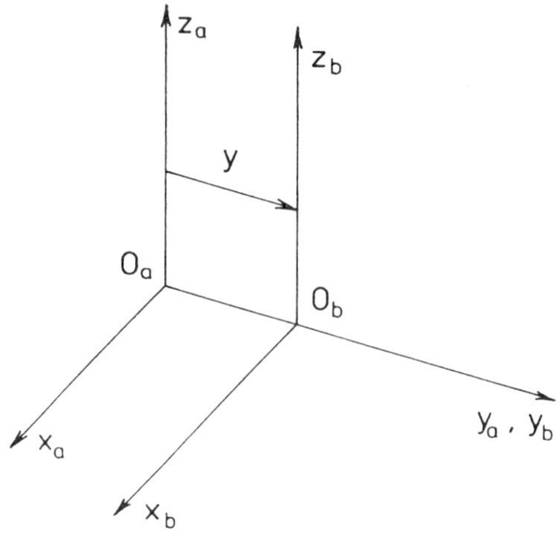

Figure 4.15. Basic motion Z2.

for the general parameter p

$$\mathbf{E}(p) = \begin{bmatrix} p & 0 & 0 & 0 \\ 0 & p & 0 & 0 \\ 0 & 0 & p & 0 \\ 0 & 0 & 0 & p \end{bmatrix} \tag{4.253}$$

2. Translational motion in the direction of the axis $o \equiv y_a \equiv y_b$ (Figure 4.15). Let y designate the displacement.

$$\mathbf{T}_{Z2}(y) = \begin{bmatrix} 1 & 0 & 0 & 0 \\ 0 & 1 & 0 & y \\ 0 & 0 & 1 & 0 \\ 0 & 0 & 0 & 1 \end{bmatrix} \tag{4.254a}$$

$$\mathbf{D}_{Z2}(\dot{y}) = \begin{bmatrix} 0 & 0 & 0 & 0 \\ 0 & 0 & 0 & \dot{y} \\ 0 & 0 & 0 & 0 \\ 0 & 0 & 0 & 0 \end{bmatrix} \tag{4.254b}$$

$$\mathbf{V}_{Z2} = \mathbf{D}_{Z2}(\dot{y}) = \mathbf{D}_{Z2}(1)\mathbf{E}(\dot{y}) \tag{4.254c}$$

$$\mathbf{A}_{Z2} = \mathbf{D}_{Z2}(\ddot{y}) = \mathbf{D}_{Z2}(1)\mathbf{E}(\ddot{y}) \tag{4.254d}$$

$$\mathbf{B}_{Z2} = \mathbf{A}_{Z2} \tag{4.254e}$$

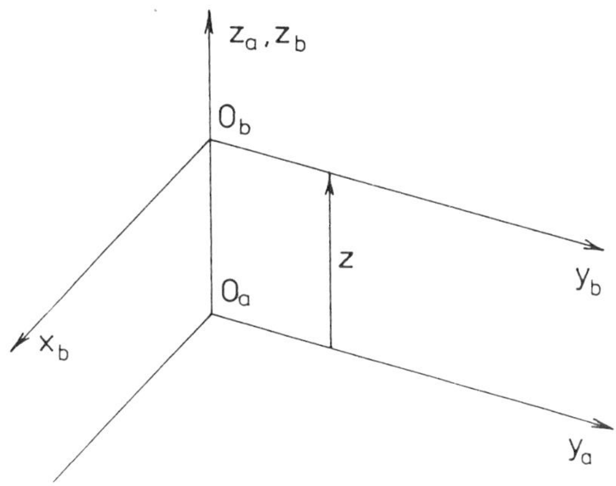

Figure 4.16. Basic motion Z3.

3. Translational motion in the direction of the axis $o = z_a = z_b$ (Figure 4.16). Let z designate the displacement.

$$\mathbf{T}_{Z3}(z) = \begin{bmatrix} 1 & 0 & 0 & 0 \\ 0 & 1 & 0 & 0 \\ 0 & 0 & 1 & z \\ 0 & 0 & 0 & 1 \end{bmatrix} \tag{4.255a}$$

$$\mathbf{D}_{Z3}(\dot{z}) = \begin{bmatrix} 0 & 0 & 0 & 0 \\ 0 & 0 & 0 & 0 \\ 0 & 0 & 0 & \dot{z} \\ 0 & 0 & 0 & 0 \end{bmatrix} \tag{4.255b}$$

$$\mathbf{V}_{Z3} = \mathbf{D}_{Z3}(\dot{z}) = \mathbf{D}_{Z3}(1)\mathbf{E}(\dot{z}) \tag{4.255c}$$

$$\mathbf{A}_{Z3} = \mathbf{D}_{Z3}(\ddot{z}) = \mathbf{D}_{Z3}(1)\mathbf{E}(\ddot{z}) \tag{4.255d}$$

$$\mathbf{B}_{Z3} = \mathbf{A}_{Z3} \tag{4.255e}$$

4. Revolute motion around the axis $o \equiv x_a \equiv x_b$ (Figure 4.17). Let φ_x be the angle of rotation.

$$\mathbf{T}_{Z4}(\varphi_x) = \begin{bmatrix} 1 & 0 & 0 & 0 \\ 0 & \mathrm{c}\varphi_x & -\mathrm{s}\varphi_x & 0 \\ 0 & \mathrm{s}\varphi_x & \mathrm{c}\varphi_x & 0 \\ 0 & 0 & 0 & 1 \end{bmatrix} \tag{4.256a}$$

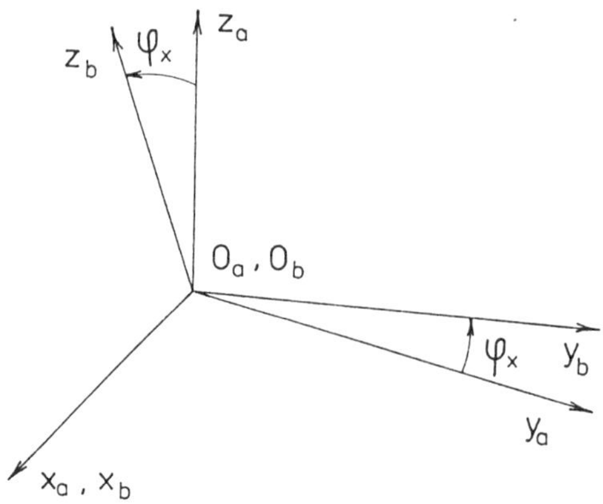

Figure 4.17. Basic motion Z4.

$$\mathbf{D}_{Z4}(\dot{\varphi}_x) = \begin{bmatrix} 0 & 0 & 0 & 0 \\ 0 & 0 & -\dot{\varphi}_x & 0 \\ 0 & \dot{\varphi}_x & 0 & 0 \\ 0 & 0 & 0 & 0 \end{bmatrix} \tag{4.256b}$$

$$\mathbf{V}_{Z4} = \mathbf{D}_{Z4}(\dot{\varphi}_x) = \mathbf{D}_{Z4}(1)\mathbf{E}(\dot{\varphi}_x) \tag{4.256c}$$

$$\mathbf{A}_{Z4} = \mathbf{D}_{Z4}(\ddot{\varphi}_x) = \mathbf{D}_{Z4}(1)\mathbf{E}(\ddot{\varphi}_x) \tag{4.256d}$$

$$\mathbf{B}_{Z4} = \mathbf{D}_{Z4}(\ddot{\varphi}_x) + \mathbf{D}^2_{Z4}(\dot{\varphi}_x) = \mathbf{D}_{Z4}(1)\mathbf{E}(\ddot{\varphi}_x) + \mathbf{D}^2_{Z4}(1)\mathbf{E}(\dot{\varphi}^2_x) \tag{4.256e}$$

5. Revolute motion around the axis $o \equiv y_a \equiv y_b$ (Figure 4.18). Let φ_y be the angle of rotation.

$$\mathbf{T}_{Z5}(\varphi_y) = \begin{bmatrix} \mathrm{c}\varphi_y & 0 & \mathrm{s}\varphi_y & 0 \\ 0 & 1 & 0 & 0 \\ -\mathrm{s}\varphi_y & 0 & \mathrm{c}\varphi_y & 0 \\ 0 & 0 & 0 & 1 \end{bmatrix} \tag{4.257a}$$

$$\mathbf{D}_{Z5}(\dot{\varphi}_y) = \begin{bmatrix} 0 & 0 & \dot{\varphi}_y & 0 \\ 0 & 0 & 0 & 0 \\ -\dot{\varphi}_y & 0 & 0 & 0 \\ 0 & 0 & 0 & 0 \end{bmatrix} \tag{4.257b}$$

$$\mathbf{V}_{Z5} = \mathbf{D}_{Z5}(\dot{\varphi}_y) = \mathbf{D}_{Z5}(1)\mathbf{E}(\dot{\varphi}_y) \tag{4.257c}$$

$$\mathbf{A}_{Z5} = \mathbf{D}_{Z5}(\ddot{\varphi}_y) = \mathbf{D}_{Z5}(1)\mathbf{E}(\ddot{\varphi}_y) \tag{4.257d}$$

$$\mathbf{B}_{Z5} = \mathbf{D}_{Z5}(\ddot{\varphi}_y) + \mathbf{D}^2_{Z5}(\dot{\varphi}_y) = \mathbf{D}_{Z5}(1)\mathbf{E}(\ddot{\varphi}_y) + \mathbf{D}^2_{Z5}(1)\mathbf{E}(\dot{\varphi}^2_y) \tag{4.257e}$$

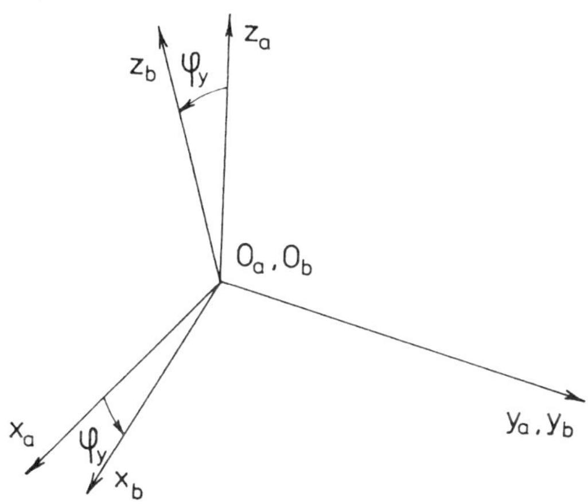

Figure 4.18 Basic motion Z5.

6. Revolute motion around the axis $o \equiv z_a \equiv z_b$ (Figure 4.19). Let φ_z be the angle of rotation.

$$\mathbf{T}_{Z6}(\varphi_z) = \begin{bmatrix} c\varphi_z & -s\varphi_z & 0 & 0 \\ s\varphi_z & c\varphi_z & 0 & 0 \\ 0 & 0 & 1 & 0 \\ 0 & 0 & 0 & 1 \end{bmatrix} \tag{4.258a}$$

$$\mathbf{D}_{Z6}(\dot{\varphi}_z) = \begin{bmatrix} 0 & -\dot{\varphi}_z & 0 & 0 \\ \dot{\varphi}_z & 0 & 0 & 0 \\ 0 & 0 & 0 & 0 \\ 0 & 0 & 0 & 0 \end{bmatrix} \tag{4.258b}$$

$$\mathbf{V}_{Z6} = \mathbf{D}_{Z6}(\dot{\varphi}_z) = \mathbf{D}_{Z6}(1)\mathbf{E}(\dot{\varphi}_z) \tag{4.258c}$$

$$\mathbf{A}_{Z6} = \mathbf{D}_{Z6}(\ddot{\varphi}_z) = \mathbf{D}_{Z6}(1)\mathbf{E}(\ddot{\varphi}_z) \tag{4.258d}$$

$$\mathbf{B}_{Z6} = \mathbf{D}_{Z6}(\ddot{\varphi}_z) + \mathbf{D}^2_{Z6}(\dot{\varphi}_z) = \mathbf{D}_{Z6}(1)\mathbf{E}(\ddot{\varphi}_z) + \mathbf{D}^2_{Z6}(1)\mathbf{E}(\dot{\varphi}^2_z) \tag{4.258e}$$

For composing basic motions, the transformation matrices of basic motions relative to the same axis, or transformation matrices of basic translations, are commutable. This can be demonstrated by a direct computation, e.g.

$$\mathbf{T}_{Z1}(x)\mathbf{T}_{Z4}(\varphi_x) = \mathbf{T}_{Z4}(\varphi_x)\mathbf{T}_{Z1}(x) \tag{4.259a}$$

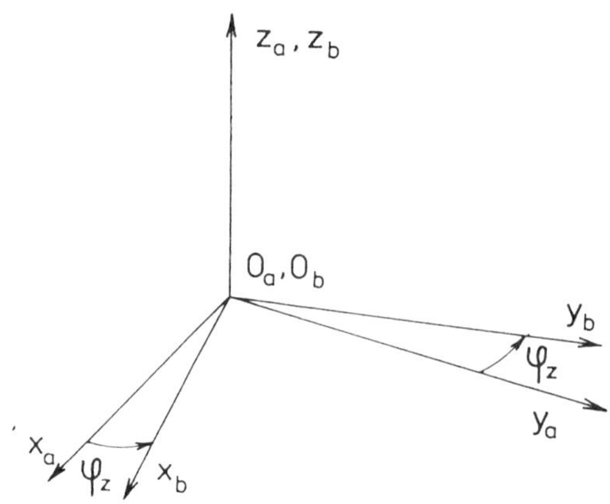

Figure 4.19. Basic motion Z6.

$$\mathbf{T}_{Z1}(x_1)\mathbf{T}_{Z1}(x_2) = \mathbf{T}_{Z1}(x_2)\mathbf{T}_{Z1}(x_1) \tag{4.259b}$$
$$\mathbf{T}_{Z4}(\varphi_{x1})\mathbf{T}_{Z4}(\varphi_{x2}) = \mathbf{T}_{Z4}(\varphi_{x2})\mathbf{T}_{Z4}(\varphi_{x1}) \tag{4.259c}$$
$$\mathbf{T}_{Z1}(x)\mathbf{T}_{Z2}(y) = \mathbf{T}_{Z2}(y)\mathbf{T}_{Z1}(x) \tag{4.259d}$$

In specific cases we shall use as parameters of basic motions and matrix differential operators various coordinates or constants. We shall often write for parameter p

$$\mathbf{T}_{Zj}(p) \quad \text{or} \quad \mathbf{T}_{Zk_j}(p)$$

where j in the former case, k_j in the latter case, designate a number from 1 to 6.

The above relations for basic matrices can be summarised as follows

$$\mathbf{V}_{Zj} = \mathbf{D}_{Zj}(\dot{q}) = \mathbf{D}_{Zj}(1)\mathbf{E}(\dot{q}) \tag{4.260a}$$
$$\mathbf{A}_{Zj} = \mathbf{D}_{Zj}(\ddot{q}) = \mathbf{D}_{Zj}(1)\mathbf{E}(\ddot{q}) \tag{4.260b}$$
$$\mathbf{B}_{Zj} = \mathbf{D}_{Zj}(\ddot{q}) + \mathbf{D}^2_{Zj}(\dot{q}) = \mathbf{D}_{Zj}(1)\mathbf{E}(\ddot{q}) + \mathbf{D}^2_{Zj}(1)\mathbf{E}(\dot{q}^2), \quad j = 1, \ldots, 6 \tag{4.260c}$$

Further, it is clear that

$$\dot{\mathbf{T}}_{Zj}(q) = \mathbf{T}_{Zj}(q)\mathbf{D}_{Zj}(\dot{q}) = \mathbf{T}_{Zj}(q)\mathbf{D}_{Zj}(1)\mathbf{E}(\dot{q}) \tag{4.261a}$$
$$\ddot{\mathbf{T}}_{Zj}(q) = \mathbf{T}_{Zj}(q)(\mathbf{D}_{Zj}(\ddot{q}) + \mathbf{D}^2_{Zj}(\dot{q})) = \mathbf{T}_{Zj}(q)(\mathbf{D}_{Zj}(1)\mathbf{E}(\ddot{q}) + \mathbf{D}^2_{Zj}(1)\mathbf{E}(\dot{q}^2)) \tag{4.261b}$$

For matrix differential operators

$$\mathbf{D}^2_{Z4}(\dot{q}) = \lceil 0, -\dot{q}^2, -\dot{q}^2, 0 \rfloor \tag{4.262a}$$

$$\begin{aligned}
\mathbf{D}^2_{Z4}(1) &= \lceil 0, -1, -1, 0 \rfloor & (4.262b)\\
\mathbf{D}^2_{Z5}(\dot{q}) &= \lceil -\dot{q}^2, 0, -\dot{q}^2, 0 \rfloor & (4.262c)\\
\mathbf{D}^2_{Z5}(1) &= \lceil -1, 0, -1, 0 \rfloor & (4.262d)\\
\mathbf{D}^2_{Z6}(\dot{q}) &= \lceil -\dot{q}^2, -\dot{q}^2, 0, 0 \rfloor & (4.262e)\\
\mathbf{D}^2_{Z6}(1) &= \lceil -1, -1, 0, 0 \rfloor & (4.262f)\\
\mathbf{D}^2_{Z1}(\dot{q}) &= \mathbf{D}^2_{Z2}(\dot{q}) = \mathbf{D}^2_{Z3}(\dot{q}) = \mathbf{0} & (4.262g)\\
\mathbf{T}_{Zj}(q)\mathbf{D}_{Zj}(\dot{q}) &= \mathbf{D}_{Zj}(\dot{q})\mathbf{T}_{Zj}(q) & (4.262h)
\end{aligned}$$

General chain rules for basic matrix derivatives are important. Let $g = g(p)$ be a function of parameter p; then

$$\frac{\partial}{\partial p}\mathbf{T}_{Zj}(g(p)) = \mathbf{T}_{Zj}(g)\mathbf{D}_{Zj}\left(\frac{\partial g}{\partial p}\right) = \mathbf{T}_{Zj}(g)\mathbf{D}_{Zj}(1)\mathbf{E}\left(\frac{\partial g}{\partial p}\right) \quad (4.263)$$

$$\frac{\partial}{\partial p}\mathbf{D}_{Zj}(g(p)) = \mathbf{D}_{Zj}\left(\frac{\partial g}{\partial p}\right) = \mathbf{D}_{Zj}(1)\mathbf{E}\left(\frac{\partial g}{\partial p}\right), \quad j = 1, \ldots, 6 \quad (4.264)$$

Other relations are easy to obtain from the above. The last two show the possibility of turning the computation of the transformation matrix derivative, expressed as a basic matrix product, into multiplication by certain special matrices – the matrix differential operators. We shall often use this in the following chapters. The last important relation describing the special properties of basic matrices is

$$\mathbf{T}^{-1}_{Zj}(p) = \mathbf{T}_{Zj}(-p), \qquad j = 1, \ldots, 6 \quad (4.265)$$

4.5.2 Permutation Matrices

We have assumed the free choice of coordinate systems for motion description. In some cases, their choice is given by the problem specification. The change in designation of axes in both systems is followed by a change of transformation matrix. To avoid having to exclude those cases or to give the form of transformation, velocity and acceleration matrices for all variants of coordinate systems, we introduce special cases of transformation matrices, the so-called permutation matrices. These enable an easy description to be made of the transformation matrix change after altering the coordinate systems a, b axis designation.

Let us redesignate coordinate system axes $O_a x_a y_a z_a$ and $O_b x_b y_b z_b$, respectively, designating the new systems $O_a x'_a y'_a z'_a$ and $O_b x'_b y'_b z'_b$, respectively (Figure 4.20). Let us determine the transformation matrix $\mathbf{T}_{a'b'}$, assuming that matrix $\mathbf{T}_{ab}$ is known. According to relation (4.115) we obtain

$$\mathbf{T}_{a'b'} = \mathbf{T}_{a'a}\mathbf{T}_{ab}\mathbf{T}_{bb'} \quad (4.266)$$

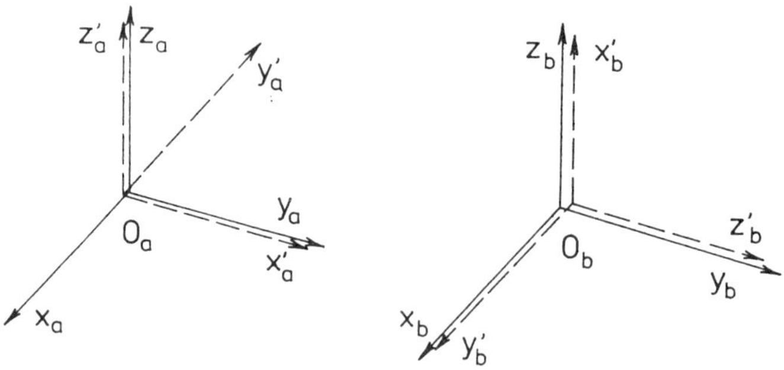

Figure 4.20. Redesignation of coordinate systems.

Systems a,a' and b,b' difer in general only in designations and perhaps orientations of their axes. This means that only direction cosinus matrix is influenced. Using (4.29) we can write by inspection of Figure 4.20 directly

$$\mathbf{S}_{a',a} = \begin{bmatrix} 0 & 1 & 0 \\ -1 & 0 & 0 \\ 0 & 0 & 1 \end{bmatrix}, \qquad \mathbf{S}_{bb'} = \begin{bmatrix} 0 & 1 & 0 \\ 0 & 0 & 1 \\ 1 & 0 & 0 \end{bmatrix}$$

Using (4.266) and (4.49) we finally obtain

$$\begin{aligned} \mathbf{T}_{a'b'} &= \left[\begin{array}{ccc|c} 0 & 1 & 0 & 0 \\ -1 & 0 & 0 & 0 \\ 0 & 0 & 1 & 0 \\ \hline 0 & 0 & 0 & 1 \end{array}\right] \left[\begin{array}{ccc|c} a_{11} & a_{12} & a_{13} & a_1 \\ a_{21} & a_{22} & a_{23} & a_2 \\ a_{31} & a_{32} & a_{33} & a_3 \\ \hline 0 & 0 & 0 & 1 \end{array}\right] \left[\begin{array}{ccc|c} 0 & 1 & 0 & 0 \\ 0 & 0 & 1 & 0 \\ 1 & 0 & 0 & 0 \\ \hline 0 & 0 & 0 & 1 \end{array}\right] = \\ &= \begin{bmatrix} a_{23} & a_{21} & a_{22} & a_2 \\ -a_{13} & -a_{11} & -a_{12} & -a_1 \\ a_{33} & a_{31} & a_{32} & a_3 \\ 0 & 0 & 0 & 1 \end{bmatrix} \end{aligned}$$

The above statements are particular cases of transformation matrices describing a displacement (basic rotation by $\pm\pi/2, \pm\pi, \pm 3\pi/2,$) in which the coordinate axes transform into others and/or change their orientation. This completely complies with the case of permutation matrices with det $= +1$. Permutation matrices with det $= -1$ describe reflection according to Section 4.2.4.

The above statements follow also from mathematical theorems on obtaining interchanged lines and or/columns of a matrix by multiplying it by another (permutation) matrix.

The relations for velocity and acceleration matrix transformation are easy to obtain using the definition and (4.266).

$$\begin{aligned}\mathbf{V}_{a'b'} = \mathbf{T}_{a'b'}^{-1}\dot{\mathbf{T}}_{a'b'} &= \mathbf{T}_{bb'}^{-1}\mathbf{T}_{ab}^{-1}\mathbf{T}_{a'a}^{-1}\mathbf{T}_{a'a}\dot{\mathbf{T}}_{ab}\mathbf{T}_{bb'} = \mathbf{T}_{bb'}^{-1}\mathbf{T}_{ab}^{-1}\dot{\mathbf{T}}_{ab}\mathbf{T}_{bb'} = \\ &= \mathbf{T}_{bb'}^{-1}\mathbf{V}_{ab}\mathbf{T}_{bb'}\end{aligned}$$

Using the orthogonality of the permutation matrix

$$\mathbf{T}_{bb'}^{-1} = \mathbf{T}_{bb'}^{T}$$

Hence

$$\mathbf{V}_{a'b'} = \mathbf{T}_{bb'}^{T}\mathbf{V}_{ab}\mathbf{T}_{bb'} \tag{4.267}$$

Analogously

$$\mathbf{A}_{a'b'} = \mathbf{T}_{bb'}^{T}\mathbf{A}_{ab}\mathbf{T}_{bb'} \tag{4.268}$$

$$\begin{aligned}\mathbf{B}_{a'b'} &= \mathbf{A}_{a'b'} + \mathbf{V}_{a'b'}^{2} = \mathbf{T}_{a'b'}^{T}\mathbf{A}_{ab}\mathbf{T}_{bb'} + \mathbf{T}_{bb'}^{T}\mathbf{V}_{ab}\mathbf{T}_{bb'}\mathbf{T}_{bb'}^{T}\mathbf{V}_{ab}\mathbf{T}_{bb'} = \\ &= \mathbf{T}_{bb'}^{T}(\mathbf{A}_{ab} + \mathbf{V}_{ab}^{2})\mathbf{T}_{bb'} = \mathbf{T}_{bb'}^{T}\mathbf{B}_{ab}\mathbf{T}_{bb'}\end{aligned} \tag{4.269}$$

A change in axis designation of system a causes no change in the velocity and acceleration matrices. This is in accordance with the fact that those matrices express vectors in space b.

4.5.3 Basic Quaternions

Expressing basic motion transformation matrices yielded basic matrices. We shall in a similar way express basic motion SDQ's and thus obtain basic quaternions. According to (4.195), (4.163) we obtain the basic quaternions $\hat{\mathbf{K}}_{Zj}(p)$, $j = 1, \ldots, 6$, as direct analogies to basic matrices $\mathbf{T}_{Zj}(p)$

$$\hat{\mathbf{K}}_{Z1}(x) = [1, 0, 0, 0; x, 0, 0] \tag{4.270a}$$

$$\hat{\mathbf{K}}_{Z2}(y) = [1, 0, 0, 0; 0, y, 0] \tag{4.270b}$$

$$\hat{\mathbf{K}}_{Z3}(z) = [1, 0, 0, 0; 0, 0, z] \tag{4.270c}$$

$$\hat{\mathbf{K}}_{Z4}(\varphi_x) = \left[\mathrm{c}\frac{\varphi_x}{2}, \mathrm{s}\frac{\varphi_x}{2}, 0, 0; 0, 0, 0\right] \tag{4.270d}$$

$$\hat{\mathbf{K}}_{Z5}(\varphi_y) = \left[\mathrm{c}\frac{\varphi_y}{2}, 0, \mathrm{s}\frac{\varphi_y}{2}, 0; 0, 0, 0\right] \tag{4.270e}$$

$$\hat{\mathbf{K}}_{Z6}(\varphi_z) = \left[\mathrm{c}\frac{\varphi_z}{2}, 0, 0, \mathrm{s}\frac{\varphi_z}{2}; 0, 0, 0\right] \tag{4.270f}$$

We can introduce differential quaternion operators $\hat{\mathbf{L}}_{Zj}^{R}(\dot{p})$ by expressions

$$\hat{\mathbf{L}}_{Z1}^{R}(\dot{x}) = [0, 0, 0, 0; \dot{x}, 0, 0] \tag{4.271a}$$

$$
\begin{aligned}
\hat{\mathbf{L}}^R_{Z2}(\dot{y}) &= [0,0,0,0;0,\dot{y},0] && (4.271\text{b})\\
\hat{\mathbf{L}}^R_{Z3}(\dot{z}) &= [0,0,0,0;0,0,\dot{z}] && (4.271\text{c})\\
\hat{\mathbf{L}}^R_{Z4}(\dot{\varphi}_x) &= \left[0,\frac{\dot{\varphi}_x}{2},0,0;0,0,0\right] && (4.271\text{d})\\
\hat{\mathbf{L}}^R_{Z5}(\dot{\varphi}_y) &= \left[0,0,\frac{\dot{\varphi}_y}{2},0;0,0,0\right] && (4.271\text{e})\\
\hat{\mathbf{L}}^R_{Z6}(\dot{\varphi}_z) &= \left[0,0,0,\frac{\dot{\varphi}_z}{2};0,0,0\right] && (4.271\text{f})
\end{aligned}
$$

Then

$$\frac{\mathrm{d}}{\mathrm{dt}}(\hat{\mathbf{K}}_{Zj}(p)) = \hat{\mathbf{K}}^L_{\varepsilon,Zj}(p)\hat{\mathbf{L}}^R_{Zj}(\dot{p}) \qquad (4.272)$$

and, for $j = 1,2,3$,

$$\frac{\mathrm{d}}{\mathrm{dt}}(\hat{\mathbf{K}}_{Zj}(p)) = \hat{\mathbf{L}}^R_{Zj}(\dot{p}) \qquad (4.273)$$

Further analogies with basic matrices hold

$$
\begin{aligned}
\hat{\mathbf{K}}^{-1}_{Zj}(p) &= \hat{\mathbf{K}}_{Zj}(-p) && (4.274\text{a})\\
\frac{\partial}{\partial p}\hat{\mathbf{K}}_{Zj}(g(p)) &= \hat{\mathbf{K}}^L_{\varepsilon,Zj}(g(p))\hat{\mathbf{L}}^R_{Zj}\left(\frac{\partial g}{\partial p}\right) && (4.274\text{b})\\
\frac{\partial}{\partial p}\hat{\mathbf{L}}^R_{Zj}(g(p)) &= \hat{\mathbf{L}}^R_{Zj}\left(\frac{\partial g}{\partial p}\right) && (4.274\text{c})\\
\hat{\mathbf{W}}^R_{Zj} &= \hat{\mathbf{L}}^R_{Zj}(\dot{p}) && (4.274\text{d})\\
\hat{\mathbf{Y}}^R_{Zj} &= \hat{\mathbf{L}}^R_{Zj}(\ddot{p}) && (4.274\text{e})\\
\hat{\mathbf{Z}}^R_{Zj} &= \hat{\mathbf{L}}^R_{Zj}(\ddot{p}) + \hat{\mathbf{L}}^R_{\varepsilon,Zj}(\dot{p})\hat{\mathbf{L}}^R_{\varepsilon,Zj}(\dot{p}) && (4.274\text{f})
\end{aligned}
$$

4.5.4 Basic Screws

Also using the screw description of general spatial motion, we can describe basic motions and obtain basic screw affinors

$$
\mathbf{A}_{Z1}(x) = \begin{vmatrix} 1 & 0 & 0\\ 0 & 1 & -\varepsilon x\\ 0 & \varepsilon x & 1 \end{vmatrix}, \quad \mathbf{A}_{Z4}(\varphi_x) = \begin{vmatrix} 1 & 0 & 0\\ 0 & \mathrm{c}\varphi_x & -\mathrm{s}\varphi_x\\ 0 & \mathrm{s}\varphi_x & \mathrm{c}\varphi_x \end{vmatrix}
$$

$$
\mathbf{A}_{Z2}(y) = \begin{vmatrix} 1 & 0 & \varepsilon y\\ 0 & 1 & 0\\ -\varepsilon y & 0 & 1 \end{vmatrix}, \quad \mathbf{A}_{Z5}(\varphi_y) = \begin{vmatrix} \mathrm{c}\varphi_y & 0 & \mathrm{s}\varphi_y\\ 0 & 1 & 0\\ -\mathrm{s}\varphi_y & 0 & \mathrm{c}\varphi_y \end{vmatrix}
$$

$$\mathbf{A}_{Z3}(z) = \begin{vmatrix} 1 & -\varepsilon z & 0 \\ \varepsilon z & 1 & 0 \\ 0 & 0 & 1 \end{vmatrix}, \quad \mathbf{A}_{Z6}(\varphi_z) = \begin{vmatrix} c\varphi_z & -s\varphi_z & 0 \\ s\varphi_z & c\varphi_z & 0 \\ 0 & 0 & 1 \end{vmatrix} \tag{4.275}$$

According to (4.249) we obtain basic dual vectors of motion

$$\begin{aligned} \mathbf{\Theta}_{Z1}(x) &= (0,0,0) + \varepsilon\left(\frac{x}{2},0,0\right) \\ \mathbf{\Theta}_{Z2}(y) &= (0,0,0) + \varepsilon\left(0,\frac{y}{2},0\right) \\ \mathbf{\Theta}_{Z3}(z) &= (0,0,0) + \varepsilon\left(0,0,\frac{z}{2}\right) \\ \mathbf{\Theta}_{Z4}(\varphi_x) &= \left(\tan\frac{\varphi_x}{2},0,0\right) + \varepsilon(0,0,0) \\ \mathbf{\Theta}_{Z5}(\varphi_y) &= \left(0,\tan\frac{\varphi_y}{2},0\right) + \varepsilon(0,0,0) \\ \mathbf{\Theta}_{Z6}(\varphi_z) &= \left(0,0,\tan\frac{\varphi_z}{2}\right) + \varepsilon(0,0,0) \end{aligned} \tag{4.276}$$

According to (4.251) we obtain basic dual quaternions

$$\begin{aligned} \mathbf{P}_{Z1}(x) &= \left[1;\varepsilon\frac{x}{2},0,0\right] = \left[1,0,0,0;0,\frac{x}{2},0,0\right] \\ \mathbf{P}_{Z2}(y) &= \left[1;0,\varepsilon\frac{y}{2},0\right] \\ \mathbf{P}_{Z3}(z) &= \left[1;0,0,\varepsilon\frac{z}{2}\right] \\ \mathbf{P}_{Z4}(\varphi_x) &= \left[c\frac{\varphi_x}{2},s\frac{\varphi_x}{2},0,0;0\right] = \left[c\frac{\varphi_x}{2},s\frac{\varphi_x}{2},0,0;0,0,0,0\right] \\ \mathbf{P}_{Z5}(\varphi_y) &= \left[c\frac{\varphi_y}{2},0,s\frac{\varphi_y}{2},0;0\right] \\ \mathbf{P}_{Z6}(\varphi_z) &= \left[c\frac{\varphi_z}{2},0,0,s\frac{\varphi_z}{2};0\right] \end{aligned} \tag{4.277}$$

The composition of screw operators is treated in Section 4.4.

4.6 Computational Complexity of Different Notations

In the above sections, the most important possible ways of describing point and body motion kinematics are dealt with. Notational compactness and low computational complexity are relevant requirements for our choice of formalism in the following chapters of this book.

For the first of the two requirements, notational compactness, vector notation is not convenient since it does not enable us to describe general spatial body motion using one symbol. Other formalisms comply with this requirement.

In order to examine the second of the two requirements, low computational complexity, we compare the number of operations necessary to compute the composition of two general spatial motions and the number of parameters required to describe the general position of the body. Since the matrix method is the compact notation of the vector method, the computational complexity of the vector and matrix formalisms is the same. The matrix formalism requires 36 multiplications and 27 additions to multiply two transformation matrices describing the composition of two motions. The general position of a body in matrix notation is determined by 12 parameters, of which only 6 are independent.

The quaternion formalism in the SDQ form requires 35 multiplications and 34 additions (or, using a different method (Dobrovodský, 1986), 31 multiplications and 38 additions) and 7 parameters to describe the position of the body. This follows from (4.203) as the SDQ multiplication requires 16 multiplications and 12 additions (or 12 multiplications and 16 additions); the computation of $\frac{1}{2}\mathbf{S}(\hat{\mathbf{u}})$ according to (4.171) requires 10 multiplications and 10 additions; the computation of $\frac{1}{2}\mathbf{S}(\hat{\mathbf{u}})\vec{v}$ requires 3 additions, and the addition of vector $\vec{u}$ requires 3 additions.

Computational complexity of screw description is enormous. Multiplication of screw affinors requires 81 multiplications and 63 additions, and a further 18 parameters to describe the position of the body using an affinor (4.247), or 6 parameters using a dual vector of motion according to (4.249), which is, however, singular for $\varphi = \pi$. The dual quaternion formalism is treated in the paragraph on quaternion notation.

From the above, it follows that the computational complexity of screw formalism is much higher than for other formalisms, and the description of position by 6 parameters contains a singular case which makes necessary the use of the seventh parameter and quaternion notation. Comparing the matrix and quaternion notations, the matrix formalism is computationally slightly easier, assuming that the computer time for multiplication is not longer than 2.2 times the time for addition. The description of body position by 7 parameters when using quaternions is convenient compared with the 12 parameters of matrix notation; we shall make use of this convenient property in the next chapter. We shall, however, use the matrix notation in subsequent chapters as its computational complexity is the lowest and it enables a simple and surveyable computation to be made of velocities and accelerations; also its interpretation is straightforward.

4.7 Special Cases

The general spatial motion of a body is dealt with in the above sections. However, in many partial cases we have to deal with special and, therefore, simpler cases. The most important cases are spherical and planar motions. The formalism of a simpler motion description is also simpler. From this point, we shall use only the matrix form.

4.7.1 Spherical Motion

Let us choose the origin O_a, O_b of coordinate systems a, b in the centre of spherical motion. In this case, vector $\mathbf{u}_{ab}$ in (4.51) vanishes; it is therefore possible to leave out the fourth lines and columns in all (4,4)-matrices in derived expressions and equations. We therefore deal only with (3,3)-matries and expressions derived from these when examining spherical motion. The basic equations of point and body motion are then of the form

$$\mathbf{u}_{aM} = \mathbf{S}_{ab}\mathbf{u}_{bM} \tag{4.278a}$$
$$\dot{\mathbf{u}}_{aM} = \mathbf{S}_{ab}\boldsymbol{\Omega}_{b,ab}\mathbf{u}_{bM} \tag{4.278b}$$
$$\boldsymbol{\Omega}_{b,ab} = \mathbf{S}_{ab}^{T}\dot{\mathbf{S}}_{ab} \tag{4.278c}$$
$$\ddot{\mathbf{u}}_{aM} = \mathbf{S}_{ab}(\mathcal{A}_{b,ab} + \boldsymbol{\Omega}_{b,ab}^{2}) \tag{4.278d}$$
$$\mathcal{A}_{b,ab} = \dot{\boldsymbol{\Omega}}_{b,ab} \tag{4.278e}$$

Accordingly, the equations of simultaneous spherical motion are then simpler.

4.7.2 Planar Motion

Planar motion is usually examined in the xy-plane. The z-coordinates vanish, and the transformation equation (4.49) takes the form

$$\begin{bmatrix} x_{aM} \\ y_{aM} \\ 0 \\ 1 \end{bmatrix} = \begin{bmatrix} a_{11} & a_{12} & 0 & a_1 \\ a_{21} & a_{22} & 0 & a_2 \\ 0 & 0 & 1 & 0 \\ 0 & 0 & 0 & 1 \end{bmatrix} \begin{bmatrix} x_{bM} \\ y_{bM} \\ 0 \\ 1 \end{bmatrix} \tag{4.279}$$

making it possible to leave out the third columns and lines in all matrices and the third elements in all extended vectors. Furthermore, one angle will suffice to describe the orientation of the body. Let us choose angle φ of the rotational displacement of axis x_b relative to x_a. Equation (4.279) then takes the form

$$\begin{bmatrix} x_{aM} \\ y_{aM} \\ 1 \end{bmatrix} = \begin{bmatrix} \mathrm{c}\varphi & -\mathrm{s}\varphi & a_1 \\ \mathrm{s}\varphi & \mathrm{c}\varphi & a_2 \\ 0 & 0 & 1 \end{bmatrix} \begin{bmatrix} x_{bM} \\ y_{bM} \\ 1 \end{bmatrix} \tag{4.280}$$

It is evident that the transformation matrix can be obtained as a product of "reduced" basic matrices:

$$\mathbf{T}^r_{ab} = \mathbf{T}^r_{Z1}(a_1)\mathbf{T}^r_{Z2}(a_2)\mathbf{T}^r_{Z6}(\varphi) \tag{4.281}$$

All derived relations also maintain their validity for reduced matrices.

Chapter 5

Kinematic Analysis of MBSs with Lower Kinematic Pairs

The determination of position, velocity and acceleration of all significant points and elements of an examined MBS based on given position, velocity and acceleration of driving elements is the task of complete kinematic analysis of MBS. The quantities sought are, in general, functions of time. In special cases, however, we may have to solve the problem of position regardless of time. As we are going to study multibody systems, the constraints of which are realized by kinematic pairs, we must first describe their properties. We shall treat MBSs with lower kinematic pairs in this chapter; higher kinematic pairs will be treated in Chapter 6.

5.1 Lower Kinematic Pairs

We shall always designate bodies constrained by the kinematic pair being examined as a and b. The coordinate systems chosen in the respective bodies and firmly connected to them shall be designated $O_a x_a y_a z_a$, $O_b x_b y_b z_b$. Their origins differ in general. The relative position of bodies a and b constrained by a lower kinematic pair is characterised by one to three variable parameters which we are going to call kinematic pair coordinates. Their number equals the degree of freedom allowed by the kinematic pair being examined.

We shall derive transformation, velocity and acceleration matrices using the results of the previous chapter for the chosen coordinate systems. These

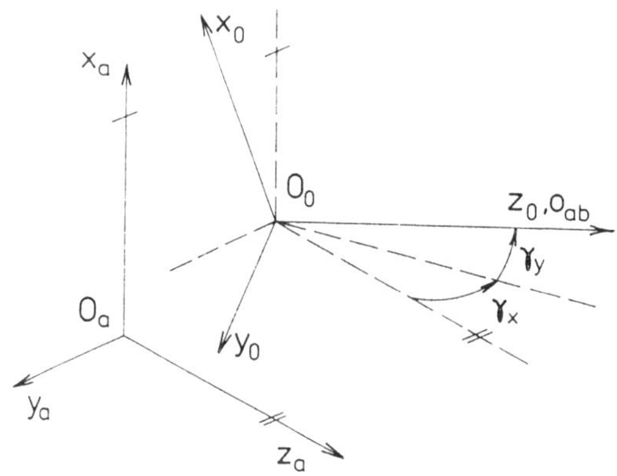

Figure 5.1. Subsidiary system $O_0x_0y_0z_0$.

matrices will be denoted by superscripts designating the kinematic pair being examined. For the sake of descriptional simplicity, it is convenient to choose the system $O_bx_by_bz_b$ so that, for example, axis $z_b \equiv o_{ab}$ is the axis of rotation or translation, etc., and we shall also assume the kinematic pair coordinates at the beginning of motion to be zero. This requires the introduction of a coordinate system $O_0x_0y_0z_0$, which is, at the beginning of the motion, identical with system $O_bx_by_bz_b$.

We shall describe the relative position of systems $O_ax_ay_az_a$ and $O_0x_0y_0z_0$ by a constant matrix $\mathbf{T}_0$ which we shall compute directly from definition (4.51)

$$\mathbf{T}_0 = \begin{bmatrix} a_{11} & a_{12} & a_{13} & x_0 \\ a_{21} & a_{22} & a_{23} & y_0 \\ a_{31} & a_{32} & a_{33} & z_0 \\ 0 & 0 & 0 & 1 \end{bmatrix} \tag{5.1}$$

where a_{ij} are the direction cosines of axes $x_0y_0z_0$ in system $O_ax_ay_az_a$ and x_0, y_0, z_0 are coordinates of origin O_0 in the same system. Using basic matrices,

$$\begin{aligned} \mathbf{T}_0 &= \mathbf{T}_{Z1}(x_0)\mathbf{T}_{Z2}(y_0)\mathbf{T}_{Z3}(z_0)\mathbf{T}_{Z4}(\gamma_x)\mathbf{T}_{Z5}(\gamma_y) = \\ &= \begin{bmatrix} \mathrm{c}\gamma_y & 0 & \mathrm{s}\gamma_y & x_0 \\ \mathrm{s}\gamma_x\mathrm{s}\gamma_y & \mathrm{c}\gamma_x & -\mathrm{s}\gamma_x\mathrm{c}\gamma_y & y_0 \\ -\mathrm{c}\gamma_x\mathrm{s}\gamma_y & \mathrm{s}\gamma_y & \mathrm{c}\gamma_x\mathrm{c}\gamma_y & z_0 \\ 0 & 0 & 0 & 1 \end{bmatrix} \end{aligned} \tag{5.2}$$

By forming the last equation, we assumed system $O_0x_0y_0z_0$ to be initially

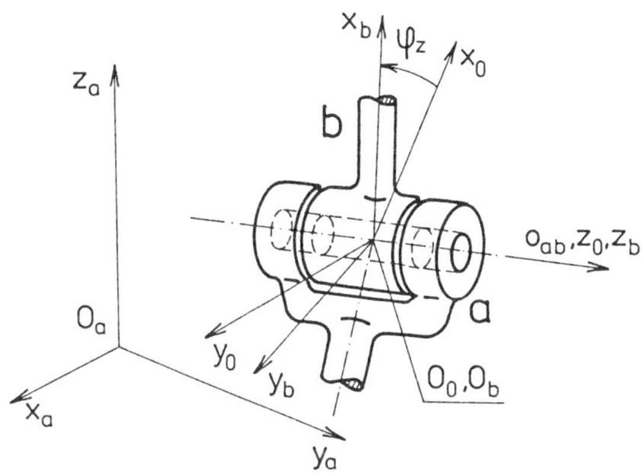

Figure 5.2. Revolute kinematic pair.

identical with system $O_a x_a y_a z_a$, and that it reached its position by (Figure 5.1

– translational displacement by x_0, y_0, z_0 in the direction of axes x_a, y_a, z_a,

– rotational displacement around the translated axis x_a by an angle γ_x such that the actual position of axis $y_a \equiv y_1 \perp o_{ab}$,

– rotational displacement around axis y_1 by angle γ_y.

In specific cases, the position of both systems is almost always a special one, so that the form of matrix $\mathbf{T}_0$ is much simpler.

5.1.1 Revolute Pair

Let us choose a coordinate system in body b so that $z_b \equiv o_{ab}$, Figure 5.2. The pair has one degree of freedom; we shall choose the angle φ_z of rotation around axis z_b for its coordinate. To obtain the transformation matrix we shall assume that system b is initially identical with a. We shall further assume that the system has reached its position by

– general displacement into position $O_0 x_0 y_0 z_0$,

– rotational displacement around axis o_{ab} by angle φ_z. Hence

$$\mathbf{T}^R = \mathbf{T}_0 \mathbf{T}_{Z6}(\varphi_z) \tag{5.3}$$

The velocity and acceleration matrices are

$$\mathbf{V}^R = (\mathbf{T}^R)^{-1}\dot{\mathbf{T}}^R = \mathbf{T}_{Z6}^{-1}(\varphi_z)\mathbf{T}_0^{-1}\mathbf{T}_0\mathbf{T}_{Z6}(\varphi_z)\mathbf{D}_{Z6}(\dot{\varphi}_1) = \mathbf{D}_{Z6}(\dot{\varphi}_z) \tag{5.4}$$

$$\mathbf{A}^R = \mathbf{D}_{Z6}(\ddot{\varphi}_z) \tag{5.5}$$

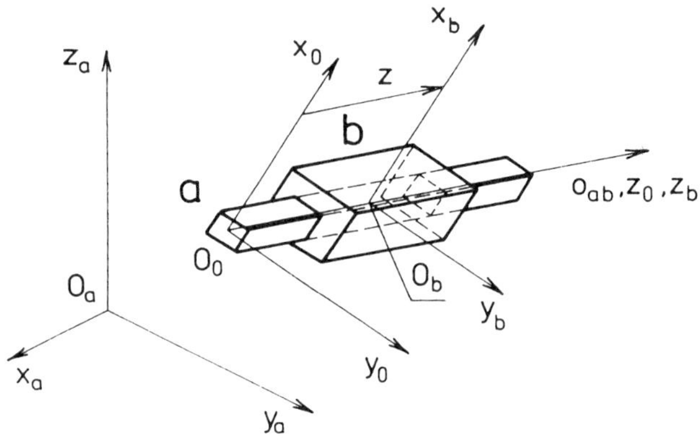

Figure 5.3. Prismatic kinematic pair.

and, using the results of Chapter 4 further, e.g.

$$\boldsymbol{\omega}_{b,ab}^{R} = [0, 0, \dot{\varphi}_z] \tag{5.6a}$$

$$\boldsymbol{\omega}_{a,ab}^{R} = \mathbf{S}^{R}\boldsymbol{\omega}_{b,ab}^{R} \tag{5.6b}$$

$$\boldsymbol{\alpha}_{b,ab}^{R} = [0, 0, \ddot{\varphi}_z] \tag{5.7a}$$

$$\boldsymbol{\alpha}_{a,ab}^{R} = \mathbf{S}^{R}\boldsymbol{\alpha}_{b,ab}^{R} \tag{5.7b}$$

5.1.2 Prismatic Pair

Let us choose the coordinate system in body b in such a way that $z_b \equiv o_{ab}$, Figure 5.3. The pair has one degree of freedom. Let us choose the displacement z in the direction of axis z_b for the coordinate. Let system b initially be identical with a. Its actual position has been reached by

– general displacement into position $O_0x_0y_0z_0$,

– rectilinear translation by z in direction z_b.

Hence

$$\mathbf{T}^{P} = \mathbf{T}_0\mathbf{T}_{Z3}(z) \tag{5.8}$$

The velocity and acceleration matrices are

$$\mathbf{V}^{P} = \mathbf{D}_{Z3}(\dot{z}) \tag{5.9}$$

$$\mathbf{A}^{P} = \mathbf{D}_{Z3}(\ddot{z}) \tag{5.10}$$

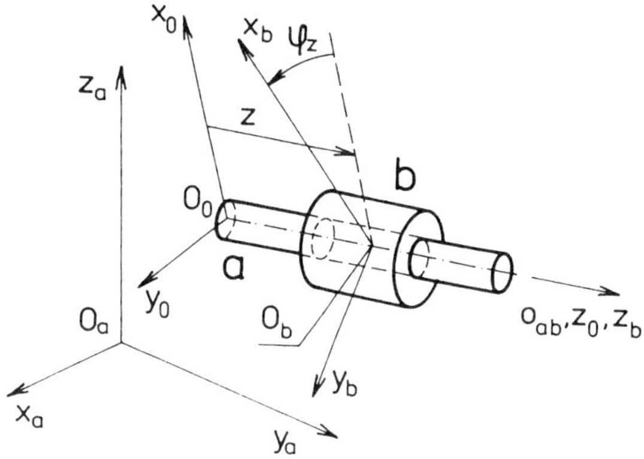

Figure 5.4. Cylindrical kinematic pair.

5.1.3 Cylindrical Pair

Let us choose a coordinate system in body b such that $z_b \equiv o_{ab}$, Figure 5.4. The pair has two degrees of freedom. We shall choose angular displacement φ_z and rectilinear translational displacement z as coordinates. To obtain the transformation matrix we assume that system b is initially identical with a and it has reached its actual position by

- general displacement into $O_0x_0y_0z_0$,
- displacement by z in direction z_b,
- angular displacement by φ_z around z_b.

Then

$$\mathbf{T}^C = \mathbf{T}_0\mathbf{T}_{Z3}(z)\mathbf{T}_{Z6}(\varphi_z) = \mathbf{T}_0\mathbf{T}_{Z6}(\varphi_z)\mathbf{T}_{Z3}(z) \tag{5.11}$$

as both basic matrices are commutable. For velocity and acceleration matrices we obtain:

$$\mathbf{V}^C = \mathbf{D}_{Z6}(\dot{\varphi}_z) + \mathbf{D}_{Z3}(\dot{z}) \tag{5.12}$$

$$\mathbf{A}^C = \mathbf{D}_{Z6}(\ddot{\varphi}_z) + \mathbf{D}_{Z3}(\ddot{z}) \tag{5.13}$$

This is the composition of independent rotation and rectilinear translation around, and along the same axis, respectively.

5.1.4 Helical Pair

A helical pair may be understood as a special case of a cylindrical pair, where the coordinates of translation and rotation are linked by the relation

$$z = k\varphi_z \tag{5.14}$$

where $k = \text{const}$ is the helix parameter. Then

$$\mathbf{T}^H = \mathbf{T}_0\mathbf{T}_{Z3}(k\varphi_z)\mathbf{T}_{Z6}(\varphi_z) = \mathbf{T}_0\mathbf{T}_{Z3}(z)\mathbf{T}_{Z6}(z/k) \tag{5.15}$$

and the pair has only one degree of freedom.

5.1.5 Spherical Pair

We shall pay more attention to the spherical kinematic pair than to other ones as it is the realization of relative spherical motion of a body. For the same reason the results of kinematic description of a spherical pair are directly applicable for the description of general spatial motion of a body, assuming we use the basic decomposition, where relative spherical motion determines the angular position (orientation) of a body in space, its angular velocity and angular acceleration.

A body carrying out spherical motion has three degrees of freedom. Its position may therefore be determined by three independent parameters. From the Euler Theorem (Section 4.1.1), we know that spherical motion may be obtained as rotation through a certain angle around an axis passing through the centre of spherical motion (a fixed point). The position of the axis of rotation may be described in a fixed coordinate system by two angles; the third parameter is the angular displacement. It is tempting to try to characterize spherical motion by a vector whose three components would be represented by these three angles. Unfortunately this is not possible, as vector components must comply with the parallelogram rule; finite rotations, however, cannot be summed in this way – they are not even commutative. It is therefore necessary to choose more complex descriptions. We shall seek a unique and intuitive description with the smallest possible number of parameters, with transformations which are as simple as possible. The requirements of simplicity (judged by the number of parameters) and of uniqueness are contradictory. As may be shown (Stuelpnagel, 1964), it is not possible to find a three-dimensional parameterization without singular points; four-dimensional parameterization is unique, and five- and more-dimensional parameterization may be one-to-one unique.

Obviously all the above requirements cannot be complied with completely and simultaneously. Therefore, in practice, a series of different descriptions is used (Rooney, 1977, 1978; Spring, 1986), e.g. Euler angles, Cardan (Bryant) angles, normed quaternions, real (3,3)-matrices, special unitary (2,2)-matrices, special unitary (3,3)-matrices and Pauli spin matrices. Each of the above descriptions has its advantages and disadvantages. We shall describe in detail only those which we shall use in subsequent chapters.

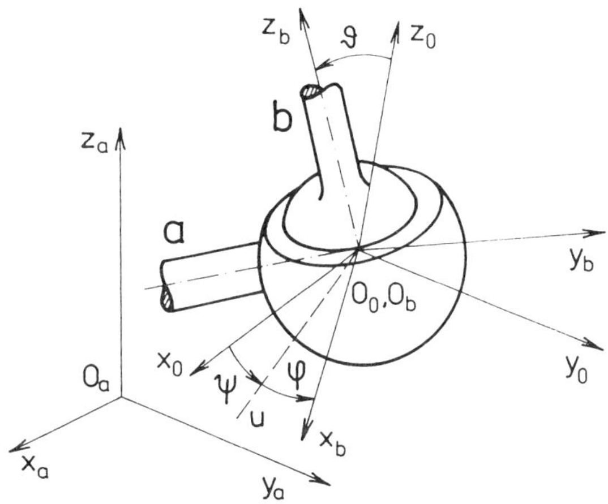

Figure 5.5. Spherical kinematic pair.

5.1.5.1 Euler Angles

There are many ways of defining Euler angles because there is a number of rotation sequences using a set of three angles for the description of the position of a body carrying out a spherical motion (Rongwed and Fletcher, 1964). Let us have two coordinate systems: $Ox_ay_az_a$ fixed in space and $Ox_by_bz_b$ attached to the body carrying out a spherical motion with a fixed point O. For the description of the body position we can use a sequence of three rotations around any axes of the system a, system b or both. In fact, there exist 24 different sequences (Kane et al., 1983). All of those sequences can be named Euler angles. However, only several sequences are in common use. They are Euler, Cardan angles and roll-pitch-yaw system.

We shall use the so-called 3–1–3 Euler angles. The designation 3–1–3 means that the first rotation is around z_b axis, the second rotation is around the instantaneous position of x_b axis and the third rotation is around z_b axis.

These Euler angles are defined as follows (Figure 5.5). Let us assume again that system a and b are initially identical and that system b has reached its position by

– general displacement to position $O_0x_0y_0z_0$,

– angular displacement by the angle ψ around instantaneous axis $z \equiv z_0$ (precession),

– angular displacement by the angle ϑ around instantaneous axis $x \equiv u$ (nutation),

– angular displacement by the angle φ around instantaneous axis $z \equiv z_b$ (rotation).
Then

$$\mathbf{T}^S = \mathbf{T}_0 \mathbf{T}_{Z6}(\psi)\mathbf{T}_{Z4}(\vartheta)\mathbf{T}_{Z6}(\varphi) = \tag{5.16}$$

$$= \mathbf{T}_0 \begin{bmatrix} c\psi c\varphi - s\psi c\vartheta s\varphi, & -c\psi s\varphi - s\psi c\vartheta c\varphi, & s\psi s\vartheta, & 0 \\ s\psi c\varphi + c\psi c\vartheta s\varphi, & -s\psi s\varphi + c\psi c\vartheta c\varphi, & -c\psi s\vartheta, & 0 \\ s\vartheta s\varphi, & s\vartheta c\varphi, & c\vartheta, & 0 \\ 0, & 0, & 0, & 1 \end{bmatrix} \tag{5.17}$$

The velocity matrix

$$\mathbf{V}^S = (\mathbf{T}^S)^{-1}\dot{\mathbf{T}}^S = \begin{bmatrix} \boldsymbol{\Omega}^S & \dot{\mathbf{u}} \\ \mathbf{0} & 0 \end{bmatrix} =$$

$$= \begin{bmatrix} 0, & -\dot{\psi} c\vartheta - \dot{\varphi}, & \dot{\psi} s\vartheta c\varphi - \dot{\vartheta} s\varphi, & 0 \\ \dot{\psi} c\vartheta + \dot{\varphi}, & 0, & -\dot{\psi} s\vartheta s\varphi - \dot{\vartheta} c\varphi, & 0 \\ -\dot{\psi} s\vartheta c\varphi + \dot{\vartheta} s\varphi, & \dot{\psi} s\vartheta s\varphi + \dot{\vartheta} c\varphi, & 0, & 0 \\ 0, & 0, & 0, & 0 \end{bmatrix} \tag{5.18}$$

Hence $\dot{\mathbf{u}} = \mathbf{0}$ and $\boldsymbol{\Omega}^S = (\mathbf{S}^S)^T \dot{\mathbf{S}}^S$ may be computed using (4.125) from

$$\boldsymbol{\Omega}^S = \mathbf{S}_\varphi^T \mathbf{S}_\vartheta^T \boldsymbol{\Omega}_\varphi \mathbf{S}_\vartheta \mathbf{S}_\varphi + \mathbf{S}_\varphi^T \boldsymbol{\Omega}_\vartheta \mathbf{S}_\varphi + \boldsymbol{\Omega}_\varphi \tag{5.19}$$

where $\mathbf{S}^S, \mathbf{S}_\psi, \mathbf{S}_\vartheta, \mathbf{S}_\varphi$ are submatrices corresponding to matrices $\mathbf{T}^S$, $\mathbf{T}_{Z6}(\psi)$, $\mathbf{T}_{Z4}(\vartheta)$, $\mathbf{T}_{Z6}(\varphi)$ from equation (5.16), and $\boldsymbol{\Omega}_\psi, \boldsymbol{\Omega}_\vartheta, \boldsymbol{\Omega}_\varphi$ are angular velocity matrices of partial motions.

In matrix $\boldsymbol{\Omega}^S$ we recognize the right hand side elements of the Euler kinematic equations.

$$\boldsymbol{\omega}_{b,ab}^S = \begin{bmatrix} \omega_x \\ \omega_y \\ \omega_z \end{bmatrix}_b = \begin{bmatrix} \dot{\psi} s\vartheta s\varphi + \dot{\vartheta} c\varphi \\ \dot{\psi} s\vartheta c\varphi - \dot{\vartheta} s\varphi \\ \dot{\psi} c\vartheta + \dot{\varphi} \end{bmatrix} = \begin{bmatrix} s\vartheta s\varphi, & c\varphi, & 0, \\ s\vartheta c\varphi, & -s\varphi, & 0, \\ c\vartheta, & 0, & 1 \end{bmatrix} \begin{bmatrix} \dot{\psi} \\ \dot{\vartheta} \\ \dot{\varphi} \end{bmatrix} \tag{5.20}$$

From this viewpoint, relations (4.64) and (4.125) may be used for generalized Euler kinematic equations. For the recomputation into space a we have

$$\boldsymbol{\omega}_{a,ab}^S = \mathbf{S}^S \boldsymbol{\omega}_{b,ab}^S \tag{5.21}$$

where $\mathbf{S}^S$ is the corresponding sub-matrix of the matrix defined by equation (5.16).

The angular acceleration matrix is

$$\mathcal{A}^S = \dot{\boldsymbol{\Omega}}^S = \begin{bmatrix} 0 & a_{12} & a_{13} \\ a_{21} & 0 & a_{23} \\ a_{31} & a_{32} & 0 \end{bmatrix} \tag{5.22}$$

where

$$\begin{aligned}
a_{12} &= -\ddot{\psi}\mathrm{c}\vartheta + \dot{\psi}\dot{\vartheta}\mathrm{s}\vartheta - \ddot{\varphi} \\
a_{13} &= \ddot{\psi}\mathrm{s}\vartheta\mathrm{c}\varphi + \dot{\psi}\dot{\vartheta}\mathrm{c}\vartheta\mathrm{c}\varphi - \dot{\psi}\dot{\varphi}\mathrm{s}\vartheta\mathrm{s}\varphi - \ddot{\vartheta}\mathrm{s}\varphi - \dot{\vartheta}\dot{\varphi}\mathrm{c}\varphi \\
a_{21} &= \ddot{\psi}\mathrm{c}\vartheta - \dot{\psi}\dot{\vartheta}\mathrm{s}\vartheta + \ddot{\varphi} \\
a_{23} &= -\ddot{\psi}\mathrm{s}\vartheta\mathrm{s}\varphi - \dot{\psi}\dot{\vartheta}\mathrm{c}\vartheta\mathrm{s}\varphi - \dot{\psi}\dot{\varphi}\mathrm{s}\vartheta\mathrm{c}\varphi - \ddot{\vartheta}\mathrm{c}\varphi + \dot{\vartheta}\dot{\varphi}\mathrm{s}\varphi \\
a_{31} &= -\ddot{\psi}\mathrm{s}\vartheta\mathrm{c}\varphi - \dot{\psi}\dot{\vartheta}\mathrm{c}\vartheta\mathrm{c}\varphi + \dot{\psi}\dot{\varphi}\mathrm{s}\vartheta\mathrm{s}\varphi + \ddot{\vartheta}\mathrm{s}\varphi + \dot{\vartheta}\dot{\varphi}\mathrm{c}\varphi \\
a_{32} &= \ddot{\psi}\mathrm{s}\vartheta\mathrm{s}\varphi + \dot{\psi}\dot{\vartheta}\mathrm{c}\vartheta\mathrm{s}\varphi + \dot{\psi}\dot{\varphi}\mathrm{s}\vartheta\mathrm{c}\varphi + \ddot{\vartheta}\mathrm{c}\varphi - \dot{\vartheta}\dot{\varphi}\mathrm{s}\varphi
\end{aligned}$$

so that

$$\boldsymbol{\alpha}^S_{b,ab} = \begin{bmatrix} \alpha_x \\ \alpha_y \\ \alpha_z \end{bmatrix}_b = \begin{bmatrix} \ddot{\psi}\mathrm{s}\vartheta\mathrm{s}\varphi + \dot{\psi}\dot{\vartheta}\mathrm{c}\vartheta\mathrm{s}\varphi + \dot{\psi}\dot{\varphi}\mathrm{s}\vartheta\mathrm{c}\varphi + \ddot{\vartheta}\mathrm{c}\varphi - \dot{\vartheta}\dot{\varphi}\mathrm{s}\varphi \\ \ddot{\psi}\mathrm{s}\vartheta\mathrm{c}\varphi + \dot{\psi}\dot{\vartheta}\mathrm{c}\vartheta\mathrm{c}\varphi - \dot{\psi}\dot{\varphi}\mathrm{s}\vartheta\mathrm{s}\varphi - \ddot{\vartheta}\mathrm{s}\varphi - \dot{\vartheta}\dot{\varphi}\mathrm{c}\varphi \\ \ddot{\psi}\mathrm{c}\vartheta - \dot{\psi}\dot{\vartheta}\mathrm{s}\vartheta + \ddot{\varphi} \end{bmatrix} \tag{5.23}$$

and again

$$\boldsymbol{\alpha}^S_{a,ab} = \mathbf{S}^S \boldsymbol{\alpha}^S_{b,ab} \tag{5.24}$$

The above relations enable the computation of matrices $\mathbf{S}^S, \mathbf{V}^S, \mathbf{A}^S$ to be made from known angles ψ, ϑ, φ and their derivatives.

The solution of the inverse problem is also important, i.e. from known matrices $\mathbf{S}^S, \mathbf{V}^S, \mathbf{A}^S$ to determine Euler angles and their derivatives. We shall, however, instead of matrix $\mathbf{T}^S$, use the matrix $\mathbf{T}_0^{-1}\mathbf{T}^S$. Using the designation $\mathbf{T}_0^{-1}\mathbf{T}^S = [s_{ij}], \mathbf{V}^S = [v_{ij}]$ and $\mathbf{A}^S = [a_{ij}]$, according to (5.16) we can write

$$\begin{aligned}
\mathrm{c}\vartheta &= s_{33} \\
\mathrm{s}\vartheta\mathrm{s}\varphi &= s_{31} \\
\mathrm{s}\vartheta\mathrm{c}\varphi &= s_{32} \\
\mathrm{s}\psi\mathrm{s}\vartheta &= s_{13} \\
-\mathrm{c}\psi\mathrm{s}\vartheta &= s_{23}
\end{aligned} \tag{5.25}$$

We could try to find the solution as follows

$$\begin{aligned}
\vartheta &= \cos^{-1}(s_{33}) \\
\varphi &= \cos^{-1}(s_{32}/\mathrm{s}\vartheta) \\
\psi &= \cos^{-1}(-s_{23}/\mathrm{s}\vartheta)
\end{aligned} \tag{5.26}$$

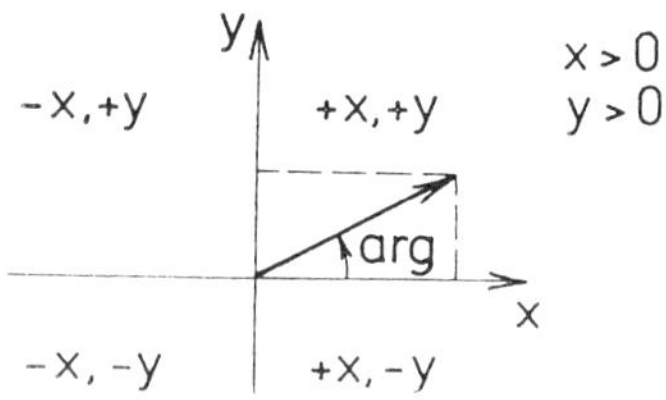

Figure 5.6. Determination of arg using function atan2.

However, this relation has inconvenient numerical properties (Paul, 1981) given by the uncertainty of the sign of the inverse cosine and its numerical inaccuracy at 0° and 180°. We therefore use as a matter of principle the function atan which is implemented on computers using the function atan2. Its advantage, apart from numerical accuracy, is in determining the angle inclusive quadrant from a value proportional to its sine and its cosine (Figure 5.6)

$$\arg = \operatorname{atan2}(\varkappa \sin(\arg), \varkappa \cos(\arg)) = \operatorname{atan2}(y, x) \tag{5.27}$$

where $\varkappa$ is an arbitrary positive constant. We shall seek the solution from equation (5.16) modified as follows

$$\mathbf{T}_0^{-1}\mathbf{T}^S = \mathbf{T}_{Z6}(\psi)\mathbf{T}_{Z4}(\vartheta)\mathbf{T}_{Z6}(\varphi) \tag{5.28}$$

$$\mathbf{T}_{Z6}^{-1}(\psi)\mathbf{T}_0^{-1}\mathbf{T}^S = \mathbf{T}_{Z4}(\vartheta)\mathbf{T}_{Z6}(\varphi) \tag{5.28a}$$

$$\begin{bmatrix} c\psi & s\psi & 0 & 0 \\ -s\psi & c\psi & 0 & 0 \\ 0 & 0 & 1 & 0 \\ 0 & 0 & 0 & 1 \end{bmatrix} \begin{bmatrix} s_{11} & s_{12} & s_{13} & s_{14} \\ s_{21} & s_{22} & s_{23} & s_{24} \\ s_{31} & s_{32} & s_{33} & s_{34} \\ 0 & 0 & 0 & 1 \end{bmatrix} = \begin{bmatrix} c\varphi & -s\varphi & 0 & 0 \\ c\vartheta s\varphi & c\vartheta c\varphi & -s\vartheta & 0 \\ s\vartheta s\varphi & s\vartheta c\varphi & c\vartheta & 0 \\ 0 & 0 & 0 & 1 \end{bmatrix} \tag{5.28b}$$

Hence, on the one hand, $s_{14} = s_{24} = s_{34} = 0$ must hold for spherical motion, and, on the other hand, for the (1,3)-element

$$c\psi s_{13} + s\psi s_{23} = 0 \tag{5.30}$$

This has two possible solutions

$$\psi = \text{atan2}(\pm s_{13}, \mp s_{23}) \tag{5.30a}$$

Hence from (2,3), (3,3) and (1,1), (1,2) elements

$$\begin{aligned} \vartheta &= \text{atan2}(s_{13}\text{s}\psi - s_{23}\text{c}\psi, s_{33}) && (5.30\text{b}) \\ \varphi &= \text{atan2}(-s_{12}\text{c}\psi - s_{22}\text{s}\psi, s_{11}\text{c}\psi + s_{21}\text{s}\psi) && (5.30\text{c}) \end{aligned}$$

Another solution may be found directly from equation (5.24):

$$\begin{aligned} \vartheta &= \text{atan2}(k\sqrt{1 - s_{33}^2}, s_{33}) && (5.31\text{a}) \\ \psi &= \text{atan2}(s_{13}k, -s_{23}k) && (5.31\text{b}) \\ \varphi &= \text{atan2}(s_{31}k, s_{32}k) && (5.31\text{c}) \end{aligned}$$

where $k = \pm 1$.
Similarly, according to (5.19)

$$\begin{bmatrix} \dot{\psi} \\ \dot{\vartheta} \\ \dot{\varphi} \end{bmatrix} = \begin{bmatrix} \text{s}\varphi/\text{s}\vartheta & \text{c}\varphi/\text{s}\vartheta & 0 \\ \text{c}\varphi & -\text{s}\varphi & 0 \\ -\text{s}\varphi/\text{t}\vartheta & -\text{c}\varphi/\text{t}\vartheta & 1 \end{bmatrix} \begin{bmatrix} v_{32} \\ v_{13} \\ v_{21} \end{bmatrix} \tag{5.32}$$

when $\text{t}\vartheta = \tan\vartheta$. Finally, due to (5.22)

$$\begin{aligned} \begin{bmatrix} \ddot{\psi} \\ \ddot{\vartheta} \\ \ddot{\varphi} \end{bmatrix} &= \begin{bmatrix} \text{s}\varphi/\text{s}\vartheta & \text{c}\varphi/\text{s}\vartheta & 0 \\ \text{c}\varphi & -\text{s}\varphi & 0 \\ -\text{s}\varphi/\text{t}\vartheta & -\text{c}\varphi/\text{t}\vartheta & 1 \end{bmatrix} \left(\begin{bmatrix} a_{32} \\ a_{13} \\ a_{21} \end{bmatrix} + \right. \\ &+ \left. \begin{bmatrix} \text{c}\vartheta\text{s}\varphi & \text{s}\vartheta\text{c}\varphi & -\text{s}\varphi \\ \text{c}\vartheta\text{c}\varphi & -\text{s}\vartheta\text{s}\varphi & -\text{c}\varphi \\ -\text{s}\vartheta & 0 & 0 \end{bmatrix} \begin{bmatrix} \dot{\psi}\dot{\vartheta} \\ \dot{\psi}\dot{\varphi} \\ \dot{\vartheta}\dot{\varphi} \end{bmatrix} \right) \end{aligned} \tag{5.33}$$

Euler angles are often used as parameters of a spherical pair although they are actually not natural for the given case, as they do not offer a simple way of finding either the position of the axis of rotation or the angle of angular displacement. Their traditional usage is apparently a consequence of the widespread knowledge of them from the theory of gyroscopes. They are, however, not very convenient for the description of spherical pair motion or of relative spherical motion by decomposition of general spatial motion of a body. They lead to non-linear transformation relations containing ambiguous trigonometric functions, and they contain singularities. From (5.24) it is evident that for $\vartheta = 0, \pm\pi$, etc., angles ψ and φ are not defined. This singularity is taken over into the calculation of Euler angle derivatives, as

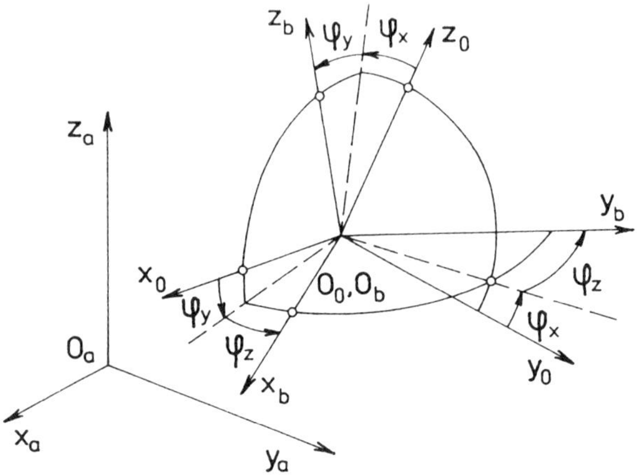

Figure 5.7. Cardan angles.

seen in (5.32), (5.33). The simplest way out of this difficulty is to select for $\vartheta = 0, \pm\pi, \ldots$ always for example $\psi = 0$; angle φ is then already determined uniquely. This modification, however, leads to discontinuity in the time behaviour of the angles. There are also other possibilities (Rongwed and Fletcher, 1964).

5.1.5.2 Cardan Angles

Cardan angles may be taken for 1–2–3 Euler angles. We obtain them as follows (Figure 5.7).

System b, initially identical with system a, has reached its position by

– general displacement into position $O_0 x_0 y_0 z_0$,

– angular displacement by angle φ_x around the instantaneous axis $x \equiv x_0$,

– angular displacement by angle φ_y around the instantaneous axis y,

– angular displacement by angle φ_z around the instantaneous axis z.

Then

$$\mathbf{T}^S = \mathbf{T}_0 \mathbf{T}_{Z4}(\varphi_x)\mathbf{T}_{Z5}(\varphi_y)\mathbf{T}_{Z6}(\varphi_z) =$$

$$= \mathbf{T}_0 \begin{bmatrix} \mathrm{c}\varphi_y \mathrm{c}\varphi_z, & -\mathrm{c}\varphi_y \mathrm{s}\varphi_z, & \mathrm{s}\varphi_y, & 0 \\ \mathrm{c}\varphi_x \mathrm{s}\varphi_z + \mathrm{s}\varphi_x \mathrm{s}\varphi_y \mathrm{c}\varphi_z, & \mathrm{c}\varphi_x \mathrm{c}\varphi_z - \mathrm{s}\varphi_x \mathrm{s}\varphi_y \mathrm{s}\varphi_z, & -\mathrm{s}\varphi_x \mathrm{c}\varphi_y, & 0 \\ \mathrm{s}\varphi_x \mathrm{s}\varphi_z - \mathrm{c}\varphi_x \mathrm{s}\varphi_y \mathrm{c}\varphi_z, & \mathrm{s}\varphi_x \mathrm{c}\varphi_z + \mathrm{c}\varphi_x \mathrm{s}\varphi_y \mathrm{s}\varphi_z, & \mathrm{c}\varphi_x \mathrm{c}\varphi_y, & 0 \\ 0, & 0, & 0, & 1 \end{bmatrix} \tag{5.34}$$

Using similar denotation as in the previous section, the velocity matrix may be written in the form

$$\mathbf{V}^S =$$

$$= \begin{bmatrix} 0, & -\dot{\varphi}_x \mathrm{s}\varphi_y - \dot{\varphi}_z, & -\dot{\varphi}_x \mathrm{c}\varphi_y \mathrm{s}\varphi_z + \dot{\varphi}_y \mathrm{c}\varphi_z, & 0 \\ \dot{\varphi}_x \mathrm{s}\varphi_y + \dot{\varphi}_z, & 0, & -\dot{\varphi}_x \mathrm{c}\varphi_y \mathrm{c}\varphi_z - \dot{\varphi}_y \mathrm{s}\varphi_z, & 0 \\ \dot{\varphi}_x \mathrm{c}\varphi_y \mathrm{s}\varphi_z - \dot{\varphi}_y \mathrm{c}\varphi_z, & \dot{\varphi}_x \mathrm{c}\varphi_y \mathrm{c}\varphi_z + \dot{\varphi}_y \mathrm{s}\varphi_z, & 0, & 0 \\ 0, & 0, & 0, & 0 \end{bmatrix} \tag{5.35}$$

Hence for the angular velocity matrix,

$$\mathbf{\Omega}^S = \mathbf{S}^T_{\varphi_z} \mathbf{S}^T_{\varphi_y} \mathbf{\Omega}_x \mathbf{S}_{\varphi_y} \mathbf{S}_{\varphi_z} + \mathbf{S}^T_{\varphi_z} \mathbf{\Omega}_y \mathbf{S}_{\varphi_z} + \mathbf{\Omega}_z \tag{5.36}$$

so that

$$\boldsymbol{\omega}^S_{b,ab} = \begin{bmatrix} \omega_x \\ \omega_y \\ \omega_z \end{bmatrix}_b = \begin{bmatrix} \dot{\varphi}_x \mathrm{c}\varphi_y \mathrm{c}\varphi_z + \dot{\varphi}_y \mathrm{s}\varphi_z \\ -\dot{\varphi}_x \mathrm{c}\varphi_y \mathrm{s}\varphi_z + \dot{\varphi}_y \mathrm{c}\varphi_z \\ \dot{\varphi}_x \mathrm{s}\varphi_y + \dot{\varphi}_z \end{bmatrix} =$$

$$= \begin{bmatrix} \mathrm{c}\varphi_y \mathrm{c}\varphi_z, & \mathrm{s}\varphi_z, & 0 \\ -\mathrm{c}\varphi_y \mathrm{s}\varphi_z, & \mathrm{c}\varphi_z, & 0 \\ \mathrm{s}\varphi_y, & 0, & 1 \end{bmatrix} \begin{bmatrix} \dot{\varphi}_x \\ \dot{\varphi}_y \\ \dot{\varphi}_z \end{bmatrix} \tag{5.37}$$

and

$$\boldsymbol{\omega}^S_{a,ab} = \mathbf{S}^S \boldsymbol{\omega}^S_{b,ab} \tag{5.38}$$

The angular acceleration matrix

$$\mathcal{A}^S = \begin{bmatrix} 0 & a_{12} & a_{13} \\ a_{21} & 0 & a_{23} \\ a_{31} & a_{32} & 0 \end{bmatrix} \tag{5.39}$$

where

$$\begin{aligned}
a_{12} &= -\ddot{\varphi}_x \mathrm{s}\varphi_y - \dot{\varphi}_x \dot{\varphi}_y \mathrm{c}\varphi_y - \ddot{\varphi}_z \\
a_{13} &= -\ddot{\varphi}_x \mathrm{c}\varphi_y \mathrm{s}\varphi_z + \dot{\varphi}_x \dot{\varphi}_y \mathrm{s}\varphi_y \mathrm{s}\varphi_z - \dot{\varphi}_x \dot{\varphi}_z \mathrm{c}\varphi_y \mathrm{c}\varphi_z + \ddot{\varphi}_y \mathrm{c}\varphi_z - \dot{\varphi}_y \dot{\varphi}_z \mathrm{s}\varphi_z \\
a_{21} &= \ddot{\varphi}_x \mathrm{s}\varphi_y + \dot{\varphi}_x \dot{\varphi}_y \mathrm{c}\varphi_y + \ddot{\varphi}_z \\
a_{23} &= -\ddot{\varphi}_x \mathrm{c}\varphi_y \mathrm{c}\varphi_z + \dot{\varphi}_x \dot{\varphi}_y \mathrm{s}\varphi_y \mathrm{c}\varphi_z + \dot{\varphi}_x \dot{\varphi}_z \mathrm{c}\varphi_y \mathrm{s}\varphi_z - \ddot{\varphi}_y \mathrm{s}\varphi_z - \dot{\varphi}_y \dot{\varphi}_z \mathrm{c}\varphi_z \\
a_{31} &= \ddot{\varphi}_x \mathrm{c}\varphi_y \mathrm{s}\varphi_z - \dot{\varphi}_x \dot{\varphi}_y \mathrm{s}\varphi_y \mathrm{s}\varphi_z - \dot{\varphi}_x \dot{\varphi}_z \mathrm{c}\varphi_y \mathrm{c}\varphi_z - \ddot{\varphi}_y \mathrm{c}\varphi_z + \dot{\varphi}_y \dot{\varphi}_z \mathrm{s}\varphi_z \\
a_{32} &= \ddot{\varphi}_x \mathrm{c}\varphi_y \mathrm{c}\varphi_z - \dot{\varphi}_x \dot{\varphi}_y \mathrm{s}\varphi_y \mathrm{c}\varphi_z - \dot{\varphi}_x \dot{\varphi}_z \mathrm{c}\varphi_y \mathrm{s}\varphi_z + \ddot{\varphi}_y \mathrm{s}\varphi_z + \dot{\varphi}_y \dot{\varphi}_z \mathrm{c}\varphi_z
\end{aligned}$$

Hence

$$\boldsymbol{\alpha}^S_{b,ab} = \begin{bmatrix} \alpha_x \\ \alpha_y \\ \alpha_z \end{bmatrix}_b =$$

$$= \begin{bmatrix} \ddot{\varphi}_x c\varphi_y c\varphi_z - \dot{\varphi}_x \dot{\varphi}_y s\varphi_y c\varphi_z - \dot{\varphi}_x \dot{\varphi}_z c\varphi_y s\varphi_z + \ddot{\varphi}_y s\varphi_z + \dot{\varphi}_y \dot{\varphi}_z c\varphi_z \\ -\ddot{\varphi}_x c\varphi_y s\varphi_z + \dot{\varphi}_x \dot{\varphi}_y s\varphi_y s\varphi_z - \dot{\varphi}_x \dot{\varphi}_z c\varphi_y c\varphi_z + \ddot{\varphi}_y c\varphi_z - \dot{\varphi}_y \dot{\varphi}_z s\varphi_z \\ \ddot{\varphi}_x s\varphi_y + \dot{\varphi}_x \dot{\varphi}_y c\varphi_y + \ddot{\varphi}_z \end{bmatrix} \tag{5.40}$$

and again

$$\boldsymbol{\alpha}^S_{a,ab} = \mathbf{S}^S \boldsymbol{\alpha}^S_{b,ab} \tag{5.41}$$

For solving the inverse problem using the designation $\mathbf{T}_0^{-1}\mathbf{T}^S = [s_{ij}]$, $\mathbf{V}^S = [v_{ij}], \mathbf{A}^S = [a_{ij}]$, in analogy to Euler angles,

$$\varphi_y = \operatorname{atan2}(s_{13}, k\sqrt{1 - s_{13}^2}) \tag{5.42a}$$
$$\varphi_x = \operatorname{atan2}(-s_{23}k, s_{33}k) \tag{5.42b}$$
$$\varphi_z = \operatorname{atan2}(-s_{12}k, s_{11}k) \tag{5.42c}$$

where $k = \pm 1$.
According to (5.37)

$$\begin{bmatrix} \dot{\varphi}_x \\ \dot{\varphi}_y \\ \dot{\varphi}_z \end{bmatrix} = \begin{bmatrix} c\varphi_z/c\varphi_y & -s\varphi_z/c\varphi_y & 0 \\ s\varphi_z & c\varphi_z & 0 \\ -c\varphi_z t\varphi_y & s\varphi_z t\varphi_z & 1 \end{bmatrix} \begin{bmatrix} v_{32} \\ v_{13} \\ v_{21} \end{bmatrix} \tag{5.43}$$

when $t\varphi_i = \tan \varphi_i$ and finally

$$\begin{bmatrix} \ddot{\varphi}_x \\ \ddot{\varphi}_y \\ \ddot{\varphi}_z \end{bmatrix} = \begin{bmatrix} c\varphi_z/c\varphi_y & -s\varphi_z/c\varphi_y & 0 \\ s\varphi_z & c\varphi_z & 0 \\ -c\varphi_z t\varphi_y & s\varphi_z t\varphi_z & 1 \end{bmatrix} \left(\begin{bmatrix} a_{32} \\ a_{13} \\ a_{21} \end{bmatrix} + \right.$$
$$\left. + \begin{bmatrix} -s\varphi_y c\varphi_z & -c\varphi_y s\varphi_z & c\varphi_z \\ s\varphi_z & -c\varphi_y c\varphi_z & -s\varphi_z \\ c\varphi_y & 0 & 0 \end{bmatrix} \begin{bmatrix} \dot{\varphi}_x \dot{\varphi}_y \\ \dot{\varphi}_x \dot{\varphi}_z \\ \dot{\varphi}_y \dot{\varphi}_z \end{bmatrix} \right) \tag{5.44}$$

Similar statements as for Euler angles may be made about Cardan angles. There is the same singularity for $\varphi_y = \pm\pi/2$ whereby neither φ_x nor φ_z is defined. In a similar way, the singularity arises when computing Cardan angle derivatives.

The use of Cardan angles is advantageous for expressing small motions, for example, when solving small spatial oscillations of a body or when solving accuracy problems. Then

$$\mathbf{T}^S = \begin{bmatrix} 1 & -\varphi_z & \varphi_y & x_0 \\ \varphi_z & 1 & -\varphi_x & y_0 \\ -\varphi_y & \varphi_x & 1 & z_0 \\ 0 & 0 & 0 & 1 \end{bmatrix} \tag{5.45}$$

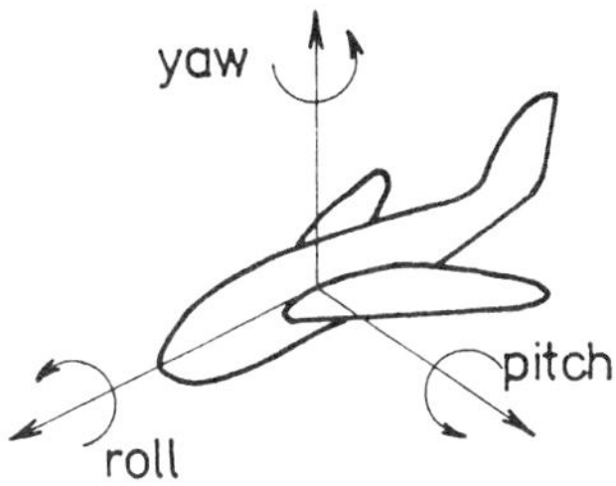

Figure 5.8. Roll, pitch and yaw definition.

Furthermore,

$$\mathbf{V}^S = \begin{bmatrix} 0 & -\dot{\varphi}_z & \dot{\varphi}_y & 0 \\ \dot{\varphi}_z & 0 & -\dot{\varphi}_x & 0 \\ -\dot{\varphi}_y & \dot{\varphi}_x & 0 & 0 \\ 0 & 0 & 0 & 0 \end{bmatrix}, \quad \mathbf{A}^S = \begin{bmatrix} 0 & -\ddot{\varphi}_z & \ddot{\varphi}_y & 0 \\ \ddot{\varphi}_z & 0 & -\ddot{\varphi}_x & 0 \\ -\ddot{\varphi}_y & \ddot{\varphi}_x & 0 & 0 \\ 0 & 0 & 0 & 0 \end{bmatrix} \tag{5.46}$$

and also

$$\begin{aligned} \boldsymbol{\varphi} &= \boldsymbol{\varphi}_a = \boldsymbol{\varphi}_b = [\varphi_x, \varphi_y, \varphi_z] \\ \boldsymbol{\omega} &= \boldsymbol{\omega}_{a,ab} = \boldsymbol{\omega}_{b,ab} = [\dot{\varphi}_x, \dot{\varphi}_y, \dot{\varphi}_z] \\ \boldsymbol{\alpha} &= \boldsymbol{\alpha}_{a,ab} = \boldsymbol{\alpha}_{b,ab} = [\ddot{\varphi}_x; \ddot{\varphi}_y, \ddot{\varphi}_z] \end{aligned} \tag{5.47}$$

5.1.5.3 Roll-Pitch-Yaw

Motion of ships, planes and cars is popularly described by similar angles designated as roll, pitch and yaw, Figure 5.8. These angles are different from previous ones because they are 1–2–3 Euler angles around the axes of the fixed coordinate system. In order to describe them let us start with the description of a rotation around a general axis o in the space. It is decribed as the sequence of three transformations. The first one is the general transformation $\mathbf{T}_0$ (5.2) to the coordinate system with one coordinate axis (e.g. z axis) identical with the axis of rotation o. The second one is the rotation $\mathbf{T}_{Z6}(\varphi)$ around the axis o and the third one is the transformation to the original system

$$\mathbf{T} = \mathbf{T}_0 \mathbf{T}_{Z6}(\varphi) \mathbf{T}_0^{-1}$$

The roll-pitch-yaw is then derived as the sequence of these rotations

- around the axis x_a by an angle α (roll),
- around the axis y_a by an angle β (pitch),
- around the axis z_a by an angle γ (yaw).

The complete transformation matrix is

$$\begin{aligned}\mathbf{T} &= \mathbf{T}_{Z4}(\alpha). \\ &\quad .(\mathbf{T}_{Z4}(-\alpha)\mathbf{T}_{Z5}(\beta)\mathbf{T}_{Z4}(\alpha)). \\ &\quad .(\mathbf{T}_{Z4}(-\alpha)\mathbf{T}_{Z5}(-\beta)\mathbf{T}_{Z6}(\gamma)\mathbf{T}_{Z5}(\beta)\mathbf{T}_{Z4}(\alpha)) = \\ &= \mathbf{T}_{Z6}(\gamma)\mathbf{T}_{Z5}(\beta)\mathbf{T}_{Z4}(\alpha)\end{aligned}$$

The resulting transformation is composed from the same matrices as for Cardan angles but in the reverse order.

5.1.5.4 Euler Parameters

Euler parameters are elements of a normed quaternion (4.163)

$$\hat{\mathbf{e}}^S = \cos\frac{\Theta}{2} + \mathbf{n}\sin\frac{\Theta}{2} \tag{5.48}$$

which may be used to describe spherical motion. In (5.48), $\mathbf{n}$ is the unit vector of the axis according to Euler theorem and Θ is the angle of angular displacement, (Figure 5.9a). We shall assume for the sake of simplicity in this section that $o_a \equiv o_b$ and we shall also leave out the superscript "S" for quantities characterizing spherical motion.

These elements of quaternion $\hat{\mathbf{e}}$ are

$$\underline{\mathbf{e}} = \lceil e_0, e_1, e_2, e_3 \rceil = \lceil \mathrm{c}\frac{\Theta}{2}, \mathrm{c}\alpha\ \mathrm{s}\frac{\Theta}{2}, \mathrm{c}\beta\ \mathrm{s}\frac{\Theta}{2}, \mathrm{c}\gamma\ \mathrm{s}\frac{\Theta}{2} \rceil \tag{5.49}$$

and they characterize position, (Figure 5.9a) i.e. they describe the large rotation according to Euler theorem (Section 4.1.1) from the initial position to the final instantaneous position.

Velocity is, according to (4.216), described by velocity SDQ, which has in this case the form

$$\hat{\mathbf{W}}^R_{b,ab} = [\hat{\mathbf{e}}^*_{ab}\dot{\hat{\mathbf{e}}}_{ab}; \vec{0}]^R \tag{5.50}$$

whereas, according to (4.225)

$$\boldsymbol{\omega}_{b,ab} = 2\hat{\mathbf{e}}^*_{ab}\dot{\hat{\mathbf{e}}}_{ab} \tag{5.51}$$

This vector describes the instantaneous rotation axis of spherical motion which is, however, different from the axis of large rotation described by $\mathbf{n}$. The last equation may be written in a way which can more readily be surveyed in the form of a matrix

$$\begin{bmatrix} 0 \\ \boldsymbol{\omega}_{b,ab} \end{bmatrix} = \begin{bmatrix} 0 \\ \omega_x \\ \omega_y \\ \omega_z \end{bmatrix}_{b,ab} = 2\begin{bmatrix} e_0 & e_1 & e_2 & e_3 \\ -e_1 & e_0 & e_3 & -e_2 \\ -e_2 & -e_3 & e_0 & e_1 \\ -e_3 & e_2 & -e_1 & e_0 \end{bmatrix}\begin{bmatrix} \dot{e}_0 \\ \dot{e}_1 \\ \dot{e}_2 \\ \dot{e}_3 \end{bmatrix} \tag{5.52}$$

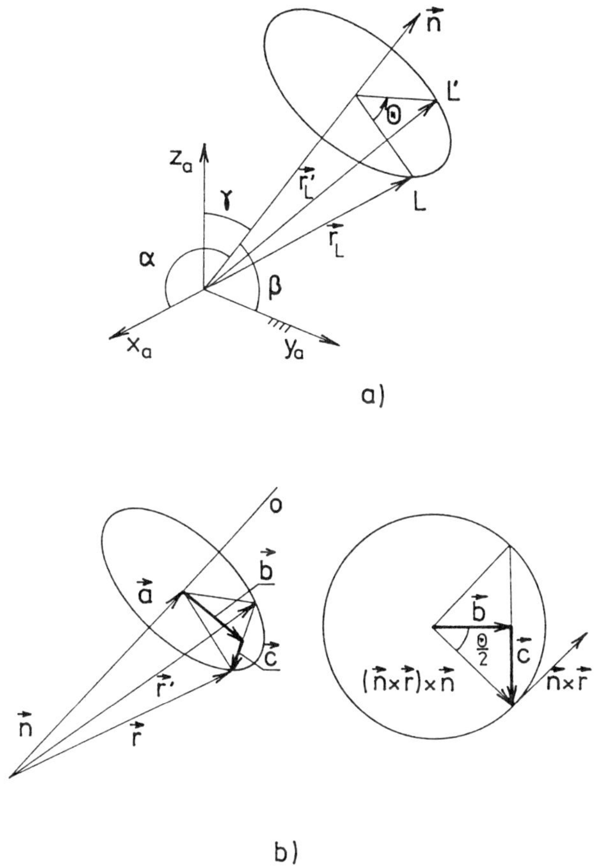

Figure 5.9. Vector description of spherical motion.

The inverse problem equation is

$$\begin{bmatrix} \dot{e}_0 \\ \dot{e}_1 \\ \dot{e}_2 \\ \dot{e}_3 \end{bmatrix} = \frac{1}{2} \left[\begin{array}{c|c} 0 & -\boldsymbol{\omega}^T \\ \hline \boldsymbol{\omega} & -\boldsymbol{\Omega} \end{array} \right] \begin{bmatrix} e_0 \\ e_1 \\ e_2 \\ e_3 \end{bmatrix} \tag{5.53}$$

Acceleration is described by the acceleration SDQ according to (4.229) in the form

$$\hat{\mathbf{Y}}^R_{b,ab} = [\hat{\mathbf{e}}^*_{ab}\ddot{\hat{\mathbf{e}}}_{ab} + \dot{\hat{\mathbf{e}}}^*_{ab}\dot{\hat{\mathbf{e}}}_{ab}; \vec{0}]^R \tag{5.54}$$

The first quaternion is

$$\boldsymbol{\alpha}_{b,ab} = 2(\hat{\mathbf{e}}^*_{ab}\ddot{\hat{\mathbf{e}}}_{ab} + \dot{\hat{\mathbf{e}}}^*_{ab}\dot{\hat{\mathbf{e}}}_{ab}) \tag{5.55}$$

It can be written in matrix form as $\boldsymbol{\omega}_{b,ab}$ in (5.52)

$$\begin{bmatrix} 0 \\ \boldsymbol{\alpha}_{b,ab} \end{bmatrix} = \begin{bmatrix} 0 \\ \alpha_x \\ \alpha_y \\ \alpha_z \end{bmatrix}_{b,ab} = 2 \begin{bmatrix} 0 & 0 & 0 & 0 \\ -e_1 & e_0 & e_3 & -e_2 \\ -e_2 & -e_3 & e_0 & e_1 \\ -e_3 & e_2 & -e_1 & e_0 \end{bmatrix} \begin{bmatrix} \ddot{e}_0 \\ \ddot{e}_1 \\ \ddot{e}_2 \\ \ddot{e}_3 \end{bmatrix} \tag{5.56}$$

To solve the inverse problem we way use for the postion the algorithm (4.177) – (4.178), then for the velocity the equation (5.53) and for the acceleration

$$\begin{bmatrix} \ddot{e}_0 \\ \ddot{e}_1 \\ \ddot{e}_2 \\ \ddot{e}_3 \end{bmatrix} = \frac{1}{2} \left[\begin{array}{c|c} 0 & -\boldsymbol{\alpha}^T \\ \hline \boldsymbol{\alpha} & -\mathcal{A} \end{array} \right] \begin{bmatrix} e_0 \\ e_1 \\ e_2 \\ e_3 \end{bmatrix} + \frac{1}{2} \left[\begin{array}{c|c} 0 & -\boldsymbol{\omega}^T \\ \hline \boldsymbol{\omega} & -\boldsymbol{\Omega} \end{array} \right] \begin{bmatrix} \dot{e}_0 \\ \dot{e}_1 \\ \dot{e}_2 \\ \dot{e}_3 \end{bmatrix} \tag{5.57}$$

The correspondence between, for example, the three-parameter description using Cardan angles and the quaternion description using the basic quaternions introduced in Section 4.5.3 may easily be obtained. Using the same basic motions (except translation $O_a \longrightarrow O_b$) as in Section 5.1.5.2 we obtain

$$\begin{aligned} \hat{\mathbf{K}}^S &= \hat{\mathbf{K}}_{Z4}(\varphi_x)\hat{\mathbf{K}}_{Z5}(\varphi_y)\hat{\mathbf{K}}_{Z6}(\varphi_z) = [\hat{\mathbf{e}}_x; \vec{0}][\hat{\mathbf{e}}_y; \vec{0}][\hat{\mathbf{e}}_z; \vec{0}] = \\ &= [\hat{\mathbf{e}}_x\hat{\mathbf{e}}_y\hat{\mathbf{e}}_z; \vec{0}] = [\hat{\mathbf{e}}^S; \vec{0}] \end{aligned} \tag{5.58}$$

where

$$\underline{\mathbf{e}_x} = [c\frac{\varphi_x}{2}, s\frac{\varphi_x}{2}, 0, 0; 0] \tag{5.59a}$$

$$\underline{\mathbf{e}_y} = [c\frac{\varphi_y}{2}, 0, s\frac{\varphi_y}{2}, 0; 0] \tag{5.59b}$$

$$\underline{\mathbf{e}_z} = [c\frac{\varphi_z}{2}, 0, 0, s\frac{\varphi_z}{2}; 0] \tag{5.59c}$$

Therefore

$$\mathbf{e}^S = \begin{bmatrix} e_0^S \\ e_1^S \\ e_2^S \\ e_3^S \end{bmatrix} = \begin{bmatrix} c\frac{\varphi_x}{2}c\frac{\varphi_y}{2}c\frac{\varphi_z}{2} - s\frac{\varphi_x}{2}s\frac{\varphi_y}{2}s\frac{\varphi_z}{2} \\ s\frac{\varphi_x}{2}c\frac{\varphi_y}{2}c\frac{\varphi_z}{2} + c\frac{\varphi_x}{2}s\frac{\varphi_y}{2}s\frac{\varphi_z}{2} \\ c\frac{\varphi_x}{2}s\frac{\varphi_y}{2}c\frac{\varphi_z}{2} - s\frac{\varphi_x}{2}c\frac{\varphi_y}{2}s\frac{\varphi_z}{2} \\ s\frac{\varphi_x}{2}s\frac{\varphi_y}{2}c\frac{\varphi_z}{2} + c\frac{\varphi_x}{2}c\frac{\varphi_y}{2}s\frac{\varphi_z}{2} \end{bmatrix} \tag{5.60}$$

The direction cosines matrix $\mathbf{S}$ computed from (5.60) using (4.171) is identical with matrix $\mathbf{S}^S$ from Section 5.1.5.2; this may be proved by direct computation. However, the matrix $\mathbf{S}$ may also be obtained directly from the vector description according to Figure 5.9b (Corben and Stehle, 1950). For these vectors, it is evident that

$$\vec{r} = \vec{a} + \vec{b} + \vec{c} \tag{5.61a}$$

$$\vec{r}\,' = \vec{a} + \vec{b} - \vec{c} \tag{5.61b}$$

where

$$\vec{a} = (\vec{r} \cdot \vec{n})\vec{n} \tag{5.62a}$$

$$\vec{b} = \cos^2\frac{\Theta}{2}(\vec{n} \times \vec{r}) \times \vec{n} + \cos\frac{\Theta}{2}\sin\frac{\Theta}{2}(\vec{n} \times \vec{r}) \tag{5.62b}$$

$$\vec{c} = -\cos\frac{\Theta}{2}\sin\frac{\Theta}{2}(\vec{n} \times \vec{r}) + \sin^2\frac{\Theta}{2}(\vec{n} \times \vec{r}) \times \vec{n} \tag{5.62c}$$

Substituting into (5.61b) and modifying gives

$$\begin{aligned}
\mathbf{r}' &= (1 - \cos\Theta)(\mathbf{r}.\mathbf{n})\mathbf{n} + \cos\Theta\mathbf{r} + \sin\Theta(\mathbf{n} \times \mathbf{r}) = \\
&= (1 - \cos\Theta)((\mathbf{r}.\mathbf{n})\mathbf{n} - \mathbf{r}) + \mathbf{r} + \sin\Theta(\mathbf{n} \times \mathbf{r}) = \\
&= 2\sin^2\frac{\Theta}{2}((\mathbf{r}.\mathbf{n})\mathbf{n} - (\mathbf{n}.\mathbf{n})\mathbf{r}) + \mathbf{r} + \sin\Theta(\mathbf{n} \times \mathbf{r}) = \\
&= 2\sin^2\frac{\Theta}{2}(\mathbf{n} \times (\mathbf{n} \times \mathbf{r})) + \mathbf{r} + \sin\Theta(\mathbf{n} \times \mathbf{r}) = \\
&= (\mathbf{E} + \sin\Theta\tilde{\mathbf{n}} + 2\sin^2\frac{\Theta}{2}\tilde{\mathbf{n}}^2)\mathbf{r}
\end{aligned} \tag{5.63}$$

Rewriting the last equation into matrix form

$$\mathbf{r}_{aL} = \mathbf{S}_{ab}\mathbf{r}_{bL}$$

we obtain the direction cosines transformation matrix

$$\mathbf{S}_{ab} = \begin{bmatrix} c^2\alpha(1 - c\Theta) + c\Theta, & c\alpha c\beta(1 - c\Theta) - c\gamma s\Theta, & c\alpha c\gamma(1 - c\Theta) + c\beta s\Theta \\ c\alpha c\beta(1 - c\Theta) + c\gamma s\Theta, & c^2\beta(1 - c\Theta) + c\Theta, & c\beta c\gamma(1 - c\Theta) + c\alpha s\Theta \\ c\alpha c\gamma(1 - c\Theta) - c\beta s\Theta & c\beta c\gamma(1 - c\Theta) + c\alpha s\Theta, & c^2\gamma(1 - c\Theta) + c\Theta \end{bmatrix} \tag{5.64}$$

whose elements contain quantities characterizing the axis of rotation and angular displacement angle.

The inverse problem, i.e. finding the axis of rotation and angle of displacement corresponding to the known matrix $\mathbf{S}$, is to be solved by the algorithm from Section 4.3.1.

Comparing the sum and difference of equations (5.61) we obtain, after modification, the relation

$$\vec{r}\,' - \vec{r} = \vec{n} \times (\vec{r} + \vec{r}') \tan \frac{\Theta}{2} \tag{5.65a}$$

which is the well known Rodrigues formula (Angeles, 1982) or in another form from (5.63)

$$\begin{aligned} \mathbf{r}' &= \left(\mathbf{E} + \sin\Theta\tilde{\mathbf{n}} + 2\sin^2\frac{\Theta}{2}\tilde{\mathbf{n}}^2\right)\mathbf{r} = \\ &= \left(\mathbf{E} + \frac{2\tan\frac{\Theta}{2}}{1+\tan^2\frac{\Theta}{2}}\tilde{\mathbf{n}} + 2\frac{\tan\frac{\Theta}{2}}{1+\tan^2\frac{\Theta}{2}}\tilde{\mathbf{n}}^2\right)\mathbf{r} = \\ &= \left(\mathbf{E} + \frac{2}{1+\gamma^2}(\tilde{\gamma} + \tilde{\gamma}^2)\right)\mathbf{r} \end{aligned} \tag{5.65b}$$

where $\boldsymbol{\gamma} = \mathbf{n}\tan(\Theta/2)$ are so-called Rodrigues parameters (Shabana, 1989). From (5.48) follows the relation between Euler and Rodrigues parameters

$$\gamma_i = \frac{e_i}{e_0}, \qquad i = 1, 2, 3 \tag{5.66}$$

However, Rodrigues parameters have the same disadvantages of singularities as Euler or Cardan angles (Shabana, 1989).

In these descriptions of spherical motion the method of Euler parameters is the only one which describes this motion in a natural way, i.e. using quantities characterizing the position of the axis and angular displacement. This is why we obtain simpler relations that can easily be written in matrix form. The description has no singularities, and is not unique (a single position resulting for two different values $\Theta_1 = \pi, \Theta_2 = -\pi$ of angular displacement). It is easy to eliminate the ambiguity by the requirement $e_0 \geq 0$. The only disadvantage of the quaternion description is the redundancy of coordinates and thus a constraint equation.

5.1.6 Flat Pair

A flat pair involves the contact of bodies a and b in such a way that the planar surface of one body moves (slides) along the planar surface of the other body. This kind of contact allows three degrees of freedom. Let us choose translational displacement x and y of origin O_b, and angular displacement φ of the body in space $O_0x_0y_0z_0$ as coordinates, (Figure 5.10).

Assuming that system b is initially identical with a, having reached its position by

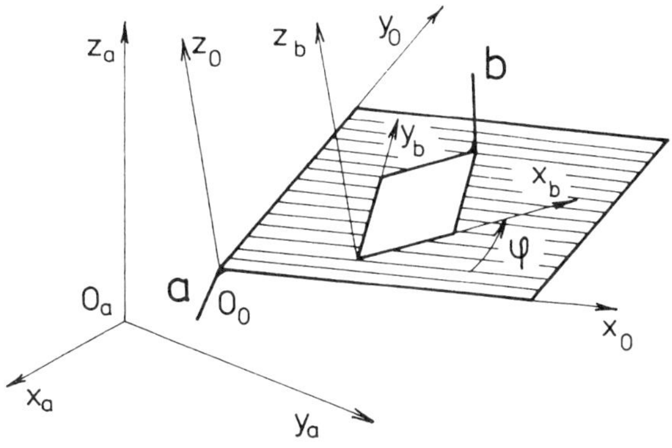

Figure 5.10. Flat kinematic pair.

- general displacement into position $O_0x_0y_0z_0$,
- translation by displacement x, y in the direction of axes x_0, y_0,
- rotation by displacement angle φ around axis z_0,

$$\mathbf{T}^F = \mathbf{T}_0\mathbf{T}_{Z1}(x)\mathbf{T}_{Z2}(y)\mathbf{T}_{Z6}(\varphi) = \mathbf{T}_0 \begin{bmatrix} c\varphi & -s\varphi & 0 & x \\ s\varphi & c\varphi & 0 & y \\ 0 & 0 & 1 & 0 \\ 0 & 0 & 0 & 1 \end{bmatrix} \tag{5.67}$$

The velocity matrix is

$$\mathbf{V}^F = \begin{bmatrix} 0 & -\dot{\varphi} & 0 & \dot{x}c\varphi + \dot{y}s\varphi \\ \dot{\varphi} & 0 & 0 & -\dot{x}s\varphi + \dot{y}c\varphi \\ 0 & 0 & 0 & 0 \\ 0 & 0 & 0 & 0 \end{bmatrix} \tag{5.68}$$

The acceleration matrix is

$$\mathbf{A}^F = \begin{bmatrix} 0 & -\ddot{\varphi} & 0 & \ddot{x}c\varphi + \ddot{y}s\varphi + (-\dot{x}s\varphi + \dot{y}c\varphi)\dot{\varphi} \\ \ddot{\varphi} & 0 & 0 & -\ddot{x}s\varphi + \ddot{y}c\varphi + (-\dot{x}c\varphi - \dot{y}s\varphi)\dot{\varphi} \\ 0 & 0 & 0 & 0 \\ 0 & 0 & 0 & 0 \end{bmatrix} \tag{5.69}$$

5.2 Simple Open Kinematic Chains

A diagram of a single open kinematic chain with n elements is presented in Figure 5.11. The motion of the last body may be understood as composed

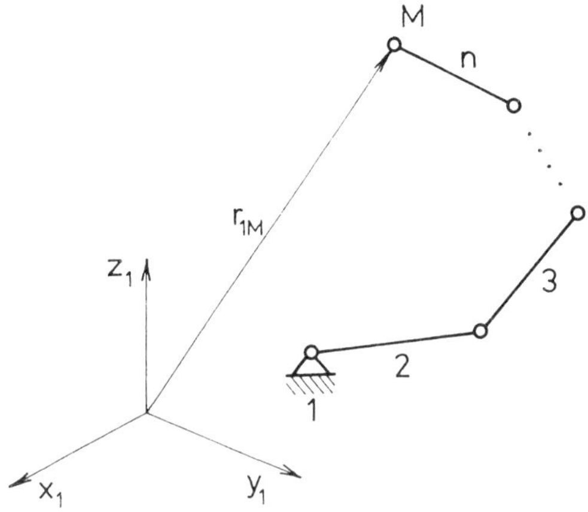

Figure 5.11. Simple open kinematic chain.

of a series of simultaneous motions 12, 23, ..., $(n-1)n$. We may use all the results of Section 4.2.3 for its description.

For the motion of a general point M, for example,

$$\mathbf{r}_{1M} = \mathbf{T}_{12}\mathbf{T}_{23}\ldots\mathbf{T}_{n-1,n}\mathbf{r}_{nM} = \mathbf{T}_{1n}\mathbf{r}_{nM} \tag{5.70}$$

where

$$\mathbf{T}_{1n} = \prod_{i=1}^{n-1} \mathbf{T}_{i,i+1} \tag{5.71}$$

In this way, the transformation matrices of single motions are, depending on the kind of link between respective bodies, equal to the matrices of lower kinematic pairs derived in Section 5.1. When the connection of, for example, bodies 2 and 3 is realized by a cylindrical pair, then $\mathbf{T}_{23} = \mathbf{T}^C$. We use the set of time variable coordinates of the kinematic pairs for the kinematic chain coordinates.

We compute the velocity and acceleration of point M from equations (4.122), (4.132). Velocity and acceleration matrices of individual bodies are defined by formulae (4.124), (4.133).

5.3 Applications to Kinematic Chains

As an example of the analysis of a simple kinematic chain, we shall outline the solution of robot PUMA 560, shown (without gripper) in Figure 5.12 in its basic position where the coordinate values are zero.

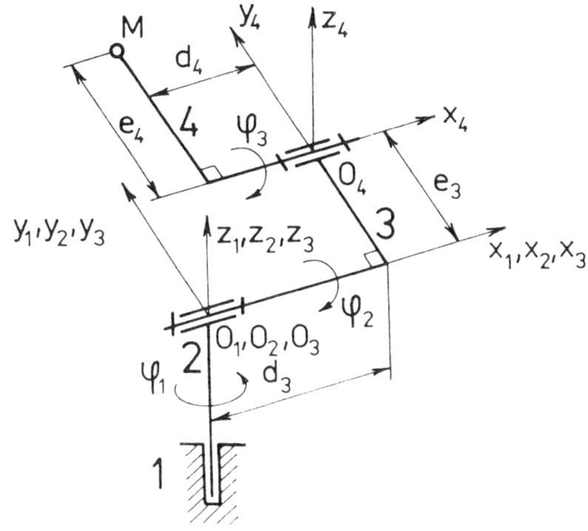

Figure 5.12. Diagram of robot PUMA 560.

The position of point M in basic space $O_1x_1y_1z_1$ is given by the relation

$$\mathbf{r}_{1M} = \mathbf{T}_{12}\mathbf{T}_{23}\mathbf{T}_{34}\mathbf{r}_{4M} \tag{a}$$

where

$$\mathbf{T}_{12} = \mathbf{T}_{Z6}(\varphi_1) \tag{b}$$

$$\mathbf{T}_{23} = \mathbf{T}_{Z4}(\varphi_2) \tag{c}$$

$$\mathbf{T}_{34} = \mathbf{T}_{Z1}(d_3)\mathbf{T}_{Z2}(e_3)\mathbf{T}_{Z4}(\varphi_3) = \begin{bmatrix} 1 & 0 & 0 & d_3 \\ 0 & c\varphi_3 & -s\varphi_3 & e_3 \\ 0 & s\varphi_3 & c\varphi_3 & 0 \\ 0 & 0 & 0 & 1 \end{bmatrix} \tag{d}$$

Hence

$$\mathbf{T}_{14} = \mathbf{T}_{12}\mathbf{T}_{23}\mathbf{T}_{34} = \begin{bmatrix} c\varphi_1 & -s\varphi_1 c\varphi_{23} & s\varphi_1 s\varphi_{23} & d_3 c\varphi_1 - e_3 s\varphi_1 c\varphi_2 \\ s\varphi_1 & c\varphi_1 c\varphi_{23} & -c\varphi_1 s\varphi_{23} & d_3 s\varphi_1 + e_3 c\varphi_1 c\varphi_2 \\ 0 & s\varphi_{23} & c\varphi_{23} & e_3 s\varphi_2 \\ 0 & 0 & 0 & 1 \end{bmatrix} \tag{e}$$

where

$$\varphi_{23} = \varphi_2 + \varphi_3 \tag{f}$$

$$\mathbf{r}_{1M} = [x_{1M}, y_{1M}, z_{1M}, 1] \tag{g}$$

$$\mathbf{r}_{4M} = [-d_4, e_4, 0, 1] \tag{h}$$

The velocities of individual bodies are described by the velocity matrices

$$\mathbf{V}_{12} = \mathbf{T}_{12}^{-1}\dot{\mathbf{T}}_{12} = \begin{bmatrix} 0 & -\dot{\varphi}_1 & 0 & 0 \\ \dot{\varphi}_1 & 0 & 0 & 0 \\ 0 & 0 & 0 & 0 \\ 0 & 0 & 0 & 0 \end{bmatrix} \tag{i}$$

$$\mathbf{V}_{13} = \mathbf{T}_{13}^{-1}\dot{\mathbf{T}}_{13} = \begin{bmatrix} 0 & -\dot{\varphi}_1 \mathrm{c}\varphi_2 & \dot{\varphi}_1 \mathrm{s}\varphi_2 & 0 \\ \dot{\varphi}_1 \mathrm{c}\varphi_2 & 0 & -\dot{\varphi}_2 & 0 \\ -\dot{\varphi}_1 \mathrm{s}\varphi_2 & \dot{\varphi}_2 & 0 & 0 \\ 0 & 0 & 0 & 0 \end{bmatrix} \tag{j}$$

$$\mathbf{V}_{14} = \mathbf{T}_{14}^{-1}\dot{\mathbf{T}}_{14} =$$
$$= \begin{bmatrix} 0 & -\dot{\varphi}_1 \mathrm{c}\varphi_{23} & \dot{\varphi}_1 \mathrm{s}\varphi_{23} & -\dot{\varphi}_1 e_3 \mathrm{c}\varphi_2 \\ \dot{\varphi}_1 \mathrm{c}\varphi_{23} & 0 & -\dot{\varphi}_2 - \dot{\varphi}_3 & \dot{\varphi}_1 d_3 \mathrm{c}\varphi_{23} + \dot{\varphi}_2 e_3 \mathrm{s}\varphi_3 \\ -\dot{\varphi}_1 \mathrm{s}\varphi_{23} & \dot{\varphi}_2 + \dot{\varphi}_3 & 0 & -\dot{\varphi}_1 d_3 \mathrm{s}\varphi_{23} + \dot{\varphi}_2 e_3 \mathrm{c}\varphi_3 \\ 0 & 0 & 0 & 0 \end{bmatrix} \tag{k}$$

Therefore

$$\boldsymbol{\omega}_{12} = [0, 0, \dot{\varphi}_1] \tag{l}$$
$$\boldsymbol{\omega}_{13} = [\dot{\varphi}_2, \dot{\varphi}_1 \mathrm{s}\varphi_{23}, \dot{\varphi}_1 \mathrm{c}\varphi_{23}] \tag{m}$$
$$\boldsymbol{\omega}_{14} = [\dot{\varphi}_2 + \dot{\varphi}_3, \dot{\varphi}_1 \mathrm{s}\varphi_{23}, \dot{\varphi}_1 \mathrm{c}\varphi_{23}] \tag{n}$$

and the velocity of point M is given by relation

$$\mathbf{v}_{1M} = \mathbf{T}_{14}\mathbf{V}_{14}\mathbf{r}_{4M} = \begin{bmatrix} a_{11} & a_{12} & a_{13} & a_{14} \\ a_{21} & a_{22} & a_{23} & a_{24} \\ a_{31} & a_{32} & a_{33} & a_{34} \\ 0 & 0 & 0 & 0 \end{bmatrix} \mathbf{r}_{4M}$$

where

$$\begin{aligned}
a_{11} &= -\dot{\varphi}_1 \mathrm{s}\varphi_1 \\
a_{12} &= -\dot{\varphi}_1 \mathrm{c}\varphi_1 \mathrm{c}\varphi_{23} + \dot{\varphi}_2 \mathrm{s}\varphi_1 \mathrm{s}\varphi_{23} + \dot{\varphi}_3 \mathrm{s}\varphi_1 \mathrm{s}\varphi_{23} \\
a_{13} &= \dot{\varphi}_1 \mathrm{c}\varphi_1 \mathrm{s}\varphi_{23} + \dot{\varphi}_2 \mathrm{s}\varphi_1 \mathrm{c}\varphi_{23} + \dot{\varphi}_3 \mathrm{s}\varphi_1 \mathrm{c}\varphi_{23} \\
a_{14} &= -\dot{\varphi}_1 d_3 \mathrm{s}\varphi_1 - \dot{\varphi}_1 e_3 \mathrm{c}\varphi_1 \mathrm{c}\varphi_2 + \dot{\varphi}_2 e_3 \mathrm{s}\varphi_1 \mathrm{s}\varphi_2 \\
a_{21} &= \dot{\varphi}_1 \mathrm{c}\varphi_1 \\
a_{22} &= -\dot{\varphi}_1 \mathrm{s}\varphi_1 \mathrm{c}\varphi_{23} - (\dot{\varphi}_2 + \dot{\varphi}_3) \mathrm{c}\varphi_1 \mathrm{s}\varphi_{23} \\
a_{23} &= \dot{\varphi}_1 \mathrm{s}\varphi_1 \mathrm{s}\varphi_{23} - (\dot{\varphi}_2 + \dot{\varphi}_3) \mathrm{c}\varphi_1 \mathrm{c}\varphi_{23} \\
a_{24} &= \dot{\varphi}_1 d_3 \mathrm{c}\varphi_1 - \dot{\varphi}_1 e_3 \mathrm{s}\varphi_1 \mathrm{c}\varphi_2 - \dot{\varphi}_2 e_3 \mathrm{c}\varphi_1 \mathrm{s}\varphi_2
\end{aligned}$$

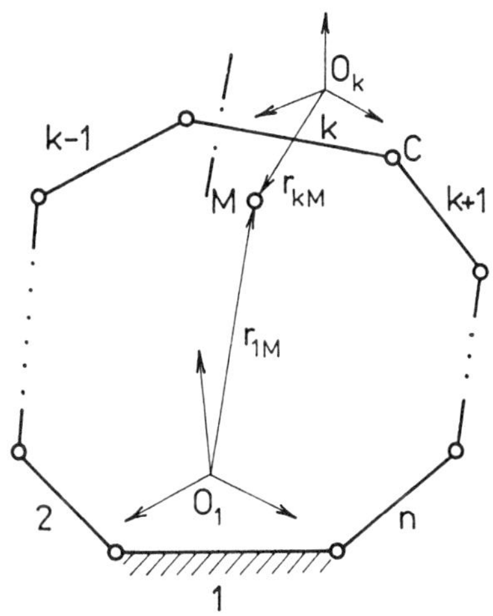

Figure 5.13. Simple mechanism.

$$
\begin{aligned}
a_{31} &= 0 \\
a_{32} &= (\dot{\varphi}_2 + \dot{\varphi}_3)c\varphi_{23} \\
a_{33} &= -(\dot{\varphi}_2 + \dot{\varphi}_3)s\varphi_{23} \\
a_{34} &= \dot{\varphi}_2 e_3 c\varphi_2
\end{aligned}
$$

In a similar way, based on the results from previous chapters, the expressions for acceleration may be found. For the sake of conciseness we do not present them here.

5.4 Simple Mechanisms

A kinematic diagram of a simple mechanism is shown in Figure 5.13. It contains a total of n elements, the first of which is the stationary frame.

To describe the mechanism being examined we shall use the coordinates that form the mechanism coordinates. Using the results from Chapter 2, we see that such a mechanism, with m degrees of freedom, has a total of $m + 6$ corresponding coordinates. We shall distinguish the independent coordinates of driving kinematic pairs, which we shall denote with the letter q, from dependent coordinates of driven pairs, for which we shall use the letter z.

To obtain a kinematic solution of a simple mechanism, we must determine the functions

$$z_i(t),\ \dot{z}_i(t),\ \ddot{z}_i(t), \qquad i = 1 \text{ to } 6$$

using the known functions

$$q_j(t),\ \dot{q}_j(t),\ \ddot{q}_j(t), \qquad j = 1, 2, \dots, m$$

In a special case we have to find only

$$z_i = z_i(q), \qquad i = 1 \text{ to } 6 \tag{5.72}$$

We shall call the above problem the position problem. It is the most fundamental and at the same time the most difficult problem in kinematic analysis.

5.4.1 Matrix Methods of Solving the Position Problem

The matrix description of motion derived in the previous text may be used for the solution of simple mechanisms in several ways. We shall briefly present the basic methods.

5.4.1.1 The Closed Loop Method

As the diagram of the mechanism to be solved is in the form of a closed loop, we may, according to (4.115), carry out such transformations that

$$\mathbf{T}_{12}\mathbf{T}_{23}\dots\mathbf{T}_{n-1,n}\mathbf{T}_{n,1} = \mathbf{T}_{11} = \mathbf{E}_4 \tag{5.73}$$

This equation says that, after carrying out all transformations in the required order, we come back to the original coordinate system. As individual transformation matrices are functions of coordinates $\mathbf{q}$ and $\mathbf{z}$, the equation (5.73) represents a system of 12 nonlinear equations of the type

$$f_k(\mathbf{z}, \mathbf{q}) = 0, \qquad k = 1 \text{ to } 12 \tag{5.74}$$

and 4 identities. From equations in (5.74), only 6 are, according to conditions (4.30), independent, but this will exactly suffice to solve the position problem, namely to determine 6 functions (5.72).

The loop closure condition in the form of (5.73), especially in general solutions, yields complicated expressions; it is therefore convenient to modify it by post-multiplying it by matrix $\mathbf{T}_{k+1,1}^{-1}$ into the form

$$\mathbf{T}_{12}\mathbf{T}_{23}\dots\mathbf{T}_{k-1,k} = \mathbf{T}_{1n}\mathbf{T}_{n,n-1}\dots\mathbf{T}_{k+1,k} \tag{5.75}$$

where $1 < k < n$ is the index of an arbitrary chain element. From the viewpoint of computational complexity, the condition (5.75) is most advantageous, when the number of transformation matrices on one side of the equation equals the number of transformation matrices on the other side, or when they differ by one.

5.4.1.2 The Disconnected Loop Method

In many problems, the coordinates of some or all internal loop pairs are of no interest to us. This leads to the possibility of disconnecting the loop in a chosen uninteresting kinematic pair. Let this be the pair linking the k-th and $(k+1)$-th elements. The coordinates of this pair do not appear at all in the position problem. The basic equation must however be written so as to express the original loop closure. It is not difficult to formulate the respective conditions for individual kinematic pairs. We shall show only the simplest case of disconnecting a loop in a spherical kinematic pair. Let C be its centre (Figure 5.13); then

$$\mathbf{T}_{12}\mathbf{T}_{23}\ldots\mathbf{T}_{k-1,k}\mathbf{r}_{kC} = \mathbf{T}_{1n}\mathbf{T}_{n,n-1}\ldots\mathbf{T}_{k+2,k+1}\mathbf{r}_{k+1,C} \tag{5.76}$$

The condition (5.76) yields three scalar equations. As the number of all functions being sought is reduced from six to three (the three coordinates corresponding to the disconnected spherical pair having been left out), and the equations suffice for a solution. In addition the relative motion $k, k+1$ may be examined, and

$$\mathbf{T}_{k,k+1} = \mathbf{T}_{1k}^{-1}\mathbf{T}_{1,k+1} \tag{5.77}$$

It is evident that the loop does not necessarily have to be disconnected in a pair. The imaginary cut may break the chain in any element of the mechanism. Let it be the k-th element. The position of point M of this element (Figure 5.13) may then be determined using two simple open chains formed by the imaginary cut. As the position of point M is the same in both cases,

$$\mathbf{T}_{12}\mathbf{T}_{23}\ldots\mathbf{T}_{k-1,k}\mathbf{r}_{kM} = \mathbf{T}_{1n}\mathbf{T}_{n,n-1}\ldots\mathbf{T}_{k+1,k}\mathbf{r}_{kM} \tag{5.78}$$

(Brát, 1981). With regard to arbitrary $\mathbf{r}_{kM}$, this gives a condition identical to (5.75). Here we use the statement that, if equation

$$\mathbf{A}\mathbf{r} = \mathbf{B}\mathbf{r} \tag{5.79}$$

is an identity for all vectors $\mathbf{r}$, then

$$\mathbf{A} = \mathbf{B} \tag{5.80}$$

Substituting coordinates of various vectors into equation (5.79), we obtain a system of linear equations for the coefficients of matrix $\mathbf{A}, \mathbf{B}$. From the condition of its solubility, the given statement follows. If equation (5.79) is an identity, we may reduce the vector $\mathbf{r}$. This cannot be done if it holds only for some values of $\mathbf{r}$.

The procedure described here may be modified so that, instead of point M, we choose in the space of body k three non-colinear points A, B, C. For each of these, a condition of the type (5.78) holds. As these are specific points, after substituting into (5.78), we obtain a total of 9 equations for 6 unknown functions z. Of the nine equations, only six are independent. From the viewpoint of geometry, it is possible to interpret this procedure as computation of the position of the ABC plane firmly connected to the body k in basic space. By suitable choice of the position of point A, B, C, we may obtain a simple form of the resultant scalar equations.

Should the points A, B, C belong to one body k, or should they not change their mutual distance, their distances are known constants. Then

$$|\mathbf{r}_{1A} - \mathbf{r}_{1B}| = d_{AB} \qquad |\mathbf{r}_{1A} - \mathbf{r}_{1C}| = d_{AC} \qquad |\mathbf{r}_{1B} - \mathbf{r}_{1C}| = d_{BC} \tag{5.81}$$

where $d_{AB} = \overline{AB}, d_{AC} = \overline{AC}, d_{BC} = \overline{BC}$. In addition to nine equations from (5.78) we obtain another three scalar equations (5.81), giving again 12 equations as in the case of a closed loop.

Should the problem of kinematic analysis contain only one unknown, one of the equations (5.81) would suffice for the solution. This is the case for four element mechanisms, where the input and output elements are linked by a draw rod with two ball journals.

5.4.2 The Matrix Method of Determining Velocities and Accelerations

Various formulations of the position problem may serve as the starting point for the computation of velocities and accelerations. We shall use here the equation (5.75). Its time derivative yields the equation for computing the velocities

$$\begin{aligned} &\sum_{j=1}^{k-1} \mathbf{T}_{12}\mathbf{T}_{23}\dots\mathbf{T}_{j,j+1}\mathbf{V}_{j,j+1}\dots\mathbf{T}_{k-1,k} = \\ = \quad &\sum_{j=n}^{k} \mathbf{T}_{1n}\mathbf{T}_{n,n-1}\dots\mathbf{T}_{j+1,j}\mathbf{V}_{j+1,j}\dots\mathbf{T}_{k+1,k} \end{aligned} \tag{5.82}$$

The time derivative of (5.82) yields for computing the accelerations

$$\begin{aligned}
& \sum_{j=1}^{k-1}(\mathbf{T}_{12}\mathbf{T}_{23}\ldots\mathbf{T}_{j,j+1}\mathbf{A}_{j,j+1}\ldots\mathbf{T}_{k-1,k} + \\
+ \;& \mathbf{T}_{12}\mathbf{T}_{23}\ldots\mathbf{T}_{j,j+1}\mathbf{V}^2_{j,j+1}\ldots\mathbf{T}_{k-1,k} + \\
+ \;& \sum_{\substack{l=1 \\ (l \neq j)}}^{k-1} \mathbf{T}_{12}\mathbf{T}_{23}\ldots\mathbf{T}_{j,j+1}\mathbf{V}_{j,j+1}\ldots\mathbf{T}_{l,l+1}\mathbf{V}_{l,l+1}\ldots\mathbf{T}_{k-1k}) = \\
= \;& \sum_{j=n}^{k}(\mathbf{T}_{1n}\mathbf{T}_{n,n-1}\ldots\mathbf{T}_{j+1,j}\mathbf{A}_{j+1,j}\ldots\mathbf{T}_{k+1,k} + \\
+ \;& \mathbf{T}_{1n}\mathbf{T}_{n,n-1}\ldots\mathbf{T}_{j+1,j}\mathbf{V}^2_{j+1,j}\ldots\mathbf{T}_{k+1,k} + \\
+ \;& \sum_{\substack{l=n \\ (l \neq j)}}^{k} \mathbf{T}_{1n}\mathbf{T}_{n,n-1}\ldots\mathbf{T}_{j+1,j}\mathbf{V}_{j+1,j}\ldots\mathbf{T}_{l+1,l}\mathbf{V}_{l+1,l}\ldots\mathbf{T}_{k+1,k})
\end{aligned} \tag{5.83}$$

Equations (5.82) are linear in velocities $\dot{\mathbf{z}}$, and equations (5.83) are linear in accelerations $\ddot{\mathbf{z}}$. They both always represent twelve scalar equations, only six of which are independent. Due to the complicated relations to which the equations lead, it is necessary to solve them using numerical methods. We shall treat them in Section 5.12. Analytical solution is suitable only in the simplest cases, when it is possible to express dependent coordinates $\mathbf{z}$ from the position equations explicitly. In such cases, however, it is more convenient, instead of using relations (5.82), (5.83), to use directly the time derivative of those coordinates. See also Section 5.8.

5.4.3 The Solution of General Point and Body Motion

After a kinematic solution of a mechanism, all its coordinates and their derivatives are known as time functions. To determine the position, velocity and acceleration of a general point of any body is then a simple matter. For this purpose, all relations derived in Section 4.2.3 may be used. It suffices to determine the chain elements connecting the frame with the body containing the point. For body k,

$$\mathbf{r}_{1M} = \mathbf{T}_{12}\mathbf{T}_{23}\ldots\mathbf{T}_{k-1,k}\mathbf{r}_{kM} = \mathbf{T}_{1k}\mathbf{r}_{kM} \tag{5.84}$$

$$\mathbf{v}_{1M} = \mathbf{T}_{1k}\mathbf{V}_{1k}\mathbf{r}_{kM} \tag{5.85}$$

$$\mathbf{a}_{1M} = \mathbf{T}_{1k}(\mathbf{A}_{1k} + \mathbf{V}^2_{1k})\mathbf{r}_{kM} \tag{5.86}$$

The motion of body k is characterized by matrices $\mathbf{T}_{1k}, \mathbf{V}_{1k}, \mathbf{A}_{1k}$, whose elements determine the position, velocity and acceleration of the reference

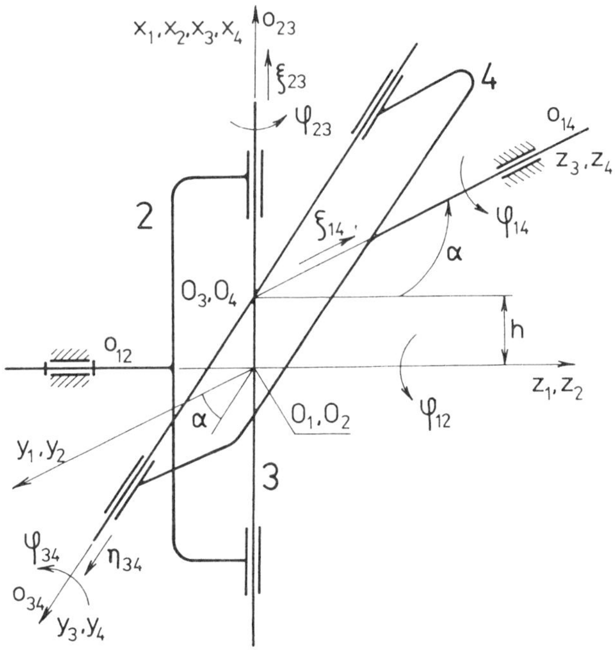

Figure 5.14. A universal joint.

point and angular velocity and angular acceleration of body k in the manner as described in Section 4.2.

5.5 Applications to Simple Mechanisms

5.5.1 A Universal Joint

To clarify the above procedures, we shall now outline the general solution of the kinematics of a universal joint (Brát, 1981).

A universal joint is a simple spatial mechanism which may be used to transfer motion from the driving shaft 2 on to the driven shaft 4 performing a helical motion (Figure 5.14). The two shafts may be skewlined, parallel, concurrent or coaxial. It is a four-element spatial mechanism of the type RCCC, which, apart from special positions, has one degree of freedom. It is formed by one closed loop; its position is described by an independent coordinate, for which we choose the angular displacement φ_{12} of fork 3 relative to frame 1, and six dependent coordinates $\varphi_{23}, \xi_{23}, \varphi_{34}, \eta_{34}, \varphi_{14}, \zeta_{14}$. The structure of the joint is of the following form: shafts 2 and 4 carry perpendicular forks, both forks being conected by a rectangular cross 3. The skewline property of shafts is characterized by constants h, α.

Function $\varphi_{12}(t)$ is given, and the complete kinematic solution is to be obtained. The choice of coordinate systems is evident from Figure 5.14, where the universal joint is drawn in its initial position, for which the values of all coordinates are zero. The axes of individual kinematic pairs are designated by the letter o and the respective index. We shall carry out the solution first using basic matrices, as introduced in Section 4.5.1.

Choosing $k = 4$, we may, according to (5.75), write

$$\mathbf{T}_{12}\mathbf{T}_{23}\mathbf{T}_{34} = \mathbf{T}_{14} \tag{a}$$

where

$$\begin{aligned}
\mathbf{T}_{12} &= \mathbf{T}_{Z6}(\varphi_{12}) && \text{(b1)}\\
\mathbf{T}_{23} &= \mathbf{T}_{Z1}(\xi_{23})\mathbf{T}_{Z4}(\varphi_{23}) && \text{(b2)}\\
\mathbf{T}_{34} &= \mathbf{T}_{Z2}(\eta_{34})\mathbf{T}_{Z5}(\varphi_{34}) && \text{(b3)}\\
\mathbf{T}_{14} &= \mathbf{T}_{Z1}(h)\mathbf{T}_{Z4}(\alpha)\mathbf{T}_{Z3}(\xi_{14})\mathbf{T}_{Z6}(\varphi_{14}) && \text{(b4)}
\end{aligned}$$

Substituting (b) into (a) and performing the multiplication leads to the result which we shall write in the form

$$\mathbf{A} = \mathbf{T}_{12}\mathbf{T}_{23}\mathbf{T}_{34} - \mathbf{T}_{14} = [a_{ij}] \tag{c}$$

Then

$$\begin{aligned}
a_{11} &\equiv \mathrm{c}\varphi_{34}\mathrm{c}\varphi_{12} - \mathrm{s}\varphi_{34}\mathrm{s}\varphi_{23}\mathrm{s}\varphi_{12} - \mathrm{c}\varphi_{14} = 0 && \text{(d1)}\\
a_{12} &\equiv -\mathrm{c}\varphi_{23}\mathrm{s}\varphi_{12} + \mathrm{s}\varphi_{14} = 0 && \text{(d2)}\\
a_{13} &\equiv \mathrm{s}\varphi_{34}\mathrm{c}\varphi_{12} + \mathrm{c}\varphi_{34}\mathrm{s}\varphi_{23}\mathrm{s}\varphi_{12} = 0 && \text{(d3)}\\
a_{14} &\equiv \xi_{23}\mathrm{c}\varphi_{12} - \eta_{34}\mathrm{c}\varphi_{23}\mathrm{s}\varphi_{12} - h = 0 && \text{(d4)}\\
a_{21} &\equiv \mathrm{c}\varphi_{34}\mathrm{s}\varphi_{12} + \mathrm{s}\varphi_{34}\mathrm{s}\varphi_{23}\mathrm{c}\varphi_{12} - \mathrm{s}\varphi_{14}\mathrm{c}\alpha = 0 && \text{(d5)}\\
a_{22} &\equiv \mathrm{c}\varphi_{23}\mathrm{c}\varphi_{12} - \mathrm{c}\varphi_{14}\mathrm{c}\alpha = 0 && \text{(d6)}\\
a_{23} &\equiv \mathrm{s}\varphi_{34}\mathrm{s}\varphi_{12} - \mathrm{c}\varphi_{34}\mathrm{s}\varphi_{23}\mathrm{c}\varphi_{12} + \mathrm{s}\alpha = 0 && \text{(d7)}\\
a_{24} &\equiv \xi_{23}\mathrm{s}\varphi_{12} + \eta_{34}\mathrm{c}\varphi_{23}\mathrm{c}\varphi_{12} + \zeta_{14}\mathrm{s}\alpha = 0 && \text{(d8)}\\
a_{31} &\equiv -\mathrm{s}\varphi_{34}\mathrm{c}\varphi_{23} - \mathrm{s}\varphi_{14}\mathrm{s}\alpha = 0 && \text{(d9)}\\
a_{32} &\equiv \mathrm{s}\varphi_{23} - \mathrm{c}\varphi_{14}\mathrm{s}\alpha = 0 && \text{(d10)}\\
a_{33} &\equiv \mathrm{c}\varphi_{34}\mathrm{c}\varphi_{23} - \mathrm{c}\alpha = 0 && \text{(d11)}\\
a_{34} &\equiv \eta_{34}\mathrm{s}\varphi_{23} - \zeta_{14}\mathrm{c}\alpha = 0 && \text{(d12)}
\end{aligned}$$

This is a system of 12 transcendental equations for six unknown dependent coordinates. In the given case the general solution may be performed using

φ_{14} from equations (d2), (d6), φ_{23} from (d6), (d10), φ_{34} from (d2), (d10) and the rest from (d4), (d8), (d12)

$$\begin{aligned}
\xi_{23} &= \frac{\mathrm{c}\varphi_{12}}{1-\mathrm{s}^2\alpha\mathrm{s}^2\varphi_{12}}h && \text{(e1)}\\
\varphi_{14} &= \operatorname{atan2}(\pm\mathrm{s}\varphi_{12}\mathrm{c}\alpha, \pm\mathrm{c}\varphi_{12}) && \text{(e2)}\\
\eta_{34} &= -\frac{\mathrm{c}\alpha\mathrm{s}\varphi_{12}}{\sqrt{1-\mathrm{s}^2\alpha\mathrm{s}^2\varphi_{12}}}h && \text{(e3)}\\
\varphi_{23} &= \operatorname{atan2}\left(\mathrm{s}\alpha\mathrm{c}\varphi_{14}, \frac{\mathrm{c}\alpha\mathrm{c}\varphi_{14}}{\mathrm{c}\varphi_{12}}\right) && \text{(e4)}\\
\zeta_{14} &= -\frac{\mathrm{s}\alpha\mathrm{s}2\varphi_{12}}{1-\mathrm{s}^2\alpha\mathrm{s}^2\varphi_{12}}h && \text{(e5)}\\
\varphi_{34} &= \operatorname{atan2}(-\mathrm{s}\alpha\mathrm{s}\varphi_{14}\mathrm{c}\varphi_{23}, \mathrm{c}\alpha\mathrm{c}\varphi_{23}) && \text{(e6)}
\end{aligned}$$

The solution of the position problem is completed by finding the functions (e) or by numerically solving equations (d).

If we wish to know, for example, the position of the cross centre S, it can be computed easily from the equation

$$\begin{aligned}
\mathbf{r}_{1S} &= \mathbf{T}_{12}\mathbf{T}_{23}\mathbf{r}_{3S} = \mathbf{T}_{13}\mathbf{r}_{3S} = \\
&= \begin{bmatrix} \mathrm{c}\varphi_{12}, & -\mathrm{s}\varphi_{12}\mathrm{c}\varphi_{23}, & \mathrm{s}\varphi_{12}\mathrm{s}\varphi_{23}, & \xi_{23}\mathrm{c}\varphi_{12} \\ \mathrm{s}\varphi_{12}, & \mathrm{c}\varphi_{12}\mathrm{c}\varphi_{23}, & -\mathrm{c}\varphi_{12}\mathrm{s}\varphi_{23}, & \xi_{23}\mathrm{s}\varphi_{12} \\ 0, & \mathrm{s}\varphi_{23}, & \mathrm{c}\varphi_{23}, & 0 \\ 0, & 0, & 0, & 1 \end{bmatrix} \mathbf{r}_{3S} && \text{(f)}
\end{aligned}$$

where $\mathbf{r}_{1S} = [x_{1S}, y_{1S}, z_{1S}, 1], \mathbf{r}_{3S} = [0, 0, 0, 1]$. We obtain

$$\begin{aligned}
x_{1S} &= \xi_{23}\mathrm{c}\varphi_{12} = \frac{c^2\varphi_{12}}{1-\mathrm{s}^2\alpha\mathrm{s}^2\varphi_{12}}h && \text{(g1)}\\
y_{1S} &= \xi_{23}\mathrm{s}\varphi_{12} = \frac{\mathrm{s}\varphi_{12}\mathrm{c}\ \varphi_{12}}{1-\mathrm{s}^2\alpha\mathrm{s}^2\varphi_{12}}h && \text{(g2)}\\
z_{1S} &= 0 && \text{(g3)}
\end{aligned}$$

In this case the velocities may be computed analytically by differentiating relations (e). We obtain

$$\begin{aligned}
\dot{\xi}_{23} &= \frac{\mathrm{s}^2\alpha c^2\varphi_{12} - c^2\alpha}{(1-\mathrm{s}^2\alpha\mathrm{s}^2\varphi_{12})^2}h\mathrm{s}\varphi_{12}\dot{\varphi}_{12} && \text{(h1)}\\
\dot{\varphi}_{23} &= -\frac{\mathrm{s}\varphi_{12}\mathrm{tg}\alpha}{1+\tan^2\alpha c^2\varphi_{12}}\dot{\varphi}_{12} && \text{(h2)}\\
\dot{\zeta}_{14} &= -\frac{\mathrm{s}\alpha(c^2\varphi_{12} - c^2\alpha\mathrm{s}^2\varphi_{12})}{(1-\mathrm{s}^2\alpha\mathrm{s}^2\varphi_{12})^2}h\dot{\varphi}_{12} && \text{(h3)}
\end{aligned}$$

$$\begin{aligned}
\dot{\varphi}_{14} &= \frac{\mathrm{c}\alpha}{1-\mathrm{s}^2\alpha\mathrm{s}^2\varphi_{12}}\dot{\varphi}_{12} && \text{(h4)}\\
\dot{\eta}_{34} &= \frac{-h\mathrm{c}\alpha\mathrm{c}\varphi_{12}}{\sqrt{(1-\mathrm{s}^2\alpha\mathrm{s}^2\varphi_{12})^3}}\dot{\varphi}_{12} && \text{(h5)}\\
\dot{\varphi}_{34} &= -\frac{\mathrm{s}\alpha\mathrm{c}\varphi_{12}}{\sqrt{1-\mathrm{s}^2\alpha\mathrm{s}^2\varphi_{12}}}\dot{\varphi}_{12} && \text{(h6)}
\end{aligned}$$

Similarly we may derive expressions for the cross 3 mass centre velocity components.

To clarify further, let us present the velocity matrix $\mathbf{V}_{13}$ expressing the velocity of absolute motion of body 3.

$$\begin{aligned}
\mathbf{V}_{3,13} &= \mathbf{T}_{13}^{-1}\dot{\mathbf{T}}_{13} = (\mathbf{T}_{Z6}(\varphi_{12})\mathbf{T}_{Z1}(\xi_{23})\mathbf{T}_{Z4}(\varphi_{23}))^{-1}.\\
&\quad .(\mathbf{T}_{Z6}(\varphi_{12})\mathbf{D}_{Z6}(\dot{\varphi}_{12})\mathbf{T}_{Z1}(\xi_{23})\mathbf{T}_{Z4}(\varphi_{23})+\\
&\quad +\mathbf{T}_{Z6}(\varphi_{12})\mathbf{T}_{Z1}(\xi_{23})\mathbf{D}_{Z1}(\dot{\xi}_{23})\mathbf{T}_{Z4}(\varphi_{23})+\\
&\quad +\mathbf{T}_{Z6}(\varphi_{12})\mathbf{T}_{Z1}(\xi_{23})\mathbf{T}_{Z4}(\varphi_{23})\mathbf{D}_{Z4}(\dot{\varphi}_{23}))=\\
&= \mathbf{T}_{Z4}(-\varphi_{23})\mathbf{T}_{Z1}(-\xi_{23})\mathbf{D}_{Z6}(\dot{\varphi}_{12})\mathbf{T}_{Z1}(\xi_{23})\mathbf{T}_{Z4}(\varphi_{23})+\\
&\quad +\mathbf{T}_{Z4}(-\varphi_{23})\mathbf{D}_{Z1}(\dot{\xi}_{23})\mathbf{T}_{Z4}(\varphi_{23})+\mathbf{D}_{Z4}(\dot{\varphi}_{23})=\\
&= \begin{bmatrix} 0, & -\dot{\varphi}_{12}\mathrm{c}\varphi_{23}, & \dot{\varphi}_{12}\mathrm{s}\varphi_{23}, & \dot{\xi}_{23}\\ \dot{\varphi}_{12}\mathrm{c}\varphi_{23}, & 0, & -\dot{\varphi}_{23}, & \xi_{23}\dot{\varphi}_{12}\mathrm{c}\varphi_{23}\\ -\dot{\varphi}_{12}\mathrm{s}\varphi_{23}, & \dot{\varphi}_{23}, & 0, & -\xi_{23}\dot{\varphi}_{12}\mathrm{s}\varphi_{23}\\ 0, & 0, & 0, & 0 \end{bmatrix} && \text{(i)}
\end{aligned}$$

Hence, for example

$$\begin{aligned}
\boldsymbol{\omega}_{3,13} &= [\dot{\varphi}_{23}, \dot{\varphi}_{12}\mathrm{s}\varphi_{23}, \dot{\varphi}_{12}\mathrm{c}\varphi_{23}] && \text{(j)}\\
\boldsymbol{\omega}_{1,13} &= \mathbf{S}_{13}\boldsymbol{\omega}_{3,13} = [\dot{\varphi}_{23}\mathrm{c}\varphi_{12}, \dot{\varphi}_{23}\mathrm{s}\varphi_{12}, \dot{\varphi}_{12}] && \text{(k)}\\
\dot{\mathbf{u}}_{3,13} &= [\dot{\xi}_{23}, \xi_{23}\dot{\varphi}_{12}\mathrm{c}\varphi_{23}, -\xi_{23}\dot{\varphi}_{12}\mathrm{s}\varphi_{23}] && \text{(l)}
\end{aligned}$$

Differentiating again with respect to time we may obtain expressions for acceleration and the acceleration matrix. We present only the quantities concerning the output element:

$$\begin{aligned}
\ddot{\xi}_{14} &= -h\mathrm{s}\alpha\left(\frac{\mathrm{c}^2\varphi_{12}-\mathrm{c}^2\alpha\mathrm{s}^2\varphi_{12}}{(1-\mathrm{s}^2\alpha\mathrm{s}^2\varphi_{12})^2}\ddot{\varphi}_{12}+2\frac{\mathrm{s}\varphi_{12}\mathrm{c}\varphi_{12}}{(1-\mathrm{s}^2\alpha\mathrm{s}^2\varphi_{12})^3}.\right.\\
&\quad \left..(-1-\mathrm{c}^2\alpha+2\mathrm{s}^2\alpha(\mathrm{c}^2\varphi_{12}-\mathrm{c}^2\alpha\mathrm{s}^2\varphi_{12}))\dot{\varphi}_{12}^2\right) && \text{(m)}\\
\ddot{\varphi}_{14} &= 2\mathrm{s}^2\alpha\mathrm{c}\alpha\mathrm{s}\varphi_{12}\mathrm{c}\varphi_{12}\frac{1}{(1-\mathrm{s}^2\alpha\mathrm{s}^2\varphi_{12})^2}\dot{\varphi}_{12}^2+\\
&\quad +\frac{\mathrm{c}\alpha}{1-\mathrm{s}^2\alpha\mathrm{s}^2\varphi_{12}}\ddot{\varphi}_{12} && \text{(n)}
\end{aligned}$$

and matrix

$$\mathcal{A}_{13} = \begin{bmatrix} 0, & \mathrm{s}\varphi_{23}\dot{\varphi}_{23}\dot{\varphi}_{12} - \mathrm{c}\varphi_{23}\ddot{\varphi}_{12}, & \mathrm{c}\varphi_{23}\dot{\varphi}_{23}\dot{\varphi}_{12} + \mathrm{s}\varphi_{23}\ddot{\varphi}_{12} \\ -\mathrm{s}\varphi_{23}\dot{\varphi}_{23}\dot{\varphi}_{12} + \mathrm{c}\varphi_{23}\ddot{\varphi}_{12}, & 0, & -\ddot{\varphi}_{23} \\ -\mathrm{c}\varphi_{23}\dot{\varphi}_{23}\dot{\varphi}_{12} - \mathrm{s}\varphi_{23}\ddot{\varphi}_{12}, & \ddot{\varphi}_{23}, & 0 \end{bmatrix} \quad \text{(o)}$$

Hence

$$\boldsymbol{\alpha}_{3,13} = [\ddot{\varphi}_{23}, \mathrm{c}\varphi_{23}\dot{\varphi}_{23}\dot{\varphi}_{12} + \mathrm{s}\varphi_{23}\ddot{\varphi}_{12}, -\mathrm{s}\varphi_{23}\dot{\varphi}_{23}\dot{\varphi}_{12} + \mathrm{c}\varphi_{23}\ddot{\varphi}_{12}] \quad \text{(p)}$$

$$\boldsymbol{\alpha}_{1,13} = [\mathrm{c}\varphi_{12}\ddot{\varphi}_{23} - \mathrm{s}\varphi_{12}\dot{\varphi}_{23}\dot{\varphi}_{12}, \mathrm{s}\varphi_{12}\ddot{\varphi}_{23} + \mathrm{c}\varphi_{12}\dot{\varphi}_{23}\dot{\varphi}_{12}, \ddot{\varphi}_{12}] \quad \text{(q)}$$

The above results can equally well be obtained using the basic quaternions introduced in Section 4.5.3. According to (4.202)

$$\hat{\mathbf{K}}_{12}\hat{\mathbf{K}}_{23}\hat{\mathbf{K}}_{34} = \hat{\mathbf{K}}_{14} \quad \text{(r)}$$

where, using (4.270),

$$\begin{aligned} \hat{\mathbf{K}}_{12} &= \hat{\mathbf{K}}_{Z6}(\varphi_{12}) && \text{(s1)} \\ \hat{\mathbf{K}}_{23} &= \hat{\mathbf{K}}_{Z1}(\xi_{23})\hat{\mathbf{K}}_{Z4}(\varphi_{23}) && \text{(s2)} \\ \hat{\mathbf{K}}_{34} &= \hat{\mathbf{K}}_{Z2}(\eta_{34})\hat{\mathbf{K}}_{Z5}(\varphi_{34}) && \text{(s3)} \\ \hat{\mathbf{K}}_{14} &= \hat{\mathbf{K}}_{Z1}(h)\hat{\mathbf{K}}_{Z4}(\alpha)\hat{\mathbf{K}}_{Z3}(\zeta_{14})\hat{\mathbf{K}}_{Z6}(\varphi_{14}) && \text{(s4)} \end{aligned}$$

Substituting (s) into (r) and multiplying yields, by comparing elements in the same place in quaternions of the left-hand and right-hand sides of the equation,

$$\begin{aligned} \mathrm{c}\frac{\varphi_{12}}{2}\mathrm{c}\frac{\varphi_{23}}{2}\mathrm{c}\frac{\varphi_{34}}{2} - \mathrm{s}\frac{\varphi_{12}}{2}\mathrm{s}\frac{\varphi_{23}}{2}\mathrm{s}\frac{\varphi_{34}}{2} &= \mathrm{c}\frac{\varphi_{14}}{2}\mathrm{c}\frac{\alpha}{2} && \text{(t1)} \\ \mathrm{c}\frac{\varphi_{12}}{2}\mathrm{s}\frac{\varphi_{23}}{2}\mathrm{c}\frac{\varphi_{34}}{2} - \mathrm{s}\frac{\varphi_{12}}{2}\mathrm{c}\frac{\varphi_{23}}{2}\mathrm{s}\frac{\varphi_{34}}{2} &= \mathrm{c}\frac{\varphi_{14}}{2}\mathrm{s}\frac{\alpha}{2} && \text{(t2)} \\ \mathrm{s}\frac{\varphi_{12}}{2}\mathrm{s}\frac{\varphi_{23}}{2}\mathrm{c}\frac{\varphi_{34}}{2} + \mathrm{c}\frac{\varphi_{12}}{2}\mathrm{c}\frac{\varphi_{23}}{2}\mathrm{s}\frac{\varphi_{34}}{2} &= -\mathrm{s}\frac{\varphi_{14}}{2}\mathrm{s}\frac{\alpha}{2} && \text{(t3)} \\ \mathrm{s}\frac{\varphi_{12}}{2}\mathrm{c}\frac{\varphi_{23}}{2}\mathrm{c}\frac{\varphi_{34}}{2} + \mathrm{c}\frac{\varphi_{12}}{2}\mathrm{s}\frac{\varphi_{23}}{2}\mathrm{s}\frac{\varphi_{34}}{2} &= \mathrm{s}\frac{\varphi_{14}}{2}\mathrm{c}\frac{\alpha}{2} && \text{(t4)} \\ -\eta_{34}\mathrm{c}\varphi_{23}\mathrm{s}\varphi_{12} + \xi_{23}\mathrm{c}\varphi_{12} &= h && \text{(t5)} \\ \eta_{34}\mathrm{c}\varphi_{23}\mathrm{c}\varphi_{12} + \xi_{23}\mathrm{s}\varphi_{12} &= -\xi_{14}\mathrm{s}\alpha && \text{(t6)} \\ \eta_{34}\mathrm{s}\varphi_{23} &= \xi_{14}\mathrm{c}\alpha && \text{(t7)} \end{aligned}$$

The last three equations are identical in both methods, as they express directly the elements of radius vector $\mathbf{r}_{14}$. There exists an unambiguous correspondence between the first nine matrix method equations and first four quaternion method equations expressed by equation (4.171) or (4.177).

In analogy to expression (i) [page 141] the velocity quaternion may be computed using (4.274)

$$\hat{\mathbf{W}}_{13}^{R} = \hat{\mathbf{K}}_{Z4}(-\varphi_{23})\hat{\mathbf{K}}_{Z1,\varepsilon}(-\xi_{23})\hat{\mathbf{L}}_{Z6}^{R}(\dot{\varphi}_{12})\hat{\mathbf{K}}_{Z1}(\xi_{23})\hat{\mathbf{K}}_{Z4}(\varphi_{23}) + \\ +\hat{\mathbf{K}}_{Z4}(-\varphi_{23})\hat{\mathbf{L}}_{Z1}^{R}(\dot{\xi}_{23})\hat{\mathbf{K}}_{Z4}(\varphi_{23}) + \mathbf{L}_{Z4}^{R}(\dot{\varphi}_{23}) \qquad \text{(u)}$$

and the relations (j) to (l) may thus be obtained. The same holds in a similar way for acceleration.

Finally we may obtain the above results using basic screws. From the three screw descriptions in Section 4.54 we shall use the screw affinors

$$\mathbf{A}_{12}\mathbf{A}_{23}\mathbf{A}_{34} = \mathbf{A}_{14} \qquad \text{(v)}$$

where

$$\begin{aligned} \mathbf{A}_{12} &= \mathbf{A}_{Z6}(\varphi_{12}) \\ \mathbf{A}_{23} &= \mathbf{A}_{Z1}(\xi_{23})\mathbf{A}_{Z4}(\varphi_{23}) \\ \mathbf{A}_{34} &= \mathbf{A}_{Z2}(\eta_{34})\mathbf{A}_{Z5}(\varphi_{34}) \\ \mathbf{A}_{14} &= \mathbf{A}_{Z1}(h)\mathbf{A}_{Z4}(\alpha)\mathbf{A}_{Z3}(\zeta_{14})\mathbf{A}_{Z6}(\varphi_{14}) \end{aligned} \qquad \text{(w)}$$

Using (4.275) and substituting (w) into (v) we obtain nine dual equations representing 18 scalar equations. (w) may, however, be modified using the concept of dual angle. As may be simply proved by substituting from (4.275)

$$\begin{aligned} \mathbf{A}_{12} &= \mathbf{A}_{Z6}(\varphi_{12}) \\ \mathbf{A}_{23} &= \mathbf{A}_{Z4}(\varphi_{23} + \varepsilon\xi_{23}) \\ \mathbf{A}_{34} &= \mathbf{A}_{Z5}(\varphi_{34} + \varepsilon\eta_{34}) \\ \mathbf{A}_{14} &= \mathbf{A}_{Z4}(\alpha + \varepsilon h)\mathbf{A}_{Z6}(\varphi_{14} + \varepsilon\xi_{14}) \end{aligned} \qquad \text{(x)}$$

The computation after substituting (x) into (v) is now simpler. We may in this case, however, use equations (d1) to (d9), where we substitute the respective dual angles, as equation (v) after substituting from (x) has the same structure as the direction cosines submatrix in (a). We obtain finally from (d1) to (d9) 9 dual equations

$$\begin{aligned} (\mathrm{c}\varphi_{34} - \varepsilon\eta_{34}\mathrm{s}\varphi_{34})\mathrm{c}\varphi_{12} &- (\mathrm{s}\varphi_{34} + \varepsilon\eta_{34}\mathrm{c}\varphi_{34})(\mathrm{s}\varphi_{23} + \varepsilon\xi_{23}\mathrm{c}\varphi_{23})\mathrm{s}\varphi_{12} = \\ &= \mathrm{c}\varphi_{14} - \varepsilon\zeta_{14}\mathrm{s}\varphi_{14} \end{aligned} \qquad \text{(y1)}$$

$$(\mathrm{c}\varphi_{34} - \varepsilon\eta_{34}\mathrm{s}\varphi_{34}) \ \mathrm{s}\varphi_{12} + (\mathrm{s}\varphi_{34} + \varepsilon\eta_{34}\mathrm{c}\varphi_{34})(\mathrm{s}\varphi_{23} + \varepsilon\xi_{23}\mathrm{c}\varphi_{23})\mathrm{c}\varphi_{12} =$$

$$
\begin{aligned}
&= (s\varphi_{14} + \varepsilon\zeta_{14}c\varphi_{14})(c\alpha - \varepsilon h s\alpha) && \text{(y2)}\\
&\quad -(s\varphi_{34} + \varepsilon\eta_{34}c\varphi_{34})(c\varphi_{23} - \varepsilon\xi_{23}s\varphi_{23}) = \\
&= (s\varphi_{14} + \varepsilon\zeta_{14}c\varphi_{14})(s\alpha + \varepsilon h c\alpha) && \text{(y3)}
\end{aligned}
$$

$$
\begin{aligned}
-(c\varphi_{23} - \varepsilon\xi_{23}s\varphi_{23})s\varphi_{12} &= -(s\varphi_{14} + \varepsilon\zeta_{14}c\varphi_{14}) && \text{(y4)}\\
(c\varphi_{23} - \varepsilon\xi_{23}s\varphi_{23})c\varphi_{12} &= (c\varphi_{14} - \varepsilon\zeta_{14}s\varphi_{14})(c\alpha - \varepsilon h s\alpha) && \text{(y5)}\\
s\varphi_{23} + \varepsilon\xi_{23}c\varphi_{23} &= (c\varphi_{14} - \varepsilon\zeta_{14}s\varphi_{14})(s\alpha + \varepsilon h s\alpha) && \text{(y6)}\\
(s\varphi_{34} + \varepsilon\eta_{34}c\varphi_{34})c\varphi_{12} + (c\varphi_{34} &- \varepsilon\eta_{34}s\varphi_{34})(s\varphi_{23} + \varepsilon\xi_{23}c\varphi_{23})s\varphi_{12} = \\
&= 0 && \text{(y7)}\\
(s\varphi_{34} + \varepsilon\eta_{34}c\varphi_{34})s\varphi_{12} - (c\varphi_{34} &- \varepsilon\eta_{34}s\varphi_{34})(s\varphi_{23} + \varepsilon\xi_{23}c\varphi_{23})c\varphi_{12} = \\
&= -(s\alpha + \varepsilon h c\alpha) && \text{(y8)}\\
(c\varphi_{34} - \varepsilon\eta_{34}s\varphi_{34})(c\varphi_{23} &- \varepsilon\xi_{23}s\varphi_{23}) = (c\alpha - \varepsilon h t\alpha) && \text{(y9)}
\end{aligned}
$$

These equations may be used to solve the position problem; for the solution of the velocity and acceleration problems, differentiation with respect to time would have to be performed.

5.5.2 RRSRR Mechanism

In solving the position problem of an RRSRR mechanism according to Figure 5.15, only the functions $\varphi_{23}(\varphi_{12}), \varphi_{54}(\varphi_{12}), \varphi_{15}(\varphi_{12})$ are of interest, in which the angular displacement φ_{12} of the driving element is a known function of time (Brát, 1981).

The formulation of the problem enables us to use the disconnected loop method in its simplest form, i.e. disconnecting in the spherical pair S. According to (5.76),

$$\mathbf{T}_{12}\mathbf{T}_{23}\mathbf{r}_{3S} = \mathbf{T}_{15}\mathbf{T}_{54}\mathbf{r}_{4S} \tag{a}$$

where

$$
\begin{aligned}
\mathbf{T}_{12} &= \mathbf{T}_{Z2}(-a)\mathbf{T}_{Z5}(\varphi_{12}) && \text{(b1)}\\
\mathbf{T}_{23} &= \mathbf{T}_{Z4}(\varphi_{23}) && \text{(b2)}\\
\mathbf{T}_{15} &= \mathbf{T}_{Z4}(\alpha)\mathbf{T}_{Z2}(b)\mathbf{T}_{Z5}(\varphi_{15}) && \text{(b3)}\\
\mathbf{T}_{54} &= \mathbf{T}_{Z4}((\pi/2) - \varphi_{54}) && \text{(b4)}\\
\mathbf{r}_{3S} &= [0, r, 0, 1] && \text{(b5)}\\
\mathbf{r}_{4S} &= [0, 0, d, 1] && \text{(b6)}
\end{aligned}
$$

Substituting (b) into (a) yields a system of three transcendental equations

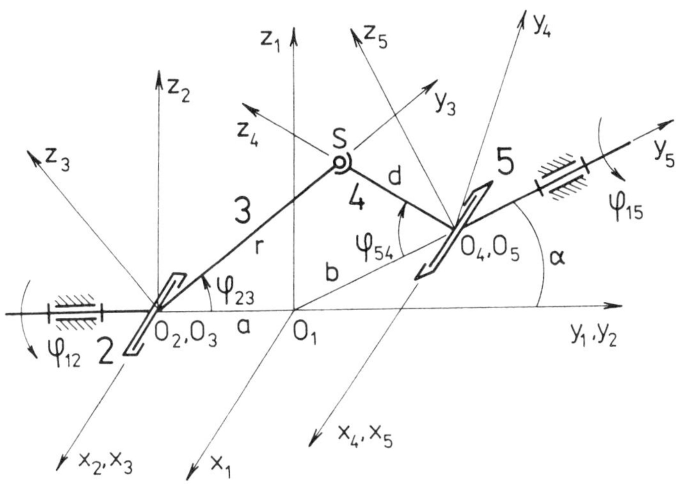

Figure 5.15. RRSRR mechanism.

$$
\begin{aligned}
rs\varphi_{23}s\varphi_{12} &= ds\varphi_{54}s\varphi_{15} && \text{(c1)}\\
rc\varphi_{23} - a &= (b - dc\varphi_{54})c\alpha - ds\varphi_{54}c\varphi_{15}s\alpha && \text{(c2)}\\
rs\varphi_{23}c\varphi_{12} &= (b - dc\varphi_{54})s\alpha + ds\varphi_{54}c\varphi_{15}c\alpha && \text{(c3)}
\end{aligned}
$$

for the three wanted functions.

In this case the position problem may also be solved analytically. Adding the squares of the left-hand and right-hand sides of equations (c) we obtain

$$r^2 + a^2 - 2arc\varphi_{23} = b^2 + d^2 - 2bdc\varphi_{54} \qquad \text{(c4)}$$

Multiplying (c2) and (c3) by $c\alpha$ and $s\alpha$ respectively, and adding, yields

$$(rc\varphi_{23} - a)c\alpha + rs\varphi_{23}c\varphi_{12}s\alpha = b - dc\varphi_{54} \qquad \text{(c5)}$$

Eliminating elements containing $c\varphi_{54}$ from (c4–c5) and modifying, we obtain

$$2(a + bc\alpha)rc\varphi_{23} + 2brc\varphi_{12}s\alpha s\varphi_{23} = a^2 + b^2 + r^2 - d^2 + 2abc\alpha \qquad \text{(c6)}$$

Denoting

$$A = 2(a + bc\alpha)r, \quad B = 2brc\varphi_{12}s\alpha,$$
$$C = a^2 + b^2 + r^2 - d^2 + 2abc\alpha$$

and rewriting (c6), we obtain

$$Ac\varphi_{23} + Bs\varphi_{23} = C$$

$$(A\mathrm{c}\varphi_{23} - C)^2 = B^2\mathrm{s}^2\varphi_{23} = B^2(1 - \mathrm{c}^2\varphi_{23})$$
$$(A^2 + B^2)\mathrm{c}^2\varphi_{23} - 2AC\mathrm{c}\varphi_{23} + C^2 - B^2 = 0$$
$$(\mathrm{c}\varphi_{23})_{1,2} = \frac{AC \pm \sqrt{A^2C^2 - (A^2 + B^2)(C^2 - B^2)}}{A^2 + B^2}$$
$$\mathrm{s}\varphi_{23} = \frac{C - A\mathrm{c}\varphi_{23}}{B}$$
$$\varphi_{23} = \mathrm{atan2}\left(\frac{C - A\mathrm{c}\varphi_{23}}{B}, \mathrm{c}\varphi_{23}\right)$$

From (c4)

$$\begin{aligned} \mathrm{c}\varphi_{54} &= \frac{b^2 + d^2 - r^2 - a^2 + 2ar\mathrm{c}\varphi_{23}}{2bd} \\ \mathrm{s}\varphi_{54} &= k\sqrt{1 - \mathrm{c}^2\varphi_{54}} \\ \varphi_{54} &= \mathrm{atan2}(k\sqrt{1 - \mathrm{c}^2\varphi_{54}}, \mathrm{c}\varphi_{54}) \end{aligned}$$

where $k = \pm 1$ and from (c1), (c3)

$$\varphi_{15} = \mathrm{atan2}(kr\mathrm{s}\varphi_{23}\mathrm{s}\varphi_{12}\mathrm{c}\alpha, k(r\mathrm{s}\varphi_{23}\mathrm{c}\varphi_{12} - (b - d\mathrm{c}\varphi_{54})\mathrm{s}\alpha))$$

The time derivative of (a) yields a system of three equations for the computation of $\dot\varphi_{23}, \dot\varphi_{54}, \dot\varphi_{15}$. Their time derivative again yields a system of linear equations for $\ddot\varphi_{23}, \ddot\varphi_{54}, \ddot\varphi_{15}$.

In the special case of $a = b$, $r = d$ the mechanism in question may serve as a homokinetic joint.

The attempt to generalize previous results leads to the question of analytical solubility of all simple mechanisms with one degree of freedom. The most general simple mechanism with one degree of freedom is the seven-element spatial mechanism where all pairs are revolute pairs, i.e. 7R-mechanism, see Figure 5.25. All simple spatial, spherical and planar mechanisms may be derived from this mechanism using special sizes or by changing a revolute pair into a prismatic pair. Its analytical solution has been until now an unsolved problem. This solution can be transformed (Lee and Liang, 1988) to finding roots of a polynomial of degree 16 which, however, cannot be decreased because there exist 16 different configurations (Lee et al., 1992).

We shall show in Section 5.13 the sufficient conditions for the analytical solution of a simple mechanism. These provide the analytical solution of a universal joint, as its three cylindrical pairs contain three translations; likewise, an RRSRR mechanism is analytically soluble, as it contains a spherical pair.

5.6 Composed Mechanisms

In this section we intend only to show that the matrix method yields a sufficient number of equations for the solution of the kinematics of a composed mechanism.

This is evident from the following inference. Let the composed mechanism being examined have m degrees of freedom and let it consist of l independent loops. Its position is then described by m independent and $6l$ dependent coordinates. Independent coordinates and perhaps their derivatives must be given; the task is to compute the $6l$ dependent coordinates. Assigning a simple chain to each loop we obtain, according to the results of Section 5.4, l matrix equations of position each of which contains 6 independent scalar equations. In non-singular cases we have exactly the right number of equations required to compute dependent coordinates.

To compute velocities and accelerations we use a similar method applying the results of Section 5.4. We examine the motion of significant points and bodies using the chain connecting the respective element with the frame in the shortest way. We use for this purpose the relations derived in Section 5.2.

Should the composed mechanism be a mixed chain, we examine first the loop motions as above. The coordinates of all connected open chains are independent and must be given. The motion of the whole complex mechanism is thus determined.

5.7 Applications to Composed Mechanisms

As an example of the kinematic solution of composed mechanisms, let us present the assembly of basic equations of motion of the front wheel suspension of a car and a robot mechanism.

5.7.1 A Wheel Suspension Mechanism

Let us solve the position problem of the spatial mechanism of the front wheel suspension of a passenger car. The kinematic diagram of the mechanism is shown in Figure 5.16. We have a composed mechanism formed by two loops, namely $O_1AO_3LBO_1$ and $O_1AO_3O_5KO_6RO_1$. In each of the two loops we may find a spherical pair, which makes solution according to (5.76) possible. In this case we are interested in the motion of significant points M, N as a function of the suspension characterized by angle φ_{12} and in the angular displacement of the steering arm given by angle φ_{17}.

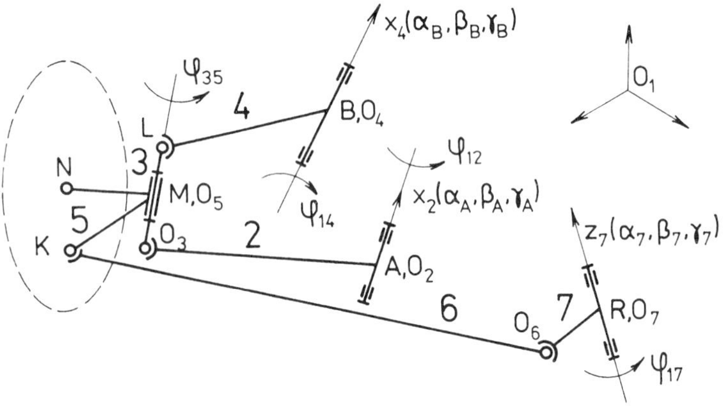

Figure 5.16. Front wheel suspension mechanism.

We shall describe the composition and position of the mechanism by two matrix equations. For the first loop disconnected at joint L

$$\mathbf{T}_{12}\mathbf{T}_{23}\mathbf{r}_{3L} = \mathbf{T}_{14}\mathbf{r}_{4L} \tag{a}$$

For the second loop disconnected at point K, similarly

$$\mathbf{T}_{12}\mathbf{T}_{23}\mathbf{T}_{35}\mathbf{r}_{5K} = \mathbf{T}_{17}\mathbf{T}_{76}\mathbf{r}_{6K} \tag{b}$$

Equations (a), (b) represent six scalar equations for six dependent coordinates, namely $\varphi_{14}, \varphi_{35}$, and four angles characterizing the position of bodies 3 and 6. After computing them we determine the position of points M, N from the relations

$$\mathbf{r}_{1M} = \mathbf{T}_{12}\mathbf{T}_{23}\mathbf{T}_{35}\mathbf{r}_{5M} \tag{c}$$
$$\mathbf{r}_{1N} = \mathbf{T}_{12}\mathbf{T}_{23}\mathbf{T}_{35}\mathbf{r}_{5N} \tag{d}$$

In both cases, before starting to compute, it is necessary to express transformation matrices of individual partial motions as a function of the dimensions of the system in question and on the chosen coordinates. For this purpose we utilize the basic matrices introduced in Section 4.5.1. Their substitution into position equations leads us to the idea of formulating the whole description in basic matrices only. We shall deal with the development of this idea in Section 5.8.

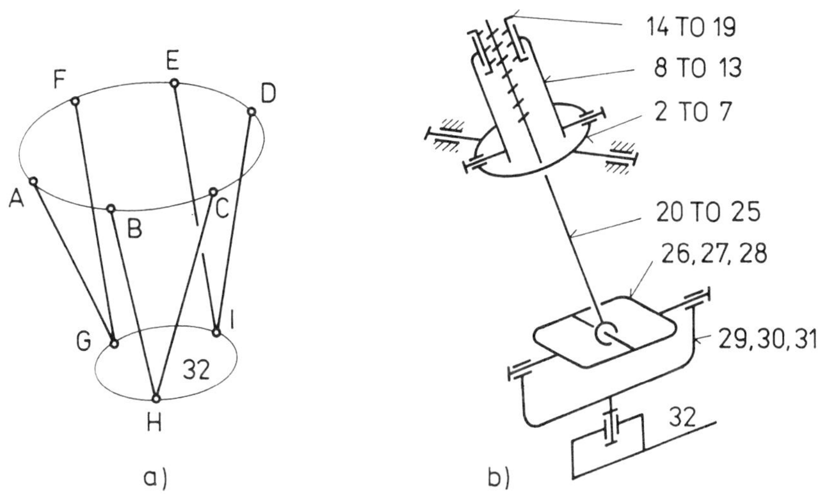

Figure 5.17a,b. Diagram of parallel robot.

5.7.2 A Robot Mechanism

Figure 5.17a shows the simplified constructional diagram of the parallel robot based on the idea of Stewart platform (Stewart, 1965). It is in principle a gripper suspended on six rods. The driving units controlling the motion of six rods with the help of a nut and screw are shown in more detail for one of the rods in Fig. 5.17b, where also the numerical designation of individual elements is given; they are seated in points A to F on the basic frame. The bridles 26, 29 placed in points G, H, I are always shared by two rods. The complete composition of the whole assembly is evident from the associated diagram in Figure 5.17c, where the designation of elements and kinematic pair types is given. The mechanism is composed of 32 bodies constrained by 6 helical and 30 revolute kinematic pairs. Using relation (3.1) we obtain for this system the value of 6 degrees of freedom in the non-singular case. As the system contains only kinematic pairs allowing motion with one degree of freedom, we use for the coordinates the designation φ_{ij}, where i, j are the numerical designations of two bodies constrained by the kinematic pair.

Let us assemble the equations for the solution of the position problem. Using relation (3.5) we find that the examined MBS contains independent loops. They are designated in diagram 5.17c. The position problem therefore requires the forming of five matrix equations of the type (5.75). Due to the symmetry of the case and to save space, we present only the equations

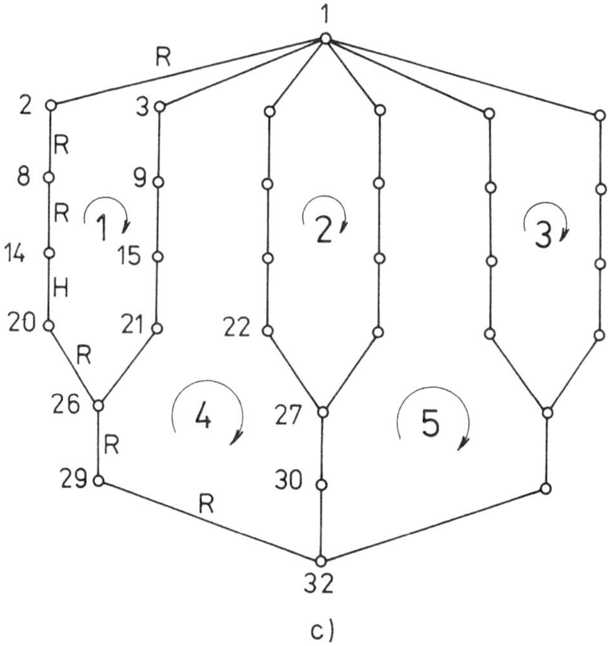

Figure 5.17c. Graph representation of parallel robot.

for loops 1 and 4. We shall write them using the designation as in Figure 5.17d. For loop 1

$$\mathbf{T}_{0102}\mathbf{T}_{0208}\mathbf{T}_{0814}\mathbf{T}_{1420}\mathbf{T}_{2026} = \mathbf{T}_{0103}\mathbf{T}_{0309}\mathbf{T}_{0915}\mathbf{T}_{1521}\mathbf{T}_{2126} \tag{a}$$

where, by indexing, the symbol i is represented by the first number pair in the subscript and symbol j by the second number pair in the subscript. Further,

$$\begin{aligned}
\mathbf{T}_{0102} &= \mathbf{T}_{Z1}(d_1)\mathbf{T}_{Z5}(\varphi_{0102}) \\
\mathbf{T}_{0208} &= \mathbf{T}_{Z4}(\varphi_{0208}) \\
\mathbf{T}_{0814} &= \mathbf{T}_{Z6}(\varphi_{0814}) \\
\mathbf{T}_{1420} &= \mathbf{T}_{Z6}(\beta_3)\mathbf{T}_{Z6}(\varphi_{1420})\mathbf{T}_{Z3}(z_{1420})\mathbf{T}_{Z3}(-d_2) \\
\mathbf{T}_{2026} &= \mathbf{T}_{Z4}(\varphi_{2026})\mathbf{T}_{Z4}(-\beta_4) \\
\mathbf{T}_{0103} &= \mathbf{T}_{Z6}(\pi/3)\mathbf{T}_{Z1}(d_1)\mathbf{T}_{Z5}(\varphi_{0103}) \\
\mathbf{T}_{0309} &= \mathbf{T}_{Z4}(\varphi_{0309}) \\
\mathbf{T}_{0915} &= \mathbf{T}_{Z6}(\varphi_{0915}) \\
\mathbf{T}_{1521} &= \mathbf{T}_{Z6}(-\beta_3)\mathbf{T}_{Z6}(\varphi_{1521})\mathbf{T}_{Z3}(z_{1521})\mathbf{T}_{Z3}(-d_2)
\end{aligned}$$

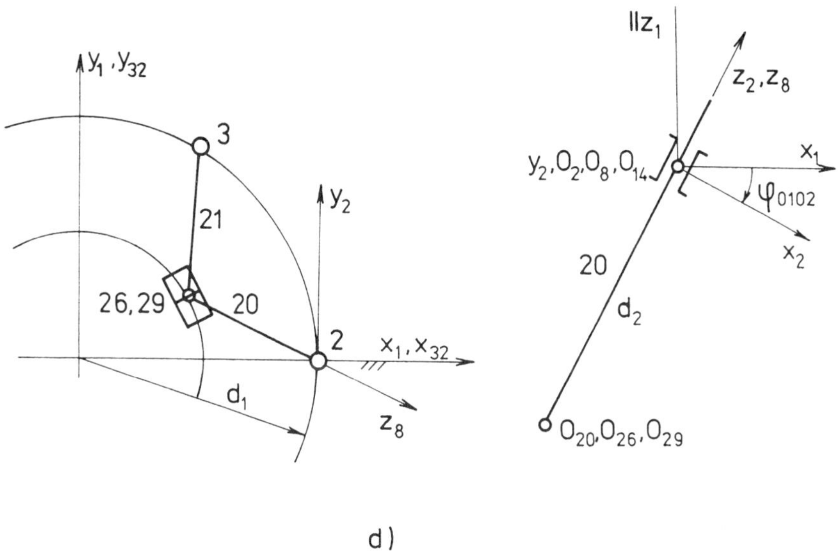

Figure 5.17d. Coordination systems.

$$\mathbf{T}_{2126} = \mathbf{T}_{Z4}(\varphi_{2126})\mathbf{T}_{Z4}(\beta_4)$$

In the above expressions we used for the indexing of coordinates the same method used for indexing transformation matrices. Further

$$z_{1420} = \varkappa\varphi_{1420}, \quad z_{1521} = \varkappa\varphi_{1521}$$

where $\varkappa$ is the helix parameter. For loop 4

$$\mathbf{T}_{0121}\mathbf{T}_{2126}\mathbf{T}_{2629}\mathbf{T}_{2932} = \mathbf{T}_{0122}\mathbf{T}_{2227}\mathbf{T}_{2730}\mathbf{T}_{3032} \tag{b}$$

where $\mathbf{T}_{0121} = \mathbf{T}_{0103}\mathbf{T}_{0309}\mathbf{T}_{0915}\mathbf{T}_{1521}$. Similarly, $\mathbf{T}_{0122}$ and the rest of the transformation matrices can be obtained:

$$\begin{aligned}
\mathbf{T}_{2629} &= \mathbf{T}_{Z5}(\varphi_{2629}) \\
\mathbf{T}_{2932} &= \mathbf{T}_{Z6}(\varphi_{2932})\mathbf{T}_{Z3}(-d_4)\mathbf{T}_{Z1}(-d_3)\mathbf{T}_{Z6}(-\pi/6) \\
\mathbf{T}_{2730} &= \mathbf{T}_{Z5}(\varphi_{2730}) \\
\mathbf{T}_{3032} &= \mathbf{T}_{Z6}(\varphi_{3032})\mathbf{T}_{Z3}(-d_4)\mathbf{T}_{Z1}(-d_3)\mathbf{T}_{Z6}(-5\pi/6)
\end{aligned}$$

In a similar way, two further equations of type (a) and one equation of type (b) may be assembled, giving altogether 5 matrix equations, representing in the general case 30 independent scalar equations. They contain 6 independent coordinates $\varphi_{0814}, \varphi_{0915}, \varphi_{1016}, \varphi_{1117}, \varphi_{1218}, \varphi_{1319}$ and 30 dependent coordinates. The position problem is soluble.

5.8 The Basic Matrix Method

Every transformation matrix may be expressed as the product of the basic matrices introduced in Section 4.5.1. We may therefore express relations (4.50), (4.59) and (4.82) for point motion and relations (4.51), (4.61) and (4.86) for body motion, using only basic matrices and their time derivatives, without distinguishing transformation matrices expressing relative motion of neighbouring bodies.

Position expressions written as basic matrices express the resultant position of a certain coordinate system body as a sequence of imagined basic motions from the initial to the final position regardless of whether or not these motions are realized by a kinematic pair.

In this way we obtain for the motion of point M in basic space

$$\mathbf{r}_{1M} = \prod_{j=1}^{N} \mathbf{T}_{Zk_j}(p_j)\mathbf{r}_{nM} \tag{5.87}$$

for its velocity

$$\mathbf{v}_{1M} = \dot{\mathbf{r}}_{1M} = \sum_{j=1}^{N} \mathbf{F}_j \mathbf{E}(\dot{p}_j)\mathbf{r}_{nM} + \prod_{j=1}^{N} \mathbf{T}_{Zkj}(p_j)\dot{\mathbf{r}}_{nM} \tag{5.88}$$

and acceleration

$$\begin{aligned}\mathbf{a}_{1M} = \ddot{\mathbf{r}}_{1M} &= \sum_{j=1}^{N}\left(\mathbf{F}_j\mathbf{E}(\ddot{p}_j) + \sum_{l=1}^{N}\mathbf{F}_{jl}\mathbf{E}(\dot{p}_j)\mathbf{E}(\dot{p}_l)\right)\mathbf{r}_{nM} + \\ &+ 2\sum_{j=1}^{N}\mathbf{F}_j\mathbf{E}(\dot{p}_j)\dot{\mathbf{r}}_{nM} + \prod_{j=1}^{N}\mathbf{T}_{Zk_j}(p_j)\ddot{\mathbf{r}}_{nM}\end{aligned} \tag{5.89}$$

where

$$\mathbf{F}_j = \mathbf{T}_{Zk_1}(p_1)\mathbf{T}_{Zk_2}(p_2)\dots\mathbf{T}_{Zk_j}(p_j)\mathbf{D}_{Zk_j}(1)\dots\mathbf{T}_{Zk_N}(p_N) \tag{5.90}$$

$$\begin{aligned}\mathbf{F}_{jl} &= \mathbf{T}_{Zk_1}(p_1)\mathbf{T}_{Zk_2}(p_2)\dots\mathbf{T}_{Zk_j}(p_j)\mathbf{D}_{Zk_j}(1)\mathbf{T}_{Zk_{j+1}}(p_{j+1})\dots \\ &\dots\mathbf{T}_{Zk_l}(p_l)\mathbf{D}_{Zk_l}(1)\mathbf{T}_{Zk_{l+1}}(p_{l+1})\dots\mathbf{T}_{Zk_N}(p_N)\end{aligned} \tag{5.91}$$

and $\mathbf{E}(\dot{p}_j)$, $\mathbf{E}(\ddot{p}_j)$ are derivatives of (4.253) with respect to time. The above holds when $j \neq l$. When $j = l$,

$$\mathbf{F}_{jj} = \mathbf{T}_{Zk_1}(p_1)\mathbf{T}_{Zk_2}(p_2)\dots\mathbf{T}_{Zk_j}(p_j)\mathbf{D}^2_{Zk_j}(1)\dots\mathbf{T}_{Zk_N}(p_N) \tag{5.92}$$

In the above relations we have assumed point M to be determined in the space of body n by radius vector $\mathbf{r}_{nM}(t) = [x_{nM}, y_{nM}, z_{nM}, 1]$ and the number of all basic motions to be N. We designate the parameter of the j-th basic matrix as p_j, whereby this quantity may be a coordinate (variable) or dimension (constant). In the latter case, of course, $\dot{p}_j = \ddot{p}_j = 0$. The type of basic matrix used is determined by the index k.

The motion of a significant body S may be expressed by relations

$$\mathbf{T}_{1S} = \prod_{j=1}^{N} \mathbf{T}_{Zk_j}(p_j) \tag{5.93a}$$

$$\mathbf{V}_{1S} = \mathbf{T}_{1S}^{-1}\dot{\mathbf{T}}_{1S} \tag{5.93b}$$

$$\mathbf{A}_{1S} = \dot{\mathbf{V}}_{1S} \tag{5.93c}$$

According to (4.260), (4.261), (4.265) we can derive

$$\mathbf{V}_{1S} = \sum_{j=1}^{N} \mathbf{G}_j \mathbf{E}(\dot{p}_j) \tag{5.93d}$$

where

$$\mathbf{G}_j = \mathbf{T}_{Zk_N}(-p_N) \dots \mathbf{T}_{Zk_{j+1}}(-p_{j+1}) \mathbf{D}_{Zk_j}(1) \mathbf{T}_{Zk_{j+1}}(p_{j+1}) \dots \mathbf{T}_{Zk_N}(p_N) \tag{5.93e}$$

The acceleration matrix can be expressed as

$$\begin{aligned}\mathbf{A}_{1S} &= \sum_{j=1}^{N} \dot{\mathbf{G}}_j \mathbf{E}(\dot{p}_j) + \sum_{j=1}^{N} \mathbf{G}_j \mathbf{E}(\ddot{p}_j) = \\ &= \sum_{j=1}^{N}\sum_{i=1}^{N} \mathbf{G}_{ji} \mathbf{E}(\dot{p}_j)\mathbf{E}(\dot{p}_i) + \sum_{j=1}^{N} \mathbf{G}_j \mathbf{E}(\ddot{p}_j)\end{aligned} \tag{5.93f}$$

where

$$\begin{aligned}\mathbf{G}_{ji} = \; & \mathbf{T}_{Zk_N}(-p_N) \dots \mathbf{T}_{Zk_i}(-p_i) \mathbf{D}_{Zk_i}(-1) \dots \\ & \dots \mathbf{T}_{Zk_{j+1}}(-p_{j+1}) \mathbf{D}_{Zkj}(1) \mathbf{T}_{Zk_{j+1}}(p_{j+1}) \dots \mathbf{T}_{Zk_N}(p_N) + \\ & + \mathbf{T}_{Zk_N}(-p_N) \dots \mathbf{T}_{Zk_{j+1}}(-p_{j+1}) \mathbf{D}_{Zk_j}(1) \mathbf{T}_{Zk_{j+1}}(p_{j+1}) \dots \\ & \dots \mathbf{T}_{Zk_i}(p_i) \mathbf{D}_{Zk_i}(1) \dots \mathbf{T}_{Zk_N}(p_N)\end{aligned} \tag{5.93g}$$

Similar formulae are valid for calculation of $\boldsymbol{\Omega}_{1S}, \mathcal{A}_{1S}$. They can be obtained from (5.93a–g) by substituting the basic matrices Z1, Z2, Z3 by unit matrix $\mathbf{E}(1)$ and keeping only the left upper submatrix, the dimension of which is (3,3).

When the mechanism in question is a simple chain, the equation (5.75) may be written in basic matrices as follows

$$\prod_{j=1}^{N_1} \mathbf{T}_{Zk_j}(p_j) = \prod_{j=N_1+1}^{N} \mathbf{T}_{Zk_j}(p_j) \tag{5.94}$$

where N_1 is the number of basic motions, in the left simple chain, which we obtained by the imaginary cut of a loop, N is the number of all basic motions. The time derivative of (5.94) yields velocity equations

$$\sum_{j=1}^{N_1} \mathbf{F}_j \mathbf{E}(\dot{p}_j) = \sum_{j=N_1+1}^{N} \mathbf{F}_j \mathbf{E}(\dot{p}_j) \tag{5.95}$$

where $\mathbf{F}_j$ is given by equation (5.90).

The next time derivative yields acceleration equations in the form

$$\sum_{j=1}^{N_1} \Big(\mathbf{F}_j \mathbf{E}(\ddot{p}_j) + \sum_{l=1}^{N_1} \mathbf{F}_{jl} \mathbf{E}(\dot{p}_j) \mathbf{E}(\dot{p}_l) \Big) =$$
$$= \sum_{j=N_1+1}^{N} \Big(\mathbf{F}_j \mathbf{E}(\ddot{p}_j) + \sum_{l=N_1+1}^{N} \mathbf{F}_{jl} \mathbf{E}(\dot{p}_j) \mathbf{E}(\dot{p}_l) \Big) \tag{5.96}$$

where we have used the symbols introduced above.

For a spherical pair with centre C in a simple mechanism between elements m and $(m+1)$, we use a shortened solution with the help of equations

$$\prod_{j=1}^{N_1} \mathbf{T}_{Zk_j}(p_j) \mathbf{r}_{mC} = \prod_{j=N_1+1}^{N} \mathbf{T}_{Zk_j}(p_j) \mathbf{r}_{m+1,C} \tag{5.97}$$

$$\Big(\sum_{j=1}^{N_1} \mathbf{F}_j \mathbf{E}(\dot{p}_j) \Big) \mathbf{r}_{mC} + \prod_{j=1}^{N_1} \mathbf{T}_{Zk_j}(p_j) \dot{\mathbf{r}}_{mC} = \Big(\sum_{j=N+1}^{N} \mathbf{F}_j \mathbf{E}(\dot{p}_j) \Big) \mathbf{r}_{m+1,C} +$$
$$+ \prod_{j=N_1+1}^{N} \mathbf{T}_{Zk_j}(p_j) \dot{\mathbf{r}}_{m+1,C} \tag{5.98}$$

$$\Big(\sum_{j=1}^{N_1} \Big(\mathbf{F}_j \mathbf{E}(\ddot{p}_j) + \sum_{l=1}^{N_1} \mathbf{F}_{jl} \mathbf{E}(\dot{p}_j) \mathbf{E}(\dot{p}_l) \Big) \Big) \mathbf{r}_{m,C} +$$
$$+ \prod_{j=1}^{N_1} \mathbf{T}_{Zk_j}(p_j) \ddot{\mathbf{r}}_{mC} + 2 \left(\sum_{j=1}^{N_1} \mathbf{F}_j \mathbf{E}(\dot{p}_j) \right) \dot{\mathbf{r}}_{mC} =$$

$$= \left(\sum_{j=N_1+1}^{N} \left(\mathbf{F}_j \mathbf{E}(\ddot{p}_j) + \sum_{j=N_1+1}^{N} \mathbf{F}_{jl} \mathbf{E}(\dot{p}_j) \mathbf{E}(\dot{p}_l) \right) \right) \mathbf{r}_{m+1,C} +$$

$$+ \prod_{j=N_1+1}^{N} \mathbf{T}_{Zk_j}(p_j) \ddot{\mathbf{r}}_{m+1,C} + 2 \left(\sum_{j=N_1+1}^{N} \mathbf{F}_j \mathbf{E}(\dot{p}_j) \right) \dot{\mathbf{r}}_{m+1,C} \tag{5.99}$$

In more general cases, where the system in question is a composed mechanism, we apply the procedure according to the results of Section 5.6 using the above expressions.

The advantage of the basic matrix method is its conceptual understandability, enabling the user to describe even a very complex MBS in a relatively simple way and without special knowledge. The further advantage is the easy way in which the whole procedure may be expressed in algorithms; this enables the complete kinematic solution to be performed on the basis of the known matrix equation of position expressed in basic matrices. We shall say more about realizing this procedure in Chapter 10.

5.9 Planar and Spherical Mechanisms

In Section 4.7 we saw that the matrix description of a spherical and of a planar motion are special cases of general spatial motion. In the same way as in Section 4.7 we obtained from the transformation matrix of general spatial motion the transformation matrices of spherical or planar motion, we obtain from the matrix methods of solution of spatial MBS kinematics the matrix methods of solution of spherical or planar MBS kinematics.

5.10 The Transmission Matrix (Velocity Transformation Matrix)

The MBS kinematics may also with advantage be described using so-called transmission matrices and transmission derivatives matrices. Their main advantage is that they are functions only of the dimensions and instantaneous position of the examined MBS. That makes it possible to perform once and forever the computation of matrices for specific cases (perhaps in symbolic form) and store them using computer. Their use then allows us to derive much more quickly the kinematic and therefore also the dynamic solution. The transmission matrices are of particular advantage when computing kinematic quantities of significant points and bodies.

We shall introduce transmission matrices for a spatial MBS, the position of which is determined by m independent coordinates $\mathbf{q}$. We define the position of j-th MBS element by basic decomposition using three coordinates $\mathbf{u}_{1,1j} = \lceil x_{1j}, y_{1j}, z_{1j} \rceil = \mathbf{u}_{1,1j}(\mathbf{q})$ of a reference point and, for example, three Cardan angles $\boldsymbol{\varphi}_j = \lceil \varphi_{xj}, \varphi_{yj}, \varphi_{zj} \rceil = \boldsymbol{\varphi}_j(\mathbf{q})$.

The velocity of the j-th body point of reference may be expressed by relations

$$\dot{\mathbf{u}}_{1,1j} = \sum_{\nu} \frac{\partial \mathbf{u}_{1,1j}}{\partial q_\nu} \dot{q}_\nu = \mathbf{P}_j^P \dot{\mathbf{q}} \tag{5.100}$$

where, using the relation

$$\mathbf{P}_j^P = [p_{j1}^P, p_{j2}^P, \ldots, p_{jm}^P] \tag{5.101}$$

we have introduced the transmission matrix of the j-th body carrying translational motion expressing the effect of general velocities $\dot{\mathbf{q}}$ on the velocity $\dot{\mathbf{u}}_{1,1j}$. The individual columns of this matrix are

$$\mathbf{p}_{j\nu}^P = \frac{\partial \mathbf{u}_{1,1j}}{\partial q_\nu} = \lceil p_{j\nu x}^P, p_{j\nu y}^P, p_{j\nu z}^P \rceil \tag{5.102}$$

and they express the effect of the ν-th independent velocity on the x, y and z components of the absolute velocity of the reference point.

In a similar way the relative spherical motion angular velocity of the j-th element may be expressed by the relation

$$\boldsymbol{\omega}_{j,1j} = \mathbf{P}_j^R \dot{\mathbf{q}} \tag{5.103}$$

The computation of matrix $\mathbf{P}_j^R$ columns $p_{j\nu}^R$ is no longer simple. We start from the defining relation

$$\boldsymbol{\Omega}_{j,1j} = \mathbf{S}_{1j}^T \dot{\mathbf{S}}_{1j} = \sum_{\nu} \mathbf{S}_{1j}^T \frac{\partial \mathbf{S}_{1j}}{\partial q_\nu} \dot{q}_\nu \tag{5.104}$$

where coefficients for individual $\dot{q}_\nu$'s are skew-symmetric (antisymmetric) matrices

$$\mathbf{S}_{1j}^T \frac{\partial \mathbf{S}_{1j}^T}{\partial q_\nu} = \begin{bmatrix} 0 & -p_{j\nu z}^R & p_{j\nu y}^R \\ p_{j\nu z}^R & 0 & -p_{j\nu x}^R \\ -p_{j\nu y}^R & p_{j\nu x}^R & 0 \end{bmatrix} \tag{5.105}$$

We then determine vectors

$$\mathbf{p}_{j\nu}^R = \lceil p_{j\nu x}^R, p_{j\nu y}^R, p_{j\nu z}^R \rceil \tag{5.106}$$

which in their turn determine the matrix

$$\mathbf{P}_j^R = [p_{j1}^R, p_{j2}^R, , p_{jm}^R] \tag{5.107}$$

For the j-th body velocity quantities we may summarize

$$\mathbf{v}_j = \begin{bmatrix} \dot{\mathbf{u}}_{1,1j} \\ \boldsymbol{\omega}_{j,1j} \end{bmatrix} = \begin{bmatrix} \mathbf{P}_j^P \\ \mathbf{P}_j^R \end{bmatrix} \dot{\mathbf{q}} \tag{5.108}$$

where we designated by indices the space in which the individual subvector components are expressed.

Equations for reference point acceleration are obtained by differentiating (5.100) with respect to time

$$\ddot{\mathbf{u}}_{1,1j} = \left(\sum_{\nu} \frac{\partial \mathbf{P}_j^P}{\partial q_\nu} \dot{q}_\nu \right) \dot{\mathbf{q}} + \mathbf{P}_j^P \ddot{\mathbf{q}} = \bar{\mathbf{P}}_j^P \dot{\bar{\mathbf{q}}} \dot{\mathbf{q}} + \mathbf{P}_j^P \ddot{\mathbf{q}} \tag{5.109}$$

where matrix

$$\bar{\mathbf{P}}_j^P = \left[\frac{\partial \mathbf{P}_j^P}{\partial q_1}, \frac{\partial \mathbf{P}_j^P}{\partial q_2}, \ldots, \frac{\partial \mathbf{P}_j^P}{\partial q_m} \right] \tag{5.110}$$

has dimensions $(3, m^2)$ and is called the transmission derivatives matrix, and the matrix

$$\dot{\bar{\mathbf{q}}} = \lceil \mathbf{E}_m \dot{q}_1, \mathbf{E}_m \dot{q}_2, \ldots, \mathbf{E}_m \dot{q}_m \rceil \tag{5.111}$$

has dimensions (m^2, m). (Using the formalism introduced in Section 8.3 it is possible to express relation (5.109) in a simpler way).

Analogously for angular acceleration of relative spherical motion, we obtain by differentiating (5.103) with respect to time the relation

$$\boldsymbol{\alpha}_{j,1j} = \left(\sum_{\nu} \frac{\partial \mathbf{P}_j^R}{\partial q_\nu} \dot{q}_\nu \right) \dot{\mathbf{q}} + \mathbf{P}_j^R \ddot{\mathbf{q}} = \bar{\mathbf{P}}_j^R \dot{\bar{\mathbf{q}}} \dot{\mathbf{q}} + \mathbf{P}_j^R \ddot{\mathbf{q}} \tag{5.112}$$

where similarly

$$\bar{\mathbf{P}}_j^R = \left[\frac{\partial \mathbf{P}_j^R}{\partial q_1}, \frac{\partial \mathbf{P}_j^R}{\partial q_2}, \ldots, \frac{\partial \mathbf{P}_j^R}{\partial q_m} \right] \tag{5.113}$$

In a similar way it is possible to introduce transmission matrices of a general point of a body, transmission matrices of all or any of the dependent coordinates, etc. Their advantage is the consistent separation of input quantities from output quantities and their independence of time. Once computed for a particular MBS, they are universally applicable for any change of input quantities, and the computation when using them is very fast.

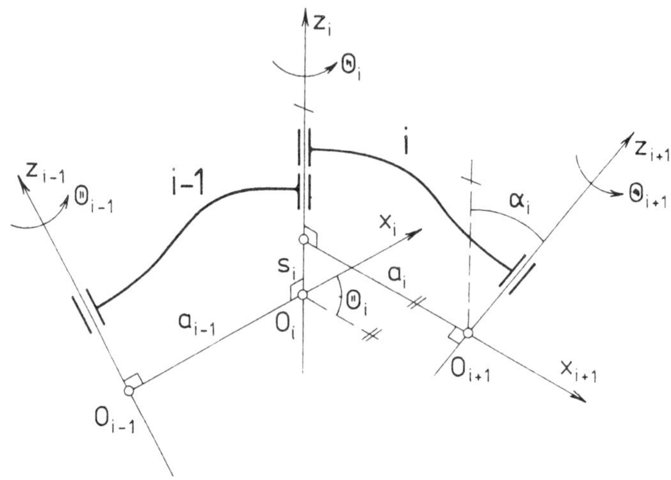

Figure 5.18. Denavit–Hartenberg's notation.

5.11 Other Approaches

Apart from the above matrix methods of solution of problems of MBS kinematic analysis, other methods have been created. We shall briefly describe their principles.

5.11.1 Denavit–Hartenberg's Method

One of the first matrix methods of solving the kinematic analysis of MBS of the open chain type or of the simple mechanism type is the Denavit–Hartenberg method (Denavit and Hartenberg, 1955; Uicker et al., 1964). This method is based on systematic introduction of coordinate systems in individual MBS elements and of parameters describing the mutual position of those coordinate systems. Using this designation, a transformation matrix analogous to (4.51) describing motion between two MBS elements linked by a lower kinematic pair (revolute or prismatic) is then introduced. Equations describing MBS are then special cases of equations (5.73) and (5.75).

The individual MBS elements are numbered from 0 to n, starting from the frame. Elements $i-1$ and i are linked by the i-th kinematic pair. With simple mechanisms, the $(n+1)$-th pair connects element n with the frame 0. In every element i we choose a coordinate system $O_i x_i y_i z_i$, whose axis z_i lies in the axis of pair i (Figure 5.18) and axis x_i is the common normal to axes z_i and z_{i-1}. The distance between axes z_i and z_{i+1} is a_i and their angle α_i. Transformation $\mathbf{A}_i$ between coordinate systems $O_i x_i y_i z_i$ and $O_{i+1} x_{i+1} y_{i+1} z_{i+1}$ is given by rotation around axis z_i by angle Θ_i, translation along axis z_i by distance s_i, translation along axis $x_i = x_{i+1}$ by distance a_i and rotation around axis x_{i+1} by angle α_i. We shall express the motions by basic matrices

$$\mathbf{A}_i = \mathbf{T}_{Z6}(\Theta_i)\mathbf{T}_{Z3}(s_i)\mathbf{T}_{Z1}(a_i)\mathbf{T}_{Z4}(\alpha_i) \tag{5.114}$$

$$\mathbf{A}_i = \begin{bmatrix} \mathrm{c}\Theta_i & -\mathrm{s}\Theta_i \mathrm{c}\alpha_i & \mathrm{s}\Theta_i \mathrm{s}\alpha_i & a_i \mathrm{c}\Theta_i \\ \mathrm{s}\Theta_i & \mathrm{c}\Theta_i \mathrm{c}\alpha_i & -\mathrm{c}\Theta_i \mathrm{s}\alpha_i & a_i \mathrm{s}\Theta_i \\ 0 & \mathrm{s}\alpha_i & \mathrm{c}\alpha_i & s_i \\ 0 & 0 & 0 & 1 \end{bmatrix}^{1)} \tag{5.115}$$

For a revolute pair the angle Θ_i is variable, for a prismatic pair the displacement s_i is variable, and for the cylindrical pair the quantities Θ_i and s_i are variables. Other kinematic pairs are not mentioned in the original formulation (Denavit and Hartenberg, 1955); we may easily obtain them, however, by composing transformations $\mathbf{A}_i$. The helical pair is a special case of the cylindrical pair, the spherical pair is described by three Cardan angles for three revolute pairs with axes intersecting in one common point, and a flat pair is to be obtained by composing two prismatic pairs with a revolute pair with mutually perpendicular axes.

The transformation matrix of the end element of an open kinematic chain is determined as follows

$$\mathbf{T}_{0n} = \prod_{i=1}^{n} \mathbf{A}_i \tag{5.116}$$

The equation describing a simple mechanism is to be obtained from a matrix equation similar to (5.73)

$$\prod_{i=1}^{n} \mathbf{A}_i = \mathbf{E}_4 \tag{5.117}$$

The Denavit–Hartenberg method is very popular in robotics, thanks to the works of R. Paul (Paul, 1981). This method is a special case of the matrix method described in this book. The disadvantage of Denavit–Hartenberg's

[1]) Note: In the Denavit–Hartenberg formulation the transformation matrix is written

$$\mathbf{T}_{DH} = \begin{bmatrix} 1 & \mathbf{0} \\ \mathbf{r} & \mathbf{S} \end{bmatrix}$$

which is a different form from that used in this book (4.51). It is evident that this is a mere formal difference in notation.

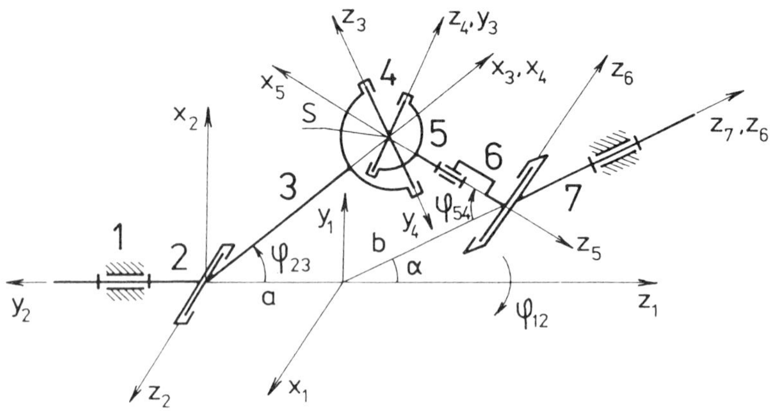

Figure 5.19. RRSRR mechanism with Denavit–Hartenberg's description.

method is the necessity of defining coordinate systems following a convention instead of fitting the specific problem. Serious problems may result (Bazerghi et al., 1984, and see Section 11.1.5). We therefore consider the basic matrix method to be more suitable as it allows complete adjustment to the individual requirements of the problem.

Let us now solve the example from Section 5.5.2 using Denavit–Hartenberg's method. Let us introduce the designation according to the conventions of the method (Figure 5.19). The parameters of individual transformations are in the table

Kinematic pair	Θ_i	s_i	a_i	α_i
1	φ_{12}	$-a$	0	$-\pi/2$
2	φ_{23}	0	r	$-\pi/2$
3	Θ_3	0	0	$-\pi/2$
4	Θ_4	0	0	$-\pi/2$
5	Θ_5	d	0	$\pi/2$
6	φ_{54}	0	0	$-\pi/2$
7	φ_{15}	$-b$	$-\alpha$	$\pi/2$

We must substitute the spherical pair by three rotations with axes intersecting in point S. This mechanism is then described by the equation

$$\mathbf{A}_1\mathbf{A}_2\mathbf{A}_3\mathbf{A}_4\mathbf{A}_5\mathbf{A}_6\mathbf{A}_7 = \mathbf{E}_4 \tag{a}$$

The equation yields 12 equations for 6 unknowns $\varphi_{23}, \Theta_3, \Theta_4, \Theta_5, \varphi_{54}, \varphi_{15}$ depending on φ_{12}.

5.11.2 The Generalized Vector Method

The vector method used for planar MBS kinematic analysis may be generalized for spatial MBS (Zinovev, 1952). In a spatial MBS with lower kinematic pairs, significant points are defined on (revolute, prismatic, cylindrical, helical) kinematic pair axes or in the centre of a spherical pair or in the plane of a flat pair. Significant points on kinematic pair axes are points of intersection of transversals of generally skew axes of kinematic pairs, of the same MBS element with those axes. With flat pairs, the significant point is either the point of intersection of the axis of the next kinematic pair of the element with the pair plane or the foot of the normal dropped on the pair plane from the significant point on the next kinematic pair of the element.

A spherical pair may be substituted by three revolute pairs with intersecting axes. A flat pair may be substituted by two prismatic pairs and one revolute pair.

Significant points of a MBS, passing them in the order in which the MBS elements are linked by kinematic pairs, form a closed polygon. The easiest description of this broken line is by the use of vectors. The vector end points are the significant points, and vectors coincide either with kinematic pair axes or with their transversals.

We shall express each vector $\mathbf{u}_i$ by its length $l_i = |\mathbf{u}_i|$, using direction cosines x_i, y_i, z_i of the respective unit vector $\mathbf{e}_i$

$$\mathbf{u}_i = l_i \mathbf{e}_i \tag{5.118}$$

$$\mathbf{e}_i = [x_i, y_i, z_i] \tag{5.119}$$

For direction cosines we can write

$$x_i^2 + y_i^2 + z_i^2 = 1 \tag{5.120}$$

If the contour is formed by n vectors $\mathbf{r}_i$, the closure condition is

$$\sum_{i=1}^{n} \mathbf{r}_i = \sum_{i=1}^{n} l_i \mathbf{e}_i = \mathbf{0} \tag{5.121}$$

i.e.

$$\sum_{i=1}^{n} l_i x_i = 0, \quad \sum_{i=1}^{n} l_i y_i = 0, \quad \sum_{i=1}^{n} l_i z_i = 0 \tag{5.122}$$

The kinematic pair axes $\mathbf{e}_i, \mathbf{e}_j$ linked with the same MBS element form a given fixed angle γ_{ij}. Hence

$$\mathbf{e}_i . \mathbf{e}_j = x_i x_j + y_i y_j + z_i z_j = \cos \gamma_{ij} \tag{5.123}$$

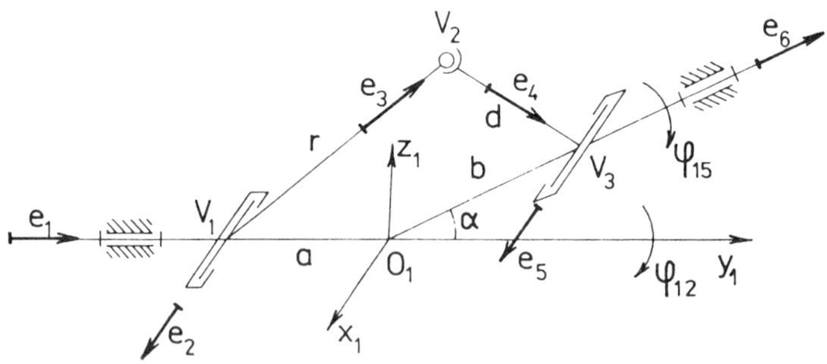

Figure 5.20. RRSRR mechanism solved by generalized vector method.

Vector $\mathbf{e}_k$ of the skew axes $\mathbf{e}_i, \mathbf{e}_j$ transversal may be expressed as

$$\mathbf{e}_k = \mathbf{e}_i \times \mathbf{e}_j = \lceil y_i z_j - y_j z_i, z_i x_j - z_j x_i, x_i y_j - x_j y_i \rceil \tag{5.124}$$

The length l_k of this transversal is also firmly given, as the kinematic pair axes firmly connected with the respective MBS element. Using the known lengths and equation (5.124) we may express all transversals in (5.121), (5.122).

In specific cases, the equations (5.120), (5.122) and (5.124) form a sufficient system of equations for solving the respective position problem. The solution for velocities and accelerations is to be obtained from the linear equation systems obtained from time derivatives of the position solution equations.

Let us solve again the example of an RRSRR mechanism from Section 5.5.2. Let us use the designations given in Figure 5.20. $\mathbf{e}_1, \mathbf{e}_2, \mathbf{e}_3, \mathbf{e}_4, \mathbf{e}_5, \mathbf{e}_6$ are unit vectors here. Condition (5.121) has the form

$$\begin{aligned} \overrightarrow{O_1V_1} + \overrightarrow{V_1V_2} + \overrightarrow{V_2V_3} + \overrightarrow{V_3O_1} &= \vec{0} \\ -a\mathbf{e}_1 + r\mathbf{e}_3 + d\mathbf{e}_4 - b\mathbf{e}_6 &= \mathbf{0} \end{aligned} \tag{a}$$

for components

$$\begin{aligned} rx_3 + dx_4 &= 0 \\ -a + ry_3 + dy_4 - b\cos\alpha &= 0 \\ rz_3 + dz_4 - b\sin\alpha &= 0 \end{aligned} \tag{b}$$

For angles of vectors $\mathbf{e}_i$

$$\begin{aligned} \mathbf{e}_1.\mathbf{e}_2 &= 0, \quad \mathbf{e}_2.\mathbf{e}_3 = 0, \quad \mathbf{e}_4.\mathbf{e}_5 = 0 \\ \mathbf{e}_5.\mathbf{e}_6 &= 0, \quad \mathbf{e}_1.\mathbf{e}_6 = \cos\alpha \end{aligned} \tag{c}$$

The components of vector $\mathbf{e}_1, \mathbf{e}_2, \mathbf{e}_6$ are known:

$$\begin{aligned} \mathbf{e}_1 &= [0,1,0], \quad \mathbf{e}_2 = [\cos\varphi_{12}, 0, -\sin\varphi_{12}], \\ \mathbf{e}_6 &= [0, \cos\alpha, \sin\alpha] \end{aligned} \tag{d}$$

Then, from (c)–(d)

$$x_3\cos\varphi_{12} - z_3\sin\varphi_{12} = 0 \tag{e1}$$
$$x_4x_5 + y_4y_5 + z_4z_5 = 0 \tag{e2}$$
$$y_5\cos\alpha + z_5\sin\alpha = 0 \tag{e3}$$

Vectors $\mathbf{e}_i$ are unit vectors, hence

$$x_3^2 + y_3^2 + z_3^2 = 1 \tag{f1}$$
$$x_4^2 + y_4^2 + z_4^2 = 1 \tag{f2}$$
$$x_5^2 + y_5^2 + z_5^2 = 1 \tag{f3}$$

We obtain a set of 9 equations (b), (e), (f) for 9 unknowns x_3, y_3, z_3, x_4, y_4, z_4, x_5, y_5, z_5 depending on φ_{12}. As we are looking for $\varphi_{15} = \varphi_{15}(\varphi_{12})$ we must introduce the angle φ_{15} into the problem by expressing

$$\begin{aligned} [x_5, y_5, z_5, 0] &= \mathbf{T}_{Z4}(\alpha)\mathbf{T}_{Z2}(b)\mathbf{T}_{Z5}(\varphi_{15})[1,0,0,0] \\ [x_5, y_5, z_5, 0] &= [\mathrm{c}\varphi_{15}, \mathrm{s}\varphi_{15}\mathrm{s}\alpha, -\mathrm{s}\varphi_{15}\mathrm{c}\alpha, 0] \end{aligned}$$

and substituting into (b), (e), (f). Equations (e3) and (f3) turn into identities. We shall then have 7 equations (b), (e1, 2), (f1, 2) for 7 unknowns x_3, y_3, z_3, x_4, y_4, z_4, φ_{15} depending on φ_{12}.

We have here an efficient alternative to the matrix method of solution of MBS kinematics. Its main advantages, which are the absence of transcendental trigonometric functions and the fact that the Jacobian is linear in unknowns, are further developed in the method of natural coordinates. Its disadvantages are the impossibility of direct expression of body motion and the need for relatively difficult geometrical computations, which are expressed in a natural way in the basic matrix method (see, for example, the conclusion of the case being considered).

From this section and from Section 5.11.4 it follows that, from the numerical solution point of view, it would be advantageous to solve the kinematic equations given by the matrix method as a function of $\sin(arg)$ and $\cos(arg)$ together with the constraint $\sin^2(arg) + \cos^2(arg) = 1$ instead of as a function of their angular arguments (arg). It removes the evaluation of transcendental functions from numerical iterations. Only the transcendental function atan2 would have to be used at the end of the solution. The evaluation of both kinematic equations and the Jacobian is performed simply by multiplications and additions. Further research into this problem may be useful.

5.11.3 The Generalized Method of Disconnecting a Loop

Kinematic equations containing neither variables of a chosen kinematic pair nor variables of kinematic pairs of a chosen MBS element may be formed for the solution of the closed chain position (Litvin, 1970, Haug et al., 1982, Woernle, 1988).

In the first variant, we imagine disconnecting a chosen kinematic pair of the closed chain and expressing the chain closure conditions from the remaining chains.

When the disconnected pair is a spherical pair, the chain closure condition is the equality of radius vectors of centres S of the spherical pair on the disconnected elements m and n

$$\mathbf{u}_{1Sm} = \mathbf{u}_{1Sn} \tag{5.125}$$

The closure condition for revolute pair R is according to Figure 5.21a

$$\mathbf{u}_{1Rm} = \mathbf{u}_{1Rn} \tag{5.126a}$$

$$\mathbf{e}_m = \mathbf{e}_n \quad \text{or} \quad (\mathbf{u}_{1Rm} - \mathbf{u}_{1Pm}) \times (\mathbf{u}_{1Rn} - \mathbf{u}_{1Qn}) = \mathbf{0} \tag{5.126b}$$

where $\mathbf{e}_m, \mathbf{e}_n$ are unit vectors and P, Q are fixed points of the revolute pair axes of elements m and n. For a prismatic pair P as in Figure 5.21b, it must be true that

$$\mathbf{e}_m = \mathbf{e}_n \quad \text{or} \quad (\mathbf{u}_{1Bm} - \mathbf{u}_{1Cm}) \times (\mathbf{u}_{1Dn} - \mathbf{u}_{1En}) = \mathbf{0} \tag{5.127a}$$

$$(\mathbf{u}_{1Bm} - \mathbf{u}_{1Dn}) \times \mathbf{e}_m = \mathbf{0} \tag{5.127b}$$

$$(\mathbf{u}_{1Am} - \mathbf{u}_{1Bm}).(\mathbf{u}_{1Dn} - \mathbf{u}_{1Fn}) = \text{const} \tag{5.127c}$$

where $\mathbf{e}_m, \mathbf{e}_n$ are unit vectors of pair axes, points B, C, D, E lie on the pair axis, and points $A, B, C(D, E, F)$ belong to element $m(n)$. For a cylindrical pair C as in Figure 5.21c, it must be true that

$$\mathbf{e}_m = \mathbf{e}_n \quad \text{or} \quad (\mathbf{u}_{1Am} - \mathbf{u}_{1Cm}) \times (\mathbf{u}_{1Bn} - \mathbf{u}_{1Dn}) = \mathbf{0} \tag{5.128a}$$

$$(\mathbf{u}_{1Am} - \mathbf{u}_{1Bn}) \times \mathbf{e}_m = \mathbf{0} \tag{5.128b}$$

where $\mathbf{e}_m, \mathbf{e}_n$ are the unit vectors of the pair axis, points A, B, C, D lie on the axis, and $A, C(B, D)$ belong to element $m(n)$. For a flat pair F as in Figure 5.21d, it must be true that

$$\mathbf{e}_m = \mathbf{e}_n \tag{5.129a}$$

$$(\mathbf{u}_{1Am} - \mathbf{u}_{1Bn}).\mathbf{e}_m = 0 \tag{5.129b}$$

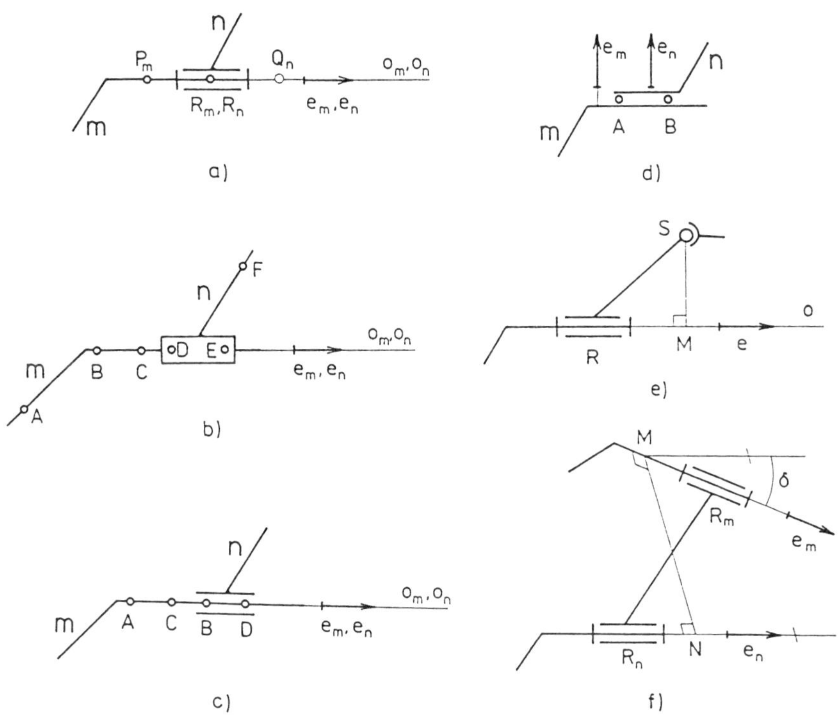

Figure 5.21. Constraints conditions expressed by vectors.

where $\mathbf{e}_m, \mathbf{e}_n$ are unit vectors normal to the pair plane, and points $A(B)$ lie in the pair plane in element $m(n)$.

We shall express vectors in conditions (5.125)–(5.129) using either the matrix method or the generalized vector method. We shall obtain equations whose number suffices to determine the remaining variables outside the disconnected pair.

In the second variant we imagine eliminating from the closed chain one element with two kinematic pairs through which the pair is linked to other chain elements, and we again express the chain closure conditions from the remaining chains.

Should the eliminated element contain two spherical pairs with centres S_1 and S_2, the closure condition is then

$$|\mathbf{u}_{1S_1} - \mathbf{u}_{1S_2}| = l \tag{5.130}$$

where $l = \overline{S_1 S_2}$ is the distance of the spherical pair centres in the element. Should the eliminated element contain a spherical pair S and a revolute pair R, the closure conditions according to Figure 5.21e are

$$|\mathbf{u}_{1S} - \mathbf{u}_{1M}| = h \tag{5.131a}$$
$$(\mathbf{u}_{1S} - \mathbf{u}_{1M}) \cdot \mathbf{e} = 0 \tag{5.131b}$$

where $h = \overline{SM}$, M is the foot of the perpendicular dropped from S on to axis o of pair R, and $\mathbf{e}$ is the unit vector of axis o. Should the eliminated element contain two revolute pairs R_m and R_n, the closure conditions according to Figure 5.21f are

$$\mathbf{u}_{1M} - \mathbf{u}_{1N} = \frac{h}{\sin\delta}(\mathbf{e}_m \times \mathbf{e}_n) \tag{5.132a}$$
$$\mathbf{e}_m . \mathbf{e}_n = \cos\delta \tag{5.132b}$$

where δ is the angle of axes o_1, o_2, MN is their transversal, h is the length of the transversal, and $\mathbf{e}_m, \mathbf{e}_n$ are unit vectors of axes. The same relations hold by substituting a revolute pair by a prismatic or cylindrical pair. The relations for planar contact pairs may also be written in a similar way.

We shall again express vectors in conditions (5.130) to (5.132), using either the matrix method or the generalized vector method. We shall again obtain sufficient equations to determine the remaining variables; these equations have the advantage of containing no variables of the eliminated pairs.

Let us solve the example from Section 5.5.2 again. We eliminate from the mechanism according to Figure 5.15 element 3 with a revolute and a spherical pair. The foot M of the perpendicular dropped from point S onto axis x_2 of the revolute pair is point O_2. For conditions (5.131) according to Section 5.5.2

$$\begin{aligned}
\lceil \mathbf{u}_{1S}, 1 \rceil &= \mathbf{T}_{15}\mathbf{T}_{54}\mathbf{r}_{4S} = \lceil d s\varphi_{54} s\varphi_{15}, (b - d c\varphi_{54}) c\alpha - \\
&\quad - d s\varphi_{54} c\varphi_{15} s\alpha, (b - d c\varphi_{54}) s\alpha + d s\varphi_{54} c\varphi_{15} c\alpha, 1 \rceil \\
\lceil \mathbf{u}_{1M}, 1 \rceil &= \mathbf{T}_{12}\mathbf{r}_{4M} = \mathbf{T}_{12}\mathbf{r}_{4O_2} = \lceil 0, -a, 0, 1 \rceil \\
\lceil \mathbf{e}, 0 \rceil &= \mathbf{T}_{12}\lceil 1, 0, 0, 0 \rceil = \lceil c\varphi_{12}, 0, -s\varphi_{12}, 0 \rceil
\end{aligned} \tag{a}$$

Hence

$$\begin{aligned}
(d s\varphi_{54} s\varphi_{15})^2 + ((b - d c\varphi_{54}) c\alpha - d s\varphi_{54} c\varphi_{15} s\alpha + a)^2 + \\
+ ((b - d c\varphi_{54}) s\alpha + d s\varphi_{54} c\varphi_{15} c\alpha)^2 = r^2
\end{aligned} \tag{b1}$$
$$d s\varphi_{54} s\varphi_{15} c\varphi_{12} - ((b - d c\varphi_{54}) s\alpha + d s\varphi_{54} c\varphi_{15} c\alpha) s\varphi_{12} = 0 \tag{b2}$$

Equations (b1, 2) are two equations for 2 unknowns $\varphi_{54}, \varphi_{15}$ depending on φ_{12}.

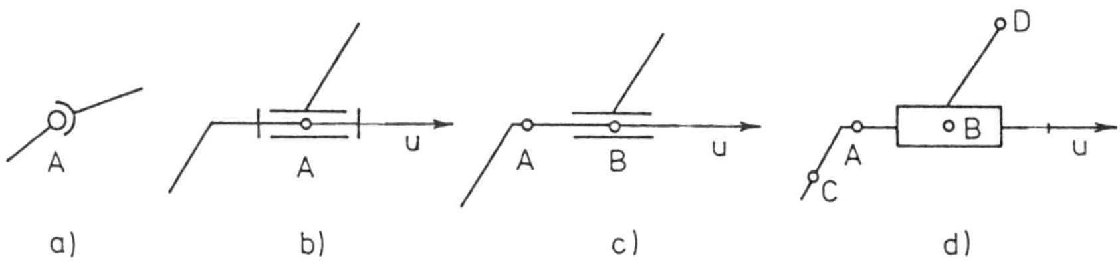

Figure 5.22. Constraints condition expressed by natural coordinates.

The great advantage of this method is the possibility of reducing the number of unknowns and the number of equations. In its formulation, however, complicated geometrical computations and the expressing of various vectors are again required, and the simplest way to obtain these is the basic matrix method. Also the method does not allow the direct expression of the motion of a body.

5.11.4 The Natural Coordinates Method

A method for solving problems of kinematics has been developed which neither requires angular coordinates to describe MBS position nor needs to determine loops in the MBS topological description. The MBS position is described by so-called natural coordinates (Garcia de Jalón et al., 1986).

The natural coordinates are formed by Cartesian coordinates of some MBS points and some unit vectors firmly linked with the MBS. The points and unit vectors are chosen on the kinematic pairs.

Individual MBS elements are described by the set of points and unit vectors. These points and unit vectors are shared by those MBS elements which are linked by respective kinematic pairs.

The spherical pair is described by a common point in the pair centre (Figure 5.22a). The revolute pair is again described by a common point on the pair axis and the common unit vector of the pair axis (Figure 5.22b). The cylindrical pair (Figure 5.22c) is described by a common unit vector of the pair axis and two points on the pair axis, each of which belongs to one MBS element connected by this pair. The prismatic pair (Figure 5.22d) is described in the same way as a cylindrical pair; each MBS element linked by this pair must always contain at least two points.

Points and unit vectors firmly connected with the given body must fulfill certain conditions expressing the rigidity of the body. Examples for this body-element MBS description using natural coordinates are shown in

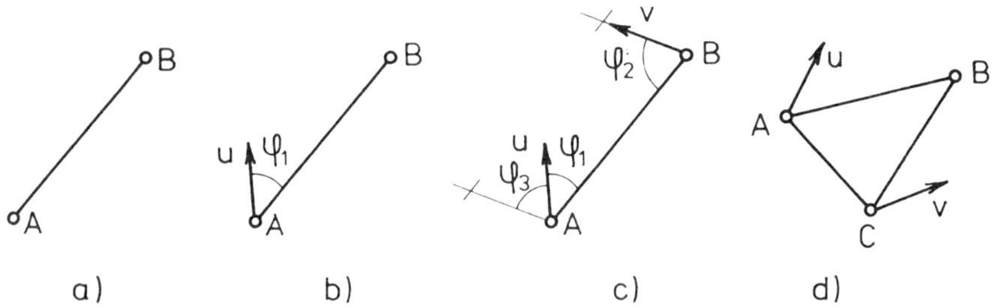

Figure 5.23. Rigid body description using natural coordinates.

Figure 5.23.

Points on the body must fulfil the constant distance condition $l_{AB} = \overline{AB}$ (Figure 5.23a)

$$(x_{1B} - x_{1A})^2 + (y_{1B} - y_{1A})^2 + (z_{1B} - z_{1A})^2 = l_{AB}^2 \tag{5.133}$$

Unit vectors must form constant angles (Figure 5.23b, c, d)

$$(x_{1B} - x_{1A})u_1 + (y_{1B} - y_{1A})u_2 + (z_{1B} - z_{1A})u_3 = l_{AB}\cos\varphi_1 \tag{5.134}$$

$$(x_{1B} - x_{1A})v_1 + (y_{1B} - y_{1A})v_2 + (z_{1B} - z_{1A})v_3 = l_{AB}\cos\varphi_2 \tag{5.135}$$

$$u_1v_1 + u_2v_2 + u_3v_3 = \cos\varphi_3 \tag{5.136}$$

Further, it is evident that

$$\sum_{i=1}^{3} u_i^2 = 1 \tag{5.137}$$

$$\sum_{i=1}^{3} v_i^2 = 1 \tag{5.138}$$

Should **u** be colinear with AB, then equation (5.134) is substituted by two of the three relations

$$\begin{aligned} l_{AB}u_1 - (x_{1B} - x_{1A}) &= 0 \\ l_{AB}u_2 - (y_{1B} - y_{1A}) &= 0 \\ l_{AB}u_3 - (z_{1B} - z_{1A}) &= 0 \end{aligned} \tag{5.139}$$

If $\mathbf{u}, \mathbf{v}$ and AB are coplanar, then one of the equations (5.134)–(5.136) is substituted by one of the equations

$$\begin{aligned} k_1u_1 + k_2v_1 - (x_{1B} - x_{1A}) &= 0 \\ k_1u_2 + k_2v_2 - (y_{1B} - y_{1A}) &= 0 \\ k_1u_3 + k_2v_3 - (z_{1B} - z_{1A}) &= 0 \end{aligned} \tag{5.140}$$

where k_1, k_2 are easily determinable constants.

A free body in space has 6 degrees of freedom. Therefore for a body described by $3p$ natural coordinates, $3p-6$ conditions of rigidity of the body must be determined. For example, if the body contains 2 points, condition (5.133) suffices. If the body contains 2 points and a unit vector, conditions (5.133), (5.134) and (5.137) apply. If the body contains 2 points and 2 unit vectors, conditions (5.133)–(5.138) apply, etc.

The connection of MBS elements by kinematic pairs is also accompanied by constraint conditions. With spherical and revolute pairs, the sharing of common natural coordinates of the pairs by elements linked by this pair suffices. A cylindrical pair is described on the one hand by sharing a common unit vector; on the other hand vectors $\overrightarrow{AB}$ and $\vec{\mathbf{u}}$ must be colinear

$$\overrightarrow{AB} \times \vec{u} = \vec{0} \tag{5.141}$$

where only two equations from (5.141) are independent. A prismatic pair is described similarly to the cylindrical pair, although vectors $\overrightarrow{AB}$ and $\overrightarrow{BD}$ do not rotate with respect to each other.

$$\begin{aligned}(x_{1C} - x_{1A})(x_{1D} - x_{1B}) &+ (y_{1C} - y_{1A})(y_{1D} - y_{1B}) + \\ +(z_{1C} - z_{1A})(z_{1D} - z_{1B}) &= l_{AC} l_{BD} \cos(\overrightarrow{AC}, \overrightarrow{BD})\end{aligned} \tag{5.142}$$

Let us solve the example from Section 5.5.2 once more. We introduce natural coordinates $\mathbf{e}_1, \mathbf{u}_{1V_1}, \mathbf{e}_2, \mathbf{u}_{1V_2}, \mathbf{e}_5, \mathbf{u}_{1V_3}, \mathbf{e}_6$ as in Figure 5.20. From those $x_2, y_2, z_2, x_{1V_2}, y_{1V_2}, z_{1V_2}, x_5, y_5, z_5$ are unknown. As the mechanism contains only revolute and spherical pairs, the connection of mechanism bodies is carried out only by sharing those natural coordinates with linked mechanism elements. It remains to express the rigidity conditions of bodies 3 and 4. First

$$\begin{aligned}\mathbf{u}_{1V_1} &= [0, -a, 0], & \mathbf{u}_{1V_3} &= [0, bc\alpha, bs\alpha] \\ \mathbf{e}_1 &= [0, 1, 0], & \mathbf{e}_6 &= [0, c\alpha, s\alpha]\end{aligned} \tag{a}$$

Then, according to (5.133), (5.134) and (5.137) for body 3

$$x_{1V_2}^2 + (y_{1V_2} + a)^2 + z_{1V_2}^2 = r^2 \tag{b}$$

$$x_{1V_2} x_2 + (y_{1V_2} + a) y_2 + z_{1V_2} z_2 = 0 \tag{c}$$

$$x_2^2 + y_2^2 + z_2^2 = 1 \tag{d}$$

Similarly for body 4

$$x_{1V_2}^2 + (y_{1V_2} - bc\alpha)^2 + (z_{1V_2} - bs\alpha)^2 = d^2 \tag{e}$$

$$x_{1V_2} x_5 + (y_{1V_2} - bc\alpha) y_5 + (z_{1V_2} - bs\alpha) z_5 = 0 \tag{f}$$

$$x_5^2 + y_5^2 + z_5^2 = 1 \tag{g}$$

From the conditions for bodies 2 and 5, only the equations (5.136) remain

$$\begin{aligned} \mathbf{e}_1.\mathbf{e}_5 &= y_5 = 0 && \text{(h)} \\ \mathbf{e}_5.\mathbf{e}_6 &= y_5 \mathrm{c}\alpha + z_5 \mathrm{s}\alpha = 0 && \text{(i)} \end{aligned}$$

We have altogether 8 equations (e)–(i) for 9 unknowns. We may then choose one more of them or introduce the angle φ_{12} into the problem. Analogously to Section 5.11.2 we know that

$$\mathbf{e}_2 = \lceil \mathrm{c}\varphi_{12}, 0, -\mathrm{s}\varphi_{12} \rceil \quad \text{(j)}$$

Substituting into (b)–(i) we obtain 6 equations (b)–(c), (e)–(g), (i) for 6 unknowns $x_{1V_2}, y_{1V_2}, z_{1V_2}, x_5, y_5, z_5$ in dependence on φ_{12}. We may also introduce angle φ_{15} into the problem. We express in the same way as in Section 5.11.2

$$\mathbf{e}_5 = \lceil c\varphi_{15}, s\varphi_{15} s\alpha, -s\varphi_{15} c\alpha \rceil \quad \text{(k)}$$

and substitute it into (b, c), (e, g), (i) obtaining four equations (b, c), (e, f) for four unknowns $x_{1V_2}, y_{1V_2}, z_{1V_2}, \varphi_{15}$ depending on φ_{12}.

This method develops advantageous properties of the generalized vector method, which enables the finding of a numerical solution by a very effective algorithm. The disadvantages are the same as in Section 5.11.2, i.e. the impossibility of directly expressing body motion, the need for difficult geometrical calculations (for which the simplest way to express them is the basic matrix method, anyway), and the choice of particular coordinates for which recomputation into the original coordinates of the problem, e.g. by open chains (robots), is not simple.

5.11.5 The Compartment Method in Kinematics

There exists a general method for assembling equations for the solution of kinematics of complex MBS's for reaching connections with MBS statics and dynamics solutions. The MBS in question is divided into parts which we shall call compartments. The MBS division into parts is, in principle, arbitrary; in practical application, however, parts which occur in more MBS's and from which the MBS may be assembled as from building elements are mostly chosen.

Let an MBS be composed of compartments $a, b, c, d, \ldots$ (Figure 5.24). These compartments are linked by constraints $H_{ab}, H_{ac}, H_{ad}, \ldots$. These constraints may be of a various character: kinematic constraints, firm connection of bodies. The position of each compartment in space is described by coordinates $s_{a_i}, s_{b_j}, s_{c_k}, s_{d_l}, \ldots$. These coordinates must enable the position and orientation to be determined for each body of the compartment in

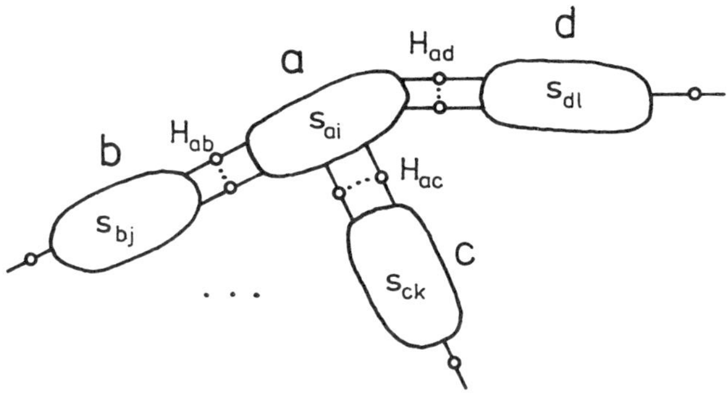

Figure 5.24. MBS composed of compartments.

space. They may be arbitrary and their number may be arbitrary (however constraints then exist between them), but they must not form relative coordinates of compartments with respect to each other, i.e. the coordinates of a compartment determine its position regardless of the position of other compartments of the MBS.

The connection of compartments by constraints H is described by equations, e.g.

$$H_{ab}(s_{a_i}, s_{b_j}, t) = 0 \tag{5.143}$$

It could be shown that these equations may always be written as the identity of boundary properties expressed by coordinates of both compartments

$$H_a(s_{a_i}, t) - H_b(s_{b_j}, t) = 0 \tag{5.144}$$

Apart from compartment constraints, internal constraints in a compartment between its coordinates may also be present, e.g.

$$g_a(s_{a_i}, t) = 0 \tag{5.145}$$

The total number of constraint equations (5.143)–(5.145) may vary.

We assemble the equations describing the MBS position problem in such a way that we take the MBS compartment constraint equations (5.145) and, according to compartment link constraints in MBS, we write the respective constraint equations (5.144). In such a way we obtain a system of equations sufficient for solving the position problem. It is then easy to determine velocities and accelerations with the help of time derivatives.

Let us solve the example from Section 5.5.2 again. Let us choose individual elements 2, 3, 4, 5 for compartments. Let us choose angle φ_{12}

for the compartment 2 coordinate, and angle φ_{15} for compartment 5. Let us further choose coordinates x_{1S}, y_{1S}, z_{1S} of spherical pair centre S in the frame coordinate system for compartment 3 coordinates, and Cardan angles $\varphi_{3x}, \varphi_{3y}, \varphi_{3z}$ (or equivalent Euler parameters with norm condition) describing the orientation of coordinate system $O_3x_3y_3z_3$ relative to frame coordinate system $O_1x_1y_1z_1$ (Figure 5.15). In a similar way we choose coordinates $x_{1S}, y_{1S}, z_{1S}, \varphi_{4x}, \varphi_{4y}, \varphi_{4z}$ for compartment 4. The condition for the connection of compartments 3 and 4 in a spherical pair is directly fulfilled by the common usage of the coordinates of its centre. For connecting compartments 2, 3 and 4, 5 by revolute pairs,

we use conditions (5.126a, b) for the identity of the pair centre and for the unit vector of the pair axis. We obtain (Figure 5.15)

$$\begin{aligned}
\mathbf{T}_{13}\mathbf{T}_{Z2}(-r)[0,0,0,1] &= \mathbf{T}_{Z2}(-a)[0,0,0,1] && \text{(a)}\\
\mathbf{T}_{13}\mathbf{T}_{Z2}(-r)[1,0,0,0] &= \mathbf{T}_{Z2}(-a)\mathbf{T}_{Z5}(\varphi_{12})[1,0,0,0] && \text{(b)}\\
\mathbf{T}_{14}\mathbf{T}_{Z2}(d)[0,0,0,1] &= \mathbf{T}_{Z4}(\alpha)\mathbf{T}_{Z2}(b)[0,0,0,1] && \text{(c)}\\
\mathbf{T}_{14}\mathbf{T}_{Z2}(d)[1,0,0,0] &= \mathbf{T}_{Z4}(\alpha)\mathbf{T}_{Z2}(b)\mathbf{T}_{Z5}(\varphi_{15})[1,0,0,0] && \text{(d)}
\end{aligned}$$

where

$$\begin{aligned}
\mathbf{T}_{13} &= \mathbf{T}_{Z1}(x_{1S})\mathbf{T}_{Z2}(y_{1S})\mathbf{T}_{Z3}(z_{1S})\mathbf{T}_{Z4}(\varphi_{3x})\mathbf{T}_{Z5}(\varphi_{3y})\mathbf{T}_{Z6}(\varphi_{3z})\\
\mathbf{T}_{14} &= \mathbf{T}_{Z1}(x_{1S})\mathbf{T}_{Z2}(y_{1S})\mathbf{T}_{Z3}(z_{1S})\mathbf{T}_{Z4}(\varphi_{4x})\mathbf{T}_{Z5}(\varphi_{4y})\mathbf{T}_{Z6}(\varphi_{4z})
\end{aligned}$$

Equations (a)–(d) yield 12 equations (whereby in both (b) and (d) there is one redundant equation given by the unit vector) for 10 unknowns x_{1S}, y_{1S}, z_{1S}, $\varphi_{3x}, \varphi_{3y}, \varphi_{3z}$, $\varphi_{4x}, \varphi_{4y}, \varphi_{4z}, \varphi_{15}$ with known φ_{12}. Equations (a)–(d) have the form (5.79), but the vectors may not be reduced, as they are specific vectors, not general vectors.

In this simple example, equations (a)–(d) are relatively complex. The compartment method is, however, a very general method allowing the solution of complicated composed mechanisms even with higher pairs with an equation complexity analogous to that of equations (a)–(d). For further details see also Section 9.7.

5.11.6 The Geometric Diagram and the Parametric Method in Kinematics

The geometric diagram method and the parametric method are both still under development; however their properties are very important and they promise to provide a highly efficient approach to the dynamic analysis of MBS's. The original methods of solving kinematic analysis, especially in velocities and accelerations, were derived at the end of the last century and the beginning of this century, and were graphical methods, e.g. (Wittenbauer, 1923). These methods can be generalized and treated in an analytical way and they can then provide advantageous methods for solving MBS kinematic analysis. These possibilities have now been investigated (Valášek and Stejskal, 1990; Valášek, 1990a,b). The advantage of these methods is the theoretical ability to achieve the linear computational complexity of MBS differential kinematic analysis in comparison with the cubic computational complexity which is achieved by all other methods. This is very important because of the possible linear computational complexity of dynamic analysis of MBS with loops, as we shall see in Chapter 9.

The generalization of old graphical methods in kinematics is based on the combination of the geometric diagram method and the parametric method. The geometric diagram method uses the geometric relation among velocities of different points of a MBS which are constructed in the so-called velocity or acceleration diagram. We fix in space a point O as origin. We shall describe the velocities (or accelerations) of points of interest of the MBS being investigated as radius vectors with the same origin O. The velocity (or acceleration) diagram is a diagram of velocity (or acceleration) vectors and their end points. In this way we obtain a geometrical interpretation of relations between velocities (or accelerations) in the MBS being investigated. The end points of velocities (or accelerations) of important points in the MBS lie in different geometrical sets (e.g. plane, straight line) and hence they can be determined by the intersections of these sets. The velocity (or acceleration) diagram method is the fundamental method of graphical solution of planar MBS's (Wittenbauer, 1923), but we can use its generalization for the spatial MBS's.

The parametric method uses other facts. In many cases, if we knew the velocity (or acceleration) of some point of the MBS, we could complete the kinematic solution. We can choose the value of this velocity (or acceleration) and thus complete the kinematic solution. However, such a solution is not in general correct because it must fulfil some conditions. From these conditions we determine the correct value of the chosen velocity (or acceleration). The chosen values are treated as parameters in the kinematic solution and thus give rise to the name "parametric method". The origin of the parametric method is also to be found in graphical methods (Wittenbauer, 1923); there have been some attempts to generalize this method (Brát, 1960) but we now use a different variant.

The geometric diagram and the parametric method are primarily methods of velocity solution of a MBS. In Section 5.10 we shall see that the matrix of the system of linear algebraic equations which are to be solved for the solution of position, velocity or acceleration of MBS are identical. Therefore the geometric diagram and the parametric method can be applied to the solution of position, velocity as well as acceleration of MBS. In this text we shall describe only the solution of MBS velocity.

The solution of MBS position is generally achieved by Newton's method of numerical solution of the nonlinear closure conditions of kinematic loops. The small corrections of dependent variables can be computed as the dependent velocities for a special velocity solution of MBS. It is the velocity solution with non-zero velocities of the ends of kinematic loops which are equal to the values of nonconditions of these loops.

The accelerations of MBS can be divided into two parts, one part depending only on velocities and the other part depending only on second time derivatives of variables (for details see Chapter 9). The total acceleration is the addition of both parts. The solution of MBS acceleration can be achieved as the solution of MBS velocity where the velocities are substituted by those parts of accelerations which depend only on the second time derivatives of variables; the parts of accelerations which depend only on velocities are then added in all appropriate points as the known components.

We start by once more solving the example from Section 5.5.2 by the geometric diagram method. Let us describe the velocity of point S from Figure 5.15 in the velocity diagram. We conclude from the analysis of the motion of member 3 that the end-point of velocity of point S in the velocity diagram lies on a straight line with the parameter $\dot{\varphi}_{23}$, and from the analysis of the motion of member 4 that the end-point of velocity of point S in the velocity diagram lies on a plane with the parameters $\dot{\varphi}_{15}$ and $\dot{\varphi}_{54}$. Hence the end-point of the velocity of point S lies in the intersection of the straight line with the plane and we obtain the velocities $\dot{\varphi}_{23}, \dot{\varphi}_{15}, \dot{\varphi}_{54}$. Similarly, we can find the corrections $\Delta\varphi_{23}, \Delta\varphi_{15}, \Delta\varphi_{54}$ for the position solution or the accelerations $\ddot{\varphi}_{23}, \ddot{\varphi}_{15}, \ddot{\varphi}_{54}$ for the acceleration solution. The idea of the geometric diagram method can also be applied directly to the position solution. The end-point of the radius vector of point S according to the origin of the frame system can be determined in two ways: from the analysis of the motion of member 3 it lies on a circle, and from the analysis of the motion of member 4 it lies on a sphere. Hence the end-point of the radius vector of point S lies in the intersection of the circle and the sphere.

Let us now describe the velocity solution of the most general 7R mechanism (Figure 5.25) where the angular velocity ω_1 is given. This solution is the basis of the kinematic solutions of all general MBS by the geometric diagram and the parametric methods. We use the second variant of the generalized method of disconnecting a loop and we remove the member 45 with the revolute joints 4 and 5. According to equations (5.132) we obtain

$$\mathbf{u}_{1B_4} - \mathbf{u}_{1B_5} = \frac{h_{45}}{\sin\delta_{45}}(\mathrm{e}_4 \times \mathrm{e}_5) \qquad (5.146)$$

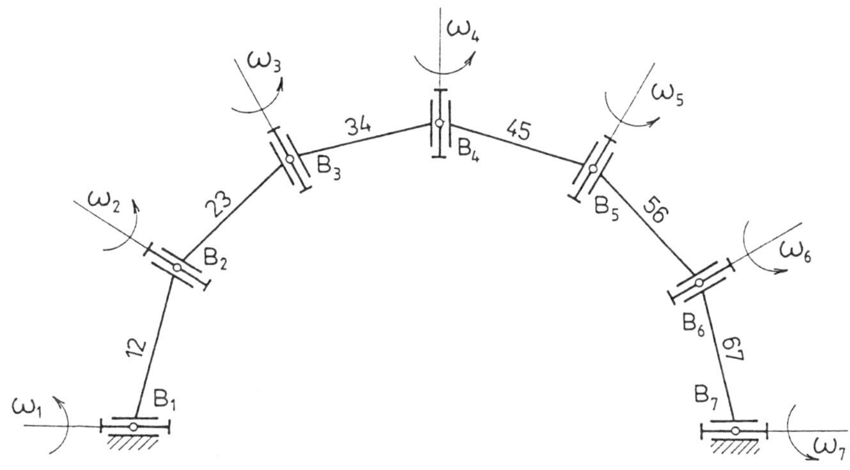

Figure 5.25. General 7R mechanism.

$$\mathbf{e}_4.\mathbf{e}_5 \quad = \quad \cos\delta_{45} \tag{5.147}$$

We differentiate these equations with respect to time

$$\dot{\mathbf{u}}_{1B_4} - \dot{\mathbf{u}}_{1B_5} = \frac{h_{45}}{\sin\delta_{45}}((\boldsymbol{\omega}_{14} \times \mathbf{e}_4) \times \mathbf{e}_5 + \mathbf{e}_4 \times (\boldsymbol{\omega}_{15} \times \mathbf{e}_5)) \tag{5.148}$$

$$(\boldsymbol{\omega}_{14} \times \mathbf{e}_4).\mathbf{e}_5 + \mathbf{e}_4.(\boldsymbol{\omega}_{15} \times \mathbf{e}_5) = \mathbf{0} \tag{5.149}$$

and after modifications of (5.148) we have

$$\begin{aligned} & \dot{\mathbf{u}}_{1B_4} - \frac{h_{45}}{\sin\delta_{45}}\mathbf{e}_5 \times (\mathbf{e}_4 \times \boldsymbol{\omega}_{14}) = \\ = \quad & \dot{\mathbf{u}}_{1B_5} - \frac{h_{45}}{\sin\delta_{45}}\mathbf{e}_4 \times (\mathbf{e}_5 \times \boldsymbol{\omega}_{15}) \end{aligned} \tag{5.150}$$

Here $\dot{\mathbf{u}}_{1B_4}$ and $\boldsymbol{\omega}_{14}$ are functions of ω_2 and ω_3 only, and, similarly, $\dot{\mathbf{u}}_{1B_5}$ and $\boldsymbol{\omega}_{15}$ are functions of ω_6 and ω_7 only. We can consider the expression on the left side of (5.150) as the velocity of a point depending on two parameters ω_2 and ω_3. The end-point of this velocity lies in the velocity diagram on a plane. Similarly the expression on the right hand of (5.150) can also be considered as the velocity of a point whose end-point lies in the velocity diagram on a plane with two parameters ω_6 and ω_7. The intersection of these two planes is a straight line with some parameter λ . Hence the angular velocities $\omega_2, \omega_3, \omega_6$ and ω_7 as well as $\boldsymbol{\omega}_{14}$ and $\boldsymbol{\omega}_{15}$ can be expressed as linear functions of the parameter λ . We substitute these linear functions in the equation (5.149) and we compute the parameter λ and then the velocities $\omega_2, \omega_3, \omega_6$ and ω_7. This type of solution uses only the geometric diagram method.

The solution of kinematics (e.g. velocities) of a general MBS is realized according to the following algorithm. An MBS with dependent variables consists of several kinematic loops. We solve these kinematic loops sequentially. In each loop we can determine 6 dependent variables. However, because the loops have common bodies and kinematic joints, the loops can contain more that 6 dependent variables. We must choose the values of these dependent variables; their correct values will be determined from further (e.g. last) kinematic loops which will be solved and where some conditions must be fulfilled for the problem to be solvable.

Let us solve the example from Section 5.7.2 using this algorithm. The variables of the helical joints (14–20, 15–21, 16–22, 17–23, 18–24, 19–25 see Figure 5.17b, c) are given and therefore these helical joints are not considered further. We solve the kinematic loops in the sequence loop 1, loop 4, loop 2, loop 5, loop 3. In loop 1 we choose the variables of the joints 1–2 and 2–8, and the variables of other joints 8–14, 20–26, 26–21, 9–15, 3–9, 1–3 are then determined. In loop 4 we choose the variables of the joints 26–29 and 29–32, and the variables of all other joints are then determined. Loop 2 then contains only 4 unknown variables. Hence, for this loop to be solvable, we obtain two conditions from which we can determine two chosen variables, e.g. 26–29 and 29–32. If the loop contains only 4 joints, then the left side of the equation (5.150) depends only on one unknown variable and also the right-hand side of the equation (5.150) depends only on one unknown variable. The end-points of velocities on both sides of the equation (5.150) lie on straight lines which may not intersect. If we change two chosen variables then we obtain the fact that the intersection points from the equation (5.150) lie on a straight line (from the combination of conditions of both sides plane – plane or space – straight line). Then we can finish the solution according to the above described algorithm from the equation (5.149). The loop 5 is directly solvable because it then contains 6 unknown variables. Finally the loop 3 contains only 4 unknown variables and from this we determine the last two chosen variables 1–2 and 2–8.

The open question is the number of these choices which are necessary for the solution of general MBS. If the number of these necessary choices is limited to some fixed finite number, this algorithm has linear computational complexity.

5.12 Numerical Solution

Parameters appearing in MBS kinematic equations are divided into constants giving various dimensions and variables. Variables may by divided into independent variables $\mathbf{q} = \lceil q_1, q_2, \ldots, q_m \rceil$ and dependent variables $\mathbf{z} = \lceil z_1, z_2, \ldots, z_n \rceil$. We rewrite all equations given by the respective method with zeros on the right-hand side; we then write them down in an arbitrarily chosen order (but fixed from now on) into components of vector $\mathbf{f}$. We thus obtain a system of non-linear equations describing the position of MBS

$$\mathbf{f}(\mathbf{z}, \mathbf{q}) = \mathbf{0} \tag{5.151}$$

For example, in the matrix method we write in the selected MBS loop order the elements of matrix equations (5.94) and (5.97), modified in such a way that the right-hand side is zero, according to lines; we then obtain from each equation (5.94) 12 equations and from equation (5.97) 3 equations.

Equations (5.151) are non-linear and their analytical solution in closed form is possible only in some specific cases – see the following Section 5.13. In general we must use a numerical method (Rektorys, 1968), of which the most suitable is Newton's method. The method is based on the Taylor expansion (5.151) to the first order in the neighbourhood of initial value $\mathbf{z}^{(k)} = \lceil z_1^{(k)}, z_2^{(k)}, \ldots, z_n^{(k)} \rceil$

$$\mathbf{f}(\mathbf{z}, \mathbf{q}) = \mathbf{f}(\mathbf{z}^{(k)}, \mathbf{q}) + \frac{\partial \mathbf{f}}{\partial \mathbf{z}^T} \Delta \mathbf{z}^{(k)} = \mathbf{0} \tag{5.152}$$

Let us denote the Jacobian of the system (5.151) as

$$\begin{aligned} \mathbf{\Phi}_z(\mathbf{z}, \mathbf{q}) &= \frac{\partial \mathbf{f}(\mathbf{z}, \mathbf{q})}{\partial \mathbf{z}^T} \\ \mathbf{\Phi}_z &= \mathbf{\Phi}_z(\mathbf{z}^{(k)}, \mathbf{q}) \end{aligned} \tag{5.153}$$

From (5.152) we obtain a system of linear equations for $\Delta \mathbf{z}^{(k)}$

$$\mathbf{\Phi}_z \Delta \mathbf{z}^{(k)} = -\mathbf{f}(\mathbf{z}^{(k)}, \mathbf{q}) \tag{5.154}$$

If the Jacobian $\mathbf{\Phi}_z$ is a non-singular matrix, it is possible to solve the system (5.154) directly

$$\Delta \mathbf{z}^{(k)} = -\mathbf{\Phi}_z^{-1} \mathbf{f}(\mathbf{z}^{(k)}, \mathbf{q}) \tag{5.155}$$

setting

$$\mathbf{z}^{(k+1)} = \mathbf{z}^{(k)} + \Delta \mathbf{z}^{(k)} \tag{5.156}$$

We consider vector $\mathbf{z}^{(k+1)}$ a more precise estimate for step $k + 1$. We substitute again into equation (5.154) and repeat the whole procedure until

$|\Delta \mathbf{z}^{(k)}| \leq \varepsilon$ for small positive ε . We then consider $\mathbf{z}^{(k)}$ to be the solution sought (5.151).

The Jacobian $\mathbf{\Phi}_z$ is generally, however, a singular matrix. The reason for this is that, in problems of kinematics (and similarly in statics and dynamics), the examined system of linear equations has different numbers of equations and unknowns. We have seen that the matrix method yields, for an MBS loop, 12 equations but only 6 unknowns. Of the 12 equations only 6 are independent. In that case, matrix $\mathbf{\Phi}_z$ would be rectangular and from 12 linear equations 6 would be independent; using this regular core we would find the solution $\Delta \mathbf{z}^{(k)}$.

We shall see further in Chapter 7 that there are the so-called singular cases, where the MBS has redundant degrees of freedom and even the square matrix $\mathbf{\Phi}_z$ is singular.

It is therefore necessary for the solution of equation (5.154) to use special algorithm, which lead to the solution of a general rectangular system of linear algebraic equations. The solution is found from the optimal choice of regular core of the system (5.154) (see Chapter 7).

Another approach follows from the connection between quaternion and matrix notations. Let us remind ourselves once more that the matrix approach leads in the formalism of (5.75) to the equality of two (4,4) matrices. Leaving out identically equal elements, the condition (5.75) leads to the equality of 12 elements in corresponding places on both sides of the equation. The significance of individual elements is, however, not the same. The elements of the last column describe a vector and they must always be included in the calculations. There remain 9 elements s_{ij} of direction cosines matrix $\mathbf{S}$ describing the orientation, of which only 3 are independent. The quaternion notation of orientation leads to equality of only 4 elements, 3 of which are independent. This fact means a tremendous saving of numerical calculations when using the quaternion approach. This saving may be obtained also in the matrix approach by utilizing the correspondence between matrix $\mathbf{S}$ and the respective normed quaternion as described in Section 4.3.1 by relations (4.177) and (4.178).

Selecting, for example, $k = 0$, we obtain the following equations expressing constraints between elements s_{ij} of the direction cosines matrix and elements e_j of the normed quaternion

$$\begin{aligned} e_0' &= \frac{1 + \operatorname{tr} \mathbf{S}}{2\sqrt{b_{00}'}}, \quad e_1' = \frac{-s_{23} + s_{32}}{2\sqrt{b_{00}'}} \\ e_2' &= \frac{-s_{31} + s_{13}}{2\sqrt{b_{00}'}}, \quad e_3' = \frac{-s_{12} + s_{21}}{2\sqrt{b_{00}'}} \end{aligned} \tag{5.157}$$

where $b_{00}' = 1 + \operatorname{tr}\mathbf{S} = 1 + s_{11} + s_{22} + s_{33}$.

Thus, when two matrices of direction cosines express the same orientation, the expressions

$$1+\mathrm{tr}\mathbf{S}, \quad -s_{23}+s_{32}, \quad -s_{31}+s_{13}, \quad -s_{12}+s_{21} \tag{5.158}$$

assembled from corresponding elements of those matrices must be equal. If in a special case $b'_{00}=0$, then also $e'_0=0$ and the last three expressions in (5.158) would be linearly dependent. At least two expressions out of

$$s_{23}+s_{32}, \qquad s_{31}+s_{13}, \qquad s_{12}+s_{21} \tag{5.159}$$

will, however, be linearly independent. These expressions together with the condition $1+\mathrm{tr}\mathbf{S}=0$ yield three independent equations for the determination of orientation.

It may be shown (Valášek and Stejskal, 1988) that a set of four expressions exists, three of which are always linearly independent. These are the expressions

$$\begin{aligned} &-\frac{1+\mathrm{tr}\mathbf{S}}{4}s_{23}+s_{32}, \qquad -\frac{1+\mathrm{tr}\mathbf{S}}{4}s_{31}+s_{13},\\ &-\frac{1+\mathrm{tr}\mathbf{S}}{4}s_{12}+s_{21}, \qquad 1+\mathrm{tr}\mathbf{S} \end{aligned} \tag{5.160}$$

where

$$0 \le \frac{1+\mathrm{tr}\mathbf{S}}{4} \le 1 \tag{5.161}$$

The above result enables us to reduce substantially the number of equations, by showing how to choose out of the equations produced by the matrix method a set of equations with the minimum degree of dependence. This is a reduction from 12 to 7 equations for every loop. Of those 7 equations only one is still linearly dependent.

Similar method to (5.158) for the description of the orientation is the usage of three expressions of vector represented by the skew-symmetric matrix

$$\frac{1}{2}(\mathbf{S}-\mathbf{S}^T)$$

which, however, could be singular.

The Newton method described above converges if we begin with a good initial estimate. Using a coarse initial estimate may lead to divergence or to oscillation around the solution. A very effective modification of the Newton method is to calculate the next iteration with the relation

$$\mathbf{z}^{(k+1)}=\mathbf{z}^{(k)}+\varkappa\Delta\mathbf{z}^{(k)} \tag{5.162}$$

instead of (5.156), where $\varkappa$ is a scalar parameter. Its value is to be obtained from a one-dimensional minimization of

$$|\mathbf{f}(\mathbf{z}^{(k+1)}, \mathbf{q})| = \sqrt{\sum_{i=1}^{n} f_i^2(\mathbf{z}^{(k+1)}, \mathbf{q})} \tag{5.163}$$

which is a function of parameter $\varkappa$ only. A simpler calculation procedure is as follows. We start with $\varkappa = 1$. We compute from (5.162) $\mathbf{z}^{(k+1)}$ and $\mathbf{f}(\mathbf{z}^{(k+1)}, \mathbf{q})$. If $|\mathbf{f}(\mathbf{z}^{\mathbf{k+1})}, \mathbf{q})| \geq |\mathbf{f}(\mathbf{z}^{(k)}, \mathbf{q})|$ we set $\varkappa = \varkappa/2$ and compute again $\mathbf{z}^{(k+1)}$ and $\mathbf{f}(\mathbf{z}^{(k+1)}, \mathbf{q})$. This is to be repeated until $|\mathbf{f}(\mathbf{z}^{(k+1)}, \mathbf{q})| < |\mathbf{f}(\mathbf{z}^{(k)}, \mathbf{q})|$. Both computational procedures are equivalent.

For the Newton method in this modification the following convergence theorem holds (Fedorenko, 1978):

Theorem: In some bounded area let $|\mathbf{f}(\mathbf{z})| \leq R$; further, let function $\mathbf{f}(\mathbf{z})$ have uniformly bounded first and second partial derivatives. In that area let $\mathbf{x} = \mathbf{f}(\mathbf{z})$ be a one-to-one mapping and $|\mathbf{\Phi}_z^{-1}(\mathbf{z})| \leq C$. Further, in that area, let a point $\mathbf{z}^*$ exist (according to the above assumptions only one point), so that $\mathbf{f}(\mathbf{z}^*) = \mathbf{0}$. The modified Newton method then converges from any estimate $|\mathbf{f}(\mathbf{z}^{(0)}| \leq R$, i.e. $\mathbf{z}^{(k)} \longrightarrow \mathbf{z}^*$ for $k \longrightarrow \infty$.

However, in some cases even the modified method does not converge or converges only slowly. The non-convergence is caused by a singularity of mapping $\mathbf{f}(\mathbf{z})$, i.e. $\det \mathbf{\Phi}_z(\mathbf{z})) = 0$ on some hypersurface in the area $|\mathbf{f}(\mathbf{z})| \leq R$. It is then necessary to use the Newton method always on one function f_i and consider the function as a function of one variable z_i. In a cycle we interchange the individual variables and the functions f_i assigned to them, while keeping the other functions in a range $\varepsilon < |f_i| < 2\varepsilon$ which is being made consecutively smaller. Should this not be the case, the calculated increment must be made smaller.

If the mapping $\mathbf{f}(\mathbf{z})$ is not singular, but the convergence is slow, we may use a scale factor modification

$$f_i(\mathbf{z}) \longrightarrow \eta_i f_i(\mathbf{z}) \tag{5.164}$$

for which the computation of increment $\Delta\mathbf{z} = \mathbf{\Phi}_z^{-1}\mathbf{f}(\mathbf{z})$ is invariant, unlike the condition

$$|\mathbf{f}(\mathbf{z} + \varkappa\Delta\mathbf{z})| < |\mathbf{f}(\mathbf{z})| \tag{5.165}$$

We therefore modify the function scale factor according to (5.164) so that quantities

$$\frac{1}{\eta_i} = |\mathbf{\Phi}_{iz}(\mathbf{z})| = \sqrt{\sum_{k=1}^{n} \left(\frac{\partial f_i(z_k)}{\partial z_k}\right)^2} \tag{5.166}$$

would be of the same order. Alternatively, instead of condition (5.165) we use for the choice of parameter $\varkappa$ the condition

$$|\mathbf{F}(\mathbf{z} + \varkappa\Delta\mathbf{z})| < |\mathbf{F}(\mathbf{z})| \tag{5.167}$$

where

$$\mathbf{F}(\mathbf{z}) = \sqrt{\sum_{i=1}^{N}(\eta_i f_i(\mathbf{z}))^2} \tag{5.168}$$

In that case however the calculation of scale factors (5.166) is dynamic, i.e. it is a function of $\mathbf{z}$ and $\varkappa$. Therefore the theorem no longer applies to it, as the norm is variable, but the practical experience is good (Fedorenko, 1978). Other possible modifications exist (Dennis and Schnabel, 1983).

Having solved the position problem the solution of velocities and accelerations is easy. The time derivative of equation (5.151) yields

$$\begin{aligned} \mathbf{\Phi}_z(\mathbf{z},\mathbf{q})\dot{\mathbf{z}} + \mathbf{\Phi}_q(\mathbf{z},\mathbf{q})\dot{\mathbf{q}} &= \mathbf{0} \\ \mathbf{\Phi}_z(\mathbf{z},\mathbf{q})\dot{\mathbf{z}} &= -\mathbf{\Phi}_q(\mathbf{z},\mathbf{q})\dot{\mathbf{q}} \end{aligned} \tag{5.169}$$

where

$$\mathbf{\Phi}_q(\mathbf{z},\mathbf{q}) = \frac{\partial \mathbf{f}(\mathbf{z},\mathbf{q})}{\partial \mathbf{q}^T} \tag{5.170}$$

Knowing $\mathbf{z}, \mathbf{q}, \dot{\mathbf{q}}$ we determine the required velocities $\dot{\mathbf{z}}$ from the system of linear algebraic equations (5.169). The matrix of system (5.169) is equal to the Jacobian of system (5.151), and to obtain a solution we must again use special algorithms for solving systems of linear algebraic equations (see Chapter 7).

The time derivative of equation (5.169) yields

$$\begin{aligned} \mathbf{\Phi}_z(\mathbf{z},\mathbf{q})\ddot{\mathbf{z}} + \dot{\mathbf{\Phi}}_z\dot{\mathbf{z}} &= -\mathbf{\Phi}_q(\mathbf{z},\mathbf{q})\ddot{\mathbf{q}} - \dot{\mathbf{\Phi}}_q\dot{\mathbf{q}} \\ \mathbf{\Phi}_z(\mathbf{z},\mathbf{q})\ddot{\mathbf{z}} &= -\mathbf{\Phi}_q(\mathbf{z},\mathbf{q})\ddot{\mathbf{q}} - \dot{\mathbf{\Phi}}_q\dot{\mathbf{q}} - \dot{\mathbf{\Phi}}_z\dot{\mathbf{z}} \end{aligned} \tag{5.171}$$

where

$$\dot{\mathbf{\Phi}}_z = \frac{\mathrm{d}}{\mathrm{dt}}\mathbf{\Phi}_z(\mathbf{z},\mathbf{q}), \qquad \dot{\mathbf{\Phi}}_q = \frac{\mathrm{d}}{\mathrm{dt}}\mathbf{\Phi}_q(\mathbf{z},\mathbf{q}) \tag{5.172}$$

Knowing $\mathbf{q}, \mathbf{z}, \dot{\mathbf{q}}, \dot{\mathbf{z}}, \ddot{\mathbf{q}}$, we determine the required accelerations $\ddot{\mathbf{z}}$ from the system of linear algebraic equations (5.171). The matrix of system (5.171) is again equal to the Jacobian of system (5.151), and special algorithms must be used for the solution.

After having solved the problem of MBS position, velocity and acceleration, we may, according to Section 5.4.3, solve the motion of an arbitrary MBS point or body.

5.13 Analytical Solution

The system of equations (5.151) for solving the MBS position contains nonlinear equations, and its analytical solution is known only for some special cases. In general it is always necessary to apply the numerical solution. However, the analytical solution has gained in significance with the development of robotics.

We describe the position of the robot end effector by the transformation matrix $\mathbf{T}_{1G}$, where 1 is the frame. This matrix is a function of coordinates of the kinematic pairs of the robot. The position of the robot end element may also be expressed as the required motion from the viewpoint of robot function with the help of the transformation matrix $\mathbf{T}_{1M}$. Then

$$\mathbf{T}_{1G} = \mathbf{T}_{1M} \tag{5.173}$$

Matrix $\mathbf{T}_{1M}$ elements are known, for example, as time functions. From equation (5.173) we may determine the unknown values of coordinates of kinematic pairs of the robot. The solution of equation (5.173) and the solution of the problem of determining robot pair coordinates for the realization of a required motion, respectively, is called the inverse robot kinematic problem.

The possibility of its analytical solution is important for sophisticated controlling of robot motion (Paul, 1981; Brady, 1982). The solution may be found directly if the degree of freedom of the robot is not more than 6. If it is more than 6, i.e. by redundant degrees of freedom, the equation (5.173) must be extended by further conditions. The robot may further contain several loops. It is then necessary to assemble the equations for those loops. Designating $\mathbf{z} = \lceil z_i \rceil$ as unknown robot pair coordinates and $\mathbf{q} = \lceil q_j \rceil$ as given coordinates for expressing the required robot motion, the equation (5.173) may be rewritten into the form (5.151) for the solution of the MBS position. The solution of the inverse robot kinematic problem has a direct connection with the solution of the MBS position problem. The requirement on the robot end element motion is equivalent to a closed loop of a mechanism.

A complete algorithm for finding the analytical solution of the inverse kinematic problem is not known. Some conditions have been found which lead to the finding of the solution (Yuchen and Wenhan, 1986). We shall further derive the pieces of knowledge which are the basis for solubility conditions and procedures for finding an analytical solution in some cases.

We have described some formalisms for the description of MBS kinematics in previous chapters. We have compared them with respect to computational complexity of deriving kinematic equations and with respect to the number of redundant equations which they yield. From these viewpoints the most advantageous formalism is the matrix formalism combined with SDQ formalism as in the previous section. The requirements on the formalism used are, from the viewpoint of analytical solution, different. We need equations which contain only a part of the unknown coordinates to be able to solve them analytically. Redundant equations do not cause any harm – the contrary is true. Therefore matrix and screw notations rather than quaternions are convenient for an analytical solution. The screw notation represented by screw affinors does yield 18 equations, but in specific cases it has not shown any advantage against the matrix notation. The reason is probably the fact that affinor dual components contain all the variables of affinor real components, as follows from (4.237) and we therefore obtain from the dual components no analytically soluble equations (see, for example, equations for the solution of the universal joint in Section 5.5.1 obtained by all three formalisms). We shall in the following use only the matrix formalism.

The kinematic equations for the description of position may be divided into equations describing the identity of translational displacement coordinates and equations describing the identity of direction cosines. The coordinates of prismatic pairs and of constant translational displacements, i.e. of transformation matrices $\mathbf{T}_{Z1}, \mathbf{T}_{Z2}, \mathbf{T}_{Z3}$ in (5.173), appear only in equations of translations. On the contrary, the coordinates of revolute pairs and constant rotations, i.e. of transformation matrices $\mathbf{T}_{Z4}, \mathbf{T}_{Z5}, \mathbf{T}_{Z6}$ in (5.173), appear generally in all equations.

The following holds:

A. If in (5.173) the unknown coordinate is a translational coordinate (i.e. an argument of $\mathbf{T}_{Z1}, \mathbf{T}_{Z2}, \mathbf{T}_{Z3}$), it does not appear in equations of direction cosines.

B. For each unknown rotation (angular) coordinate in (5.173) (i.e. argument of $\mathbf{T}_{Z4}, \mathbf{T}_{Z5}, \mathbf{T}_{Z6}$) it is possible to assemble from (5.173) 4 equations (3 of direction cosines, 2 of which are independent, and 1 translation) or 6 equations (3 of direction cosines, 2 of which are independent, and 3 translations), which do not contain this coordinate.

C. If axes of 2 or more consecutivelly following rotations described by unknown coordinates intersect, it is possible to assemble from (5.173) 3 equations of translation, which do not contain unknown coordinates of those rotations.

D. If axes of 2 or more rotations following each other, described by unknown coordinates, are parallel to each other, the direction cosines equations may be expressed in such a way that they contain, instead of 2 or more of those unknowns, only one unknown – their sum.

E. If the kinematic pairs consecutively following each other, described by unknown coordinates, form the equivalent of a flat pair, it is possible to assemble 1 equation of translations in which their unknown coordinates do not appear.

The validity of statement A follows from the fact that dividing a transformation matrix into submatrices is invariant with respect to multiplication (see Section 4.2.3.1).

The validity of statement B follows from a modification of (5.173), for example with an unknown rotation coordinate φ as the argument of the basic matrix $\mathbf{T}_{Z4}$. Equation (5.173) may be written in general

$$\mathbf{T}_A \mathbf{T}_{Z4}(\varphi) \mathbf{T}_B = \mathbf{T}_{1M} \tag{5.174}$$

$$\mathbf{T}_{Z4}(\varphi) \mathbf{T}_B = \mathbf{T}_A^{-1} \mathbf{T}_{1M} \tag{5.175}$$

$$\mathbf{T}_A \mathbf{T}_{Z4}(\varphi) = \mathbf{T}_{1M} \mathbf{T}_B^{-1} \tag{5.176}$$

The equations sought correspond to equality of elements 11, 12, 13 and 14 in (5.175) or 11, 21, 31, 14, 24 and 34 in (5.176).

The validity of statement C follows from a modification of (5.173) together with the consideration that the part of the transformation corresponding to intersecting axes of rotation is described by a sequence of multiplications of basic matrices $\mathbf{T}_{Z4}, \mathbf{T}_{Z5}, \mathbf{T}_{Z6}$ which contain as non-null submatrices only direction cosines submatrices and this division into submatrices is invariant with respect to multiplication. Designating the product of those basic matrices $\mathbf{T}_S$, (5.173) may in general be written as

$$\mathbf{T}_A \mathbf{T}_S \mathbf{T}_B = \mathbf{T}_{1M} \tag{5.177}$$

$$\mathbf{T}_A \mathbf{T}_S = \mathbf{T}_{1M} \mathbf{T}_B^{-1} \tag{5.178}$$

$$\begin{bmatrix} \mathbf{S}_A & 0 \\ \mathbf{r}_A & 1 \end{bmatrix} \begin{bmatrix} \mathbf{S}_S & 0 \\ 0 & 1 \end{bmatrix} = \begin{bmatrix} \mathbf{S}_{MB} & 0 \\ \mathbf{r}_{MB} & 1 \end{bmatrix} \tag{5.179}$$

Hence

$$\mathbf{S}_A \mathbf{S}_S = \mathbf{S}_{MB} \tag{5.180a}$$

$$\mathbf{r}_A = \mathbf{r}_{MB} \tag{5.180b}$$

Equations (5.180b) do not contain coordinates of those revolute pairs whose axes intersect.

Statement D follows from the invariance of division of transformation matrices into submatrices (Section 4.2.3.1). We obtain the resultant direction cosines submatrix as the product of direction cosines submatrices of individual basic matrices. If the axes of rotation are parallel, then among their basic matrices, e.g. $\mathbf{T}_{Z4}$, are only basic matrices of translations; and their direction cosines submatrices are unit matrices. Thus the product of direction cosines submatrices of basic matrices for the part of transformation (5.173) corresponding to parallel axes of rotation is given directly as a product of direction cosine submatrices of those parallel rotations. It may be proved by elementary substitution that, for example, for

$$\mathbf{S}_{Z4}(\varphi_1)\mathbf{S}_{Z4}(\varphi_2) = \mathbf{S}_{Z4}(\varphi_1 + \varphi_2) \tag{5.181}$$

and hence statement D holds.

Finally we obtain the validity of statement E by projecting the radius vector of the resultant translation carried out by a chain into the direction perpendicular to the plane equivalent to the flat pair. As unknown coordinates of pairs equivalent to flat pair describe motion performed only in the flat pair plane, they do not appear in the direction perpendicular to this plane.

An empirical statement holds: for analytical solubility of (5.173) it is necessary to obtain from (5.173) two sets of 3 equations for at most three unknowns. Combining conditions from statements A to E to obtain from (5.173) a system of 3 equations with 3 unknowns, we obtain cases where we are able to find analytically the inverse kinematic solution of robots with 6 degrees of freedom or the analytical solution of a simple mechanism.

The combinations are as follows:

1. A chain containing 3 translations (in prismatic or cylindrical pairs) (Figure 5.26a).

2. A chain containing 3 intersecting axes of rotation (or equivalently a spherical pair) (Figure 5.26b).

3. A chain containing a flat pair equivalent, i.e. either a sequence of three following parallel axes of rotation (Figure 5.26c) or a sequence in any order of 2 parallel axes of rotation and 1 translation in the direction perpendicular to the axis of rotation (Figure 5.26d) or a sequence in any order of two translations and 1 rotation with the axis normal to the plane given by the directions of those translations (Figure 5.26e).

4. A chain containing two translations and two parallel axes of rotation (Figure 5.26f).

In cases 1 to 4 we are able to find the analytical solution. If more conditions A – E are fulfilled than in 1 to 4, the analytical solution is simpler.

The above cases of possible analytical solution are of such significance for controlling robots that robot construction would take them into account. In particular, the robot wrist would often be constructed as a spherical pair.

We shall now prove cases 1 – 4. We shall use the Denavit–Hartenberg method formulated in basic matrices for the description of individual kinematic pairs. Its advantage is the pair description and designation convention which enables the forming of general conclusions in the form of resultant kinematic equations. We shall start with the proof from case 2.

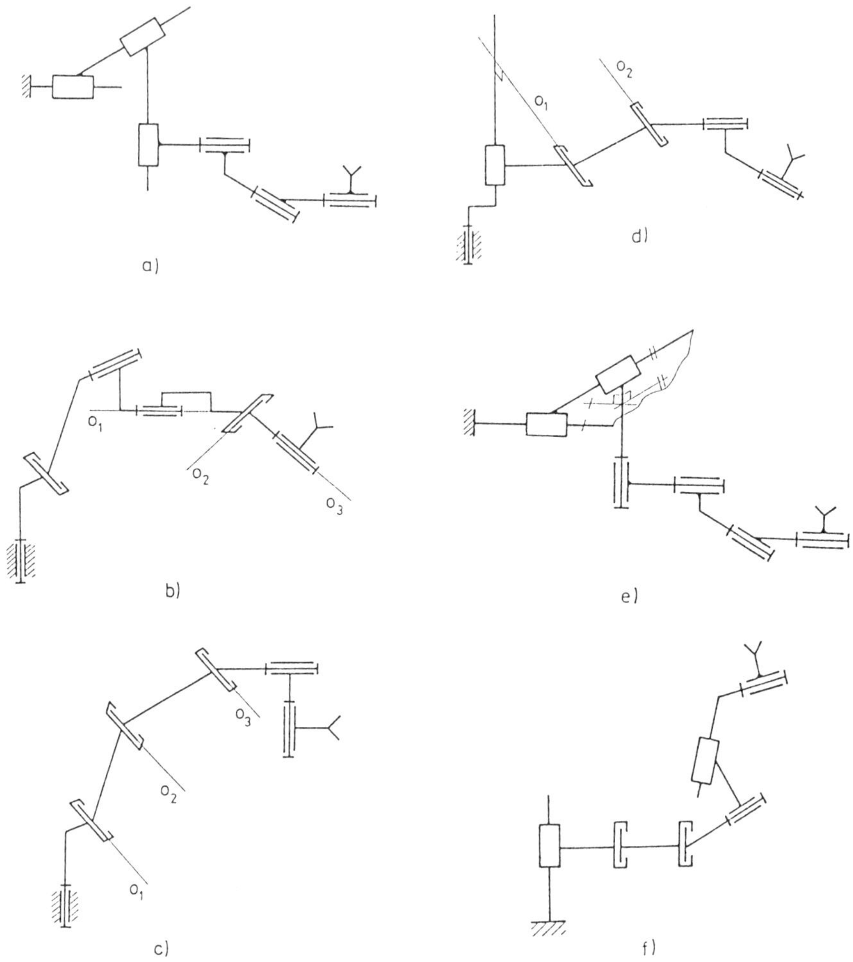

Figure 5.26. Diagram of robots with analytical solution possible.

Let us consider the most difficult and most frequent case 2, where the revolute pairs with intersecting axes are, counting from the frame, the first three or the last three in the kinematic chain. Then, according to (5.87) expressed in basic matrices for (5.180b), the following holds

$$\mathbf{T}_{Z6}(\varphi_{12})\mathbf{T}_{Z3}(s_{12})\mathbf{T}_{Z1}(a_{12})\mathbf{T}_{Z4}(\alpha_{12})\mathbf{T}_{Z6}(\varphi_{23})\mathbf{T}_{Z3}(s_{23})\mathbf{T}_{Z1}(a_{23}). \\ .\mathbf{T}_{Z4}(\alpha_{23})\mathbf{T}_{Z6}(\varphi_{34})\mathbf{T}_{Z3}(s_{34})\mathbf{T}_{Z1}(a_{34})\mathbf{T}_{Z4}(\alpha_{34})\mathbf{r}_{4S} = \mathbf{r}_{1S'} \quad (5.182)$$

Let us assume these to be only revolute pairs. The variables are then only basic motions Z6. Let us introduce constant transformation matrices $\mathbf{T}_A, \mathbf{T}_B, \mathbf{T}_C$ and let us rewrite (5.182)

$$\mathbf{T}_{Z6}(\varphi_{12})\mathbf{T}_A\mathbf{T}_{Z6}(\varphi_{23})\mathbf{T}_B\mathbf{T}_{Z6}(\varphi_{34})\mathbf{T}_C\mathbf{r}_{4S} = \mathbf{r}_{1S'} \tag{5.183}$$

Modifying,

$$\mathbf{T}_{Z6}(\varphi_{23})\mathbf{T}_B\mathbf{T}_{Z6}(\varphi_{34})\mathbf{T}_C\mathbf{r}_{4S} = \mathbf{T}_A^{-1}\mathbf{T}_{Z6}(-\varphi_{12})\mathbf{r}_{1S'} \tag{5.184}$$

Further designating

$$\begin{aligned} \mathbf{r}_1 &= \mathbf{T}_C\mathbf{r}_{4S} = \lceil \mathbf{u}_1, 1 \rceil \\ \mathbf{r}_2 &= \mathbf{T}_B\mathbf{T}_{Z6}(\varphi_{34})\mathbf{r}_1 = \lceil \mathbf{S}_B\mathbf{S}_{Z6}(\varphi_{34})\mathbf{u}_1 + \mathbf{u}_B, 1 \rceil \\ \mathbf{r}_3 &= \mathbf{T}_{Z6}(-\varphi_{12})\mathbf{r}_{1S'} = \lceil \mathbf{S}_{Z6}(-\varphi_{12})\mathbf{u}_{1S'}, 1 \rceil \end{aligned} \tag{5.185}$$

and modifying (5.184),

$$\begin{aligned} \mathbf{T}_{Z6}(\varphi_{23})\mathbf{r}_2 &= \mathbf{T}_A^{-1}\mathbf{r}_3 \\ \mathbf{S}_{Z6}(\varphi_{23})\mathbf{u}_2 &= \mathbf{S}_A^T\mathbf{u}_3 - \mathbf{S}_A^T\mathbf{u}_A \end{aligned} \tag{5.186}$$

Computing from (5.186) the square of the left-hand and right-hand side vector lengths

$$\begin{aligned} (\mathbf{S}_{Z6}(\varphi_{23})\mathbf{u}_2)^T(\mathbf{S}_{Z6}(\varphi_{23})\mathbf{u}_2) &= (\mathbf{S}_A^T\mathbf{u}_3 - \mathbf{S}_A^T\mathbf{u}_A)^T(\mathbf{S}_A^T\mathbf{u}_3 - \mathbf{S}_A^T\mathbf{u}_A) \\ \mathbf{u}_2^T\mathbf{u}_2 = \mathbf{u}_3^T\mathbf{u}_3 &+ \mathbf{u}_A^T\mathbf{u}_A - 2\mathbf{u}_A^T\mathbf{u}_3 \end{aligned} \tag{5.187}$$

Substituting (5.185) into (5.187) and modifying,

$$\begin{aligned} &\mathbf{u}_1^T\mathbf{u}_1 + \mathbf{u}_B^T\mathbf{u}_B + 2\mathbf{u}_B^T(\mathbf{S}_B\mathbf{S}_{Z6}(\varphi_{34})\mathbf{u}_1) = \\ &= \mathbf{u}_{1S'}^T\mathbf{u}_{1S'} + \mathbf{u}_A^T\mathbf{u}_A - 2\mathbf{u}_A^T(\mathbf{S}_{Z6}(-\varphi_{12})\mathbf{u}_{1S'}) \end{aligned} \tag{5.188}$$

Multiplying in (5.188) we obtain an equation with constant coefficient H_{ij}, G_{ij}

$$H_{11}\text{c}\varphi_{34} + H_{12}\text{s}\varphi_{34} + H_{13} = G_{11}\text{c}\varphi_{12} + G_{12}\text{s}\varphi_{12} + G_{13} \tag{5.189}$$

We now take the third component of the vector of the left- and right-hand side of equation (5.186). Due to the special form of the matrix $\mathbf{S}_{Z6}(\varphi_{23})$ according to (4.258), the third component of the vector of the left-hand side equals the third component of the vector $\mathbf{r}_2$; after substituting from (5.185) and performing the multiplication, we obtain an equation of a similar form to (5.189)

$$H_{21}\text{c}\varphi_{34} + H_{22}\text{s}\varphi_{34} + H_{23} = G_{21}\text{c}\varphi_{12} + G_{22}\text{s}\varphi_{12} + G_{23} \tag{5.190}$$

From the system of equations (5.189), (5.190), we calculate $c\varphi_{34}$ and $s\varphi_{34}$ with constant coefficients K_{ij}

$$\begin{aligned} c\varphi_{34} &= K_{11}c\varphi_{12} + K_{12}s\varphi_{12} + K_{13} \\ s\varphi_{34} &= K_{21}c\varphi_{12} + K_{22}s\varphi_{12} + K_{23} \end{aligned} \tag{5.191}$$

Using identity $s^2\varphi_{34} + c^2\varphi_{34} = 1$ and (5.191) we obtain

$$\begin{gathered} (K_{11}^2 + K_{21}^2)c^2\varphi_{12} + (K_{12}^2 + K_{22}^2)s^2\varphi_{12} + \\ +2(K_{11}K_{12} + K_{21}K_{22})s\varphi_{12}c\varphi_{12} + 2(K_{11}K_{13} + K_{21}K_{23})c\varphi_{12} + \\ + 2(K_{12}K_{13} + K_{22}K_{23})s\varphi_{12} + K_{13}^2 + K_{23}^2 = 1 \end{gathered} \tag{5.192}$$

In this equation, we either express $s\varphi_{12}$ and $c\varphi_{12}$ with the help of $\tan(\varphi_{12}/2)$, or we leave on one side of the equation terms with the first power of $s\varphi_{12}$; we then raise both sides of the equation to power 2 and substitute $s^2\varphi_{12} = 1 - c^2\varphi_{12}$. In both cases we obtain an algebraic equation of fourth degree, e.g. for $c\varphi_{12}$. We solve it analytically by converting it into the problem of cubic resolvent solution, and we solve the cubic equation most suitably using trigonometric substitution (Rektorys, 1968).

Let us have an algebraic equation of degree 4 with real coefficients $a \neq 0, b, c, d, e$, such that

$$ax^4 + bx^3 + cx^2 + dx + e = 0 \tag{5.193}$$

We convert it by substituting $x = y - (b/4a)$ into the equation

$$\begin{gathered} y^4 + py^2 + qy + r = 0 \\ p = -\frac{3b^2}{8a^2} + \frac{c}{a}, \quad q = \frac{b^3}{8a^3} - \frac{bc}{2a^2} + \frac{d}{a}, \\ r = -\frac{3b^4}{256a^4} + \frac{b^2c}{16a^3} - \frac{bd}{4a^2} + \frac{e}{a} \end{gathered} \tag{5.194}$$

We determine its roots $y_{1,2,3,4}$ from the solution of a cubic resolvent

$$z^3 + \frac{p}{2}z^2 + \left(\frac{p^2}{16} - \frac{r}{4}\right)z - \frac{q^2}{64} = 0 \tag{5.195}$$

using

$$\begin{aligned} y_1 &= \sqrt{z_1} + \sqrt{z_2} + \sqrt{z_3} \\ y_2 &= \sqrt{z_1} - \sqrt{z_2} - \sqrt{z_3} \\ y_3 &= -\sqrt{z_1} + \sqrt{z_2} - \sqrt{z_3} \\ y_4 &= -\sqrt{z_1} - \sqrt{z_2} + \sqrt{z_3} \end{aligned} \tag{5.196}$$

and fulfilling the condition

$$\sqrt{z_1}\sqrt{z_2}\sqrt{z_3} = -\frac{q}{8} \tag{5.197}$$

We convert the cubic equation with real coefficients $a \neq 0, b, c, d$

$$ax^3 + bx^2 + cx + d = 0 \tag{5.198}$$

by substituting $x = y - (b/3a)$ into the equation

$$y^3 + 3py + 2q = 0$$

$$3p = \frac{3ac - b^2}{3a^2}, \qquad 2q = \frac{2b^3}{27a^3} - \frac{bc}{3a^2} + \frac{d}{a} \tag{5.199}$$

We determine its solution as follows

$$r = \varepsilon\sqrt{|p|}, \qquad \varepsilon = \begin{cases} 1 \ldots q > 0 \\ -1 \ldots q < 0 \end{cases} \tag{5.200}$$

$p < 0$		$p > 0$
$p^3 + q^2 \leq 0$	$p^3 + q^2 > 0$	
$\cos\varphi = \frac{q}{r^3}$ $y_1 = -2r\cos\frac{\varphi}{3}$ $y_2 = 2r\cos(\frac{\pi}{3} - \frac{\varphi}{3})$ $y_3 = 2r\cos(\frac{\pi}{3} + \frac{\varphi}{3})$	$\cosh\varphi = \frac{q}{r^3}$ $y_1 = -r\cosh\frac{\varphi}{3}$ $y_{2,3} = r\cosh\frac{\varphi}{3}+$ $\pm i\sqrt{3}r\sinh\frac{\varphi}{3}$	$\sinh\varphi = \frac{q}{r^3}$ $y_1 = -2r\sinh\frac{\varphi}{3}$ $y_{2,3} = r\sinh\frac{\varphi}{3}+$ $\pm i\sqrt{3}\cosh\frac{\varphi}{3}$

(5.201)

From the known value $c\varphi_{12}$ we determine the angle φ_{13} with the help of computer function atan2

$$\varphi_{12} = \text{atan2}(\pm\sqrt{1 - c^2\varphi_{12}}, c\varphi_{12}) \tag{5.202}$$

There may exist more solutions to an algebraic equation of degree 4 permissible for $c\varphi_{12}$, and for each of these values there are two solutions from (5.202). The ambiguity of the solution is typical for the solution of MBS and robot position. It corresponds to the existence of various configurations of MBS or robot elements which satisfy the reaching of the required position.

We determinc $s\varphi_{34}$ and $c\varphi_{34}$ from the value φ_{12} and from (5.191); we determine the angle φ_{34} using the function atan2. Substituting into (5.186) and performing the multiplication we obtain from equalities of the first two vector components a system of two linear equations for $s\varphi_{23}$ and $c\varphi_{23}$. Hence finally, using function atan2, we determine the angle φ_{23}.

Of the three pairs, let two be revolute and one prismatic. The case where the prismatic pair is the first pair is equivalent to the case where the prismatic pair is the third pair. If it is the first pair, we seek out of equation (5.182) variables $s_{12}, \varphi_{23}, \varphi_{34}$. Multiplication of matrices $\mathbf{T}_{Z6}$ and $\mathbf{T}_{Z3}$ is commutative according to (4.259). We interchange the order of those matrices in (5.182) and, instead of equations (5.183)–(5.188), we obtain analogous equations

$$
\begin{aligned}
\mathbf{T}_{Z3}(s_{12})\mathbf{T}_A\mathbf{T}_{Z6}(\varphi_{23})\mathbf{T}_B\mathbf{T}_{Z6}(\varphi_{34})\mathbf{T}_C\mathbf{r}_{4S} &= \mathbf{r}_{1S'} \\
\mathbf{T}_{Z6}(\varphi_{23})\mathbf{T}_B\mathbf{T}_{Z6}(\varphi_{34})\mathbf{T}_C\mathbf{r}_{4S} &= \mathbf{T}_A^{-1}\mathbf{T}_{Z3}(-s_{12})\mathbf{r}_{1S'} \\
\mathbf{T}_{Z6}(\varphi_{23})\mathbf{r}_2 &= \mathbf{T}_A^{-1}(\mathbf{r}_{1S'} - \lceil 0, 0, s_{12}, 1\rceil)
\end{aligned}
$$

$$
\begin{gathered}
\mathbf{S}_{Z6}(\varphi_{23})\mathbf{u}_2 = \mathbf{S}_A^T(\mathbf{u}_{1S'} - \lceil 0, 0, s_{12}\rceil) - \mathbf{S}_A^T\mathbf{u}_A \\
(\mathbf{S}_{Z6}(\varphi_{23})\mathbf{u}_2)^T(\mathbf{S}_{Z6}(\varphi_{23})\mathbf{u}_2) = (\mathbf{S}_A^T(\mathbf{u}_{1S'} - \lceil 0, 0, s_{12}\rceil) - \\
-\mathbf{S}_A^T\mathbf{u}_A)^T(\mathbf{S}_A^T(\mathbf{u}_{1S'} - \lceil 0, 0, s_{12}\rceil) - \mathbf{S}_A^T\mathbf{u}_A) \\
\mathbf{u}_2^T\mathbf{u}_2 = \mathbf{u}_{1S'}^T\mathbf{u}_{1S'} + s_{12}^2 - 2\mathbf{u}_{1S'}^T\lceil 0, 0, s_{12}\rceil + \mathbf{u}_A^T\mathbf{u}_A - \\
- 2\mathbf{u}_A^T\mathbf{u}_{1S'} + 2\mathbf{u}_A^T\lceil 0, 0, s_{12}\rceil \qquad (5.203) \\
\mathbf{u}_1^T\mathbf{u}_1 + \mathbf{u}_B^T\mathbf{u}_B + 2\mathbf{u}_B^T(\mathbf{S}_B\mathbf{S}_{Z6}(\varphi_{34})\mathbf{u}_1) = s_{12}^2 - \\
- 2(\mathbf{u}_{1S'}^T - \mathbf{u}_A^T)\lceil 0, 0, s_{12}\rceil + \mathbf{u}_{1S'}^T\mathbf{u}_{1S'} + \mathbf{u}_A^T\mathbf{u}_A - 2\mathbf{u}_A^T\mathbf{u}_{1S'} \qquad (5.204)
\end{gathered}
$$

From (5.204) we obtain an equation analogous to (5.189)

$$H_{11}\mathrm{c}\varphi_{34} + H_{12}\mathrm{s}\varphi_{34} + H_{13} = G_{11}s_{12}^2 + G_{12}s_{12} + G_{13} \qquad (5.205)$$

Again, from the third component of the vector of the left- and right-hand side of (5.203) we obtain in analogy to (5.190)

$$H_{21}\mathrm{c}\varphi_{34} + H_{22}\mathrm{s}\varphi_{34} + H_{23} = G_{21}s_{12} + G_{22} \qquad (5.206)$$

It is possible either from equation (5.206) to express s_{12}, which we then substitute into (5.205) getting an analogy of equation (5.192) for φ_{34}, or with null G_{21} to solve the equation (5.206) for φ_{34} directly. The rest of the solution is easy to find.

If the prismatic pair is the second one, we commute again in the multiplication of matrices $\mathbf{T}_{Z6}$ and $\mathbf{T}_{Z3}$ and we include $\mathbf{T}_{Z6}$ into $\mathbf{T}_B$. Instead of equations (5.183)–(5.186) we obtain

$$
\begin{aligned}
\mathbf{T}_{Z6}(\varphi_{12})\mathbf{T}_A\mathbf{T}_{Z3}(s_{23})\mathbf{T}_B\mathbf{T}_{Z6}(\varphi_{34})\mathbf{T}_C\mathbf{r}_{4S} &= \mathbf{r}_{1S'} \\
\mathbf{T}_{Z3}(s_{23})\mathbf{T}_B\mathbf{T}_{Z6}(\varphi_{34})\mathbf{T}_C\mathbf{r}_{4S} &= \mathbf{T}_A^{-1}\mathbf{T}_{Z6}(-\varphi_{12})\mathbf{r}_{1S'} \\
\mathbf{T}_{Z3}(s_{23})\mathbf{r}_2 &= \mathbf{T}_A^{-1}\mathbf{r}_3 \\
\lceil 0, 0, s_{23}\rceil + \mathbf{u}_2 &= \mathbf{S}_A^T\mathbf{u}_3 - \mathbf{S}_A^T\mathbf{u}_A \qquad (5.207)
\end{aligned}
$$

From the equality of the first two components of vectors (5.207) we obtain directly equations analogous to (5.189)–(5.190), and it is easy to complete the solution.

We shall prove case 1 in the following way. If the mechanism contains three translations, the remaining pairs must be revolute. Should the contrary hold, the translations would be redundant as they are contained only in 3 equations in (5.173). We shall describe the remaining 3 revolute pairs using the Denavit–Hartenberg convention. We shall express the direction cosines submatrix in (5.173)

$$\mathbf{S}_A\mathbf{S}_{Z6}(\varphi_1)\mathbf{S}_B\mathbf{S}_{Z6}(\varphi_2)\mathbf{S}_C\mathbf{S}_{Z6}(\varphi_3)\mathbf{S}_D = \mathbf{S}_{1M} \tag{5.208}$$

where $\mathbf{S}_A, \mathbf{S}_B, \mathbf{S}_C, \mathbf{S}_D, \mathbf{S}_{1M}$ are constant matrices. We further modify

$$\mathbf{S}_{Z6}(\varphi_1)\mathbf{S}_B\mathbf{S}_{Z6}(\varphi_2)\mathbf{S}_C = \mathbf{S}_A^{-1}\mathbf{S}_{1M}\mathbf{S}_D^{-1}\mathbf{S}_{Z6}(-\varphi_3) \tag{5.209}$$

Elements 31, 32 and 33 of equation (5.209) are of the form (5.189). We choose two of them and solve them in the same way as equations (5.189)–(5.190). For the third coordinates of translations we obtain 3 linear equations in components of the radius vector of the resultant translation.

To prove case 3 we shall express the mechanism again using Denavit–Hartenberg's formalism. We write the equation (5.173) in the form

$$\mathbf{A}_1\mathbf{A}_2\mathbf{A}_3\mathbf{A}_4\mathbf{A}_5\mathbf{A}_6 = \mathbf{T}_{1M} \tag{5.210}$$

Let us first consider three parallel axes of rotation. We find the solution using statement E from one equation of translation and using B, D from two direction cosine equations not containing angles of three rotations with parallel axes. Let the pairs being considered be the first three pairs in the chain. We modify the equation (5.210)

$$\left[\begin{array}{c|c} \mathbf{S}_{Z6}(\varphi_1+\varphi_2+\varphi_3) & \begin{array}{c} a_1\mathrm{c}_1 + a_2\mathrm{c}_{12} + a_3\mathrm{c}_{123} \\ a_1\mathrm{s}_1 + a_2\mathrm{s}_{12} + a_3\mathrm{s}_{123} \\ d_1 + d_2 + d_3 \end{array} \\ \hline 0 \;\; 0 \;\; 0 & 1 \end{array}\right] \mathbf{T}_{Z4}(\alpha_3).$$

$$.\mathbf{T}_{Z6}(\varphi_4)\mathbf{T}_{Z3}(d_4)\mathbf{T}_{Z1}(a_4)\mathbf{T}_{Z4}(\alpha_4)\mathbf{T}_{Z6}(\varphi_5) =$$
$$= \mathbf{T}_{1M}\mathbf{T}_{Z4}^{-1}(\alpha_6)\mathbf{T}_{Z1}^{-1}(a_6)\mathbf{T}_{Z3}^{-1}(d_6)\mathbf{T}_{Z6}^{-1}(\varphi_6)\mathbf{T}_{Z4}^{-1}(\alpha_5)\mathbf{T}_{Z1}^{-1}(a_5)\mathbf{T}_{Z3}^{-1}(d_5) \tag{5.211}$$

as $\alpha_1 = \alpha_2 = 0$ for the axes to be parallel. The designation $\mathrm{s}_{123} = \mathrm{s}(\varphi_1 + \varphi_2 + \varphi_3)$, etc. is used again. We express from (5.211) element 34, which is of the form (5.189) for angles φ_4, φ_6

$$H_{11}\mathrm{c}_4 + H_{12}\mathrm{s}_4 + H_{13} = G_{11}\mathrm{c}_6 + G_{12}\mathrm{s}_6 + G_{13} \tag{5.212}$$

This is an equation of translation according to E. We express further from (5.211) element 33, which is again of the form (5.189), and (5.212) for angles φ_4 and φ_6 respectively, and we find the solution according to (5.191)–(5.201). Based on the expression of element 31 or 32 in (5.211), we determine φ_5 and for some remaining direction cosine equations the angle $\varphi_1 + \varphi_2 + \varphi_3$. Finally, we obtain from elements 14, 24 in (5.211), two equations of the type (5.189) for angles φ_1 and $\varphi_1 + \varphi_2$.

If pairs 2, 3, 4 are the revolute pairs with parallel axes in (5.210), we modify it again into a form analogous to (5.211)

$$\left[\begin{array}{c|c} \mathbf{S}_{Z6}(\varphi_2+\varphi_3+\varphi_4) & \begin{array}{c} a_2c_2 + a_3c_{23} + a_4c_{234} \\ a_2s_2 + a_3s_{23} + a_4s_{234} \\ d_2 + d_3 + d_4 \end{array} \\ \hline 0 \;\; 0 \;\; 0 & 1 \end{array}\right] \mathbf{T}_{Z4}(\alpha_4).$$

$$.\mathbf{T}_{Z6}(\varphi_5)\mathbf{T}_{Z3}(d_5)\mathbf{T}_{Z1}(a_5)\mathbf{T}_{Z6}(\varphi_6) =$$

$$= \mathbf{T}_{Z4}^{-1}(\alpha_1)\mathbf{T}_{Z1}^{-1}(a_1)\mathbf{T}_{Z3}^{-1}(d_1)\mathbf{T}_{Z6}^{-1}(\varphi_1)\mathbf{T}_{1M}\mathbf{T}_{Z4}^{-1}(\alpha_6)\mathbf{T}_{Z1}^{-1}(a_6)\mathbf{T}_{Z3}^{-1}(d_6) \tag{5.213}$$

We use the expression of elements 34 and 33 from (5.213). We obtain two equations of the type (5.189) for angles φ_1 and φ_5. We determine angle φ_6 from elements 31 or 32. From other elements of the direction cosine submatrix we determine $\varphi_2 + \varphi_3 + \varphi_4$, and from elements 14, 24 remaining φ_2 and $\varphi_3 + \varphi_3$. The solution of other positions of 3 revolute pairs with parallel axes is similar, converting into the cases solved above by inversion of both sides of equation (5.210).

If the chain contains 2 parallel axes of rotation and a translation in the direction perpendicular to it, we again modify (5.210) into the form (5.211) or (5.213), whereas the first matrix obtains the form of the type

$$\left[\begin{array}{c|c} \mathbf{S}_{Z6}(\varphi_1+\varphi_2) & \begin{array}{c} f_1(z_3,\varphi_1,\varphi_2) \\ f_2(z_3,\varphi_1,\varphi_2) \\ d \end{array} \\ \hline 0 \;\; 0 \;\; 0 & 1 \end{array}\right] \tag{5.214}$$

Here f_i designates a general function bilinear in z_3. Performing the same procedure as in the above cases we determine $\varphi_4, \varphi_6, \varphi_5, \varphi_1 + \varphi_2$, and from elements 14, 24, after eliminating z_3 and using the knowledge of $\varphi_1 + \varphi_2$ we determine φ_1 or φ_2.

If the chain contains 2 translations and a rotation the axis of which is perpendicular to the plane of translations, we again modify (5.210) into the form of (5.211) or (5.213), whereas the first matrix obtains the form of the type

$$\left[\begin{array}{c|c} & f_1(z_2, z_3, \varphi_1) \\ \mathbf{S}_{Z6}(\varphi_1) & f_2(z_2, z_3, \varphi_1) \\ & d \\ \hline 0\ 0\ 0 & 1 \end{array}\right] \tag{5.215}$$

Here f_i designates a general function for given φ_1 linear in z_2, z_3. The solution procedure is analogous. We determine $\varphi_4, \varphi_6, \varphi_5, \varphi_1$ and from elements 14, 24 we obtain 2 linear equations for z_2, z_3.

Case 4 is easy to prove by the following consideration. The direction cosine submatrix according to statement A contains no unknown coordinates of two translations and according to D contains unknown coordinates of 2 rotations with parallel axes only as their sum. Hence the direction cosine submatrix contains only three unknown angles (the sum of two angles and two remaining angles), which leads to the solution of equation (5.208) from case 1. From three equations of translations, which are linear in coordinates of translations, we eliminate those coordinates and we obtain an equation for one of the angles of rotation with parallel axes.

For real robots the finding of an analytical inverse kinematic solution is substantially simpler. On the other hand, robots are often constructed in such a way as to enable the analytical solution of their inverse kinematics to be made.

Let us present the well-known example of the Stanford manipulator (Figure 5.27).

The manipulator is described according to equation (5.173)

$$\mathbf{T}_{Z6}(\varphi_1)\mathbf{T}_{Z3}(l_1)\mathbf{T}_{Z5}(\varphi_2)\mathbf{T}_{Z2}(l_2)\mathbf{T}_{Z1}(-x)\mathbf{T}_{Z4}(\varphi_4)\mathbf{T}_{Z6}(\varphi_5)$$
$$\mathbf{T}_{Z4}(\varphi_6)\begin{bmatrix} \mathbf{S}_{6T} & \mathbf{r}_{6T} \\ \mathbf{0} & 1 \end{bmatrix} - \mathbf{T}_{1M} \tag{a}$$

where $\mathbf{S}_{6T}, \mathbf{r}_{6T}$ determines the orientation and position of the tool in the hand of robot 6. The inverse kinematical solution may be found analytically, as, in the wrist of the robot, the axes of rotation 4, 5, 6 intersect in point S. We modify equation (a) into the form (5.176)

$$\mathbf{T}_{Z6}(\varphi_1)\mathbf{T}_{Z3}(l_1)\mathbf{T}_{Z5}(\varphi_2)\mathbf{T}_{Z2}(l_2)\mathbf{T}_{Z1}(-x).$$
$$\cdot\begin{bmatrix} \mathbf{S}_{Z4}(\varphi_4)\mathbf{S}_{Z6}(\varphi_5)\mathbf{S}_{Z4}(\varphi_6), & 0 \\ \mathbf{0}, & 1 \end{bmatrix} = \mathbf{T}_{1M}\begin{bmatrix} \mathbf{S}_{6T}^T, & -\mathbf{S}_{6T}^T\mathbf{r}_{6T} \\ \mathbf{0}, & 1 \end{bmatrix} \tag{b}$$

and after designating the transformation matrix on the right-hand side $\mathbf{T}_{MT}$ we modify it into the form of (5.180b) and (5.178) respectively.

$$\mathbf{T}_{Z6}(\varphi_1)\mathbf{T}_{Z3}(l_1)\mathbf{T}_{Z5}(\varphi_2)\mathbf{T}_{Z2}(l_2)\mathbf{T}_{Z1}(-x)[0,0,0,1] = \mathbf{r}_{MT} \tag{c}$$

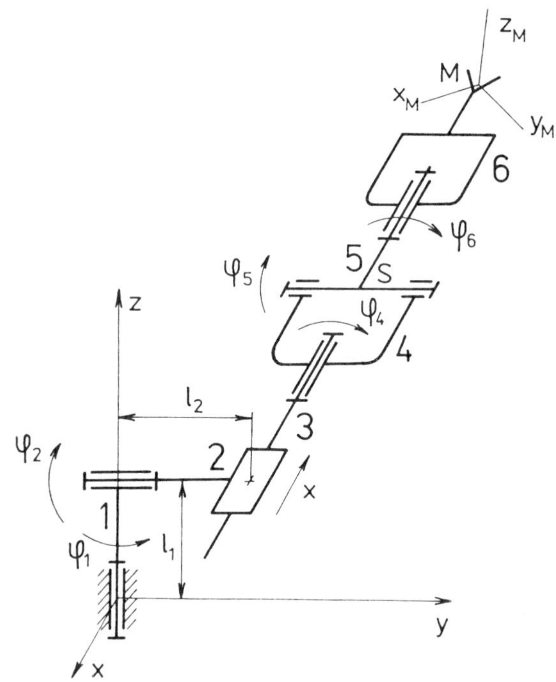

Figure 5.27. Stanford manipulator.

Further modification according to (5.180) leads to

$$\mathbf{T}_{Z5}(\varphi_2)\mathbf{T}_{Z2}(l_2)\mathbf{T}_{Z1}(-x)[0,0,0,1] = \mathbf{T}_{Z3}(-l_1)\mathbf{T}_{Z6}(-\varphi_1)\mathbf{r}_{MT} \qquad \text{(d)}$$

In the given case it suffices to perform the multiplications in (d) and we obtain

$$\begin{aligned} c\varphi_2(-x) &= c\varphi_1 r_{MT,1} + s\varphi_1 r_{MT,2} \\ l_2 &= -s\varphi_1 r_{MT,1} + c\varphi_1 r_{MT,2} \\ -s\varphi_2(-x) &= r_{MT,3} - l_1 \end{aligned} \qquad \text{(e)}$$

Now from the second equation of (e) we obtain

$$\begin{aligned} \alpha &= \operatorname{atan2}(r_{MT,1}, r_{MT,2}) \\ c(\varphi_1 + \alpha) &= l_2/\sqrt{r_{MT,1}^2 + r_{MT,2}^2} \\ \varphi_1 &= \operatorname{atan2}(\pm\sqrt{r_{MT,1}^2 + r_{MT,2}^2 - l_2^2}, l_2) - \operatorname{atan2}(r_{MT,1}, r_{MT,2}) \end{aligned} \qquad \text{(f)}$$

Then, from the first and third equation of (e)

$$\varphi_2 = \operatorname{atan2}(-(r_{MT,3} - l_1), c\varphi_1 r_{MT,1} + s\varphi_1 r_{MT,2}) \tag{g}$$

In this case only one solution exists, as $(-x) > 0$. Finally, again, from the first and third equation of (e)

$$(-x) = c\varphi_2(c\varphi_1 r_{MT,1} + s\varphi_1 r_{MT,2}) - s\varphi_2(r_{MT,3} - l_1) \tag{h}$$

We now express the equation (5.180a)

$$\mathbf{S}_{Z6}(\varphi_1)\mathbf{S}_{Z5}(\varphi_2)\mathbf{S}_{Z4}(\varphi_4)\mathbf{S}_{Z6}(\varphi_5)\mathbf{S}_{Z4}(\varphi_6) = \mathbf{S}_{MT}$$

$$\mathbf{S}_{Z4}(\varphi_4)\mathbf{S}_{Z6}(\varphi_5)\mathbf{S}_{Z4}(\varphi_6) = \mathbf{S}_{Z5}(-\varphi_2)\mathbf{S}_{Z6}(-\varphi_1)\mathbf{S}_{MT} \tag{i}$$

We know the elements of the matrix on the right-hand side and we designate the matrix $\mathbf{S}_S$. We modify the equation further according to (a) obtaining

$$\mathbf{S}_{Z4}(\varphi_4)\mathbf{S}_{Z6}(\varphi_5) = \mathbf{S}_S\mathbf{S}_{Z4}^{-1}(\varphi_6) = \mathbf{S}_S\mathbf{S}_{Z4}(-\varphi_6)$$

$$\begin{bmatrix} c\varphi_5, & -s\varphi_5, & 0 \\ c\varphi_4 s\varphi_5, & c\varphi_4 c\varphi_5, & -s\varphi_4 \\ s\varphi_4 s\varphi_5, & s\varphi_4 c\varphi_5, & c\varphi_4 \end{bmatrix} =$$

$$= \begin{bmatrix} s_{11}, & s_{12}c\varphi_6 + s_{13}s\varphi_6, & -s_{12}s\varphi_6 + s_{13}c\varphi_6 \\ s_{21}, & s_{22}c\varphi_6 + s_{23}s\varphi_6, & -s_{22}s\varphi_6 + s_{23}c\varphi_6 \\ s_{31}, & s_{32}c\varphi_6 + s_{33}s\varphi_6, & -s_{32}s\varphi_6 + s_{33}c\varphi_6 \end{bmatrix} \tag{j}$$

Hence, comparing elements 21, 31, 13, 11, 12

$$\begin{aligned} \varphi_4 &= \operatorname{atan2}(s_{31}, s_{21}) \\ \varphi_6 &= \operatorname{atan2}(\pm s_{13}, \pm s_{12}) \\ \varphi_5 &= \operatorname{atan2}(-(s_{12}c\varphi_6 + s_{13}s\varphi_6), s_{11}) \end{aligned} \tag{k}$$

Chapter 6

MBSs with Higher Kinematic Pairs

Higher kinematic pairs are, according to Section 3.1, those pairs where the two bodies have their contact in a point or curve. We may classify the types of contact as follows:

a) surface and surface, Figure 6.1a,
b) surface and curve, Figure 6.1b,
c) surface and point, Figure 6.1c,
d) curve and curve, Figure 6.1d,
e) curve and point, Figure 6.1e.

An infinite number of specific surface and curve forms could, however, be presented. That is the reason why higher kinematic pairs, with the exception of a general kinematic pair, do not have special designations. A general kinematic pair constrains the motion of two bodies only in the normal direction (assuming ideal smooth bodies) and allows relative motion with five degrees of freedom. It is therefore a kinematic pair of class 1.

From the viewpoint of MBS composition a general kinematic pair allows the realization of most general input-output relations at minimum element number. Limiting our treatment of the topic to the most frequent practical examples of MBS with one degree of freedom, it follows from the discussion of relation (3.1) in the integer domain that for the minimum possible number of three bodies a mechanism with one degree of freedom may be realized using one general pair and two pairs of class 5, (Figure 6.1a–e). Six dependent coordinates correspond to such a mechanism, as we know from Section 3.3. Five of them belong to the general pair, and the sixth is the coordinate of the driven (output) element. An exception is the mechanism

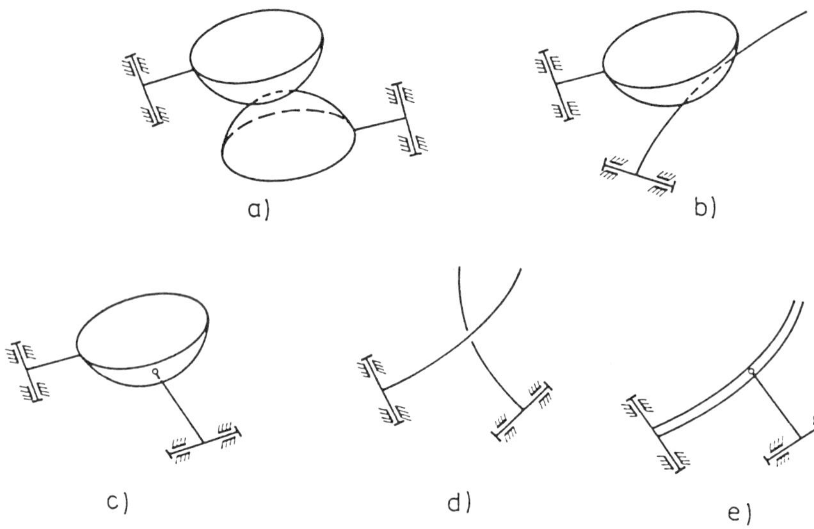

Figure 6.1. Higher kinematic pairs.

in Figure 6.1e, where only four dependent coordinates correspond to the general pair of the curve and point contact.

General curves and surfaces are only rarely used in technical applications. What we usually have are cylinders, cones, spheres and planes. A significant exception is the teeth contact of gears and the cams. The theory of gears, gear trains and cams are however special independent disciplines and we shall not treat those topics here.

In the following text we shall assume smooth surfaces unless otherwise stated.

6.1 Mechanisms with a General (Smooth) Pair

We shall examine a MBS containing one or more general smooth pairs and show how to assemble the position description equations. If the MBS being examined is a simple mechanism, the equations described in the following suffice for its solution. If the MBS is more complex, it is necessary to combine methods from the previous chapter and this chapter in order to assemble the equations for every loop of the MBS.

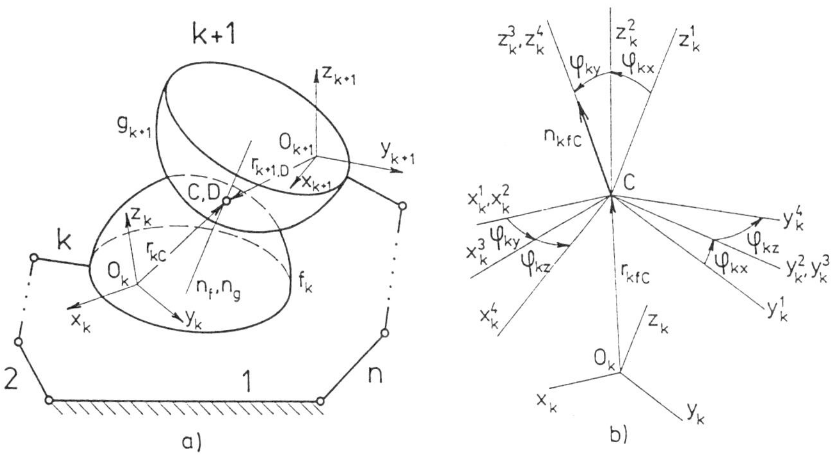

Figure 6.2. Surface and surface contact.

Let us have a general pair between elements k and $k+1$ (Figure 6.2a). Let the pair be formed by the contact of two general surfaces f_k and g_{k+1} in point $C \equiv D$, $C \in f_k$, $D \in g_{k+1}$. To solve the position problem we may use either the first variant of the generalized method of disconnecting a loop or the matrix method of loop closure.

In the former case, we disconnect a MBS loop in a general kinematic pair and express the chain closure conditions from the remaining chains. In the given case we have two conditions

$$\mathbf{r}_{1C} = \mathbf{r}_{1D} \tag{6.1}$$

$$\mathbf{n}_{1fC} = \lambda \mathbf{n}_{1gD} \qquad \text{or} \qquad \mathbf{n}_{1fC} \times \mathbf{n}_{1gD} = \mathbf{0} \tag{6.2}$$

The first expresses the common point of both surfaces, and the second the identity of the normals of both surfaces in the point of contact. The parameter equations of surfaces in spaces k and $k+1$ are

$$\begin{aligned} f_k: \qquad \mathbf{r}_{kf} &= \mathbf{r}_{kf}(u_k, v_k) \\ \mathbf{r}_{kf} &= [x_{kf}, y_{kf}, z_{kf}, 1] = \\ &= [x_{kf}(u_k, v_k), y_{kf}(u_k, v_k), z_{kf}(u_k, v_k), 1] \qquad (6.3a) \\ g_{k+1}: \qquad \mathbf{r}_{k+1,g} &= \mathbf{r}_{k+1,g}(u_{k+1}, v_{k+1}) \\ \mathbf{r}_{k+1,g} &= [x_{k+1,g}, y_{k+1,g}, z_{k+1,g}] = \end{aligned}$$

$$= [x_{k+1,g}(u_{k+1}, v_{k+1}), y_{k+1,g}(u_{k+1}, v_{k+1}), z_{k+1,g}(u_{k+1}, v_{k+1}), 1] \tag{6.3b}$$

Here u_k, v_k and u_{k+1}, v_{k+1} are surface parameters. We assume the surfaces to be regular. Their equations in space 1 are

$$\begin{aligned} \mathbf{r}_{1f}(u_k, v_k) &= \mathbf{T}_{1k}\mathbf{r}_{kf} \\ \mathbf{r}_{1g}(u_{k+1}, v_{k+1}) &= \mathbf{T}_{1,k+1}\mathbf{r}_{k+1,g} \end{aligned} \tag{6.4}$$

For normal vectors $\mathbf{n}_{1f}, \mathbf{n}_{1g}$ which are, however, not unit vectors,

$$\begin{aligned} \mathbf{n}_{1f} &= \frac{\partial \mathbf{u}_{1f}}{\partial u_k} \times \frac{\partial \mathbf{u}_{1f}}{\partial v_k} = \mathbf{S}_{1k}\left(\frac{\partial \mathbf{u}_{kf}}{\partial u_k} \times \frac{\partial \mathbf{u}_{kf}}{\partial v_k}\right) \\ \mathbf{n}_{1g} &= \frac{\partial \mathbf{u}_{1g}}{\partial u_{k+1}} \times \frac{\partial \mathbf{u}_{1g}}{\partial v_{k+1}} = \mathbf{S}_{1,k+1}\left(\frac{\partial \mathbf{u}_{k+1,g}}{\partial u_{k+1}} \times \frac{\partial \mathbf{u}_{k+1,g}}{\partial v_{k+1}}\right) \end{aligned} \tag{6.5}$$

Here $\mathbf{u}_{..}$ designates a vector corresponding to the extended vector $\mathbf{r}_{..}$. Conditions (6.1), (6.2) then have the form

$$\mathbf{T}_{1k}\mathbf{r}_{kfC} = \mathbf{T}_{1,k+1}\mathbf{r}_{k+1,gD} \tag{6.6}$$

$$\frac{\partial \mathbf{u}_{1fC}}{\partial u_k} \times \frac{\partial \mathbf{u}_{1fC}}{\partial v_k} = \lambda \frac{\partial \mathbf{u}_{1gD}}{\partial u_{k+1}} \times \frac{\partial \mathbf{u}_{1gD}}{\partial v_{k+1}} \tag{6.7}$$

Equations (6.6)–(6.7) represent six equations and contain five unknown parameters describing a general pair, i.e. parameters of surfaces u_k, v_k, u_{k+1}, v_{k+1} and parameter λ expressing the normal vector length ratio. In accordance with the function of a general pair, equations (6.6), (6.7) set one constraint on the chain coordinates, thereby reducing the degree of freedom by one.

If one of the surfaces, e.g. g_{k+1}, reduces to a curve, we obtain case b) of surface and curve contact. In such a case, description (6.3b) reduces to a one-parameter description.

$$\begin{aligned} g_{k+1}: \quad \mathbf{r}_{k+1,g} &= \mathbf{r}_{k+1,g}(u_{k+1}) \\ \mathbf{r}_{1g} &= \mathbf{T}_{1,k+1}\mathbf{r}_{k+1,g} \end{aligned} \tag{6.8}$$

The solution of the position problem using the method of disconnecting a loop is now described by conditions (6.1) and (6.6), respectively. The condition of perpendicularity of surface normal and curve tangent is described by

$$\mathbf{n}_{1fC}.\mathbf{t}_{1gD} = 0 \tag{6.9}$$

where

$$\mathbf{t}_{1gD} = \frac{\partial \mathbf{u}_{1gD}}{\partial u_{k+1}} = \mathbf{S}_{1,k+1}\frac{\partial \mathbf{u}_{k+1,gD}}{\partial u_{k+1}} \tag{6.10}$$

is the tangent vector of curve g_{k+1}. Conditions (6.6) and (6.9) then represent four equations for three unknowns u_k, v_k, u_{k+1} and reduce the degree of freedom by one.

If the curve reduces to a point, we get case c). If both surfaces reduce to curves we get case d). A further curve reduction to a point gives case e). In those cases conditions (6.1) and (6.6) respectively are sufficient for the solution, and we have here the direct usage of the simplest case of the disconnected loop method.

In the latter case, instead of disconnecting a MBS loop we express the general pair transformation matrix using pair coordinates and applying the loop closure condition. The general form of the general pair transformation matrix $\mathbf{T}_{k,k+1}$ is as follows

$$\mathbf{T}_{k,k+1} = \mathbf{T}_{k,C}\mathbf{T}_{CD}\mathbf{T}_{D,k+1} \tag{6.11}$$

where $\mathbf{T}_{k,C}$ describes the transformation from the reference coordinate system to the significant coordinate system in the point of contact C; similarly $\mathbf{T}_{D,k+1} = \mathbf{T}_{k+1,D}^{-1}$ is the transformation of a significant coordinate system in point of contact D to the reference system in body $k+1$, and $\mathbf{T}_{CD}$ express the transformation between both significant systems. The significant coordinate system of the surface is the coordinate system with axis z in the surface normal; for a curve, it is the accompanying trihedral and, for a point, the displaced reference system of the body.

For a surface we may express the transformation $\mathbf{T}_{k,C}$ as a displacement from point O_k to point C by translation (Figure 6.2b) by vector $\mathbf{r}_{kfC}$, and rotation around axes x_k^1, y_k^2 by angles $\varphi_{kx}, \varphi_{ky}$ such that axis z_k^3 coincides with normal $\mathbf{n}_{kf}$ of surface f in point C. Then:

$$\mathbf{T}_{k,C} = \mathbf{T}_{Z1}(x_{kfC})\mathbf{T}_{Z2}(y_{kfC})\mathbf{T}_{Z3}(z_{kfC})\mathbf{T}_{Z4}(\varphi_{kx})\mathbf{T}_{Z5}(\varphi_{ky}) \tag{6.12}$$

where

$$\begin{aligned} x_{kfC} &= x_{kf}(u_k, v_k), \qquad \varphi_{kx} = \varphi_{kx}(u_k, v_k) \\ y_{kfC} &= y_{kf}(u_k, v_k), \qquad \varphi_{ky} = \varphi_{ky}(u_k, v_k) \\ z_{kfC} &= z_{kf}(u_k, v_k) \end{aligned}$$

Angles $\varphi_{kx}, \varphi_{ky}$ can be determined from the condition that axis z_k^3 and its vector $[0, 0, \|\mathbf{n}_{kf}\|]$, respectively, after rotations by those angles, are identical with the normal vector $\pm\mathbf{n}_{kf}$

$$\begin{aligned} \varphi_{kx} &= \operatorname{atan2}(\mp n_{kf,2}, \pm n_{kf,3}) \\ \varphi_{ky} &= \operatorname{atan2}(\pm n_{kf,1}, \pm\sqrt{n_{kf,2}^2 + n_{kf,3}^2}) \end{aligned} \tag{6.13}$$

The sign in (6.13) must be chosen to represent the correct sides of the surfaces for their contact. We may express

$$\mathbf{T}_{Z4}(\varphi_{kx})\mathbf{T}_{Z5}(\varphi_{ky}) = \begin{bmatrix} \frac{d_2}{d_1}, & 0, & \frac{\pm n_{kf,1}}{d_1}, & 0 \\ \frac{-n_{kf,1}n_{kf,2}}{d_1 d_2}, & \frac{\pm n_{kf,3}}{d_2}, & \frac{\pm n_{kf,2}}{d_1}, & 0 \\ \frac{-n_{kf,1}n_{kf,3}}{d_1 d_2}, & \mp\frac{n_{kf,2}}{d_2}, & \frac{\pm n_{kf,3}}{d_1}, & 0 \\ 0, & 0, & 0, & 1 \end{bmatrix} \tag{6.14}$$

$$d_1 = \sqrt{n_{kf,1}^2 + n_{kf,2}^2 + n_{kf,3}^2}, \qquad d_2 = \sqrt{n_{kf,2}^2 + n_{kf,3}^2}$$

For a curve, we may express the transformation $\mathbf{T}_{k,C}$ (or similarly $\mathbf{T}_{D,k+1} = \mathbf{T}_{k+1,D}^{-1}$), as a displacement from point O_k to point C by translation by vector $\mathbf{r}_{kfC}$ and rotations around axes x_k^1, y_k^2, z_k^3 by angles φ_{kx}, φ_{ky}, φ_{kz} so that the coordinate system $Cx_k^4y_k^4z_k^4$ becomes identical with the accompanying trihedral of curve f_k.

$$\mathbf{T}_{k,C} = \mathbf{T}_{Z1}(x_{kfC})\mathbf{T}_{Z2}(y_{kfC})\mathbf{T}_{Z3}(z_{kfC})\mathbf{T}_{Z4}(\varphi_{kx})\mathbf{T}_{Z5}(\varphi_{ky})\mathbf{T}_{Z6}(\varphi_{kz}) \tag{6.15}$$

where

$$\begin{aligned} x_{kfC} &= x_{kf}(u_k), & \varphi_{kx} &= \varphi_{kx}(u_k) \\ y_{kfC} &= y_{kf}(u_k), & \varphi_{ky} &= \varphi_{ky}(u_k) \\ z_{kfC} &= z_{kf}(u_k), & \varphi_{kz} &= \varphi_{kz}(u_k) \end{aligned}$$

In the case of a point, the transformation $\mathbf{T}_{k,C}$ is given by a translation from point O_k to point C by a fixed vector $\overrightarrow{O_kC}$.

For transformation between significant coordinate systems, in the case of surface contact (Figure 6.1a)

$$\mathbf{T}_{CD} = \mathbf{T}_{Z6}(\varphi) \tag{6.16}$$

where φ is the angle of rotation around the common normal. In the case of surface and curve contact (Figure 6.1b)

$$\mathbf{T}_{CD} = \mathbf{T}_{Z6}(\varphi)\mathbf{T}_{Z5}(\vartheta) \tag{6.17}$$

where φ is the angle of rotation around the surface normal, perpendicular to the curve tangent, which is needed to cause the coincidence of axis y_k^4 and the curve tangent; ϑ is the angle of rotation around the curve tangent causing the coordinate system $Cx_k^5y_k^5z_k^5$ to turn identically with the accompanying trihedral of curve g_{k+1} (Figure 6.3).

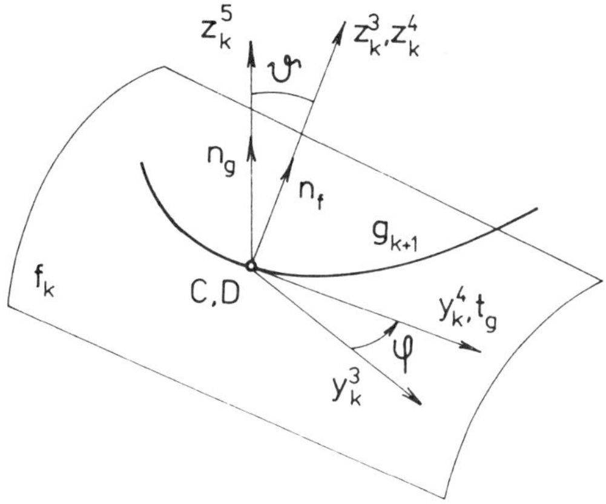

Figure 6.3. Surface and curve contact.

In the case of surface and point contact (Figure 6.1c), two curves (Figure 6.1d) or curve and point (Figure 6.1e) we have angular displacement φ, ϑ, ψ around three axes x, y, z to cause the identity of two general coordinate systems with common origin $C \equiv D$

$$\mathbf{T}_{CD} = \mathbf{T}_{Z4}(\varphi)\mathbf{T}_{Z5}(\vartheta)\mathbf{T}_{Z6}(\psi) \tag{6.18}$$

We assemble the general pair transformation matrix (6.11) from the transformation matrices in particular cases of Figure 6.1. In the case of Figure 6.1a the transformation matrix (6.11) has coordinates u_k, v_k, in the case of Figure 6.1c coordinates $u_k, v_k, \varphi, \vartheta, \psi$, in the case of Figure 6.1d coordinates $u_k, \varphi, \vartheta, \psi, u_{k+1}$, and in the case of Figure 6.1e coordinates $u_k, \varphi, \vartheta, \psi$. In cases a) to d) we have five coordinates and five degrees of freedom, and in case e) four coordinates and four degrees of freedom. The position problem is, using the transformation matrix approach, described as

$$\mathbf{T}_{1k}\mathbf{T}_{k,k+1}\mathbf{T}_{k+1,1} = \mathbf{E}_4 \tag{6.19}$$

and

$$\begin{aligned}
\mathbf{T}_{1k}\mathbf{T}_{k,k+1} &= \mathbf{T}_{1,k+1} \\
\mathbf{T}_{1k}\mathbf{T}_{k,C}\mathbf{T}_{CD} &= \mathbf{T}_{1,k+1}\mathbf{T}_{k+1,D}
\end{aligned}$$

respectively.

Having solved the position problem, the solution of velocities, accelerations and motion of a MBS point or body is easy. The equations for the

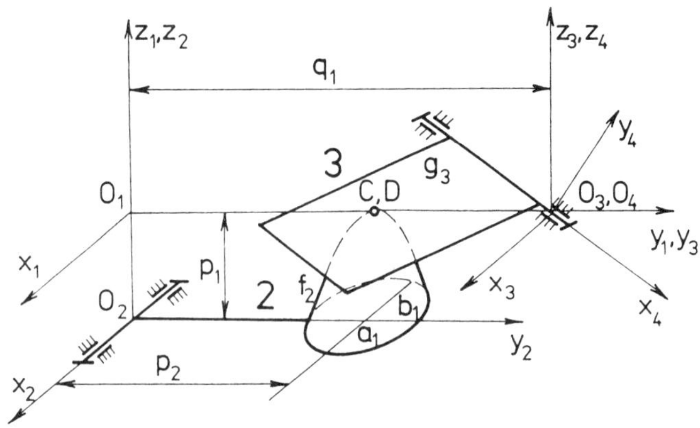

Figure 6.4. The three-element mechanism with higher kinematic pair.

solution of velocities and accelerations are to be derived by differentiating the respective conditions used in the position solution with respect to time. Let us note only the motion of the point of contact C

$$\begin{aligned}
\mathbf{r}_{1fC} &= \mathbf{T}_{1k}\mathbf{r}_{kfC} \\
\mathbf{v}_{1fC} &= \dot{\mathbf{T}}_{1k}\mathbf{r}_{kfC} + \mathbf{T}_{1k}\dot{\mathbf{r}}_{kfC} \\
\dot{\mathbf{r}}_{kfC} &= \frac{\partial \mathbf{r}_{kfC}}{\partial u_k}\dot{u}_k + \frac{\partial \mathbf{r}_{kfC}}{\partial v_k}\dot{v}_k
\end{aligned} \tag{6.20}$$

where $\mathbf{T}_{1k}\dot{\mathbf{r}}_{kfC}$ is the velocity of point C along surface f_k and $\dot{\mathbf{T}}_{1k}\mathbf{r}_{kfC}$ is the velocity of point C as the point of element k. If point C is understood as a fixed point of body k and not as a point of contact, its velocity $\dot{\mathbf{r}}_{kfC}$ in motion along surface f_k is zero.

6.2 Applications

6.2.1 A Spatial Parabolic Cam

A three-element mechanism is presented in Figure 6.4. Element 2 is formed by a pin in the shape of an elliptic paraboloid, element 3 by a plane.

The equations of surface f_2 in space 2 and surface g_3 in space 3 are

$$\begin{aligned}
\mathbf{r}_{2fC} &= \lceil u_2, v_2, p_1\left(-\frac{u_2^2}{a_1^2} - \frac{(v_2-p_2)^2}{b_1^2} + 1\right), 1\rceil \\
\mathbf{r}_{3gD} &= \lceil u_3, v_3, 0, 1\rceil
\end{aligned} \tag{a}$$

where u_2, v_2, u_3, v_3 are surface parameters. The mechanism position solution is described by equations (6.6), (6.7). In the above case

$$\begin{aligned} \mathbf{T}_{12}\mathbf{r}_{2fC} &= \mathbf{T}_{13}\mathbf{r}_{3gD} \\ \frac{\partial \mathbf{u}_{1fC}}{\partial u_2} \times \frac{\partial \mathbf{u}_{1fC}}{\partial v_2} &= \lambda \frac{\partial \mathbf{u}_{1gD}}{\partial u_3} \times \frac{\partial \mathbf{u}_{1gD}}{\partial v_3} \qquad \text{(b)} \\ \mathbf{T}_{12} &= \mathbf{T}_{Z3}(-p_1)\mathbf{T}_{Z4}(\varphi_{12}) \\ \mathbf{T}_{13} &= \mathbf{T}_{Z2}(q_1)\mathbf{T}_{Z6}(\alpha)\mathbf{T}_{Z4}(\varphi_{13}) \qquad \text{(c)} \end{aligned}$$

Performing the operations in (b) yields six equations for six unknowns u_2, v_2, $u_3, v_3, \lambda, \varphi_{13}$ with respect to φ_{12}

$$u_2 = u_3 \mathrm{c}\alpha - v_3 \mathrm{s}\alpha \mathrm{c}\varphi_{13}$$

$$v_2 \mathrm{c}\varphi_{12} - p_1 \left(-\frac{u_2^2}{a_1^2} - \frac{(v_2 - p_2)^2}{b_1^2} + 1 \right) \mathrm{s}\varphi_{12} = u_3 \mathrm{s}\alpha + v_3 \mathrm{c}\alpha \mathrm{c}\varphi_{13} + q_1$$

$$v_2 \mathrm{s}\varphi_{12} + p_1 \left(-\frac{u_2^2}{a_1^2} - \frac{(v_2 - p_2)^2}{b_1^2} + 1 \right) \mathrm{c}\varphi_{12} - p_1 = v_3 \mathrm{s}\varphi_{13}$$

$$\frac{2p_1 u_2}{a_1^2} = \lambda \mathrm{s}\alpha \mathrm{s}\varphi_{13}$$

$$-\mathrm{s}\varphi_{12} + \frac{2p_1}{b_1^2}(v_2 - p_2)\mathrm{c}\varphi_{12} = \lambda(-\mathrm{c}\alpha \mathrm{s}\varphi_{13})$$

$$\mathrm{c}\varphi_{12} + \frac{2p_1}{b_1^2}(v_2 - p_2)\mathrm{s}\varphi_{12} = \lambda \mathrm{c}\varphi_{13} \qquad \text{(d)}$$

Using for the solution equation (6.11), we obtain according to (6.8)

$$\begin{aligned} \mathbf{T}_{23} = \; & \mathbf{T}_{Z1}(u_2)\mathbf{T}_{Z2}(v_2)\mathbf{T}_{Z3}\left(p_1\left(-\frac{u_2^2}{a_1^2} - \frac{(v_2 - p_2)^2}{b_1^2} + 1\right)\right)\mathbf{T}_{Z4}(\varphi_{2x}). \\ & .\mathbf{T}_{Z5}(\varphi_{2y})\mathbf{T}_{Z6}(\varphi)\mathbf{T}_{Z5}(-\varphi_{3y})\mathbf{T}_{Z4}(-\varphi_{3x})\mathbf{T}_{Z3}(0)\mathbf{T}_{Z2}(-v_3). \\ & .\mathbf{T}_{Z1}(-u_3) \qquad \text{(e)} \end{aligned}$$

and according to (6.11)

$$\mathbf{T}_{12}\mathbf{T}_{23} = \mathbf{T}_{13}$$

$$\begin{aligned} &\mathbf{T}_{Z3}(-p_1)\mathbf{T}_{Z4}(\varphi_{12})\mathbf{T}_{Z1}(u_2)\mathbf{T}_{Z2}(v_2)\mathbf{T}_{Z3}\left(p_1\left(-\frac{u_2^2}{a_1^2} - \frac{(v_2 - p_2)^2}{b_1^2} + 1\right)\right). \\ &.\mathbf{T}_{Z4}(\varphi_{2x})\mathbf{T}_{Z5}(\varphi_{2y})\mathbf{T}_{Z6}(\varphi)\mathbf{T}_{Z5}(-\varphi_{3y})\mathbf{T}_{Z4}(-\varphi_{3x})\mathbf{T}_{Z1}(-u_3). \\ &.\mathbf{T}_{Z2}(-v_3)\mathbf{T}_{Z3}(0) = \mathbf{T}_{Z2}(q_1)\mathbf{T}_{Z6}(\alpha)\mathbf{T}_{Z4}(\varphi_{13}) \qquad \text{(f)} \end{aligned}$$

Modifying (f) according to the principle expressed by relation (5.75) yields

$$\mathbf{T}_{Z3}(-p_1)\mathbf{T}_{Z4}(\varphi_{12})\mathbf{T}_{Z1}(u_2)\mathbf{T}_{Z2}(v_2).$$

$$.\mathbf{T}_{Z3}\left(p_1\left(-\frac{u_2^2}{a_1^2}-\frac{(v_2-p_2)^2}{b_1^2}+1\right)\right)\mathbf{T}_{Z4}(\varphi_{2x})\mathbf{T}_{Z5}(\varphi_{2y})\mathbf{T}_{Z6}(\varphi)=$$

$$=\mathbf{T}_{Z2}(q_1)\mathbf{T}_{Z6}(\alpha)\mathbf{T}_{Z4}(\varphi_{13})\mathbf{T}_{Z1}(u_3)\mathbf{T}_{Z2}(v_3)\mathbf{T}_{Z3}(0)\mathbf{T}_{Z4}(\varphi_{3x})\mathbf{T}_{Z5}(\varphi_{3y}) \tag{g}$$

The equalities of radius vectors in equation (g) are identical with equations (d1–3). The equalities of the third columns of the direction cosines matrices in equation (g) are identical with equations (d4–6). Having solved these equations, angle φ can be computed easily. The identical equations yielded by the two approaches are a consequence of a general property, and this may be easily proved from equations (6.19) and (6.14).

6.2.2 A Tripod Driving Device

The diagram in Figure 6.5 presents a spatial mechanism used in the driving devices for car wheels (Brát, 1981). It consists of a tripod joint and a Hooke joint in series. It is a five-element mechanism of the type

```
        G
      /   \
 R̲̅−   G   −R − R − R̲̿
      \   /
        G
```

consisting of three general pairs A, B, C and is formed by three loops. Element 2 is the input element, element 5 is the output element, and axes z_1 and z_5 are skew lines. The general pairs are of type d); we have here the case of curve and curve contact. For the solution of kinematics, conditions (6.6) in each loop suffice

$$\begin{aligned}\mathbf{T}_{12}\mathbf{r}_{2A} &= \mathbf{T}_{13}\mathbf{r}_{3A}\\ \mathbf{T}_{12}\mathbf{r}_{2B} &= \mathbf{T}_{13}\mathbf{r}_{3B}\\ \mathbf{T}_{12}\mathbf{r}_{2C} &= \mathbf{T}_{13}\mathbf{r}_{3C}\end{aligned} \tag{a}$$

where

$$\begin{aligned}\mathbf{T}_{12} &= \mathbf{T}_{Z6}(\varphi_{12})\\ \mathbf{T}_{13} &= \mathbf{T}_{Z3}(b)\mathbf{T}_{Z2}(e)\mathbf{T}_{Z5}(\gamma)\mathbf{T}_{Z3}(l)\mathbf{T}_{Z6}(\varphi_{15})\mathbf{T}_{Z4}(\varphi_{54})\mathbf{T}_{Z5}(\varphi_{43})\end{aligned}$$

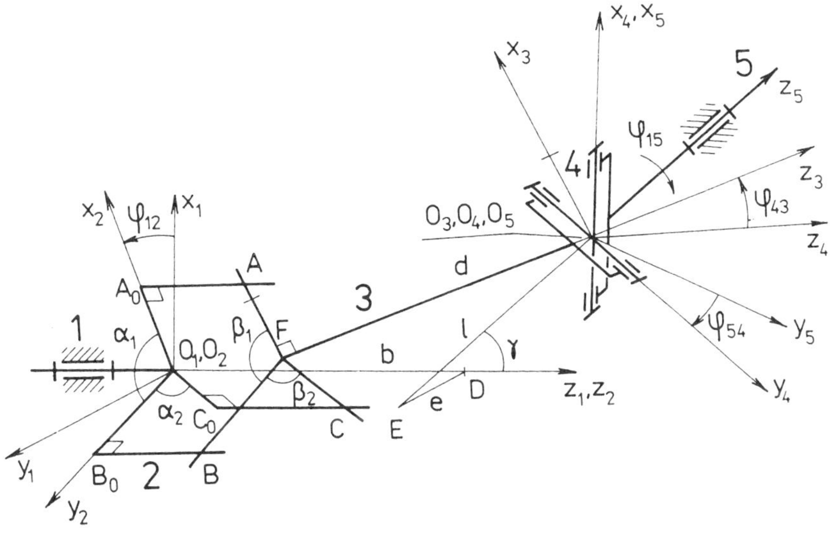

Figure 6.5. Driving mechanism with tripod and Hooke's joints.

$$\begin{aligned}
\mathbf{r}_{2A} &= [a_1, 0, \zeta_1, 1] \\
\mathbf{r}_{2B} &= [a_2 c\alpha_1, a_2 s\alpha_1, \zeta_2, 1] \\
\mathbf{r}_{2C} &= [a_3 c(\alpha_1+\alpha_2), a_3 s(\alpha_1+\alpha_2), \zeta_3, 1] \\
\mathbf{r}_{3A} &= [\xi_1, 0, -d, 1] \\
\mathbf{r}_{3B} &= [\xi_2 c\beta_1, \xi_2 s\beta_1, -d, 1] \\
\mathbf{r}_{3C} &= [\xi_3 c(\beta_1+\beta_2), \xi_3 s(\beta_1+\beta_2), -d, 1]
\end{aligned}$$

$$\begin{array}{rrr}
O_2A_0 = a_1, & O_2B_0 = a_2, & O_2C_0 = a_3 \\
A_0A = \zeta_1, & B_0B = \zeta_2, & C_0C = \zeta_3 \\
FA = \xi_1, & FB = \xi_2, & FC = \xi_3
\end{array}$$

$$O_1D = b, \qquad DE = e, \qquad EO_3 = l, \qquad FO_3 = d \tag{b}$$

Equations (a) represent nine equations for nine unknowns $\zeta_1, \zeta_2, \zeta_3, \xi_1, \xi_2, \xi_3$, $\varphi_{15}, \varphi_{54}, \varphi_{43}$ depending on φ_{12}. Unknowns ζ_i and ξ_i are curve parameters. To determine the velocities we differentiate (a) with respect to time

$$\begin{aligned}
\dot{\mathbf{T}}_{12}\mathbf{r}_{2A} + \mathbf{T}_{12}\dot{\mathbf{r}}_{2A} &= \dot{\mathbf{T}}_{13}\mathbf{r}_{3A} + \mathbf{T}_{13}\dot{\mathbf{r}}_{3A} \\
\dot{\mathbf{T}}_{12}\mathbf{r}_{2B} + \mathbf{T}_{12}\dot{\mathbf{r}}_{2B} &= \dot{\mathbf{T}}_{13}\mathbf{r}_{3B} + \mathbf{T}_{13}\dot{\mathbf{r}}_{3B} \\
\dot{\mathbf{T}}_{12}\mathbf{r}_{2C} + \mathbf{T}_{12}\dot{\mathbf{r}}_{2C} &= \dot{\mathbf{T}}_{13}\mathbf{r}_{3C} + \mathbf{T}_{13}\dot{\mathbf{r}}_{3C}
\end{aligned} \tag{c}$$

where, for example

$$\begin{aligned}\dot{\mathbf{r}}_{2B} &= [0, 0, \dot{\zeta}, 0] \\ \dot{\mathbf{r}}_{3B} &= [\dot{\xi}_2 c\beta_1, \dot{\xi}_2 s\beta_1, 0, 0]\end{aligned}$$

The motion of point F is important. It is described by the equation

$$\mathbf{r}_{1F} = \mathbf{T}_{13}\mathbf{r}_{3F} \tag{d}$$

where

$$\mathbf{r}_{3F} = [0, 0, -d, 1]$$

Its velocity and acceleration are given by

$$\begin{aligned}\mathbf{v}_{3F} &= \dot{\mathbf{T}}_{13}\mathbf{r}_{3F} \\ \mathbf{a}_{3F} &= \ddot{\mathbf{T}}_{13}\mathbf{r}_{3F}\end{aligned}$$

6.3 Mechanisms with a General Roll Pair

Let us examine the same MBS as in the previous section, but with the general pair being formed by rough surfaces.

We shall use the first variant of the generalized method for disconnecting a loop. The loop closure conditions for a general roll pair are given by conditions (6.1), (6.2) and (6.6), (6.7) respectively and by the condition of equality of absolute velocities of actual points of contact $C \equiv D$

$$\left.\frac{d\mathbf{r}_{1f}}{dt}\right|_{\substack{\dot{u}_k = 0 \\ \dot{v}_k = 0 \\ C \equiv (u_k, v_k)}} = \left.\frac{d\mathbf{r}_{1g}}{dt}\right|_{\substack{\dot{u}_{k+1} = 0 \\ \dot{v}_{k+1} = 0 \\ D \equiv (u_{k+1}, v_{k+1})}} \tag{6.21}$$

expressing the fact that, in this point, the relative velocity of both bodies is zero. The point is the pole of relative motion. Using (6.20) we may modify

$$\dot{\mathbf{T}}_{1k}\mathbf{r}_{kC} = \dot{\mathbf{T}}_{1,k+1}\mathbf{r}_{k+1,D} \tag{6.22}$$

Condition (6.21) represents a non-holonomic constraint formed by three equations, one of which is dependent as its velocity vector projection into the normal is zero. The general roll pair therefore represents eight equations containing five unknowns $u_k, v_k, \lambda, u_{k+1}, v_{k+1}$, and reduces the degree of freedom by 3. Instead of equation (6.21), it is possible in the case of rolling surfaces to use the equation

$$\left.\left.\mathbf{S}_{1k}\mathbf{S}_{kC}\frac{d\mathbf{u}_{1f}}{dt}\right|_{\substack{\dot{u}_k = 0\\ \dot{v}_k = 0\\ C \equiv (u_k, v_k)}}\right|_{\substack{x\\ y}} = \left.\left.\mathbf{S}_{1k}\mathbf{S}_{kC}\frac{d\mathbf{u}_{1g}}{dt}\right|_{\substack{\dot{u}_{k+1} = 0\\ \dot{v}_{k+1} = 0\\ D \equiv (u_{k+1}, v_{k+1})}}\right|_{\substack{x\\ y}} \tag{6.23}$$

This expresses the identity of velocity x, y components of both points expressed in the significant coordinate system of surface k (and similarly for surface $k + 1$). The x, y plane of this system coincides with the common tangent plane of both surfaces.

Conditions (6.21), (6.22) may also be used for a curve rolling against a surface in case b) or for curve and curve rolling.

For the solution of a MBS with a roll pair, the transformation matrix (6.11) and the condition (6.21) may also be used. The surface parameters in (6.3) or (6.11) are not mutually independent; they are linked by the non-holonomic constraint (6.21). If we could integrate this non-holonomic constraint and could obtain the behaviour of parameters of one surface relative to the parameters of the other surface, we could even assemble and use the general roll pair transformation matrix by substituting the dependence into (6.11).

6.4 Applications to MBSs with Roll Pairs

6.4.1 A Planar Tyre

A tyre in the form of a circle with radius r rolls in a plane along axis x (Figure 6.6).

Let us solve the case using reduced transformation matrices from Section 4.7.2 and the above methods. We shall use the method of compartments according to Section 5.11.5 to find the solution. We shall choose the tyre g as one compartment and the straight line f as another compartment. The tyre centre coordinates x_A, y_A and the angle of angular displacement ψ of the coordinate system $O_1x_1y_1$ firmly fixed to the tyre are the compartment coordinates. For the solution, we use a planar analogy of conditions (6.1), (6.2) and (6.21). The coordinate of line f is x, and the coordinate of circle g is the angle φ, so that

$$\begin{aligned}\mathbf{r}_{1fP} &= \mathbf{r}_{1gP}\\ \mathbf{n}_{1fP} &= \lambda\mathbf{n}_{1gP}\end{aligned}$$

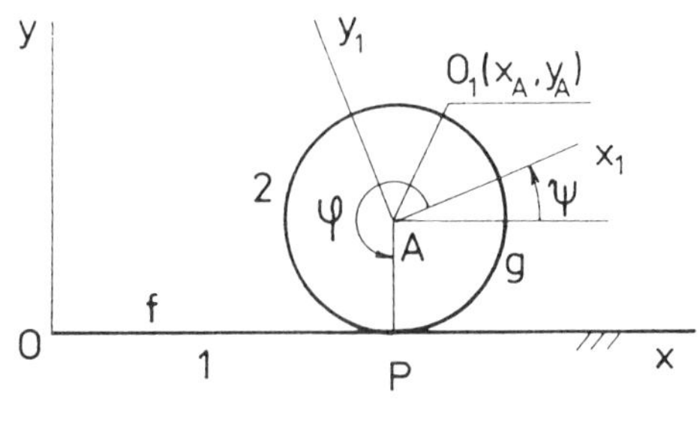

Figure 6.6. Circle rolling along axis x.

$$\left.\frac{\mathrm{d}\mathbf{r}_{1fP}}{\mathrm{d}t}\right|_{\dot{x}=0} = \left.\frac{\mathrm{d}\mathbf{r}_{1gP}}{\mathrm{d}t}\right|_{\dot{\varphi}=0} \tag{a}$$

Expressing

$$\begin{aligned}
\mathbf{r}_{1fP} &= \lceil x, 0, 1\rceil \\
\mathbf{r}_{1gP} &= \mathbf{T}^r_{12}\mathbf{r}_{2gP} \\
\mathbf{T}^r_{12} &= \mathbf{T}^r_{Z1}(x_A)\mathbf{T}^r_{Z2}(y_A)\mathbf{T}^r_{Z6}(\psi) \\
\mathbf{r}_{2gP} &= \lceil r\mathrm{c}\varphi, r\mathrm{s}\varphi, 1\rceil \\
\mathbf{n}_{1fP} &= \lceil 0, 1\rceil \\
\mathbf{n}_{1gP} &= \mathbf{S}^r_{12}\mathbf{n}_{2gP} \\
\mathbf{n}_{2gP} &= \lceil -r\mathrm{c}\varphi, -r\mathrm{s}\varphi\rceil
\end{aligned} \tag{b}$$

we obtain from equations (a)

$$\begin{aligned}
x &= r\mathrm{c}(\varphi+\psi) + x_A && \text{(c1)}\\
0 &= r\mathrm{s}(\varphi+\psi) + y_A && \text{(c2)}\\
0 &= -\lambda r\mathrm{c}(\varphi+\psi) && \text{(c3)}\\
1 &= -\lambda r\mathrm{s}(\varphi+\psi) && \text{(c4)}\\
0 &= -r\mathrm{s}(\varphi+\psi)\dot{\psi} + \dot{x}_A && \text{(c5)}\\
0 &= r\mathrm{c}(\varphi+\psi)\dot{\psi} + \dot{y}_A && \text{(c6)}
\end{aligned}$$

From (c3, 4), it follows that

$$\begin{aligned}
\mathrm{c}(\varphi+\psi) &= 0 && \text{(d1)}\\
\mathrm{s}(\varphi+\psi) &= -1 && \text{(d2)}\\
\lambda &= 1/r && \text{(d3)}\\
\varphi+\psi &= (2l-1)\frac{\pi}{2}, \qquad l = 0, \pm 1, \pm 2, \ldots && \text{(d4)}
\end{aligned}$$

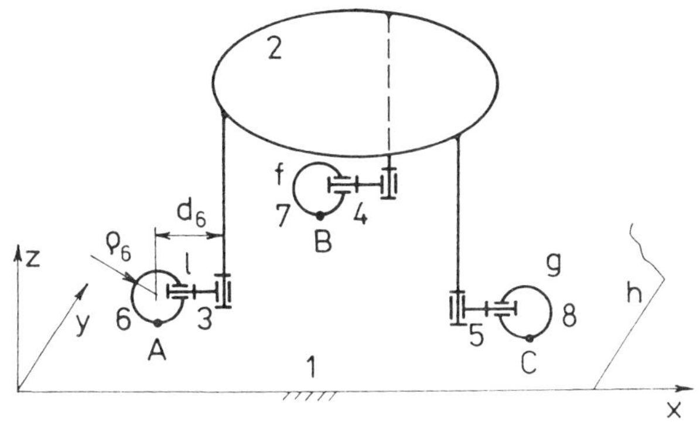

Figure 6.7. Undercarriage of a mobile robot.

Substituting into other equations we obtain

$$x = x_A \tag{e1}$$

$$y_A = r \tag{e2}$$

$$\dot{x}_A = -r\dot{\psi} \tag{e3}$$

$$\dot{y}_A = 0 \tag{e4}$$

Integrating (e3) we finally have

$$x_A = x_A(0) - r(\psi - \psi(0)) \tag{f}$$

Relations (f), (e1, 2) and (d3, 4) give the complete solution.

6.4.2 A Mobile Robot

Figure 6.7 shows the diagram of the undercarriage (bogie) of a mobile robot similar to the Stanford robot Shakey, formed by a platform 2 on three spherical wheels 6, 7, 8, each revolving about two axes. Let the wheels and the floor have an ideal rough contact (rolling).

Let both rotations 6:3 and 7:4 of wheels 6 and 7 be driven. We shall use the compartment method of Section 5.11.5 to solve the position problem. We choose compartments, namely the whole bogie, including wheels, and the frame 1. The coordinates of bogie 2 are the coordinates of the element 2, with mass centre $[x_2, y_2, z_2]$ in the frame coordinate system. Cardan angles $[\varphi_x, \varphi_y, \varphi_z]$ (or Euler parameters e_0, e_1, e_2, e_3 with the constraint $e_0^2 + e_1^2 + e_2^2 + e_3^2 = 1$) describe the orientation of the coordinate system firmly connected with the bogie 2 relative to the frame and angles of revolute pairs $\varphi_{23}, \varphi_{36}, \varphi_{24}, \varphi_{47}, \varphi_{25}, \varphi_{58}$ together with the parameters of wheel spherical surfaces given by the spherical coordinates $\varphi_6, \nu_6, \varphi_7, \nu_7, \varphi_8, \nu_8$. The coordinates of the frame 1 formed by a plane are the coordinates of points of contact A, B, C, namely $x_A, y_A, x_B, y_B, x_C, y_C$. The bogie position is determined by the transformation matrix

$$\mathbf{T}_{12} = \mathbf{T}_{Z1}(x_2)\mathbf{T}_{Z2}(y_2)\mathbf{T}_{Z3}(z_2)\mathbf{T}_{Z4}(\varphi_x)\mathbf{T}_{Z5}(\varphi_y)\mathbf{T}_{Z6}(\varphi_z) \tag{a}$$

The position problem is in this case described by equations (6.1), (6.6) and (6.21)

$$\begin{aligned}
\mathbf{r}_{1lA} &= \mathbf{T}_{12}\mathbf{T}_{23}\mathbf{T}_{36}\mathbf{r}_{6lA} = \mathbf{r}_{1hA} \\
\mathbf{r}_{1fB} &= \mathbf{T}_{12}\mathbf{T}_{24}\mathbf{T}_{47}\mathbf{r}_{7fB} = \mathbf{r}_{1hB} \\
\mathbf{r}_{1gC} &= \mathbf{T}_{12}\mathbf{T}_{25}\mathbf{T}_{58}\mathbf{r}_{8gC} = \mathbf{r}_{1hC} && \text{(b)} \\
\mathbf{n}_{1lA} &= \mathbf{S}_{12}\mathbf{S}_{23}\mathbf{S}_{36}\mathbf{n}_{6lA} = \lambda_A\mathbf{n}_{1hA} \\
\mathbf{n}_{1fB} &= \mathbf{S}_{12}\mathbf{S}_{24}\mathbf{S}_{47}\mathbf{n}_{7fB} = \lambda_B\mathbf{n}_{1hB} \\
\mathbf{n}_{1gC} &= \mathbf{S}_{12}\mathbf{S}_{25}\mathbf{S}_{58}\mathbf{n}_{8gC} = \lambda_C\mathbf{n}_{1hC} && \text{(c)}
\end{aligned}$$

$$\begin{aligned}
\left.\frac{\mathrm{d}\mathbf{r}_{1lA}}{\mathrm{d}t}\right|_{\substack{\dot\varphi_6 = 0 \\ \dot\nu_6 = 0}} &= \left.\frac{\mathrm{d}\mathbf{r}_{1hA}}{\mathrm{d}t}\right|_{\substack{\dot x_A = 0 \\ \dot y_A = 0}} \\
\left.\frac{\mathrm{d}\mathbf{r}_{1fB}}{\mathrm{d}t}\right|_{\substack{\dot\varphi_7 = 0 \\ \dot\nu_7 = 0}} &= \left.\frac{\mathrm{d}\mathbf{r}_{1hB}}{\mathrm{d}t}\right|_{\substack{\dot x_B = 0 \\ \dot y_B = 0}} \\
\left.\frac{\mathrm{d}\mathbf{r}_{1gC}}{\mathrm{d}t}\right|_{\substack{\dot\varphi_8 = 0 \\ \dot\nu_8 = 0}} &= \left.\frac{\mathrm{d}\mathbf{r}_{1hC}}{\mathrm{d}t}\right|_{\substack{\dot x_C = 0 \\ \dot y_C = 0}} && \text{(d)}
\end{aligned}$$

where

$$\begin{aligned}
\mathbf{T}_{23} &= \mathbf{T}_{\text{const}}\mathbf{T}_{Z6}(\varphi_{23}), \quad \mathbf{T}_{36} = \mathbf{T}_{Z4}(\varphi_{36}) \\
\mathbf{r}_{6lA} &= [\varrho_6 \mathrm{c}\nu_6 \mathrm{c}\varphi_6 + d_6, \varrho_6 \mathrm{c}\nu_6 \mathrm{s}\varphi_6, \varrho_6 \mathrm{s}\nu_6, 1] \\
\mathbf{r}_{1hA} &= [x_A, y_A, 0, 1]
\end{aligned}$$

and similarly for other cases. The problem has 24 unknowns $x_2, y_2, z_2, \varphi_x, \varphi_y, \varphi_z, \varphi_{24}, \varphi_{25}, \varphi_{58}, \nu_6, \varphi_6, \nu_7, \varphi_7, \nu_8, \varphi_8, x_A, y_A, x_B, y_B, x_C, y_C, \lambda_A, \lambda_B, \lambda_C$ by given dependence $\varphi_{23} = \varphi_{23}(t), \varphi_{36} = \varphi_{36}(t), \varphi_{47} = \varphi_{47}(t)$. System (b)–(d) contains 27 equations, 3 of which, namely (d), are dependent.

Chapter 7

Overconstrained and Singular Cases of MBSs

The methods for MBS kinematic solution usually yield more equations than unknowns, as follows from their description given in Chapters 3 and 4. The consequence of this is that the set of equations contains many dependent ones. The main problem in the actual mathematical solution is to find a set of independent equations and, furthermore, the set which would be most suitable (best conditioned) from the viewpoint of numerical solution. This condition makes it necessary to seek independent equations at practically each computation step, as the conditioning of equations changes with MBS position. An alternative approach to the solution of the problem is to attempt to fulfil all equations yielded by the method with a predetermined condition which may be, for example, the requirement to have the smallest sum of squared residui.

Apart from the standard situations, there are cases of MBS kinematic solutions where, usually due to the dimensions of a particular element or the orientation of a pair of axes, additive mobility or, conversely, restricted mobility is obtained in disagreement with the mobility calculation according to (3.1); in these cases, the given situation may occur only in a particular position or it may be valid for the whole range of motion. We describe such cases as singular cases. Their solution is the topic of the present chapter. Similar cases may also occur in the solution of static and dynamic problems of MBS, but they are not so frequent or significant (Stejskal et al., 1990) and are mentioned in the Section 8.1.

The difficulty of treatment of these cases is that it is caused by the influences of both physical and numerical nature and the solving procedures must correctly detect them.

7.1 Mathematical Background

In Chapter 5 we saw that the solution of a MBS position problem using the Newton approach and the solution of velocities and accelerations lead in general to the problem of finding a vector **x** of unknowns from a set of linear equations. These are of the form

$$\mathbf{A}\mathbf{x} = \mathbf{b} \tag{7.1}$$

where **A** is the matrix of a system of linear equations of the type (m, n), $m \gtreqless n$ containing known elements, and **b** is a given vector of dimension m.

The key part in solving the system (7.1) is played by matrix **A** rank r_A and extended matrix $[\mathbf{A}, \mathbf{b}]$ rank r_B. The well-known condition for the existence of the solution is $r_A = r_B$. It is also a known fact that for $r_A = m = n$, matrix **A** is regular and system (7.1) has a single solution for any vector **b**.

Further, we distinguish two singular cases which may occur simultaneously, namely, the case $r_A < n$ and the case $r_A < m$. If $r_A < n$ then the solution **x** (assuming it exists, i.e. $r_A = r_B$) is not unique. If $r_A < m$ then the system (7.1) is not solvable for each **b**. More precisely we can classify the MBS described by (7.1) from two main points of view.

A. Mathematical (the criterion being the solvability of (7.1)). This involves

a) solvable (or precisely exactly solvable) systems for $r_A = r_B$, in which case a unique or parametric solution exists;

b) unsolvable (or precisely pseudosolvable) systems for $r_A \neq r_B$, (i.e. $r_A < r_B$), in which case no (exact) solution exists.

B. Physical (the criterion being the existence of "free" unknowns). This involves

a) regular system for $r_A = n$

α) safe cases ($m = n$), in which a unique solution exists;

β) dangerous cases ($m > n$), in which a unique or no (exact) solution exists;

b) irregular system for $r_A < n$

α) safe cases ($r_A = m$), in which a parametric solution exists;

β) dangerous cases ($r_A < m$), in which a parametric or no (exact) solution exists.

The classification from the physical point of view is expressed in compact form by Table 7.1.

The cases Baα, Baβ with uniquc solution and Bbα and Bbβ with parametric solution correspond to the case Aa. The solution of the case Baα can be obtained using the classical inverse matrix $\mathbf{A}^{-1}$. However the numerical method of solution is quite different as we shall see. The solution in other cases needs more sophisticated approaches as follows.

TABLE 7.1. Classification of singular cases

MBS		solution	type of **A**
regular $r_A = n$	safe $m = n$	1	
	dangerous $m > n$	1 0	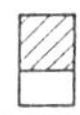
irregular $r_A < n$	safe $r_A = m$	∞	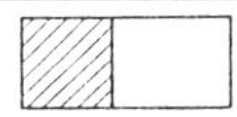
	dangerous $r_A > m$	∞ 0	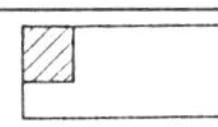

The cases Baβ and Bbβ with no (exact) solution correspond to the case Ab. There exists no solution which exactly satisfies (7.1). We can only construct some pseudosolutions which satisfy (7.1) approximately in some sense, e.g. we require the minimization of vector norm of residui $||\mathbf{Ax} - \mathbf{b}||$ or of solution $||\mathbf{x}||$. If $r_A < r_B, r_A = n$ then the solution which minimizes $||\mathbf{Ax} - \mathbf{b}||$ is

$$\mathbf{x} = (\mathbf{A}^T\mathbf{A})^{-1}\mathbf{A}^T\mathbf{b} \tag{7.2}$$

where $(\mathbf{A}^T\mathbf{A})^{-1}\mathbf{A}^T$ is Moore–Penrose generalized inverse of $\mathbf{A}$. If $r_A < r_B$, $r_A < n = \min(m, n)$ then the solution which minimizes $||\mathbf{x}||$ is

$$\mathbf{x} = \mathbf{A}^T(\mathbf{A}\mathbf{A}^T)^{-1}\mathbf{b} \tag{7.3}$$

assuming that $\mathbf{AA}^T$ is of full rank. $\mathbf{A}^T(\mathbf{AA}^T)^{-1}$ is another pseudoinverse of $\mathbf{A}$.

The most important procedures for numerical determination of above described cases and for solution of (7.1) are the following.

It is obviously vital to compute the rank r_A and to determine the linearly independent rows and columns of $\mathbf{A}$. All existing algorithms suffer from an inaccuracy due to rounding-off errors. Let us recall the so-called LU-decomposition with complete pivoting of the given matrix $\mathbf{A}$ as performed by subroutine MFGR (IBM Application Program).

After r steps of the algorithm, one can decompose the matrix $\mathbf{A}$ as follows:

$$\mathbf{RP} * \mathbf{A} * \mathbf{CP} = \mathbf{LDU} \tag{7.4}$$

where $\mathbf{L}$ is a (r, r) lower triangular matrix with units on the diagonal, $\mathbf{U}$ is a (r, r) upper triangular matrix, and $\mathbf{D}$ is a rectangular $(m - r, n - r)$ matrix, see (Figure 7.1). The symbol $\mathbf{E}$ denotes a unit matrix the size of which is clear from the context. Writing $\mathbf{0}$ signifies a rectangular matrix

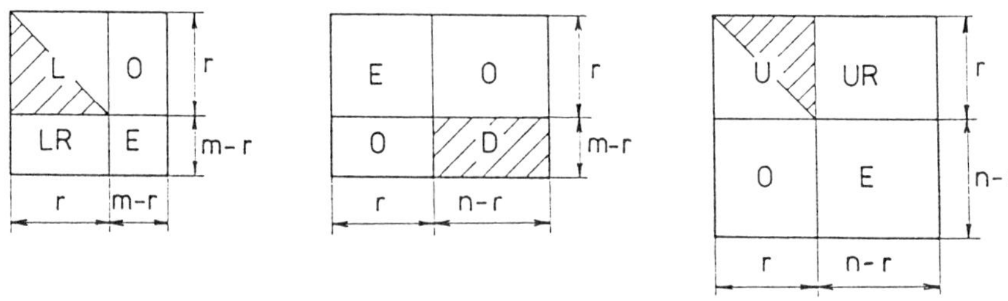

Figure 7.1. Matrices $\mathbf{L}, \mathbf{D}, \mathbf{U}$.

with zero elements.

Both **RP** and **CP** are (m, m) and (m, n) permutation matrices.

Let d_{max} be the absolutely largest element of $\mathbf{D}$. If $d_{max} \geq$ tolerance, then we proceed with the next (i.e. the $(r+1)$-st) step of the decomposition. If $d_{max} <$ tolerance, then we stop and claim that r is (numerically computed) the rank of $\mathbf{A}$.

Solving (7.1) we substitute $\mathbf{x} = \mathbf{CP} * \mathbf{y}$, $\mathbf{y} = [y_1, y_2, \ldots, y_n]$. Note that $(\mathbf{CP})^{-1} = (\mathbf{CP})^T$.

Using decomposition (7.4), equation (7.1) is

$$\mathbf{RP} * \mathbf{A} * \mathbf{x} = \mathbf{RP} * \mathbf{A} * \mathbf{CP} * \mathbf{y} = \mathbf{LDUy} = \mathbf{RP} * \mathbf{b} = \mathbf{c} \tag{7.5}$$

In order to exploit the block structure of matrices $\mathbf{L}, \mathbf{U}, \mathbf{D}$ we also split both $\mathbf{y}$ and $\mathbf{c}$ into blocks, namely:

$$\mathbf{y} = [\mathbf{y}_D, \mathbf{y}_I], \quad \mathbf{c} = [\mathbf{c}_D, \mathbf{c}_I] \tag{7.6}$$

Then (7.1) is equivalent to the following conditions upon $\mathbf{y}_D, \mathbf{y}_I$

$$\mathbf{y}_D = \mathbf{U}^{-1}\mathbf{L}^{-1}\mathbf{c}_D - \mathbf{U}^{-1} * \mathbf{UR} * \mathbf{y}_I \tag{7.7a}$$
$$\mathbf{c}_I = \mathbf{LR} * \mathbf{L}^{-1}\mathbf{c}_D + \mathbf{D}\mathbf{y}_I \tag{7.7b}$$

Since d_{max} is small enough we may write the last condition as

$$\mathbf{c}_I = \mathbf{LR} * \mathbf{L}^{-1}\mathbf{c}_D \tag{7.8}$$

This equation represents the consistency condition for (7.1) to be solvable.

We define the matrices

$$\mathbf{H} = -\mathbf{U}^{-1} * \mathbf{UR}, \quad \mathbf{C} = \mathbf{LR} * \mathbf{L}^{-1} \tag{7.9}$$

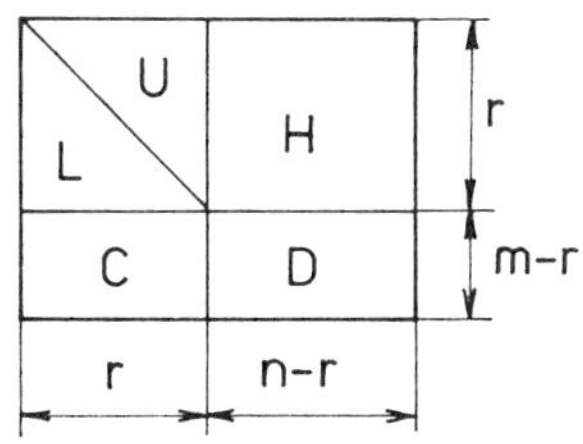

Figure 7.2. Allocation of matrices $\mathbf{L}, \mathbf{U}, \mathbf{H}, \mathbf{C}, \mathbf{D}$.

and both (7.7a) and (7.8) can be rewritten as

$$\mathbf{C}\mathbf{c}_D = \mathbf{c}_I \tag{7.10}$$

$$\mathbf{y}_D = \mathbf{U}^{-1}\mathbf{L}^{-1}\mathbf{c}_D + \mathbf{H}\mathbf{y}_I \tag{7.11}$$

The quoted subroutine MFGR returns the matrices $\mathbf{U}, \mathbf{L}, \mathbf{H}, \mathbf{C}, \mathbf{D}$, all of them being allocated in the blocks of the given matrix $\mathbf{A}$, the value of r and two vectors $IROW(i), i = 1, 2, \ldots, m$, $ICOL(j), j = 1, 2, \ldots, n$, (see Figure 7.2). These vectors define the permutation matrices $\mathbf{RP}$ and $\mathbf{CP}$ from (7.4) in the following way

$$\mathbf{RP}(i,j) = \begin{cases} 1 & \text{for} \quad j = IROW(i) \\ 0 & \text{for} \quad j \neq IROW(i) \end{cases} \tag{7.12}$$

and

$$\mathbf{CP}(i,j) = \begin{cases} 1 & \text{for} \quad i = ICOL(j) \\ 0 & \text{for} \quad i \neq ICOL(j) \end{cases} \tag{7.13}$$

There is, however, no need to construct the permutation matrices in order to construct $\mathbf{c}$ from $\mathbf{b}$ and reconstruct $\mathbf{x}$ from $\mathbf{y}$. In fact

$$c_j = b_i \quad \text{for} \quad j = IROW(i), \quad i = 1, 2, \ldots, m \tag{7.14}$$

$$x_i = y_j \quad \text{for} \quad i = ICOL(j), \quad j = 1, 2, \ldots, n \tag{7.15}$$

There are alternative approaches to a numerical solution of (7.1). For example, the singular value decomposition (SVD) of $\mathbf{A}$ can be used (Forsythe et al., 1977) which decomposes $\mathbf{A}$ as follows

$$\mathbf{A} = \mathbf{U}\mathbf{S}\mathbf{V}^T \tag{7.16}$$

$\mathbf{U}$ is (m, m) matrix and $\mathbf{V}$ is (n, n) matrix satisfying $\mathbf{U}^T\mathbf{U} = \mathbf{E}_m$ and $\mathbf{V}^T\mathbf{V} = \mathbf{E}_n$, respectively, and $\mathbf{S}$ is an (m, n) matrix with diagonal elements

$d_1, \ldots, d_k, k = \min(m, n)$.
The diagonal elements are ordered as follows

$$d_1 \geq d_2 \geq \ldots \geq d_k \geq 0 \tag{7.17}$$

The (numerically computed) rank r of $\mathbf{A}$ is defined as the integer $1 \leq r \leq k$, which satisfies the condition

$$d_r \leq \text{tolerance}, \quad d_{r-1} > \text{tolerance} \tag{7.18}$$

Having the matrix $\mathbf{A}$ in the form (7.16), the system (7.1) can easily be solved.
We substitute

$$\mathbf{x} = \mathbf{V}\mathbf{y} \tag{7.19}$$

into (7.1) and multiply this equation by the matrix $\mathbf{U}^T$; we then have

$$\mathbf{U}^T\mathbf{A}\mathbf{V}\mathbf{y} = \mathbf{U}^T\mathbf{b} \tag{7.20}$$

We set

$$\mathbf{c} = \mathbf{U}^T\mathbf{b} \tag{7.21}$$

and we use the decomposition (7.16). It is simple to deduce that

$$\mathbf{S}\mathbf{y} = \mathbf{c} \tag{7.22}$$

is equivalent to (7.1). We split both $\mathbf{y}$ and $\mathbf{c}$ into two blocks, namely

$$\mathbf{y} = [\mathbf{y}_D, \mathbf{y}_I] \tag{7.23a}$$
$$\mathbf{c} = [\mathbf{c}_D, \mathbf{c}_I] \tag{7.23b}$$

Obviously, the first r equations of (7.22) read as

$$\begin{bmatrix} d_1 & & 0 \\ & \ddots & \\ 0 & & d_r \end{bmatrix} \mathbf{y}_D = \mathbf{c}_D \tag{7.24}$$

The remaining $m - r$ equations of (7.22) are neglected, since $d_{r+1}, \ldots, d_k$ are small numbers. Thus $\mathbf{y}_I$ can be chosen arbitrarily and

$$\mathbf{c}_I = \mathbf{0} \tag{7.25}$$

is the consistency condition for (7.1) to be solvable. We resume by showing that the solution $\mathbf{x}$ of (7.1) is given by the formula

$$\mathbf{x} = \mathbf{V} \begin{bmatrix} \mathbf{y}_D \\ \mathbf{y}_I \end{bmatrix} \tag{7.26}$$

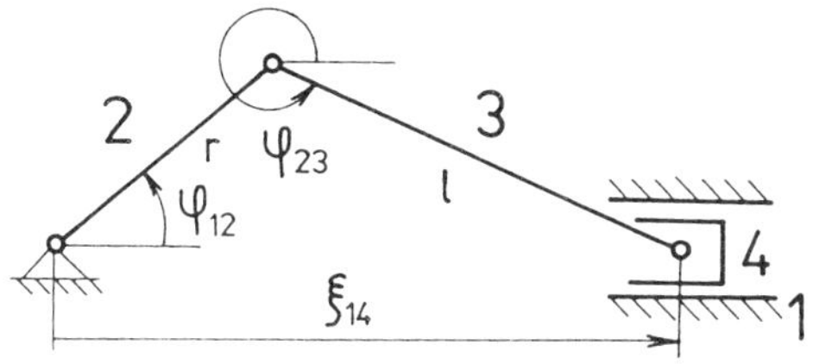

Figure 7.3. Crank mechanism.

where $\mathbf{y}_D$ solves (7.22) and the $\mathbf{y}_I$ represents the degrees of freedom. The equation (7.1) is solvable if the last $m - r$ components of the vector $\mathbf{U}^T\mathbf{b}$ equal zero.

It is worth noting that the degrees of freedom of the solution $\mathbf{x}$ are given in a transformed coordinate system (the action of $\mathbf{V}$ means a rotation of the coordinate system). From the computational point of view, the singular value decomposition (SVD) is computationally more demanding than the LU-decomposition with complete pivoting. On the other hand, the algorithms of SVD are claimed to be numerically more stable.

7.2 Physical Interpretation in Kinematics

It is evident from equations (5.154), (5.169) and (5.171) that, in the position, velocity and acceleration problem, the part of matrix $\mathbf{A}$ from the previous section is played by matrix $\mathbf{\Phi}_z$ defined by relation (5.153). Its rank $r(\mathbf{\Phi}_z)$ together with the rank of the extended system determines the solubility of the respective problem.

It is useful to realize that matrix $\mathbf{\Phi}_z$ is a submatrix of matrix

$$\mathbf{\Phi}_s = [\mathbf{\Phi}_z, \mathbf{\Phi}_q] \tag{7.27}$$

This division is, however, a subjective one. It depends on which coordinates are assumed to be independent. Through this choice we influence the solubility of the problem. It may happen that for an unsuitable choice of an independent (driving) coordinate the system is not soluble, whereas for another choice in the same position the system is soluble. We shall show this situation in the simple case of a crank mechanism as in Figure 7.3. Differentiating constraint equations

$$\begin{aligned} rc\varphi_{12} + lc\varphi_{23} - \xi_{14} &= 0 \\ rs\varphi_{12} + ls\varphi_{23} &= 0 \end{aligned} \tag{7.28}$$

with respect to time we obtain

$$\begin{bmatrix} -rs\varphi_{12}, & -ls\varphi_{23}, & -1 \\ rc\varphi_{12}, & lc\varphi_{23}, & 0 \end{bmatrix} \begin{bmatrix} \dot{\varphi}_{12} \\ \dot{\varphi}_{23} \\ \dot{\xi}_{14} \end{bmatrix} = \mathbf{0} \tag{7.29}$$

In a general position, the driving (independent) coordinate may be any one out of $\varphi_{12}, \varphi_{23}, \xi_{14}$. We have a different situation in, for example the upper turning point where $\varphi_{12} = 0, \varphi_{23} = 2\pi$.

Choosing in this position ξ_{14} for the driving coordinate, we obtain for the computation of velocities the equation

$$\begin{bmatrix} 0 & 0 \\ r & l \end{bmatrix} \begin{bmatrix} \dot{\varphi}_{12} \\ \dot{\varphi}_{23} \end{bmatrix} = \begin{bmatrix} 1 \\ 0 \end{bmatrix} \dot{\xi}_{14} \tag{7.30}$$

for which $r(\mathbf{\Phi}_z) = 1$. According to the classification from Table 7.1 we have an irregular dangerous case Bbβ which has for $\dot{\xi}_{14} = 0$ an infinite number of solutions

$$\frac{\dot{\varphi}_{12}}{\dot{\varphi}_{23}} = -\frac{l}{r} \tag{7.31}$$

and no solution for $\dot{\xi}_{14} \neq 0$.

Choosing φ_{12} for a driving coordinate, we obtain equation

$$\begin{bmatrix} 0 & -1 \\ l & 0 \end{bmatrix} \begin{bmatrix} \dot{\varphi}_{23} \\ \dot{\xi}_{14} \end{bmatrix} = \begin{bmatrix} 0 \\ -r \end{bmatrix} \dot{\varphi}_{12} \tag{7.32}$$

Here $r(\mathbf{\Phi}_z) = 2$ and we have a regular safe case Baα with a single solution

$$\dot{\xi}_{14} = 0, \quad \dot{\varphi}_{23} = -\frac{r}{l}\dot{\varphi}_{12} \tag{7.33}$$

The choice of $\varphi_{12}, \varphi_{23}$ for independent coordinates leads to a regular dangerous case Baβ, which has for

$$r\dot{\varphi}_{12} + l\dot{\varphi}_{23} = 0 \tag{7.34}$$

a single solution $\dot{\xi}_{14} = 0$ and no solution when (7.34) does not hold.

The examples presented show the dependence of the solution of kinematic problems not only on the topological properties of the MBS being solved (structure) but also on its metric properties (dimensions) which change according to position. Which element (which coordinate) is a driving element must furthermore be taken into consideration.

The above statements on solubility hold in exactly the same way for the computation of MBS mobility, which may be expressed by the relation

$$i = n - r(\mathbf{\Phi}_s) \tag{7.35}$$

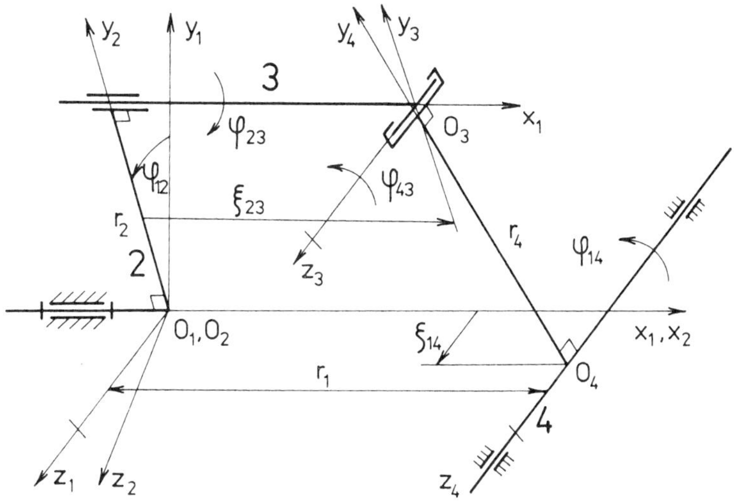

Figure 7.4. Altmann's mechanism.

where n is the number of coordinates required for a unique determination of the MBS position. This is also the reason why formula (3.1), which takes only MBS composition into account, cannot give a reliable result in all cases.

The above case of crank mechanism singularity occurs only in two positions. Cases do exist, however, where a MBS is permanently singular. This property is not rare, particularly with spatial mechanisms used in practice. This is the case for the Bennett mechanism, the Bricard mechanism, the Altmann mechanism, etc. They often have the advantage of consisting of a smaller number of elements or have simpler kinematic pairs. According to formula (3.1), from the viewpoint of composition, they are immobile. Mobility may, however, be achieved by a suitable choice of longitudinal and angular dimensions. The Turbula mechanism which we shall treat in Section 7.3 also belongs in this group.

For example, for the position of the Altman mechanism as in Figure 7.4, the following holds

$$\mathbf{T}_{12}\mathbf{T}_{23} = \mathbf{T}_{14}\mathbf{T}_{43} \qquad \text{(a)}$$

where

$$\begin{aligned}\mathbf{T}_{12} &= \mathbf{T}_{Z4}(\varphi_{12}) \\ \mathbf{T}_{23} &= \mathbf{T}_{Z2}(r_2)\mathbf{T}_{Z1}(\xi_{23})\mathbf{T}_{Z4}(\varphi_{23})\end{aligned}$$

$$
\begin{aligned}
\mathbf{T}_{14} &= \mathbf{T}_{Z1}(r_1)\mathbf{T}_{Z3}(\zeta_{14})\mathbf{T}_{Z6}(\varphi_{14}) \\
\mathbf{T}_{43} &= \mathbf{T}_{Z2}(r_4)\mathbf{T}_{Z6}(\varphi_{43})
\end{aligned} \qquad \text{(b)}
$$

Performing the multiplications in (a) we find that from 9 equations corresponding to direction cosines submatrix $\mathbf{S}$, only 2 are independent. Altogether 5 independent constraint equations are available, namely

$$
\begin{aligned}
\mathrm{s}(\varphi_{14}+\varphi_{43}) &= 0 \\
\mathrm{s}(\varphi_{12}+\varphi_{23}) &= 0 \\
\xi_{23}+r_4\mathrm{s}\varphi_{14}-r_1 &= 0 \\
r_2\mathrm{c}\varphi_{12}-r_4\mathrm{c}\varphi_{14} &= 0 \\
r_2\mathrm{s}\varphi_{12}-\zeta_{14} &= 0
\end{aligned} \qquad \text{(c)}
$$

The problem has altogether 6 variables, i.e. $\mathbf{s} = [\varphi_{12}, \varphi_{23}, \varphi_{43}, \varphi_{14}, \xi_{23}, \zeta_{14}]$ and for these

$$
\mathbf{\Phi}_s = \begin{bmatrix}
0, & 0, & \mathrm{c}(\varphi_{14}+\varphi_{43}), & \mathrm{c}(\varphi_{14}+\varphi_{43}), & 0, & 0 \\
\mathrm{c}(\varphi_{12}+\varphi_{23}), & \mathrm{c}(\varphi_{12}+\varphi_{23}), & 0, & 0, & 0, & 0 \\
0, & 0, & 0, & -r_4\mathrm{c}\varphi_{14}, & 1, & 0 \\
-r_2\mathrm{s}\varphi_{12}, & 0, & 0, & -r_4\mathrm{s}\varphi_{14}, & 0, & 0 \\
r_2\mathrm{c}\varphi_{12}, & 0, & 0, & 0, & 0, & -1
\end{bmatrix} \qquad \text{(d)}
$$

and $r(\mathbf{\Phi}_s) = 5$. Although according to (3.1) the degree of freedom of the Altman mechanism is 0, according to (7.35) the actual mobility degree of freedom $i = 6 - 5 = 1$. This is a consequence of the particular position of axes $z_1 \parallel z_3 \parallel z_4$, $x_1 \parallel x_3$.

In a similar way it is possible to examine the Bennett mechanism (Figure 7.5). It is a mechanism of the RRRR type which according to formula (3.1) has (-2) degrees of freedom, but which for

$$
\frac{d_1}{\sin\beta_3} = \pm\frac{d_2}{\sin\beta_4} = \pm\frac{d_3}{\sin\beta_1} = \pm\frac{d_4}{\sin\beta_2}
$$

is mobile (Beggs, 1983).

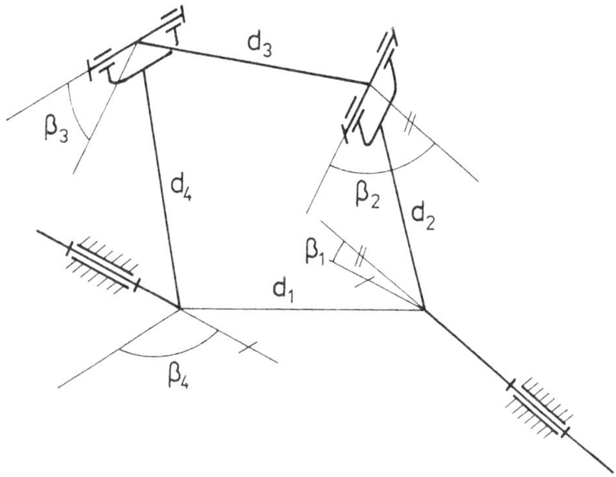

Figure 7.5. Bennett's mechanism.

7.3 Applications

Let us examine the motion of a spatial mechanism of the $\overline{R}-R-R-R-R-\underline{R}$ type as in Figure 7.6 (Brát, 1969). This mechanism is the basis of the Turbula machine, where element 4 is a drum supporting a vessel containing the material to be mixed. The input element is shaft 2. The drum performs a general spatial motion.

We have here a six element mechanism where all pairs are revolute. According to formula (3.1) we have $i = 0$; the mechanism should have no mobility.

Let us carry out the analysis of its mobility. Let us first assemble the equations for solution of the position problem. Let us perform an imaginary cut through element 4. This yields, according to (5.75), the following matrix equation

$$\mathbf{T}_{12}\mathbf{T}_{23}\mathbf{T}_{34} = \mathbf{T}_{16}\mathbf{T}_{65}\mathbf{T}_{54} \tag{7.36}$$

From Figure 7.6.

$$\begin{aligned}
\mathbf{T}_{12} &= \mathbf{T}_{Z6}(\varphi_{12}) \\
\mathbf{T}_{23} &= \mathbf{T}_{Z4}(\varphi_{23}) \\
\mathbf{T}_{34} &= \mathbf{T}_{Z3}(a)\mathbf{T}_{Z5}(\varphi_{34}) \\
\mathbf{T}_{54} &= \mathbf{T}_{54}\mathbf{T}_{44} = \mathbf{T}_{Z3}(a)\mathbf{T}_{Z4}(\varphi_{54})\mathbf{T}_{Z3}(a)\mathbf{T}_{Z4}(\pi) \\
\mathbf{T}_{65} &= \mathbf{T}_{Z5}(\varphi_{65}) \\
\mathbf{T}_{16} &= \mathbf{T}_{Z2}(-b)\mathbf{T}_{Z6}(\varphi_{16})
\end{aligned} \tag{7.37}$$

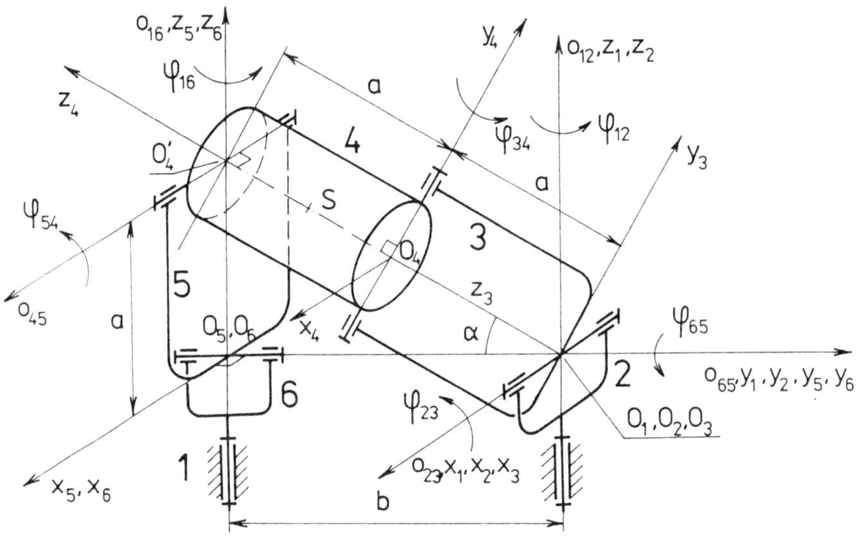

Figure 7.6. Mechanism TURBULA.

Substituting from (7.37) into (7.36) and performing the multiplications yields

$$\mathrm{c}\varphi_{34}\mathrm{c}\varphi_{12} - \mathrm{s}\varphi_{34}\mathrm{s}\varphi_{23}\mathrm{s}\varphi_{12} = \mathrm{c}\varphi_{65}\mathrm{c}\varphi_{16} \tag{7.38a}$$

$$\mathrm{c}\varphi_{34}\mathrm{s}\varphi_{12} + \mathrm{s}\varphi_{34}\mathrm{s}\varphi_{23}\mathrm{c}\varphi_{12} = \mathrm{c}\varphi_{65}\mathrm{s}\varphi_{16} \tag{7.38b}$$

$$-\mathrm{s}\varphi_{34}\mathrm{c}\varphi_{23} = -\mathrm{s}\varphi_{65} \tag{7.38c}$$

$$-\mathrm{c}\varphi_{23}\mathrm{s}\varphi_{12} = \mathrm{c}\varphi_{54}\mathrm{s}\varphi_{16} - \mathrm{s}\varphi_{54}\mathrm{s}\varphi_{65}\mathrm{c}\varphi_{16} \tag{7.38d}$$

$$\mathrm{c}\varphi_{23}\mathrm{c}\varphi_{12} = -\mathrm{c}\varphi_{54}\mathrm{c}\varphi_{16} - \mathrm{s}\varphi_{54}\mathrm{s}\varphi_{65}\mathrm{s}\varphi_{16} \tag{7.38e}$$

$$\mathrm{s}\varphi_{23} = -\mathrm{s}\varphi_{54}\mathrm{c}\varphi_{65} \tag{7.38f}$$

$$\mathrm{s}\varphi_{34}\mathrm{c}\varphi_{12} + \mathrm{c}\varphi_{34}\mathrm{s}\varphi_{23}\mathrm{s}\varphi_{12} = -\mathrm{s}\varphi_{54}\mathrm{s}\varphi_{16} - \mathrm{c}\varphi_{54}\mathrm{s}\varphi_{65}\mathrm{c}\varphi_{16} \tag{7.38g}$$

$$\mathrm{s}\varphi_{34}\mathrm{s}\varphi_{12} - \mathrm{c}\varphi_{34}\mathrm{s}\varphi_{23}\mathrm{c}\varphi_{12} = \mathrm{s}\varphi_{54}\mathrm{c}\varphi_{16} - \mathrm{c}\varphi_{54}\mathrm{s}\varphi_{65}\mathrm{s}\varphi_{16} \tag{7.38h}$$

$$\mathrm{c}\varphi_{34}\mathrm{c}\varphi_{23} = -\mathrm{c}\varphi_{54}\mathrm{c}\varphi_{65} \tag{7.38i}$$

$$a\mathrm{s}\varphi_{23}\mathrm{s}\varphi_{12} = a(\mathrm{s}\varphi_{54}\mathrm{s}\varphi_{16} + \mathrm{c}\varphi_{54}\mathrm{s}\varphi_{65}\mathrm{c}\varphi_{16} + \mathrm{s}\varphi_{65}\mathrm{c}\varphi_{16}) \tag{7.38j}$$

$$-a\mathrm{s}\varphi_{23}\mathrm{c}\varphi_{12} = a(-\mathrm{s}\varphi_{54}\mathrm{c}\varphi_{16} + \mathrm{c}\varphi_{54}\mathrm{s}\varphi_{65}\mathrm{s}\varphi_{16} + \mathrm{s}\varphi_{65}\mathrm{s}\varphi_{16}) - b \tag{7.38k}$$

$$a\mathrm{c}\varphi_{23} = a(\mathrm{c}\varphi_{54}\mathrm{c}\varphi_{65} + \mathrm{c}\varphi_{65}) \tag{7.38l}$$

In the initial position, evidently $\varphi_{12} = 0, \varphi_{23} = (\pi/2) - \alpha, \varphi_{34} = 0, \varphi_{54} = \pi + (\pi/2) - \alpha = (3\pi/2) - \alpha, \varphi_{65} = 0, \varphi_{16} = 0$, (Figure 7.6). Substituting into (7.38) we easily find that equations (7.38–1) to (7.38–10) are valid, as

$$\begin{aligned} c\varphi_{23} &= s\alpha, \quad s\varphi_{23} = c\alpha \\ c\varphi_{54} &= -s\alpha, \quad s\varphi_{54} = -c\alpha \end{aligned} \tag{7.39}$$

The last two equations are

$$-a\ c\alpha = a\ c\alpha - b \tag{7.40a}$$

$$a\ s\alpha = a(-s\alpha + 1) \tag{7.40b}$$

According to Figure 7.6. we express

$$c\alpha = \frac{a}{\sqrt{a^2+b^2}}, \quad s\alpha = \frac{b}{\sqrt{a^2+b^2}} \tag{7.41}$$

Substituting (7.41) into (7.40a) we easily obtain

$$\frac{b}{a} = \sqrt{3} \tag{7.42}$$

Hence

$$\alpha = \frac{\pi}{6} \tag{7.43}$$

which fulfills at the same time (7.40b), and the mechanism is therefore mobile. The mobility proof for the whole range of motions is performed by finding a general analytical solution of equations (7.38) assuming (7.43).

Raising to power and adding equations (7.38–6) and (7.38–12) yields

$$1 = 2c^2\varphi_{65}(1 + c\varphi_{54}) \tag{7.44}$$

Substituting (7.44) into (7.38–12) yields

$$c\varphi_{23} = \frac{1}{2c\varphi_{65}} \tag{7.45}$$

Hence

$$\begin{aligned} c\varphi_{65} &= \frac{1}{2c\varphi_{23}} \\ s\varphi_{65} &= \frac{\pm\sqrt{4c^2\varphi_{23} - 1}}{2c\varphi_{23}} \end{aligned} \tag{7.46}$$

From (7.44) and (7.45)

$$c\varphi_{54} = 2c^2\varphi_{23} - 1 \tag{7.47}$$

and from (7.38–6) and (7.45)

$$s\varphi_{54} = -2s\varphi_{23}c\varphi_{23} \tag{7.48}$$

Further, from (7.38–9), (7.47) and (7.45),

$$c\varphi_{34} = \frac{1 - 2c^2\varphi_{23}}{2c^2\varphi_{23}} \tag{7.49}$$

and hence

$$s\varphi_{34} = \frac{\pm\sqrt{4c^2\varphi_{23} - 1}}{2c^2\varphi_{23}} \tag{7.50}$$

Finally, we express $s\varphi_{16}$ and $c\varphi_{16}$ from (7.38–1,2) and substitute into equation (7.38–11). We then express all angles in this equation from (7.46) to (7.50), using angles φ_{23} and φ_{12}. Modifying, we obtain

$$s\varphi_{23} = \frac{\sqrt{3}}{2}c\varphi_{12} \tag{7.51}$$

The complete solution (one of several solutions) can easily be expressed and so

$$\begin{aligned}
s\varphi_{23} &= \tfrac{\sqrt{3}}{2}c\varphi_{12}, & c\varphi_{23} &= \tfrac{1}{2}\sqrt{4 - 3c^2\varphi_{12}} \\
s\varphi_{34} &= \frac{2\sqrt{3}s\varphi_{12}}{4 - 3c^2\varphi_{12}}, & c\varphi_{34} &= \frac{3c^2\varphi_{12} - 2}{4 - 3c^2\varphi_{12}} \\
s\varphi_{54} &= -\tfrac{\sqrt{3}}{2}\sqrt{4 - 3c^2\varphi_{12}}c\varphi_{12}, & c\varphi_{54} &= \tfrac{1}{2}(2 - 3c^2\varphi_{12}) \\
s\varphi_{65} &= \frac{-\sqrt{3}s\varphi_{12}}{\sqrt{4 - 3c^2\varphi_{12}}}, & c\varphi_{65} &= \frac{1}{\sqrt{4 - 3c^2\varphi_{12}}} \\
s\varphi_{16} &= \frac{-2s\varphi_{12}}{\sqrt{4 - 3c^2\varphi_{12}}}, & c\varphi_{16} &= \frac{c\varphi_{12}}{\sqrt{4 - 3c^2\varphi_{12}}}
\end{aligned} \tag{7.52}$$

The mechanism transmission ratio is

$$\mu_{26} = \frac{d\varphi_{16}}{d\varphi_{12}} = \frac{-2}{4 - 3c^2\varphi_{12}} \tag{7.53}$$

The diagrams of transmission ratio behaviour (7.53) and of behaviour of φ_{16} from (7.52) are presented in Figure 7.7.

For the study of mechanism properties the examination of element 4 (drum) motion is essential. We express it using a description of this motion decomposed into carrying translational motion of element 4 centre of mass S and relative spherical motion around point S. We assume the element 4 centre of mass S in the centre of line segment O_4O_4'. Then

$$\mathbf{r}_{1S} = \mathbf{T}_{12}\mathbf{T}_{23}\mathbf{T}_{34}\mathbf{r}_{4S} \tag{7.54}$$

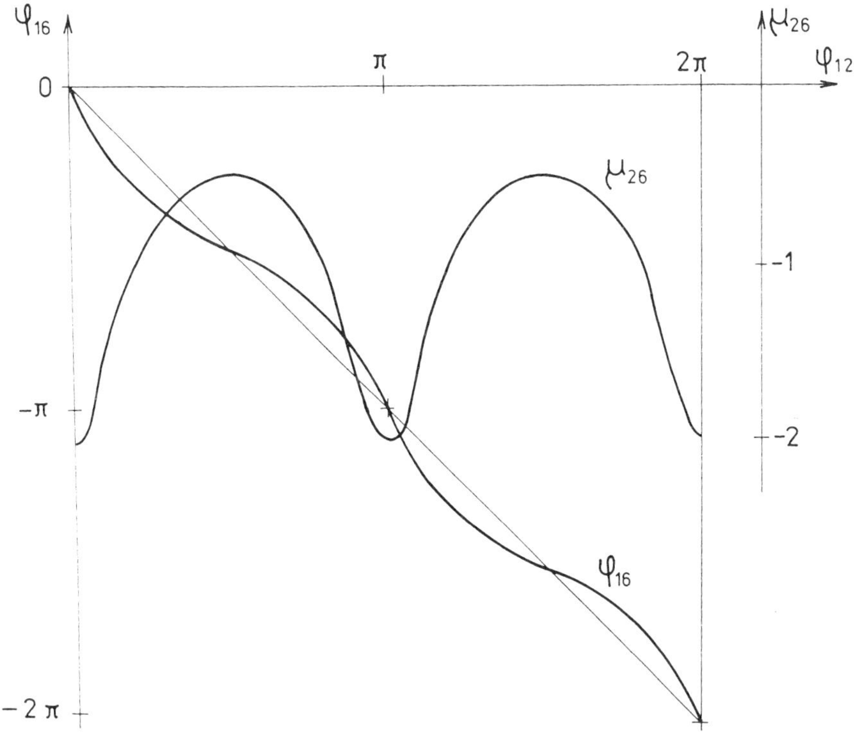

Figure 7.7. Functions $\varphi_{16}(\varphi_{12}), \mu_{26}(\varphi_{12})$.

where $\mathbf{r}_{1S} = [x_{1S}, y_{1S}, z_{1S}, 1]$ and $\mathbf{r}_{4S} = [0, 0, \frac{1}{2}a, 1]$. Substituting and computing yields

$$\begin{bmatrix} x_{1S} \\ y_{1S} \\ z_{1S} \end{bmatrix} = a \begin{bmatrix} \frac{\sqrt{3}}{8}\frac{2 - 3\mathrm{c}^2\varphi_{12}}{4 - 3\mathrm{c}^2\varphi_{12}}\mathrm{s}(2\varphi_{12}) \\ \frac{\sqrt{3}}{4}\frac{3\mathrm{c}^4\varphi_{12} - 2\mathrm{c}^2\varphi_{12} - 4}{4 - 3\mathrm{c}^2\varphi_{12}} \\ \frac{3}{4}\frac{1 + \mathrm{s}^2\varphi_{12}}{\sqrt{4 - 3\mathrm{c}^2\varphi_{12}}} \end{bmatrix} \tag{7.55}$$

In Figure 7.8 the path of point S for $a = 0.1$ is shown.

For the description of relative spherical motion we may use various means from Section 5.1.5. The problem we have is to choose a motion description as intuitively as possible. From this viewpoint, a suitable description of element 4 relative spherical motion is to use hodographs of

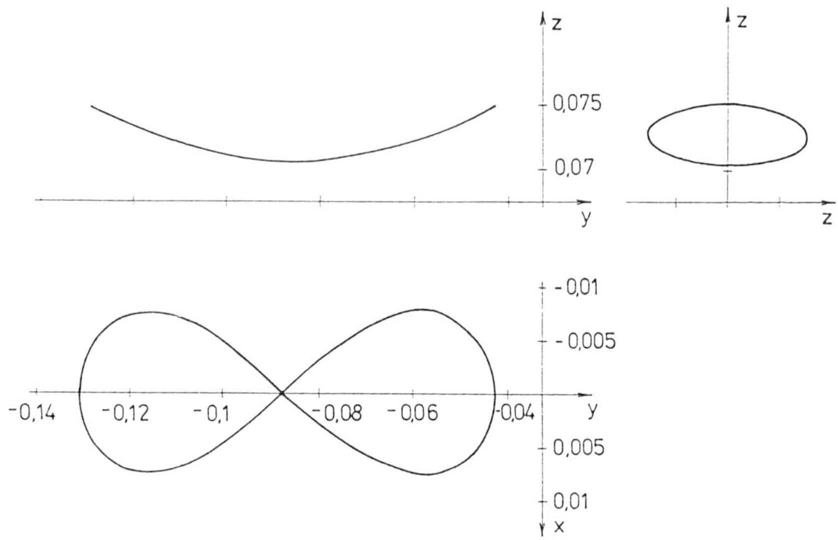

Figure 7.8. Trajectory of point S.

unit vectors of the coordinate system $O_4x_4y_4z_4$ firmly connected with the drum body. These vectors equal the columns of matrix $\mathbf{S}_{14}$, which we easily express from (7.36) and (7.37).

$$\mathbf{S}_{14} = \mathbf{S}_{12}\mathbf{S}_{23}\mathbf{S}_{34} = \mathbf{S}_{Z6}(\varphi_{12})\mathbf{S}_{Z4}(\varphi_{23})\mathbf{S}_{Z5}(\varphi_{34}) \tag{7.56}$$

$$\mathbf{S}_{14} = \begin{bmatrix} x_{1,4} & y_{1,4} & z_{1,4} \\ x_{2,4} & y_{2,4} & z_{2,4} \\ x_{3,4} & y_{3,4} & z_{3,4} \end{bmatrix} =$$

$$= \begin{bmatrix} \dfrac{\mathrm{c}\varphi_{12}(1-6\mathrm{s}^2\varphi_{12})}{D^2}, & -\frac{1}{2}\mathrm{s}\varphi_{12}D, & \dfrac{(\sqrt{3}/2)\mathrm{s}\varphi_{12}\mathrm{c}\varphi_{12}(2+3\mathrm{c}^2\varphi_{12})}{D^2} \\ \dfrac{2\mathrm{s}\varphi_{12}(2-3\mathrm{s}^2\varphi_{12})}{D^2}, & \frac{1}{2}\mathrm{c}\varphi_{12}D, & \dfrac{(\sqrt{3}/2)(2+2\mathrm{s}^2\varphi_{12}-3\mathrm{c}^4\varphi_{12})}{D^2} \\ \dfrac{\sqrt{3}\mathrm{s}\varphi_{12}}{D}, & \frac{\sqrt{3}}{2}\mathrm{c}\varphi_{12}, & \dfrac{(3\mathrm{c}^2\varphi_{12}-2)}{D} \end{bmatrix} \tag{7.57}$$

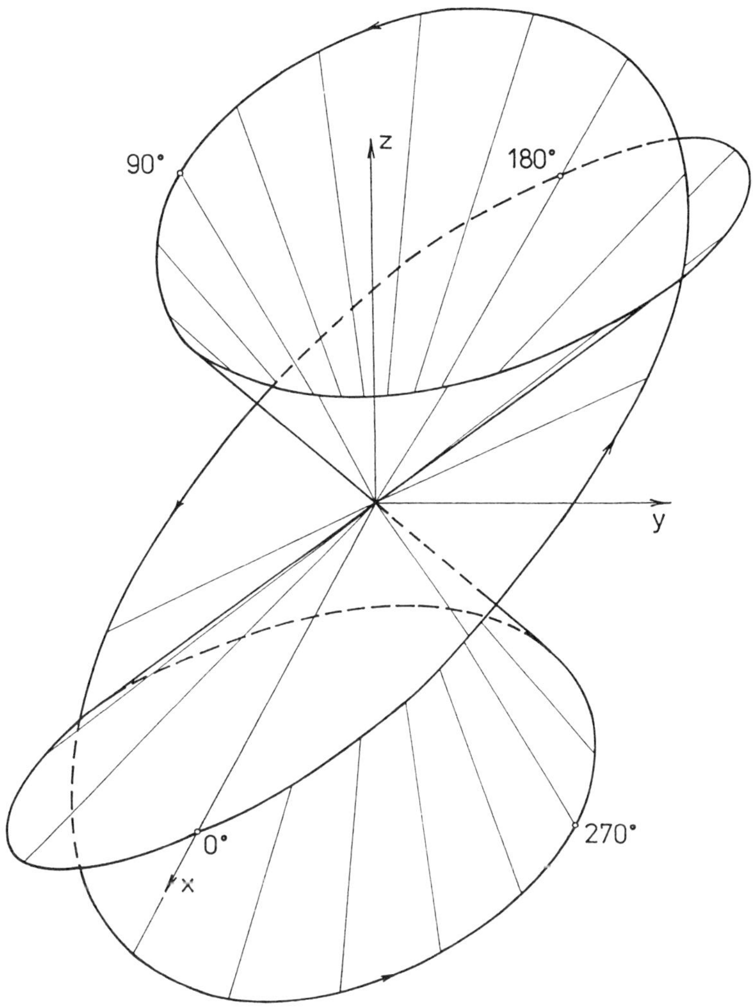

Figure 7.9. Hodograph of unit vector $\mathbf{x}_4^0$.

where $D = \sqrt{4 - 3c^2\varphi_{12}}$. The hodographs are shown in Figure 7.9. The respective velocities and accelerations are easily obtainable from relations (7.55) and (7.57) differentiating them with respect to time.

Chapter 8

Kinetostatic Analysis of MBSs

In this chapter we are going to add force effects (forces and force couples) and mass quantities (masses, moments of inertia, products of inertia) to the set of quantities describing MBS. We thus arrive at MBS dynamics, where the motion of certain elements may be prescribed and, simultaneously, given action forces may act on the system.

The most general problems in MBS dynamics are the determination of external force effects securing the required motion of certain bodies, the determination of motion of the remaining bodies, and finally the computation of reactions in all kinematic pairs. The solution of MBS dynamics is in practice only rarely performed in its full generality. Usually we treat the so-called direct problem (first problem of dynamics), where we determine the motion caused by given action forces and the inverse (indirect) problem (second problem of dynamics, kinetostatics), where we determine the action forces needed to secure the required motion. The kinetostatic solution is the topic of this chapter.

8.1 General Solution of the Problem

The MBS kinetostatic solution method is influenced to a great extent by the formalism used. The expression of force effects (forces and force couples) with six-element vectors introduced in Section 4.4 has the advantage of being concise. This is possible because a force acting on a rigid body has the same behaviour as a vector bound to a line. It is a typical line vector, whose effect (from the equivalent statics viewpoint) may be expressed in

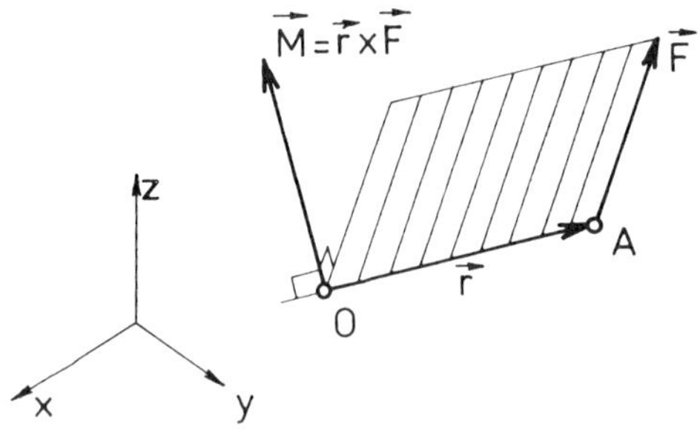

Figure 8.1. Moment of force.

any point of substitution by a vector of force $\vec{F}$ and by a vector of couple $\vec{M} = \vec{r} \times \vec{F}$, where the radius vector $\vec{r}$ determines the position of the current point A of the force fixed line of action with respect to a chosen point of substitution O, (Figure 8.1). We shall express this substitute effect in accordance with Section 4.4 by motor

$$\mathbf{F}_0 = \lceil \mathbf{F}, \mathbf{M} \rceil = \lceil F_x, F_y, F_z, M_x, M_y, M_z \rceil \tag{8.1}$$

To assemble the basic equations we use the free body method. Forces of three types act on free body j: active (applied), reaction and inertia. We shall denote them by the symbol $\mathbf{F}$ with superscript A, R, I. The effects of all of these must be in balance, according to the d'Alembert principle. In order to be able to compare them, they must be expressed in the same system of coordinates. This may be the fixed coordinate system of the frame. However the equations are simpler in a system fixed to the body. We shall choose its origin O_j in the centre of mass of this body and the axes x_j, y_j, z_j identical with the principal central axes of inertia. We shall call this system of coordinates the basic system of body j. At the same time, the origin O_j is going to be the point of substitution of all force effects acting on the body. We shall compute all of them.

If known action forces act on a body in points $\alpha = A, B, C, \ldots$ we can compute their resultant motor according to the relation

$$\bar{\mathbf{F}}^A_{jj} = \sum_\alpha \bar{\mathbf{F}}^A_{jj\alpha} \tag{8.2}$$

where the first subscript designates the system in which the motor is expressed, the second subscript designates the body on which the force acts, and index α designates the order in which the applied forces act.

As the MBS motion is determined, the kinematic quantities of all its elements may be computed. We use the basic decomposition with point of reference O_j. The resultant force of inertia of body j expressed in space 1 fixed to the frame is

$$\mathbf{F}_{1j}^{I} = -m_j \ddot{\mathbf{u}}_{10j} \tag{8.3}$$

where m_j is the mass of body j, and $\ddot{\mathbf{u}}_{10j}$ the absolute acceleration of point O_j expressed in space 1. As a vector quantity we transform it into the basic space of body j according to (4.28), using the relation

$$\mathbf{F}_{jj}^{I} = \mathbf{S}_{j1}\mathbf{F}_{1j}^{I} \tag{8.4}$$

The resultant inertial couple acting on body j is expressed in its basic system by a negated left side of the Euler dynamic equations (9.15a)

$$\mathbf{M}_{jj}^{I} = -(\mathbf{I}_{jj}\dot{\boldsymbol{\omega}}_{jj} + \boldsymbol{\Omega}_{jj}\mathbf{I}_{jj}\boldsymbol{\omega}_{jj}) \tag{8.5}$$

In this case,

$$\mathbf{I}_{jj} = \begin{bmatrix} I_{xx} & -I_{xy} & -I_{xz} \\ -I_{yx} & I_{yy} & -I_{yz} \\ -I_{zx} & -I_{zy} & I_{zz} \end{bmatrix} \tag{8.6}$$

is the constant matrix (tensor) of inertia of body j in its basic system, $\boldsymbol{\omega}_{jj} = \boldsymbol{\omega}_{j,1j}$ is the vector of absolute angular velocity of body j expressed by components in the same system, $\boldsymbol{\Omega}_{jj}$ is its expression in form (4.40) and $\dot{\boldsymbol{\omega}}_{jj}$ is its time derivative. The motor of resultant inertial effects acting on body j is then

$$\bar{\mathbf{F}}_{jj}^{I} = \lceil \mathbf{F}_{jj}^{I}, \mathbf{M}_{jj}^{I} \rceil \tag{8.7}$$

To express the reactions is more complicated. From the viewpoint of a simple interpretation of motor components of each reaction, it is convenient to use a system of coordinates corresponding to the reaction, (see Table 8.1). This leads to the problem of motor transformation from space $O_\beta x_\beta y_\beta z_\beta$ of a kinematic pair β into the basic space α of body j. We perform it by the relation

$$\bar{\mathbf{F}}_{\alpha} = \bar{\mathbf{T}}_{\alpha\beta}\bar{\mathbf{F}}_{\beta} \tag{8.8}$$

in which the transformation matrix

$$\bar{\mathbf{T}}_{\alpha\beta} = \begin{bmatrix} \mathbf{S}_{\alpha\beta} & \mathbf{0} \\ \mathbf{R}_{\alpha\beta}\mathbf{S}_{\alpha b\beta} & \mathbf{S}_{\alpha\beta} \end{bmatrix} \tag{8.9}$$

with dimension (6,6) has been introduced; $\mathbf{S}_{\alpha\beta} = [\mathbf{e}_{\beta x}, \mathbf{e}_{\beta y}, \mathbf{e}_{\beta z}]$ contains, in columns, system β unit vectors expressed in system α and

$$\mathbf{R}_{\alpha\beta} = \begin{bmatrix} 0 & -z & y \\ z & 0 & -x \\ -y & x & 0 \end{bmatrix} \tag{8.10}$$

TABLE 8.1. Components of reaction motors

kinematic pair	number of static parameters	components of reaction motor	transformation matrix
revolute	5	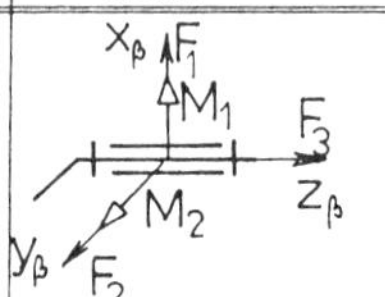	$\bar{T}^R_{\alpha\beta} = \left[\begin{array}{c\|c} S_{\alpha\beta} & 0 \\ \hline R_{\alpha\beta} S_{\alpha\beta} & e_{\beta x}, e_{\beta y} \end{array}\right]$
prismatic	5	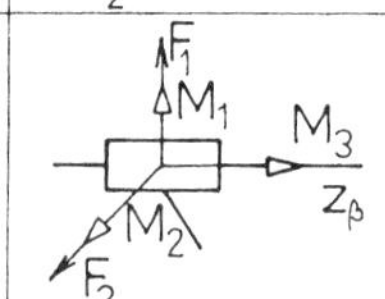	$\bar{T}^P_{\alpha\beta} = \left[\begin{array}{c\|c} e_{\beta x}, e_{\beta y} & 0 \\ \hline R_{\alpha\beta}[e_{\beta x}, e_{\beta y}] & S_{\alpha\beta} \end{array}\right]$
helical	5	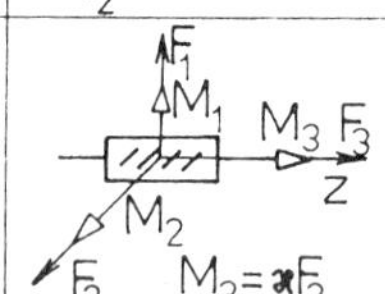	$\bar{T}^H_{\alpha\beta} = \left[\begin{array}{c\|c} S_{\alpha\beta} & 0 \\ \hline R_{\alpha\beta} S_{\alpha\beta} & e_{\beta x}, e_{\beta y} \end{array}\right]$
cylindrical	4	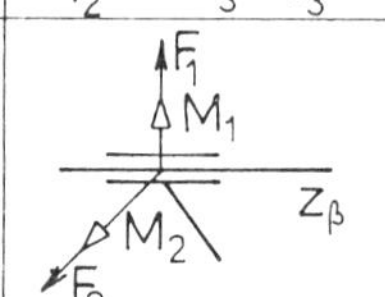	$\bar{T}^C_{\alpha\beta} = \left[\begin{array}{c\|c} e_{\beta x}, e_{\beta y} & 0 \\ \hline R_{\alpha\beta}[e_{\beta x}, e_{\beta y}] & e_{\beta x}, e_{\beta y} \end{array}\right]$
spherical	3	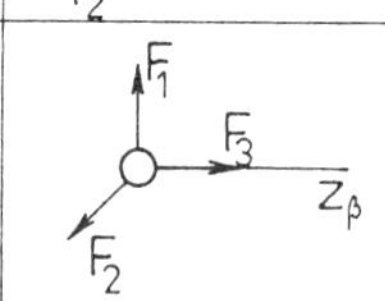	$\bar{T}^S_{\alpha\beta} = \left[\begin{array}{c} S_{\alpha\beta} \\ \hline R_{\alpha\beta} S_{\alpha\beta} \end{array}\right]$
flat	3	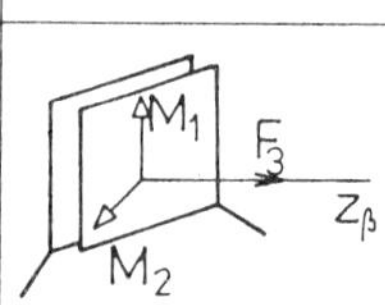	$\bar{T}^F_{\alpha\beta} = \left[\begin{array}{c\|c} e_{\beta z} & 0 \\ \hline R_{\alpha\beta} e_{\beta z} & e_{\beta x}, e_{\beta y} \end{array}\right]$

is a skew symmetric (antisymmetric) matrix expressing the radius vector of origin O_β in system α. As the number of static parameters (components of reaction forces and couples) of a given kinematic pair is determined by its class, it is always less than 6. We therefore modify the transformation matrices $\bar{\mathbf{T}}_{\alpha\beta}$ for the recomputation of the reaction motor from space β of the kinematic pair into space α in such a way that we leave out columns corresponding to null motor components. The transformation matrices are then rectangular (see Table 8.1).

The resultant motor of reaction effects acting on body j and substituted in point O_j may be expressed using the designation introduced by relation

$$\bar{\mathbf{F}}_{jj}^{R} = \sum_{k} \bar{\mathbf{F}}_{jjk}^{R} \tag{8.11}$$

where $\bar{\mathbf{F}}_{jjk}^{R}$ is the motor of reaction acting from body k on to body j substituted in point O_j. We compute it from the expression

$$\bar{\mathbf{F}}_{jj}^{R} = \bar{\mathbf{T}}_{j\beta}^{\gamma} \bar{\mathbf{F}}_{jk}^{*} \tag{8.12}$$

where $\bar{\mathbf{T}}_{j\beta}^{\gamma}$ is one of the transformation matrices introduced in Table 8.1 (kinematic pair type given by superscript γ), transforming the shortened motor $\bar{\mathbf{F}}_{jk}^{*}$ of reaction (containing only non-null components as a matter of principle) from system β into the basic system α of the body.

From all the types of forces considered in this analysis, the unknown action force effects, for example robot actuators, remain to be considered. These act mostly in kinematic pairs and in directions in which the reaction components vanish. We may conveniently include them as components in the motors of reactions of driven kinematic pairs. If this is not the case, we treat the reaction formally in the same way as a known applied force; the elements of motor $\bar{\mathbf{F}}^{AU}$ are, however, unknown.

We can now write the equilibrium equation of body j

$$\bar{\mathbf{F}}_{jj}^{A} + \bar{\mathbf{F}}_{jj}^{I} + \bar{\mathbf{F}}_{jj}^{R} + \bar{\mathbf{F}}_{jj}^{AU} = \mathbf{0} \tag{8.13}$$

For a system of n bodies including the frame, it is possible to write altogether $n-1$ equations. We must remember that each of the equations (8.13) is written in a different system of coordinates, so that the same reaction effect has a different expression in the basic system of bodies between which it acts. If the system solved has i degrees of freedom and contains d_j kinematic pairs of class j, the kinetostatic computation represents $6(n-1)$ equations. It follows from formula (3.1) that the number of equations and unknowns in a non-singular case is the same. This confirms the solubility of the problem.

Before starting with the proper computation of unknowns, we modify the equation (8.13) so that we express all elements in the form of the product of the transformation matrix (8.9) and the respective motor. In the case of applied force determined by components in space j, the matrix (8.9) has the form

$$\bar{\mathbf{T}}_{\alpha\beta} = \begin{bmatrix} \mathbf{E}_3 & \mathbf{0} \\ \mathbf{R}_{\alpha\beta} & \mathbf{E}_3 \end{bmatrix} \tag{8.14}$$

for inertial effects $\bar{\mathbf{T}}_{\alpha\beta} = \mathbf{E}_6$; in the case of reaction effects $\bar{\mathbf{T}}_{\alpha\beta} = \bar{\mathbf{T}}^{\gamma}_{\alpha\beta}$ from (8.12). Equation (8.13) may be more concisely rewritten in the form

$$\mathbf{T}_j \mathbf{f}_j = \mathbf{0} \tag{8.15}$$

where matrix $\mathbf{T}_j$ contains as sub-matrices the transformation matrices $\mathbf{T}_{\alpha\beta}$ of all force effects acting on body j, and vector $\mathbf{f}_j$ contains components of all their motors. Performing the same procedure with equations of all bodies yields

$$\mathbf{Tf} = \mathbf{0} \tag{8.16}$$

where the row dimension of matrix $\mathbf{T}$ is $6(n-1)$. Equation (8.13) is linear in the case of unknown applied forces and in reaction components, and may be written in the standard form

$$\mathbf{Ax} = \mathbf{b} \tag{8.17}$$

where $\mathbf{A}$ is the system matrix, $\mathbf{x}$ the vector of unknowns and $\mathbf{b}$ the right-hand side vector. In the solution of this equation, singular cases may occur according to Chapter 7. Their physical meaning (statically indeterminate cases or neglected mass in dynamical models) has been described in detail in (Stejskal et al., 1990).

We now consider the method of the automatic assembly of equation (8.13). There is advantage in using for this purpose the reduced incidence matrix $\mathbf{C}$ of the associated oriented graph. We introduce it in such a way that nodes correspond to bodies of the MBS to be solved and edges correspond to force effects. The edge orientation denotes the sense in which the force effect acts. If it is oriented from node i to node j, the respective force effect is understood as the effect of body i on body j. The inertial effects are at the same time understood as effects of the frame on the respective body. For the graph constructed in this way we define the reduced incidence matrix $\mathbf{C}$ in such a way that its elements

$$c_{ij} = \begin{cases} +1 & \text{if edge } j \text{ ends in node } i \\ -1 & \text{if edge } j \text{ begins in node } i \\ 0 & \text{otherwise} \end{cases}$$

whereas body 1 (frame) is not considered.

Comparing the structure of matrix **C** with the structure of matrix **T** as far as non-null elements are concerned, we find that they are identical. This conformity follows from the definition of the associated graph. Each edge corresponds to a force effect, whereby all force effects acting on a certain body coincide with the node representing this body in the graph. The given conformities may be used for automated assembly of the system (8.17) of equilibrium equations (Stejskal, 1978). We describe the algorithm as follows:

1. We enumerate the bodies and force effects.

2. We designate each force effect by a chosen code giving its type (applied, inertial, reaction effect of revolute, prismatic, etc. kinematic pair) and, for example, using the sign we distinguish given and unknown effects.

3. We create the reduced incidence matrix **C**.

4. We search column-wise the matrix **C** for non-null elements. On finding such an element, we store its row (i) and column (j) subscript and continue in the following way.

4.1. From the column index and chosen code we find out the type of force effect with which we are dealing.

4.2. In case of a given force effect, we compute the respective motor from relations (8.2) and (8.7), change the signs of all elements, and place it as a subvector into vector **b** from equation (8.17) in the position determined by index i.

4.3. In case of an unknown force effect which is not included in the reaction motor, we compute the corresponding transformation matrix (8.9) and place it as a block matrix in matrix **A** from equation (8.17), unchanged for $c_{ij} = +1$ or changing signs of all elements for $c_{ij} = -1$. The position of matrix $\bar{\mathbf{T}}$ in matrix **A** is given by the value of indices i, j and the information on the types of these force effects where the designation is less than j.

After evaluating the whole matrix **C**, we have assembled a system of equilibrium equations (8.17) which we solve numerically using a suitable algorithm.

A difficult and time-consuming part of the above procedure is the computation of elements of transformation matrices. This is a kinematic problem for the solution of which we use procedures presented in Chapters 4 and 5. We do not deal with them here in more detail. Let us only remember that this computation must be repeated for each solved MBS position. The computation may be made faster by performing once for ever the symbolic computations of the required elements of the transformation matrices for the given type of MBS and only evaluating the resultant expressions at each step.

In some cases it is not necessary to calculate the value of the reaction effect; we are concerned only with the unknown driving effects. As the

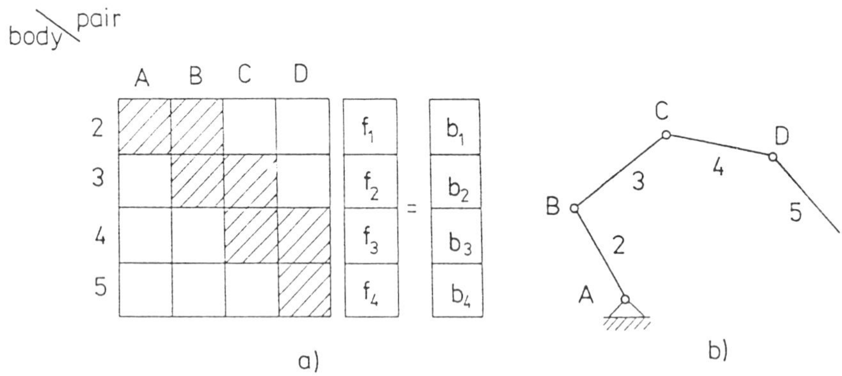

Figure 8.2. Description of open chain.

solved problem is described by a system of linear algebraical equations, it is possible to exclude reactions from the computation using the known procedure: we split the vector $\mathbf{x}$ in equation (8.17) into two parts: a subvector $\mathbf{x}_1$ containing action forces components and a subvector $\mathbf{x}_2$ containing reaction forces components. According to this division we also split the matrix $\mathbf{A}$ and vector $\mathbf{b}$ so that

$$\begin{bmatrix} \mathbf{A}_{11} & \mathbf{A}_{12} \\ \mathbf{A}_{21} & \mathbf{A}_{22} \end{bmatrix} \begin{bmatrix} \mathbf{x}_1 \\ \mathbf{x}_2 \end{bmatrix} = \begin{bmatrix} \mathbf{b}_1 \\ \mathbf{b}_2 \end{bmatrix} \tag{8.18}$$

Hence, writing it out as two matrix equations, we easily carry out the formal computation

$$\mathbf{x}_1 = (\mathbf{A}_{11} - \mathbf{A}_{12}\mathbf{A}_{22}^{-1}\mathbf{A}_{21})^{-1}(\mathbf{b}_1 - \mathbf{A}_{12}\mathbf{A}_{22}^{-1}\mathbf{b}_2) \tag{8.19}$$

Let us return once more to the structure of matrix $\mathbf{A}$ from equation (8.17). Until now, we have considered the solution of a MBS with at least one loop, where the whole system of $6(n-1)$ equations must be solved at once. However for MBS's of the simple open chain type, which is often the case with robots and manipulators, it is possible by choosing a suitable order of equations to achieve the band property of matrix $\mathbf{A}$. Let us for the sake of simplicity consider such a chain formed by five elements. The structure of equations (8.17) corresponds to Figure 8.2b.

As the kinematic pair of class j also has j static parameters and allows $6-j$ degrees of freedom on which unknown action force effects may act, we have in a general case exactly 6 unknowns. Furthermore, as six equilibrium equations are available for each body, it is possible to utilize the band structure of matrix **A** from Figure 8.2a for faster recursive computation. The following holds

$$\mathbf{f}_{i-1} = \mathbf{A}_{i-1}^{-1}(\mathbf{b}_{i-1} - \mathbf{A}_i\mathbf{f}_i), \qquad i = 2, 3, \ldots, n+1 \tag{8.20a}$$

In the case of a simple open chain, it is obviously possible to use recursive computation even for the determination of kinematic quantities required for obtaining the inertial effects. If, for example, the motion of bodies $i, i+1$ constrained by a revolute kinematic pair is specified by quantities $\boldsymbol{\omega}_{ij}, \boldsymbol{\alpha}_{ij}$, we use, according to the theory of simultaneous motions presented in Chapter 4, the relations

$$\mathbf{v}_{i+1,1,i+1} = \mathbf{S}_{i+1,i}\mathbf{v}_{i,1i} + \boldsymbol{\Omega}_{i+1,i,i+1}\mathbf{r}_{i,i+1} \tag{8.20b}$$

$$\boldsymbol{\omega}_{i+1,1,i+1} = \mathbf{S}_{i+1,i}\boldsymbol{\omega}_{i,1i} + \boldsymbol{\omega}_{i+1,i,i+1} \tag{8.20c}$$

$$\mathbf{a}_{i+1,1,i+1} = \mathbf{S}_{i+1,i}\mathbf{a}_{i,1i} + \mathbf{S}_{i+1,i}\boldsymbol{\Omega}_{i+1,i,i+1}\mathbf{v}_{i,1i} + \mathcal{A}_{i+1,1,i+1}\mathbf{r}_{i,i+1} \tag{8.20d}$$

$$\boldsymbol{\alpha}_{i+1,1,i+1} = \mathbf{S}_{i+1,i}\boldsymbol{\alpha}_{i,1i} + \boldsymbol{\alpha}_{i+1,i,i+1} \tag{8.20e}$$

The kinematic calculation is then performed in a "forward pass" (from frame to gripper), and the kinetostatic calculation in a "backward pass". A slightly different method will be described in Sections 9.5.1.8 and 9.6.

The same calculation procedure may be used for the static solution, in which we use only the position part of the kinematic calculation. The static calculation is then performed in a "gripper to frame pass".

The fact that we calculate reactions in the problem to be solved enables us easily to take into account friction which magnitudes do depend on those reactions. They must however be expressed as linear combinations of reaction components without nonlinear relations. For example, in the case of a revolute kinematic pair, we usually express the resistances in the form of an axial pin friction moment

$$M_{pa} = r_s f_a F_3 \tag{8.21}$$

and a radial pin friction moment

$$M_{pr} = r_p f_p \sqrt{F_1^2 + F_2^2}; \tag{8.22}$$

we use the reaction components notation as in Table 8.1 and symbols f_a, f_p for the respective friction coefficients (which we assume to be constant), r_s for the mean radius of the collar on which the friction occurs, and r_p for the pin radius. Using relation (8.21) causes no difficulties; either relation (8.22) must be made linear using, for example, the Poncelet approximate relation (Šrejtr, 1954)

$$M_{pr} \simeq r_p f_p (0.96F_1 + 0.4F_2), \qquad F_1 > F_2 \tag{8.23}$$

or we use an iterative solution. We start with computation of the reaction forces without friction. In the next iteration we compute friction using the reaction forces from the previous iteration. For details see also Section 9.9. We proceed in a similar way when expressing the friction of the remaining kinematic pairs.

We add the coefficients of the reaction components from the actual computation to the elements of the transformation matrices in Table 8.1.

8.2 The Parametric Method in Statics

The fundamental idea of the parametric method in statics is identical with that of the parametric method in kinematics. In many cases if we knew the reaction forces of some kinematic joint of a multibody system then we could complete the static solution. We can choose the value of this force and thus complete the static solution. However, such a solution is not in general correct because it must fulfil some conditions; from these conditions we determine the correct value of the chosen force. In the static solution, the chosen values are treated as parameters – hence the name "parametric method". The origin of the parametric method is also to be found in graphical methods (Wittenbauer, 1923); there have been some attempts to generalize this method (Brát, 1960), but we now use a different variant.

We substitute all kinematic joints by their equivalent reaction forces using the free-body method for all bodies of a multibody system. There act on every body reaction forces of its kinematic joints, applied forces and dynamic forces due to the d'Alembert principle. We formulate a number of equilibrium equations for every body of a multibody system — 6 for spatial, 3 for planar. Among these equations we find equations which are directly solvable for their unknown variables (i.e. they contain just one unknown variable). These equations are solved for their unknown variables and these variables are further substituted into the system of remaining equations. If there are no directly solvable equations, we find an equation which contains the minimum number of unknown variables. We choose the values of some unknown variables in this equation to make this equation solvable. Then we solve this equation and we substitute the result into the system of remaining equations. If we have chosen the values of some unknown variables, we later reach an equation which is unsolvable, i.e. it contains only known values and therefore it is not generally fulfilled. For this equation to be solvable, we obtain a condition for the determination of some chosen variable.

The algorithm described is that most generally used in the parametric method for the spatial multibody system. A more efficient algorithm is under development (Valášek, 1990b; Valášek and Stejskal, 1990).

The spatial (or planar) multibody system is transformed into a spatial (or planar) truss (possibly slightly modified) and the solution of the equivalent spatial (or planar) truss according to the algorithm described bellow gives the static solution of the multibody system being studied. The transformation of a spatial multibody system into a spatial truss is realized by two steps.

Firstly, every kind of kinematic joint is represented by a special truss. These trusses corresponding to particular joints of a body are connected together in such a way that they are statically determined within the body. However, the whole truss of the multibody system enables the virtual motions of particular joints of the multibody system to take place. The spherical joint is represented by a node in its centre and up to three bars in both bodies of the joint. The revolute joint is represented by two tetrahedra with two common nodes at the axis of rotation without the bar at the axis. The prismatic and cylindrical joints are represented by two oblongs parallel with the axis, each with a diagonal. They are connected by two crossing bars on two parallel sides with no bar at the axis. In the case of the prismatic joint there is one extra bar to prevent possible rotation around the axis. In this way we can prepare equivalent trusses of different kinds of bodies together with different numbers and kinds of kinematic joints; for a particular multibody system we can simply connect these trusses together.

Secondly, all forces which act on each body are decomposed in the forces acting in the nodes of the equivalent truss.

The transformation of a planar multibody system into a planar truss is similar to that described above. On the axis of a revolute joints there is just one node and from this node there come bars from the two bodies. The translational joint is represented by parallelogram with two sides parallel and the others perpendicular to the axis of translation.

The trusses are solved by the parametric method. The most important fact is that the forces act only in nodes, and the bars can transfer an axial force. We use two procedures:

– If a node contains only 3 bars in space (or 2 in a plane) with unknown forces, then these 3 (or 2) forces can be determined from the equilibrium condition with all known forces acting in the node (i.e. the equilibrium of four forces in the space or three forces in the plane).

– If none of the nodes in the truss contains only 3 (or 2) unknown forces, then we choose the values of some unknown forces in order to be able to solve some node. During the further solution there is always a node which is unsolvable, i.e. it contains less than 3 (or 2) unknown forces. Hence we have a condition which must be fulfilled and the chosen values can be determined.

This algorithm is the generalization of Cremone's diagram.

8.3 MBS Static Analysis: The Equilibrium Position

If the MBS motion is so slow that we can neglect inertial effects, the kinetostatic solution turns into a quasi-static solution. All of the above relations and procedures maintain their validity; equation terms expressing inertial effects vanish. We speak of a strict static solution when solving the equilibrium of a MBS whose elements do not carry out any motion. The purpose of an immobile MBS solution is to compute reactions in all or selected kinematic pairs. Considering a movable MBS static solution we may have the following two problems

– determination of such additional applied force effects to maintain MBS equilibrium in a prescribed position,

– determination of the equilibrium position from known applied forces. In both cases the computation of reactions may also be required.

Whereas the first problem may be solved by modifying the procedure described in Section 8.1, the problem of determining the equilibrium position is quite different. We shall treat it in more detail.

Let us consider a moving MBS in a potential force field (gravitation) and perhaps containing elastic elements. In that case we may calculate the potential energy of the solved MBS and, for the determination of its equilibrium position, use the known fact that the stable equilibrium position corresponds to minimum potential energy (Lagrange-Dirichlet theorem).

In practical cases we have to take into account the potential position energy of each body in the gravitational field and the potential energy of straight or torsion springs. We may assume constant stiffness of the springs and express potential energy by the relation

$$V = \frac{1}{2}\left(\sum_i k_i(\xi_i - \xi_{i0})^2 + \sum_j k_j \varphi_j^2\right) - \sum_k \mathbf{G}_k^T \left(\mathbf{r}_k - \mathbf{r}_{k_0}\right) \tag{8.24}$$

where k_i is the stiffness of the straight line spring, ξ_i is the instantaneous length of the spring, ξ_{i_0} the free length of the spring, k_j the torsion stiffness of the spring, φ_j the instantaneous torsion of spring j with respect to its initial position, $\mathbf{G}_k$ the weight (gravity) of body k, $\mathbf{r}_k$ the instantaneous absolute position of the centre of gravity of body k in basic space, and $\mathbf{r}_{k_0}$ the initial position of the centre of gravity.

In the same way as with gravity, for the calculation of potential energy we deal with all forces which are constant in the basic system. If the stiffnesses of some springs are not constant we must use their contribution towards the potential energy by integration. The same holds in the case of non-constant applied forces.

After expressing all variables in (8.24) as functions of coordinates s_j, $j = 1, 2, \dots, n$ uniquely determining the instantaneous position of the system, we obtain

$$V = V(\mathbf{s}) \tag{8.25}$$

The coordinates $\mathbf{s}$ are in general subjected to constraint equations

$$f_i(\mathbf{s}) = 0, \qquad i = 1, 2, \dots, m \tag{8.26}$$

which we assume to be independent.

As the search for the equilibrium position makes sense only for mobile systems, in non-singular cases we always have $m < n$.

The determination of the equilibrium position can now be formulated as the problem of finding the minimum of function (8.25) taking into account conditions (8.26), i.e.

$$V(\mathbf{s}) \longrightarrow \min, \qquad \mathbf{f}(\mathbf{s}) = \mathbf{0} \tag{8.27}$$

The solution may in general be sought using various optimizing procedures (Polak, 1971) or using necessary conditions for an extreme and subsequently solving a system of nonlinear equations.

There are two procedures available for finding the system.

a) We express from equation (8.26) m dependent coordinates $\mathbf{z}$ as a function of $n - m$ independent coordinates $\mathbf{q}$ and substitute into (8.25). Then $V = V(\mathbf{q})$ and conditions

$$\frac{\partial V}{\partial q_k} = 0, \quad k = 1, 2, \dots, n - m \tag{8.28}$$

for the minimum of this function form a system of $n - m$ equations for the unknown $\mathbf{q}$. Dependent coordinates are subsequently obtained from (8.26).

b) We use the method of Lagrange multipliers and construct the function

$$\overline{V}(\mathbf{s}, \boldsymbol{\lambda}) = V(\mathbf{s}) + \boldsymbol{\lambda}^T \mathbf{f}(\mathbf{s}) \tag{8.29}$$

The necessary conditions for its stationary value

$$\begin{aligned} \frac{\partial \bar{V}}{\partial s_j} &= 0, \qquad j = 1, 2, \dots, n \\ \frac{\partial \bar{V}}{\partial \lambda_i} &= 0, \qquad i = 1, 2, \dots, m \end{aligned} \tag{8.30}$$

form a system of $n + m$ equations for the calculation of n coordinates s_j and m multipliers λ_i.

The analytical solution of the equilibrium position is possible only in extremely simple cases. In general, numerical computation is required.

A suitable algorithm for minimizing a function of more variables with subsidiary conditions in the form of equalities may be used; alternatively, a procedure according to a) which we are going to describe in more detail may be used. Both methods, of course, lead to the same computational algorithm given by relations (8.35), (8.43).

Let our starting point be an equilibrium position estimate, which we shall make more precise using iteration. Let us present the procedure in the iteration step k, where the k-th approximation $\mathbf{s}^{(k)}$ is available and we search for the approximation $k+1$

$$\mathbf{s}^{(k+1)} = \mathbf{s}^{(k)} + \Delta\mathbf{s}^{(k)} \tag{8.31}$$

where $\Delta\mathbf{s}^{(k)}$ are the coordinate changes in step k. The expansion of function (8.26) into a Taylor series with terms of the order up to 1 in the neighbourhood of point $\mathbf{s}^{(k)}$ leads to the result

$$\mathbf{f}(\mathbf{s}) \doteq \mathbf{f}(\mathbf{s}^{(k)}) + \mathbf{\Phi}_s^{(k)}\Delta\mathbf{s}^{(k)} = \mathbf{0} \tag{8.32}$$

We factorize the matrix

$$\mathbf{\Phi}_s^{(k)} = \left.\frac{\partial\mathbf{f}(\mathbf{s})}{\partial\mathbf{s}^T}\right|_{\mathbf{s}=\mathbf{s}^{(k)}} = [\mathbf{\Phi}_z^{(k)}, \mathbf{\Phi}_q^{(k)}] \tag{8.33}$$

where $\mathbf{\Phi}_z^{(k)}$ with dimension (m, m) forms the regular core of matrix $\mathbf{\Phi}_s^{(k)}$ and $\mathbf{\Phi}_q^{(k)}$ with dimension $(m, n-m)$ corresponds to independent coordinates $\mathbf{q}$. We split the coordinate changes

$$\Delta\mathbf{s}^{(k)} = \lceil \Delta\mathbf{z}^{(k)}, \Delta\mathbf{q}^{(k)} \rceil \tag{8.34}$$

into changes $\Delta\mathbf{z}^{(k)}$ of dependent and $\Delta\mathbf{q}^{(k)}$ of independent ones. Then it is possible to write

$$\mathbf{f}(\mathbf{s}^{(k)}) + \mathbf{\Phi}_q^{(k)}\Delta\mathbf{q}^{(k)} + \mathbf{\Phi}_z^{(k)}\Delta\mathbf{z}^{(k)} = \mathbf{0}$$

$$\Delta\mathbf{z}^{(k)} = -(\mathbf{\Phi}_z^{(k)})^{-1}\left(\mathbf{f}(\mathbf{s}^{(k)}) + \mathbf{\Phi}_q^{(k)}\Delta\mathbf{q}^{(k)}\right) \tag{8.35}$$

The equation (8.35) has an infinite number of solutions. Changes $\Delta\mathbf{q}^{(k)}$ are, from the viewpoint of geometry, free parameters. Let us remember that equations (8.35) also hold in the case when one of the equations $\mathbf{f}(\mathbf{s})$ is, either permanently or in a certain position, dependent. This only means a different splitting in (8.34). From the viewpoint of statics, the changes

$\Delta\mathbf{q}^{(k)}$ must, however, comply with $n-m$ potential energy minimum conditions

$$\frac{\partial V(\mathbf{z}(\mathbf{q}),\mathbf{q})}{\partial \mathbf{q}^T} = \boldsymbol{\Gamma}(\mathbf{z}(\mathbf{q}),\mathbf{q}) = \mathbf{g}(\mathbf{q}) = \mathbf{0} \tag{8.36}$$

Further

$$\boldsymbol{\Gamma}(\mathbf{z}(\mathbf{q}),\mathbf{q}) = \frac{\partial V(\mathbf{z},\mathbf{q})}{\partial \mathbf{q}^T} + \frac{\partial V(\mathbf{z},\mathbf{q})}{\partial \mathbf{z}^T}\frac{\partial \mathbf{z}}{\partial \mathbf{q}^T} \tag{8.37}$$

and from the validity of the constraint equations $\mathbf{f}(\mathbf{z},\mathbf{q}) = \mathbf{0}$

$$\frac{\partial \mathbf{f}(\mathbf{z}(\mathbf{q}),\mathbf{q})}{\partial \mathbf{q}^T} = \boldsymbol{\Phi}_q + \boldsymbol{\Phi}_z\frac{\partial \mathbf{z}}{\partial \mathbf{q}^T} = 0 \tag{8.38}$$

Hence

$$\frac{\partial \mathbf{z}}{\partial \mathbf{q}^T} = -\boldsymbol{\Phi}_z^{-1}\boldsymbol{\Phi}_q \tag{8.39}$$

Denoting

$$\left.\frac{\partial V(\mathbf{z},\mathbf{q})}{\partial \mathbf{q}^T}\right|_{\mathbf{z}=\mathbf{z}(\mathbf{q})} = \mathbf{V}_q(\mathbf{z}(\mathbf{q}),\mathbf{q}), \quad \left.\frac{\partial V(\mathbf{z},\mathbf{q})}{\partial \mathbf{z}^T}\right|_{\mathbf{z}=\mathbf{z}(\mathbf{q})} = \mathbf{V}_z(\mathbf{z}(\mathbf{q}),\mathbf{q}) \tag{8.40}$$

we obtain the conditions (8.37) expressed as:

$$\boldsymbol{\Gamma}(\mathbf{z}(\mathbf{q}),\mathbf{q}) = \mathbf{V}_q - \mathbf{V}_z\boldsymbol{\Phi}_z^{-1}\boldsymbol{\Phi}_q = \mathbf{0} \tag{8.41}$$

The problem of finding the equilibrium position is described by two systems of nonlinear equations (8.26), by splitting the coordinates $\mathbf{s}$ into $\mathbf{z}$ and $\mathbf{q}$, using (8.41). We shall solve them simultaneously. We shall also expand (8.41) in the neighbourhood of point $\mathbf{q}^{(k)}$ in a Taylor series. Then

$$\boldsymbol{\Gamma}(\mathbf{z}(\mathbf{q}),\mathbf{q}) \doteq \boldsymbol{\Gamma}(\mathbf{z}(\mathbf{q}^{(k)}),\mathbf{q}^{(k)}) + \boldsymbol{\Gamma}_q^{(k)}\Delta\mathbf{q}^{(k)} + \boldsymbol{\Gamma}_z^{(k)}\Delta\mathbf{z}^{(k)} = \mathbf{0} \tag{8.42}$$

Substituting from (8.35) we obtain

$$\boldsymbol{\Gamma}^{(k)} + \boldsymbol{\Gamma}_q^{(k)}\Delta\mathbf{q}^{(k)} - \boldsymbol{\Gamma}_z^{(k)}(\boldsymbol{\Phi}_z^{(k)})^{-1}(\mathbf{f}^{(k)} + \boldsymbol{\Phi}_q^{(k)}\Delta\mathbf{q}^{(k)}) = \mathbf{0}$$

$$\Delta\mathbf{q}^{(k)} = \left(\boldsymbol{\Gamma}_q^{(k)} - \boldsymbol{\Gamma}_z^{(k)}(\boldsymbol{\Phi}_z^{(k)})^{-1}\boldsymbol{\Phi}_q^{(k)}\right)^{-1}\left(\boldsymbol{\Gamma}_z^{(k)}(\boldsymbol{\Phi}_z^{(k)})^{-1}\mathbf{f}^{(k)} - \boldsymbol{\Gamma}^{(k)}\right) \tag{8.43}$$

where

$$\begin{aligned}\boldsymbol{\Gamma}_q &= \left.\frac{\partial \boldsymbol{\Gamma}(\mathbf{z},\mathbf{q})}{\partial \mathbf{q}^T}\right|_{\mathbf{z}=\mathbf{z}(\mathbf{q})} = \mathbf{V}_{qq} - (\mathbf{V}_z\boldsymbol{\Phi}_z^{-1}\boldsymbol{\Phi}_q)_q \\ &= \mathbf{V}_{qq} - \mathbf{V}_{zq}\boldsymbol{\Phi}_z^{-1}\boldsymbol{\Phi}_q - \mathbf{V}_z(\boldsymbol{\Phi}_z^{-1})_q\boldsymbol{\Phi}_q - \mathbf{V}_z\boldsymbol{\Phi}_z^{-1}\boldsymbol{\Phi}_{qq}\end{aligned} \tag{8.44}$$

$$
\begin{aligned}
\mathbf{\Gamma}_z &= \left.\frac{\partial \mathbf{\Gamma}(\mathbf{z},\mathbf{q})}{\partial \mathbf{z}^T}\right|_{\mathbf{z}=\mathbf{z}(\mathbf{q})} = \mathbf{V}_{qz} - (\mathbf{V}_z \mathbf{\Phi}_z^{-1} \mathbf{\Phi}_q)_z \\
&= \mathbf{V}_{qz} - \mathbf{V}_{zz}\mathbf{\Phi}_z^{-1}\mathbf{\Phi}_q - \mathbf{V}_z(\mathbf{\Phi}_z^{-1})_z\mathbf{\Phi}_q - \mathbf{V}_z\mathbf{\Phi}_z^{-1}\mathbf{\Phi}_{qz} \qquad (8.45)
\end{aligned}
$$

$$
\begin{aligned}
\mathbf{V}_{qq} &= \left.\frac{\partial \mathbf{V}_q(\mathbf{z},\mathbf{q})}{\partial \mathbf{q}^T}\right|_{\mathbf{z}=\mathbf{z}(\mathbf{q})} = \left[\frac{\partial V}{\partial q_i \partial q_j}\right] \\
\mathbf{V}_{zz} &= \left.\frac{\partial \mathbf{V}_z(\mathbf{z},\mathbf{q})}{\partial \mathbf{z}^T}\right|_{\mathbf{z}=\mathbf{z}(\mathbf{q})} = \left[\frac{\partial V}{\partial z_i \partial z_j}\right] \\
\mathbf{V}_{qz} &= \left.\frac{\partial \mathbf{V}_q(\mathbf{z},\mathbf{q})}{\partial \mathbf{z}^T}\right|_{\mathbf{z}=\mathbf{z}(\mathbf{q})} = \left[\frac{\partial V}{\partial q_i \partial z_j}\right] = \\
= \mathbf{V}_{zq} &= \left.\frac{\partial \mathbf{V}_z(\mathbf{z},\mathbf{q})}{\partial \mathbf{q}^T}\right|_{\mathbf{z}=\mathbf{z}(\mathbf{q})} = \left[\frac{\partial V}{\partial z_i \partial q_j}\right] \\
\mathbf{\Phi}_{qq} &= \left.\frac{\partial \mathbf{\Phi}_q(\mathbf{z},\mathbf{q})}{\partial \mathbf{q}^T}\right|_{\mathbf{z}=\mathbf{z}(\mathbf{q})} = \left[\frac{\partial f_i}{\partial q_j \partial q_k}\right] \\
\mathbf{\Phi}_{zz} &= \left.\frac{\partial \mathbf{\Phi}_z(\mathbf{z},\mathbf{q})}{\partial \mathbf{z}^T}\right|_{\mathbf{z}=\mathbf{z}(\mathbf{q})} = \left[\frac{\partial f_i}{\partial z_j \partial z_k}\right] \\
\mathbf{\Phi}_{qz} &= \left.\frac{\partial \mathbf{\Phi}_q(\mathbf{z},\mathbf{q})}{\partial \mathbf{z}^T}\right|_{\mathbf{z}=\mathbf{z}(\mathbf{q})} = \left[\frac{\partial f_i}{\partial q_j \partial z_k}\right] = \\
= \mathbf{\Phi}_{zq} &= \left.\frac{\partial \mathbf{\Phi}_z(\mathbf{z},\mathbf{q})}{\partial \mathbf{q}^T}\right|_{\mathbf{z}=\mathbf{z}(\mathbf{q})} = \left[\frac{\partial f_i}{\partial z_j \partial q_k}\right] \qquad (8.46)
\end{aligned}
$$

We express $(\mathbf{\Phi}_z^{-1})_q, (\mathbf{\Phi}_z^{-1})_z$ from the identity $\mathbf{\Phi}_z^{-1}\mathbf{\Phi}_z = \mathbf{E}$ by differentiating

$$
\begin{aligned}
(\mathbf{\Phi}_z^{-1})_q\mathbf{\Phi}_z + \mathbf{\Phi}_z^{-1}\mathbf{\Phi}_{zq} &= 0 \\
(\mathbf{\Phi}_z^{-1})_q &= -\mathbf{\Phi}_z^{-1}\mathbf{\Phi}_{zq}\mathbf{\Phi}_z^{-1} \qquad (8.47)
\end{aligned}
$$

$$
\begin{aligned}
(\mathbf{\Phi}_z^{-1})_z\mathbf{\Phi}_z + \mathbf{\Phi}_z^{-1}\mathbf{\Phi}_{zz} &= 0 \\
(\mathbf{\Phi}_z^{-1})_z &= -\mathbf{\Phi}_z^{-1}\mathbf{\Phi}_{zz}\mathbf{\Phi}_z^{-1} \qquad (8.48)
\end{aligned}
$$

The computation of, for example, (8.48) is performed as follows:

$$
[(\mathbf{\Phi}_z^{-1})_z]_{i,j,k} = -\sum_{j_1}[\mathbf{\Phi}_z^{-1}]_{i,j_1}\left(\sum_{j_2}[\mathbf{\Phi}_{zz}]_{j_1,j_2,k}[\mathbf{\Phi}_z^{-1}]_{j_2,j}\right) \qquad (8.49)
$$

Similarly, for example, (8.44)

$$
[\mathbf{\Gamma}_q]_{i,j} = [\mathbf{V}_{qq}]_{i,j} - \sum_{j_1}[\mathbf{V}_{zq}]_{j_1,i}\left(\sum_{j_2}[\mathbf{\Phi}_z^{-1}]_{j_1,j_2}[\mathbf{\Phi}_q]_{j_2,j}\right) -
$$

$$-\sum_{j_1}[\mathbf{V}_z]_{j_1}\left(\sum_{j_2}[(\boldsymbol{\Phi}_z^{-1})_q]_{j_1,j_2,j}[\boldsymbol{\Phi}_q]_{j_2,i}\right)-$$

$$-\sum_{j_1}[\mathbf{V}_z]_{j_1}\left(\sum_{j_2}[(\boldsymbol{\Phi}_z^{-1})]_{j_1,j_2}[\boldsymbol{\Phi}_{qq}]_{j_2,i,j}\right) \tag{8.50}$$

and, for example, in (8.43)

$$[\boldsymbol{\Gamma}_q-\boldsymbol{\Gamma}_z\boldsymbol{\Phi}_z^{-1}\boldsymbol{\Phi}_q]_{i,j}=[\boldsymbol{\Gamma}_q]_{i,j}-\sum_{j_1}[\boldsymbol{\Gamma}_z]_{i,j_1}\left(\sum_{j_2}[\boldsymbol{\Phi}_z^{-1}]_{j_1,j_2}[\boldsymbol{\Phi}_q]_{j_2,j}\right) \tag{8.51}$$

The computation using the Newton method according to (8.35) and (8.43) may also be altered for the modified Newton method according to Section 5.10 by substituting the modified $\varkappa\Delta\mathbf{q}^{(k)}$ and $\varkappa\Delta\mathbf{z}^{(k)}$ into (8.34) and (8.31) rather than the calculated $\Delta\mathbf{q}^{(k)}$ and $\Delta\mathbf{z}^{(k)}$. In the actual computation we first split in the given position $\mathbf{s}^{(k)}$ coordinates according to (8.33), we then calculate from (8.43) the change $\Delta\mathbf{q}^{(k)}$ and after substituting them into (8.35) we find the changes $\Delta\mathbf{z}^{(k)}$. The new position $\mathbf{s}^{(k+1)}$ is determined from (8.31). Even using the calculation according to (8.43), it may happen that $r(\boldsymbol{\Gamma}_q^{(k)}-\boldsymbol{\Gamma}_z^{(k)}(\boldsymbol{\Phi}_z^{(k)})^{-1}\boldsymbol{\Phi}_q^{(k)})<(n-m)$. From the viewpoint of physics this means that some of the coordinates $\mathbf{q}$ are actually independent and the equilibrium position is independent of this coordinate. This is an indifferent equilibrium position. The quality of the resulting equilibrium position is decided by the behaviour of function $V(\mathbf{s})$ in the neighbourhood of the stationary point. In most cases, the stable equilibrium position being sought occurs when all principal minors of matrix $\boldsymbol{\Gamma}_q$ are greater then zero (Sylvester theorem), i.e. $\boldsymbol{\Gamma}_q$ must be positively definite. Should some of the principal minors be null, it is necessary to analyse the higher order differentials. In a general case, if the coordinates $\mathbf{s}$ are constrained by constraint equations (8.26), the expressing of conditions of the type of equilibrium position is more difficult. The second total differential of function $\overline{V}(\mathbf{s},\lambda)$ (8.29) is to be constructed and $\mathrm{d}\mathbf{z}=-\boldsymbol{\Phi}_z^{-1}\boldsymbol{\Phi}_q\mathrm{d}\mathbf{q}$ is to be substituted. Modifying we obtain the quadratic form expressed only in $\mathrm{d}\mathbf{q}$. Its definiteness determines the type of equilibrium position.

Evidently the problem may have several solutions determining various configurations of equilibrium position. With the procedure given above, their finding is affected by the closeness of the initial estimate. This dependence on the initial estimate is particularly significant in the procedure given above, in which the starting point is the necessary conditions for the minimum, which corresponds to second order optimizing procedures (Polak, 1971). The influence of this dependence is smaller when using first

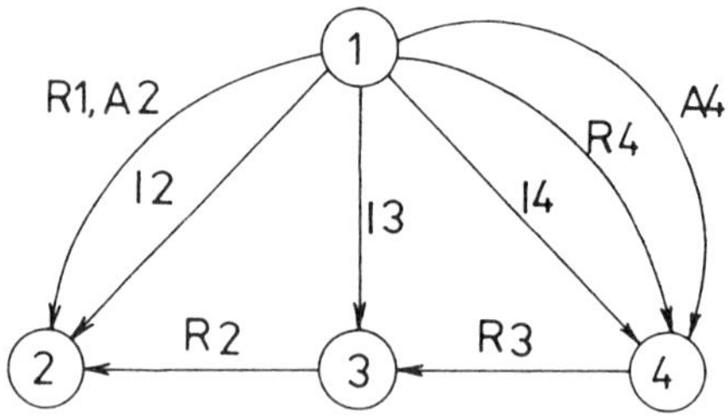

Figure 8.3. Associated graph.

order optimizing procedures based on the gradient method; however they are slower. It is possible to eliminate the dependence on the initial estimate while maintaining the speed of the second order procedure using the following method, which is faster and, from the viewpoint of physics, more straightforward. This is based on the idea of quasistatic loading (Livermore, 1967). The beginning is in an equilibrium position with no external forces acting. Load is subsequently applied using small increments, and for each value of the external forces a new equilibrium position is computed. The MBS geometry changes in the course of loading in the same way as in a real system until the resultant values of external forces and thereby the resultant equilibrium position are reached.

8.4 Applications

8.4.1 Universal Joint Kinetostatics

As an example of the above procedure, we shall briefly outline the kinetostatic solution of a universal joint (Figure 5.14) loaded by an output force Z_4 and a torque M_4 (Brát and Stejskal, 1979). We wish to calculate the driving torque M_2 required to let element 2 revolve with angular velocity ω_{12} = const. Further, we would like to know the behaviour of reaction components in all kinematic pairs in dependence on the driving element angular displacement φ_{12}. All the required dimensions and mass quantities are known.

Let us first construct the associated oriented graph of the problem, (Figure 8.3), where R1 is the reaction acting from the frame on body 2 including the input torque M2, R2 is the reaction acting from body 3 on body 2, R3 is the reaction acting from body 4 on body 3, R4 is the reaction acting from the frame on body 4, A2 is the unknown input torque (M_2)

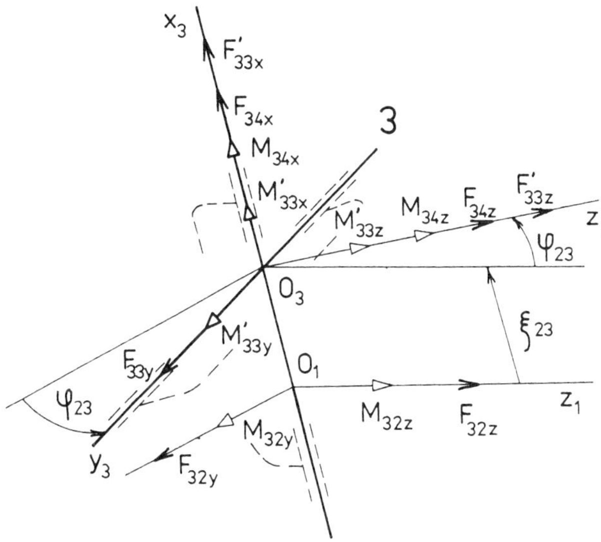

Figure 8.4a. Free-body diagram of body 3.

included in R1, A4 is the given output force and torque (M_4, Z), and I2, I3 and I4 are the inertial effects acting on bodies 2, 3 and 4. The reduced incidence matrix has the form

$$\begin{array}{cc} & \begin{array}{cccccccc} \text{edge} & R1 & R2 & R3 & R4 & A4 & I2 & I3 & I4 \\ \text{node} & & & & & & & & \end{array} \\ \mathbf{C} = & \begin{array}{c} 2 \\ 3 \\ 4 \end{array} \begin{bmatrix} 1 & 1 & 0 & 0 & 0 & 1 & 0 & 0 \\ 0 & -1 & 1 & 0 & 0 & 0 & 1 & 0 \\ 1 & 0 & -1 & 1 & 1 & 0 & 0 & 1 \end{bmatrix} \end{array} \qquad \text{(a)}$$

The centre of gravity of bodies 2, 3 lies in the origins O_2, O_3 of the basic coordinate systems, and the centre of gravity S_4 of body 4 for $\varphi_{12} = 0$ is identical with point D.

In the written-out form, we present only the motors of force effects acting on body 3, (Figure 8.4). For the inertial force according to (8.3)

$$\mathbf{F}_{13}^I = -m_3 \ddot{\mathbf{u}}_{1O_3} \qquad \text{(b)}$$

We transform this into space 3 using the relation

$$\mathbf{F}_{33}^I = \mathbf{S}_{31}\mathbf{F}_{13}^I = -m_3 \mathbf{S}_{13}^T \ddot{\mathbf{u}}_{1O_3} \qquad \text{(c)}$$

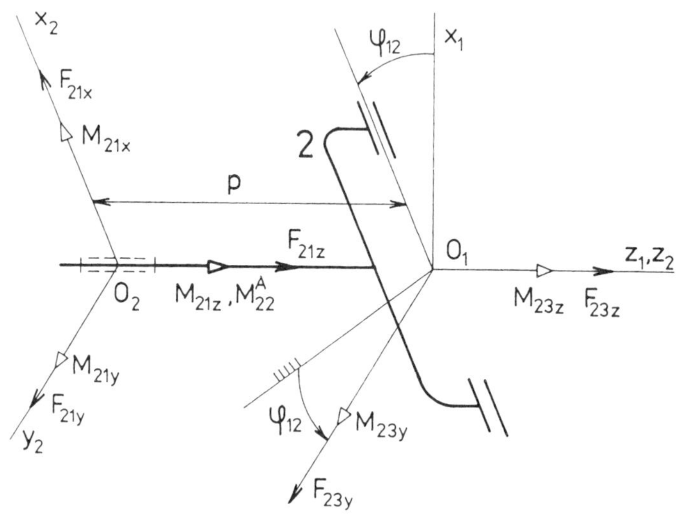

Figure 8.4b. Free-body diagram of body 2.

The inertial couple is, according to (8.5), given by the expression

$$\mathbf{M}_{33}^{T} = -(\mathbf{I}_{33}\dot{\boldsymbol{\omega}}_{3,13} + \boldsymbol{\Omega}_{3,13}\mathbf{I}_{33}\boldsymbol{\omega}_{3,13}) \tag{d}$$

where

$$\boldsymbol{\Omega}_{3,13} = \mathbf{S}_{23}^{T}\boldsymbol{\Omega}_{2,12}\mathbf{S}_{23} + \boldsymbol{\Omega}_{3,23} \tag{e}$$

Altogether, the motor of inertial effects is

$$\mathbf{F}_{33}^{I} = \lceil \mathbf{F}_{33}^{I}, \mathbf{M}_{33}^{I} \rceil \tag{f}$$

From body 3 on body 2 the reaction R2 acts in a cylindrical kinematic pair expressed constructionally as a fork with two short sleeves each of which transmits a force perpendicular to the pair axis. Each of the two unknown components of these forces consists of four unknown static parameters corresponding to a pair of class 4. To simplify the computation, we substitute the actual pair by a statically and kinematically equivalent imaginary cylindrical kinematic pair in point O_1 characterized by four other coordinates, namely two force components F_{23y} and F_{23z} and two torque components M_{23y} and M_{23z}. All of the components mentioned lie in a plane perpendicular to the kinematic pair axis. We transform them into the basic system

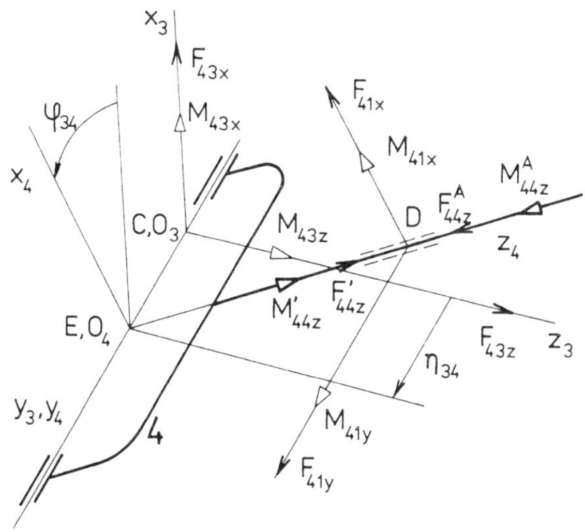

Figure 8.4c. Free-body diagram of body 4.

of body 3 using the relation

$$\bar{\mathbf{F}}_{32}^{R} = -\bar{\mathbf{T}}_{32}\bar{\mathbf{F}}_{23}^{R} \tag{g}$$

where the minus sign expresses the change of orientation. The submatrices of matrix $\bar{\mathbf{T}}_{32}$ are

$$\mathbf{S}_{32} = \begin{bmatrix} 1 & 0 & 0 \\ 0 & c\varphi_{23} & s\varphi_{23} \\ 0 & -s\varphi_{23} & c\varphi_{23} \end{bmatrix}, \quad \mathbf{R}_{32} = \begin{bmatrix} 0 & 0 & 0 \\ 0 & 0 & \xi_{23} \\ 0 & -\xi_{23} & 0 \end{bmatrix} \tag{h}$$

and the motor of reaction R2 is

$$\bar{\mathbf{F}}_{23}^{R} = \lceil 0, F_{23y}, F_{23z}, 0, M_{23y}, M_{23z} \rceil$$

In a similar way we process the reaction R3. We substitute the actual cylindrical kinematic pair combining bodies 3 and 4 by a statically and kinematically equivalent imaginary cylindrical pair directly in point O_3. Then

$$\bar{\mathbf{F}}_{34}^{R} = \lceil F_{34x}, 0, F_{34z}, M_{34x}, 0, M_{34z} \rceil \tag{i}$$

and no transformation is required.

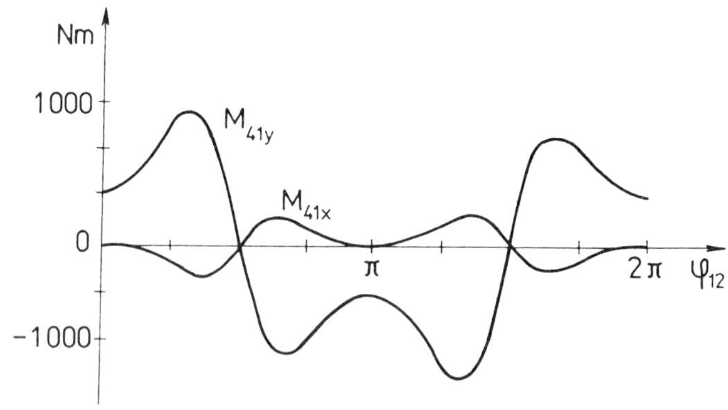

Figure 8.5. Components of the reaction couple M_{14}.

Expressing in a similar way the forces acting on freed bodies 2 and 4, (Figure 8.4b, c), we may write the equilibrium equations for the whole system in the form

$$
\begin{array}{rl}
 & \begin{array}{ccccccc} \text{edge}\; R1 & R2 & R3 & R4 & I2 & I3 & I4 \end{array} \\
\begin{array}{r} \text{node}: 2 \\ 3 \\ 4 \end{array} &
\begin{bmatrix}
\mathbf{T}_{202} & \mathbf{T}_{201} & \mathbf{0} & \mathbf{0} & \mathbf{E}_6 & \mathbf{0} & \mathbf{0} \\
\mathbf{0} & -\mathbf{T}_{32} & \mathbf{T}_{303} & \mathbf{0} & \mathbf{0} & \mathbf{E}_6 & \mathbf{0} \\
\mathbf{0} & \mathbf{0} & -\mathbf{T}_{43} & \mathbf{T}_{4D} & \mathbf{0} & \mathbf{0} & \mathbf{E}_6
\end{bmatrix}
\begin{bmatrix}
\bar{\mathbf{F}}^R_{21} \\ \bar{\mathbf{F}}^R_{23} \\ \bar{\mathbf{F}}^R_{43} \\ \bar{\mathbf{F}}^R_{41} \\ \bar{\mathbf{F}}^I_{2} \\ \bar{\mathbf{F}}^I_{32} \\ \bar{\mathbf{F}}^I_{4}
\end{bmatrix} = \mathbf{0}
\end{array}
\tag{j}
$$

The structure of the matrix of the system corresponds to the structure of matrix $\mathbf{C}$. In the problem being solved,

$$\mathbf{T}_{202} = \mathbf{T}_{303} = \mathbf{E}_6, \qquad \mathbf{F}^I_2 = \mathbf{0} \tag{k}$$

Using the algorithm described in Section 8.2 we obtain from equation (j) a system of eighteen linear algebraic equations for 17 components of reaction effects and the desired driving torque M_2. The solution has been carried out numerically for the following values $\alpha = 30^o$, $h = 0.05$ m, $\omega_{12} = 150$ rad.s^{-1}, $p = 0.1$ m, $q = 0.15$ m, $M_4 = 3$ Nm, $Z_4 = 200$ N, $m_3 = 1.65$ kg, $I_{3x} = 0.013$ kg.m^2, $I_{3y} = 0.013$ kg.m^2, $I_{3z} = 0.025$ kg.m^2, $m_4 = 3$ kg, $I_{4z} = 0.015$ kg.m^2. Of all the results, we present in Figure 8.5 only the behaviour of $M_{41x}(\varphi_{12}), M_{41y}(\varphi_{12})$.

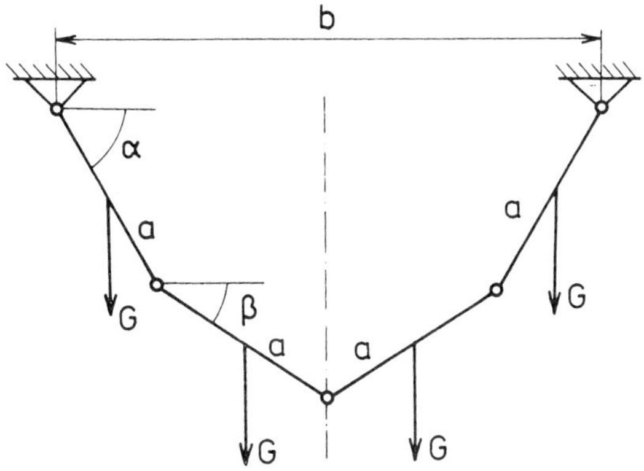

Figure 8.6. System of rods.

8.4.2 The Equilibrium Position of a System of Bars

Let us solve the equilibrium position of a system according to Figure 8.6. This is a system of four equal bars of length a all of them in the vertical plane and connected by revolute kinematic pairs. Each of the bars has weight G. Other force effects are not be considered here. Because of symmetry, it is necessary to solve only one half of the system. We describe its position by two angles α, β constrained by the condition

$$a \ \mathrm{c}\alpha + a \ \mathrm{c}\beta - \frac{b}{2} = 0 \tag{a}$$

This is a system with one degree of freedom. We determine the equilibrium position from the minimum potential energy with a subsidiary condition (a), i.e. as a minimum of function

$$V(\alpha, \beta, \lambda) = -G\frac{a}{2}\mathrm{s}\alpha - G(a \ \mathrm{s}\alpha + \frac{a}{2}\mathrm{s}\beta) + \lambda\left(\mathrm{c}\alpha + \mathrm{c}\beta - \frac{b}{2a}\right)$$

The minimum conditions for this function

$$\begin{aligned} \frac{3}{2}a \ G \ \mathrm{c}\alpha + \lambda \ \mathrm{s}\alpha &= 0 \\ \frac{a}{2}G \ \mathrm{c}\beta + \lambda \ \mathrm{s}\beta &= 0 \\ \mathrm{c}\alpha + \mathrm{c}\beta - \frac{b}{2a} &= 0 \end{aligned}$$

are the conditions for α, b, λ. Excluding λ we obtain two equations

$$\begin{aligned} \tan\alpha - 3\tan\beta &= 0 \\ c\alpha + c\beta - \frac{b}{2a} &= 0 \end{aligned}$$

for which, solving numerically for $b/a = 2.5$, we obtain $\alpha = 64.507°$, $\beta = 34.956°$. It is evident that the required conditions are also fulfilled by values $\alpha = -64.507°$, $\beta = -34.956°$. They correspond to the unstable equilibrium position.

Alternatively let us solve the same problem by iterative computation. If $z = \alpha$ and $q = \beta$, the equation (8.35) has in our case the form

$$\Delta\alpha^{(k)} = \frac{1}{s\alpha^{(k)}}(c\alpha^{(k)} + c\beta^{(k)} - 1.25 - \Delta\beta^{(k)} s\beta^{(k)})$$

and equation (8.43) has the form

$$\left(3\left(\frac{s^2\beta^{(k)}}{s^3\alpha^{(k)}} + \frac{c\beta^{(k)}}{\tan\alpha^{(k)}}\right) + s\beta^{(k)}\right)\Delta\beta^{(k)} =$$

$$= c\beta^{(k)} - 3\frac{s\beta^{(k)}}{\tan\alpha^{(k)}} + 3\frac{s\beta^{(k)}}{s^3\alpha^{(k)}}(c\alpha^{(k)} + c\beta^{(k)} - 1.25)$$

For the estimate $\alpha^{(0)} = 60°$, $\beta^{(0)} = 40°$ we obtain with the required precision 1.E-8 after four iterations the result $\alpha^{(5)} = 64.506697°$, $\beta^{(5)} = 34.955787°$.

8.4.3 The Equilibrium Position of a Wheel Suspension Mechanism

To make the presentation on equilibrium position of more complex systems more specific, we outline the principal relations required for the calculation of the equilibrium position of the front wheel suspension mechanism of a car. The suspension is formed by seven bodies (including the frame), constrained by 3 revolute, 1 prismatic and 4 spherical kinematic pairs (Figure 8.7). The mechanism is loaded by a force G acting in the wheel centre N; (this is the part of the weight of the whole car which acts on the wheel). Furthermore there is a spring with stiffness $k = \text{const}$ acting between point P of the frame and point R of the lower arm. We have to compute the equilibrium position of the mechanism.

The position problem is solved by two matrix equations written for the disconnected loops $O_1O_2O_3LO_4O_1$ and $O_1O_2O_3O_5KO_6O_7SO_1$

$$\mathbf{T}_{12}\mathbf{T}_{23}\mathbf{r}_{3L} = \mathbf{T}_{14}\mathbf{r}_{4L} \qquad \text{(a)}$$

$$\mathbf{T}_{12}\mathbf{T}_{23}\mathbf{T}_{35}\mathbf{r}_{5K} = \mathbf{T}_{17}\mathbf{T}_{76}\mathbf{r}_{6K} \qquad \text{(b)}$$

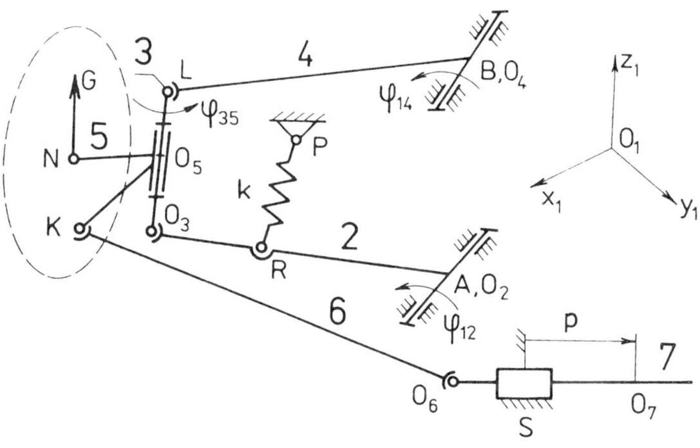

Figure 8.7. Mechanism of front wheel suspension.

They represent a system of six scalar equations for eight coordinates, not counting the passive revolution of bodies 3 and 6 around their own axes. It is thus a device with two degrees of freedom. If we consider the steering position determined by value p to be a free parameter, we have a system described by eight coordinates $s_j, j = 1, 2, \ldots, 8$, one of which is independent. We determine the equilibrium position from the potential energy minimum condition

$$V(\mathbf{s}) = (\mathbf{u}_N - \mathbf{u}_{N_0})^T \mathbf{G} + \frac{1}{2}k(l - d_{RP})^2 \tag{c}$$

where $\mathbf{u}_N(\mathbf{s})$ denotes the instantaneous position of point N, $\mathbf{u}_{N_0}$ its position for $d_{RP} = l$, and $d_{RP}(\mathbf{s})$ is the instantaneous spring length.

Equations (a), (b) and (c) form a system of seven equations for seven cordinates corresponding to the desired equilibrium position.

8.4.4 RRSRR Mechanism Kinetostatics by the Parametric Method

Let us solve the kinetostatics of an RRSRR mechanism (Figure 5.15). On the input and output rotational joints 12 and 51 there act the input and output torques M_{12} and M_{15}. We wish to calculate the driving torque M_{15} required to let element 2 revolve with angular velocity ω_{12} = const if the mechanism is loaded by the torque M_{15}. This mechanism is transformed into the equivalent truss in Figure 8.8. The forces in bars 15, 16 and 17, 18 represent the input and output torques respectively. The nodes A and B

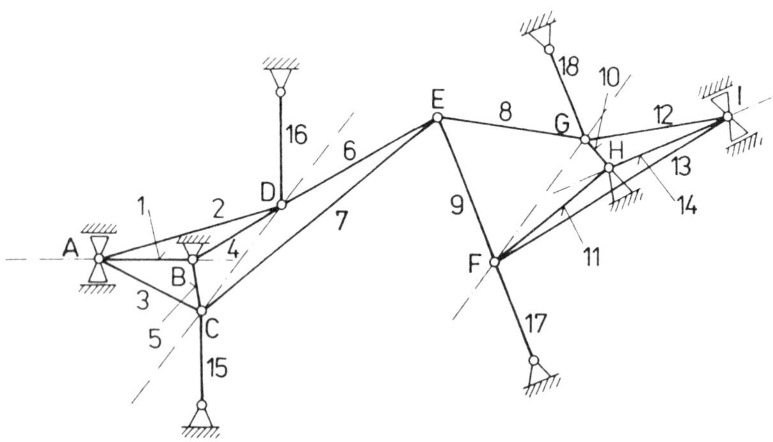

Figure 8.8. Equivalent truss of RRSRR mechanism.

represent the revolute joint 12, the nodes C and D represent the revolute joint 23, the node E represents the spherical joint 34, the nodes F and G represent the revolute joint 45 and the nodes H and I represent the revolute joint 51. The inertia effects and other applied forces are decomposed into the forces acting in the nodes of the truss. From the given torque M_{15} we know the force in the bar 18. None of the nodes is directly solvable. We choose the value of the force in the bar 8. Hence we can solve the nodes E,F,G and determine the forces in the bars 17 and 18 as a function of the chosen value f_8. The torque of these two forces must be the same as output torque. Hence we have the condition of the solvability for the determination of the chosen force in the bar 8. Then we correct the force in bar 8 and in nodes E, F, G. After that we proceed with the real forces in nodes C, D. The solution is completed by the solution of the nodes A, B, I and H which gives the reaction forces.

Chapter 9

MBS Dynamic Analysis

The knowledge of the kinematic solution of MBS is very important and often decisive in judging the suitability of its properties. However, the dynamic solution of the MBS gives more integral and comprehensive information to support the final design. The main purpose of the dynamic solution is to find a connection between motion and the forces acting. The results of the dynamic solution form the basis for testing and designing the drive of the device being studied, the constructional executions of kinematic pairs and the cross section characteristics of individual elements.

Most MBS dynamic investigation methods set out from the so-called equations of motion, the assembly of which is one of the key problems of dynamics. Using the classification method presented in Chapter 2, these equations form an operational mathematical model. With the help of this we attempt to express as well as possible the behaviour of the real object being studied from the viewpoint of mechanics.

The description of the bodies in a system as a three-dimensional continuum characterized by continuously variable parameters in space and time is the one closest to reality. These bodies are connected by discontinuous (discrete) kinematic pairs which we assume, unless other wise stated, to be ideal, i.e. geometrically precise with no backlash and no loss of energy. The mathematical model is formed in this case by partial differential equations, enabling the strain and stress in the bodies to be computed. The solution in the closed form of such a systems is not possible. It is therefore necessary for computational reasons to turn from a continuum to discontinuous sets of numerical values characterizing the object position in previously chosen (also discontinuous) time instants. Such calculation is usually realized by the procedure used in the finite element method.

To study the behaviour of individual bodies of the system as integral units, it is sufficient to use a discontinuous description, where each real body is substituted by a dynamically equivalent model body with the same mass, the same centre of mass position and the same moments of inertia as the original body. The mathematical model of such a system is formed by ordinary differential and algebraic equations. The model is discontinuous in space and continous with respect to time.

We must also in this case, because of the need for numerical solution, turn to time discontinuation. The model presented is in practice the most commonly used, and we are going to treat it in the following text in more detail.

However, let us note that MBS dynamic investigation with non-constant transmission of motions and forces between many elements is a very demanding problem from the viewpoint of both theory and application. It is not easy just to formulate the equations of motion of a spatial MBS. The solution of these equations needs extensive calculations of systems of differential and algebraic equations (DAE) which are unimaginable without the use of a computer and which are mathematically demanding. The successful solution of MBS dynamics requires good knowledge of dynamics, numerical mathematics and computer programming. The creation of general programs and the solution of more complicated problems is not possible without the cooperation of specialists in these disciplines.

We stress once more that even a successful solution is only the description of dynamic behaviour of the accepted model and not of the real system. Even the correct results of simulation of a model can be an incorrect description of the reality. This must be kept in mind when interpreting the results.

9.1 Multibody System Description

The preceding chapters dealing with the study of MBS kinematics present several ways of describing the MBS position and its state of motion as specified by velocities and accelerations. To meet the needs of dynamics, further properties and terms of the MBS description must be studied.

The position of a multibody system is described by a certain number of coordinates s_j $(j = 1, \ldots, n)$. In general we shall call these coordinates physical coordinates, as they usually originate in the natural physical description of the system. We have learned in kinematics that we may in general distinguish independent coordinates q_k $(k = 1, \ldots, i)$ and dependent coordinates z_k $(k = 1, \ldots, n - i)$. Independent coordinates q_k fully specify the MBS position, and the dependent coordinates z_k may be derived from them. This connection is implicitly described by (kinematic) constraints

(5.151) which the physical, independent and dependent coordinates must satisfy. From the viewpoint of dynamics we distinguish holonomic constraints

$$f(s_j, t) = 0 \tag{9.1}$$

which are functions only of coordinates s_j and time t, and nonholonomic constraints of order h

$$g(s_j, \dot{s}_j, \ldots\ldots s_j^{(h)}, t) = 0 \tag{9.2}$$

which are functions of time derivatives of coordinates up to order h. We distinguish nonholonomic nonintegrable constraints in the form (9.2), usually designated as proper (current) nonholonomic constraints (Leech, 1958; Dobronravov, 1976, 1970) and nonholonomic integrable (also semi-holonomic) constraints which are also in the form (9.2) but which can be obtained from holonomic constraints (9.1) by differentiating and possibly altering them, for example by reducing a common factor. The term originates, of course, from the integrability of constraints in the form (9.2) into the form (9.1). The contact of higher kinematic pairs by rolling (6.21) is an example of nonholonomic constraints. The possibility of understanding the description of the behaviour of part of the system as a general nonholonomic constraint for the whole system is an advantage of the general approach of (9.2). Feedback loops of MBS control subsystems are an example. The most frequent form of nonholonomic constraints is, however, linear and

$$\sum_{j=1}^{n} A_j \mathrm{d}s_j + A_t \mathrm{d}t = 0 \tag{9.3}$$

Both types of constraints are further distinguished into skleronomic (stationary), which are not explicitly dependent on time ($\partial f/\partial t = 0$), and rheonomic (non stationary), which are explicitly dependent on time ($\partial f/\partial t \neq 0$), see (9.1), (9.2).

According to the occurrence of types of constraints in the system under investigation, both MBS's and constraints are classified into holonomic skleronomic, holonomic rheonomic, nonholomic skleronomic and nonholomic rheonomic.

The above constraints are two-sided constraints; they are described by equations. One-sided constraints are described by inequalities. Examples are the motion of a mass point on the inside of a spherical surface described by the constraint $x^2 + y^2 + z^2 \leq R^2$, or the motion of two mass points connected by an ideally flexible fibre with no yield.

In the following text we are going to formulate the equations of motion when describing the MBS using various coordinates. The MBS in question is usually described first by physical coordinates. As the next step, independent and dependent coordinates would be applied. Independent coordinates, due to their significance for the formulation of equations of motion, are called generalized coordinates in analytical mechanics. These coordinates would be obtained from physical coordinates by a simple choice, in which case they are called simple coordinates. Generalized coordinates with a general dependence on physical coordinates are called combined coordinates.

Apart from introducing new coordinates by a simple functional relation, new coordinates may also be defined by derivatives of certain velocities generally in the form of nonholomic constraints. If this expression is integrable, the coordinates are proper coordinates. If it is not integrable the term "coordinate" is not appropriate and we apply the term "quasi-coordinate" instead (see Section 9.3).

9.2 Basic Approaches to the Dynamics of MBSs

We may in principle distinguish synthetic (vector), analytic (scalar) and special approaches.

Vector approaches have the advantage of intuitive and concise notation, the validity of which is independent of the choice of coordinate system. They yield values and orientation of the quantities being sought. In specific usage, vector relations are converted into corresponding scalar (component) expressions if a graphical solution is used. Such approaches are usually called Newton-Euler approaches.

The initial quantities for analytical approaches are the scalar quantities: work and energy. In many practical cases, they allow us to obtain an effective description of the dynamic behaviour of the system being studied on the basis of mathematical formulation without deep knowledge of dynamics. Analytical approaches are usually less intuitive and less transparent than vector approaches. The fundamental principles of dynamics apply here: the principle of virtual work, the principle of virtual power (Jourdain's principle), Gauss's principle, Hamilton's principle and above all, Langrange's equations of which equations of the second type and of mixed type are particularly used.

Under the heading of special approaches, there are first of all the recursive formalisms based on the so-called composite rigid-body inertia and articulated-body inertia, further the so-called compartment method, approaches based on graph theory and others. These approaches use above-mentioned procedures of formulation of equations of motion for a single

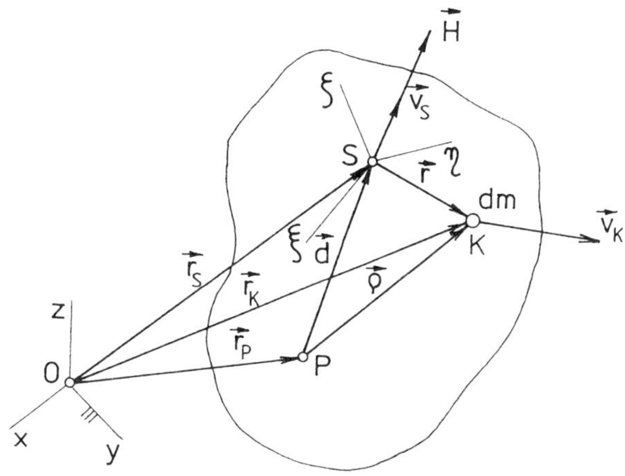

Figure 9.1. Moment and moment of momentum of a body.

body, but they use special methods for the description of dynamic behaviour of the whole MBS.

We present briefly in this chapter the fundamental relations and algorithms of individual approaches.

9.2.1 Approaches Based on Vector Mechanics

9.2.1.1 Newton-Euler Approach

The basic quantities characterizing the dynamics of a body are momentum and moment of momentum. For momentum (Figure 9.1)

$$\vec{H} = \int_{(m)} \vec{v}_K \mathrm{d}m = \int_{(m)} (\vec{v}_S + \vec{\omega} \times \vec{r})\mathrm{d}m = m\vec{v}_S \tag{9.4}$$

where from the definition of mass centre S

$$\int_{(m)} \vec{r}\mathrm{d}m = \vec{0} \tag{9.5}$$

and

$$\vec{v}_K = \frac{\mathrm{d}}{\mathrm{d}t}\vec{r}_K = \frac{\mathrm{d}}{\mathrm{d}t}(\vec{r}_S + \vec{r}) = \vec{v}_S + \vec{\omega} \times \vec{r} \tag{9.6}$$

is the velocity of a general point K of the body, m its mass and $\vec{v}_S$ the velocity of the mass centre. The body momentum is a vector bound to the body mass centre S colinear to vector $\vec{v}_S$.

The moment $\vec{L}_P$ of body momentum with respect to point P is defined by the relation

$$\begin{aligned}\vec{L}_P &= \int_{(m)} \vec{\varrho} \times \mathrm{d}\vec{H} = \int_{(m)} \vec{\varrho} \times \vec{v}_K \mathrm{d}m = \\ &= \int_{(m)} (\vec{d} + \vec{r}) \times (\vec{v}_S + \vec{\omega} \times \vec{r}) \mathrm{d}m = \\ &= \vec{d} \times m\vec{v}_S + \int_{(m)} \vec{r} \times (\vec{\omega} \times \vec{r}) \mathrm{d}m = \vec{d} \times \vec{H} + \vec{L}_S \end{aligned} \tag{9.7}$$

where $\vec{\varrho} = \vec{d} + \vec{r}$ and $\vec{L}_S$ is the moment of momentum of the body about its mass centre S. In a special case $P = S$ we have $\vec{L}_P = \vec{L}_S$. Because of the importance of this case we express $\vec{L}_S$ in more detail using matrix calculus

$$\begin{aligned}\mathbf{L}_S &= \int_{(m)} (\mathbf{r} \times (\boldsymbol{\omega} \times \mathbf{r})) \mathrm{d}m = \int_{(m)} (\mathbf{r} \times \boldsymbol{\Omega}\mathbf{r}) \mathrm{d}m = \\ &= -\int_{(m)} (\mathbf{r} \times \mathbf{R}\boldsymbol{\omega}) \mathrm{d}m = -\int_{(m)} \mathbf{R}\mathbf{R} \mathrm{d}m \boldsymbol{\omega} = \mathbf{I}_S \boldsymbol{\omega}\end{aligned} \tag{9.8}$$

where we designate $\mathbf{I}_S$ as the matrix (tensor) of inertia of the body and $\mathbf{R} = \tilde{\mathbf{r}}$ is skew-symmetric representation of vector $\mathbf{r}$. Expressing elements of $\mathbf{I}_S$in the system of coordinates $S\xi\eta\zeta$ fixed to the body, we obtain

$$\mathbf{I}_S = -\int_{(m)} \mathbf{R}^2 \mathrm{d}m = -\int_{(m)} \begin{bmatrix} 0 & -\zeta & \eta \\ \zeta & 0 & -\xi \\ -\eta & \xi & 0 \end{bmatrix}^2 \mathrm{d}m =$$

$$= \int_{(m)} \begin{bmatrix} \eta^2 + \zeta^2 & -\xi\eta & -\xi\zeta \\ -\xi\eta & \xi^2 + \zeta^2 & -\eta\zeta \\ -\xi\zeta & -\eta\zeta & \eta^2 + \xi^2 \end{bmatrix} \mathrm{d}m = \begin{bmatrix} I_\xi & -I_{\xi\eta} & -I_{\xi\zeta} \\ -I_{\eta\xi} & I_\eta & -I_{\eta\zeta} \\ -I_{\zeta\xi} & -I_{\zeta\eta} & I_\zeta \end{bmatrix} \tag{9.9}$$

The elements of the matrix of inertia are constant in this case. Should axes $\xi\eta\zeta$ be the principal axes inertia, the products of inertia $\mathbf{I}_{\alpha\beta}$, $\alpha, \beta = \xi, \eta, \zeta$ would vanish and the matrix $\mathbf{I}_S$ would be a diagonal matrix.

Using the theorem on change of momentum with respect to time

$$\frac{\mathrm{d}\vec{H}}{\mathrm{d}t} = \vec{F} \tag{9.10}$$

and the theorem on change of moment of momentum with respect to time

$$\frac{\mathrm{d}\vec{L}_P}{\mathrm{d}t} = \vec{M}_P \tag{9.11}$$

where $\vec{F}$ and $\vec{M}_P$ designate the resultant force and the resultant moment of all forces acting on the body resolved in the point P respectively, we obtain the equations of motion.

The form of the equations of motion depends on the space in which the basic quantities are expressed and what decomposition of the general spatial motion is used.

For the usual case of the basic decomposition of motion determined by the point of reference $P \equiv S$ we obtain from (9.10) and (9.11) the following special form of equations of motion. From (9.10) after substituting (9.4) we obtain

$$m\mathbf{a}_S = \mathbf{F} \tag{9.12}$$

These are well known as Newton's equations of motion. Equations (9.12) are written in the matrix form and express three scalar equations. Expressing the respective components in the fixed basic system $Oxyz$

$$m[a_{Sx}, a_{Sy}, a_{Sz}] = [F_x, F_y, F_z] \tag{9.13}$$

Similarly in the system $S\xi\eta\zeta$ fixed to the body

$$m[a_{S\xi}, a_{S\eta}, a_{S\zeta}] = [F_\xi, F_\eta, F_\zeta] \tag{9.14}$$

We obtain another three equations by substituting (9.8) into (9.11). It must be realized that $\vec{L}_S$ is a vector determined in the moving space $S\xi\eta\zeta$, the angular velocity of which with respect to the basic space $Sxyz$ is $\boldsymbol{\omega}$. Thus, for the differentiation, relation (4.3) must be used. We then obtain in the moving system

$$\mathbf{I}_S\dot{\boldsymbol{\omega}} + \boldsymbol{\omega} \times \mathbf{I}_S\boldsymbol{\omega} = \mathbf{M}_S \tag{9.15a}$$

or expanded

$$\mathbf{I}_S \begin{bmatrix} \dot{\omega}_\xi \\ \dot{\omega}_\eta \\ \dot{\omega}_\zeta \end{bmatrix} + \begin{bmatrix} 0 & -\omega_\zeta & \omega_\eta \\ \omega_\zeta & 0 & -\omega_\xi \\ -\omega_\eta & \omega_\xi & 0 \end{bmatrix} \mathbf{I}_S \begin{bmatrix} \omega_\xi \\ \omega_\eta \\ \omega_\zeta \end{bmatrix} = \begin{bmatrix} M_\xi \\ M_\eta \\ M_\zeta \end{bmatrix} \tag{9.15b}$$

where $\mathbf{I}_S$ is given by expression (9.9). For the special and frequent case where axes $\xi\eta\zeta$ are the principal central axes of inertia

$$\begin{aligned} I_\xi\dot{\omega}_\xi + (I_\zeta - I_\eta)\omega_\eta\omega_\zeta &= M_\xi \\ I_\eta\dot{\omega}_\eta + (I_\xi - I_\zeta)\omega_\zeta\omega_\xi &= M_\eta \\ I_\zeta\dot{\omega}_\zeta + (I_\eta - I_\xi)\omega_\xi\omega_\eta &= M_\zeta \end{aligned} \tag{9.15c}$$

These are the well known Euler's dynamic equations. The advantage of writing them in system $S\xi\eta\zeta$ is that the matrices of inertia are constant.

Euler's dynamic equations may also be expressed in system $Oxyz$. Although a direct derivation using relations (9.11) and (9.8) is possible, we are going to use a transformation. Denoting system $Oxyz$ as system a and system $O\xi\eta\zeta$ as system b according to (4.28)

$$\boldsymbol{\omega}_a = \mathbf{S}\boldsymbol{\omega}_b, \qquad \mathbf{I}_a = \mathbf{S}\mathbf{I}_b\mathbf{S}^T, \qquad \mathbf{M}_a = \mathbf{S}\mathbf{M}_b \tag{9.16}$$

where $\mathbf{S} = \mathbf{S}_{ab}$ is the matrix of directional cosines. After substituting (9.16) into (9.15a) and modifying, we obtain

$$\mathbf{I}_a\dot{\boldsymbol{\omega}}_a + \boldsymbol{\omega}_a \times \mathbf{I}_a\boldsymbol{\omega}_a = \mathbf{M}_a \tag{9.17}$$

This relation is formally identical to (9.15), except that it contains quantities expressed in the basic system. Its disadvantage is the variability of elements of the matrix of inertia. They are functions of body position.

For the general case of the arbitrary point of reference P we obtain from (9.10) and (9.11) the following general form of equations of motion. From (9.10) after substituting (9.4) and

$$\vec{v}_S = \frac{\mathrm{d}}{\mathrm{d}t}(\vec{r}_P + \vec{d}) = \vec{v}_P + \vec{\omega} \times \vec{d} \tag{9.18}$$

we obtain

$$m\mathbf{a}_P + \dot{\boldsymbol{\omega}} \times m\mathbf{d} + m(\boldsymbol{\omega} \times (\boldsymbol{\omega} \times \mathbf{d})) = \mathbf{F}_P \tag{9.19}$$

By substituting (9.4), (9.7) and (9.8) into (9.11) we obtain

$$\mathbf{d} \times m\mathbf{a}_S + \mathbf{I}_S\dot{\boldsymbol{\omega}} + \boldsymbol{\omega} \times \mathbf{I}_S\boldsymbol{\omega} = \mathbf{M}_P \tag{9.20}$$

From (9.18) we derive

$$\vec{a}_S = \vec{a}_P + \dot{\vec{\omega}} \times \vec{d} + (\vec{\omega} \times (\vec{\omega} \times \vec{d})) \tag{9.21}$$

We substitute (9.21) into (9.20) and using

$$\begin{aligned} \vec{d} \times (\vec{\omega} \times (\vec{\omega} \times \vec{d})) &= \vec{d} \times ((\vec{\omega}.\vec{d})\vec{\omega} - (\vec{\omega}.\vec{\omega})\vec{d}) = \\ &= -\vec{\omega} \times ((\vec{d}.\vec{\omega})\vec{d} - (\vec{d}.\vec{d})\vec{\omega}) = \\ &= -\vec{\omega} \times (\vec{d} \times (\vec{d} \times \vec{\omega})) = -\boldsymbol{\omega} \times \tilde{\mathbf{d}}^2\boldsymbol{\omega} \end{aligned}$$

we obtain

$$\mathbf{d} \times m\mathbf{a}_P + \mathbf{I}_P\dot{\boldsymbol{\omega}} + \boldsymbol{\omega} \times \mathbf{I}_P\boldsymbol{\omega} = \mathbf{M}_P \tag{9.22}$$

where

$$\mathbf{I}_P = \mathbf{I}_S - m\tilde{\mathbf{d}}^2 \tag{9.23}$$

is the tensor of inertia with respect to the point P according to Steiner's theorem. Similarly as in Section 4.3.1 vector multiplication "$\vec{d}\times$ " is expressed by multiplying skew-symmetric matrix $\tilde{\mathbf{d}}$.

Further it will be very useful to write the equations of motion (9.19) and (9.22) not in two parts for translational and rotational motions but in one compact form. Therefore we shall use vectors of the type (6,1) for the accelerations and force effects as in Chapter 8

$$\mathbf{a}^P = \begin{bmatrix} \mathbf{a}_P \\ \boldsymbol{\alpha} \end{bmatrix} \tag{9.24}$$

$$\mathbf{F}^P = \begin{bmatrix} \mathbf{F} \\ \mathbf{M}_P \end{bmatrix} \tag{9.25}$$

Then the equations of motion (9.19) and (9.22) may be written

$$\begin{bmatrix} m\mathbf{E} & m\tilde{\mathbf{d}}^T \\ m\tilde{\mathbf{d}} & \mathbf{I}_P \end{bmatrix} \begin{bmatrix} \mathbf{a}_P \\ \boldsymbol{\alpha} \end{bmatrix} = \begin{bmatrix} \mathbf{F} \\ \mathbf{M}_P \end{bmatrix} - \begin{bmatrix} m\boldsymbol{\omega}\times(\boldsymbol{\omega}\times\mathbf{d}) \\ \boldsymbol{\omega}\times\mathbf{I}_P\boldsymbol{\omega} \end{bmatrix} \tag{9.26}$$

or

$$\mathbf{I}^P\mathbf{a}^P = \mathbf{F}^P + \boldsymbol{\beta}^P \tag{9.27}$$

where

$$\mathbf{I}^P = \begin{bmatrix} m\mathbf{E} & m\tilde{\mathbf{d}}^T \\ m\tilde{\mathbf{d}} & \mathbf{I}_P \end{bmatrix} \tag{9.28}$$

$$\boldsymbol{\beta}^P = -\begin{bmatrix} m\boldsymbol{\omega}\times(\boldsymbol{\omega}\times\mathbf{d}) \\ \boldsymbol{\omega}\times\mathbf{I}_P\boldsymbol{\omega} \end{bmatrix} \tag{9.29}$$

The expression (9.28) is called composite rigid-body inertia and has been introduced in slightly different form by Featherstone (1983, 1987) and by Brandl, Johanni and Otter (1987, 1988). Featherstone tried to introduced the composite accelerations $[\boldsymbol{\alpha}, \mathbf{a}]$ and force effects $[\mathbf{F}, \mathbf{M}]$ in complete analogy with the screw notation. Then the composite rigid-body inertia is represented by a singular matrix and for its transformation a special operation of spatial transposition must be introduced. If we use the screw notation just for accelerations or just for force effects, then the composite rigid-body inertia is represented by a regular matrix and its transformation can be described by traditional matrix operations.

9.2.1.2 D'Alembert's Principle

Understanding the quantity

$$-\vec{a}\mathrm{d}m = \mathrm{d}\vec{D}$$

as a force, we may reformulate equations (9.12) and (9.15) into the form

$$\mathbf{F} + \mathbf{D} = \mathbf{0} \tag{9.30}$$

$$\mathbf{M} + \mathbf{M}^D = \mathbf{0} \tag{9.31}$$

where

$$\mathbf{D} = -m\mathbf{a}_S \tag{9.32}$$

$$\mathbf{M}^D = -\mathbf{I}_S\dot{\boldsymbol{\omega}} - \boldsymbol{\omega} \times \mathbf{I}_S\boldsymbol{\omega}_S \tag{9.33}$$

are inertial effects (inertial force and inertial couple). Equations (9.30) and (9.31) express d'Alembert's principle according to which a body moves in such a way as to satisfy at any instant the (fictitious) equilibrium of applied forces and inertial forces.

We have so far considered the case of a single body, free in space, for whose motion we have formulated the basic quantities and their relations. A MBS is, however, a system of bodies. For the purpose of assembling MBS equations of motion, the free body principle may be used. According to this principle, the constraints of the real system are to be substituted by forces with which the constraints act on the individual bodies or groups of bodies. The bodies "freed" in this way may then be considered to be free bodies, and all derived relations may be used for the description of their motion. This is the same approach as with the kinetostatic solution described in Chapter 8. The components of reaction force effects are shown in Table 8.1. The equations of motion may be written either in the Newton-Euler form (9.12), (9.15) or in the d'Alembert form (9.30), (9.31). In addition to unknown kinematic and geometric quantities, unknown reaction force effects appear in the equations. Kinematic and geometric MBS quantities are, in general, not independent. To enable their calculation (or elimination), the respective number of constraint equations must be added to the equations of motion.

We often try to eliminate all dependent quantities and, in a system with i degrees of freedom, retain only i independent coordinates. Further eliminating all constraints reactions from the system of equations of motion, we obtain directly the relations between applied forces and quantities of motion. These are called proper equations of motion.

In normal cases, general treatment of equations cannot be considered due to the great complexity of the relations. The elimination can only be performed numerically. The procedures used are dealt with in Section 9.8. In a given case, the result is a minimum number of differential equations which we integrate numerically. The determination of further kinematic quantities is then a strictly kinematic matter. Another common procedure is to solve numerically the system of differential equations of motion together with the system of algebraic constraint equations. The resulting solution yields the time behaviour of the geometric and kinematic quantities and the constraint reaction.

9.2.2 Approaches Based on Analytical Mechanics

The principles of analytical mechanics are usually formulated in such a way that simple formulation of the laws of motion may be made even for complex systems. A certain disadvantage is the usually more complicated mathematical apparatus which is required. The principles of analytical mechanics may be divided into differential (describing motion in a certain instant) and integral (observing the state of motion of the system being examined in a certain finite time interval). Several principles have their origin in variation calculus. We call these variation principles.

The basic quantities characterizing the dynamics of a body in approaches based on analytical mechanics are the kinetic energy of the body and the MBS, potential energy and the work completed by applied forces. For the kinetic energy of a body,

$$T_k = \frac{1}{2}\int_{(m)} \dot{\vec{r}}^2 \mathrm{d}m \tag{9.34a}$$

Using the decomposition with respect to the centre of mass as in Figure 9.1, we derive Koenig's theorem

$$\begin{aligned} T_k &= \frac{1}{2}\int_{(m)} (\dot{\vec{r}}_S + \vec{\omega}\times\vec{r})^2 \mathrm{d}m = \frac{1}{2}\int_{(m)} \dot{\vec{r}}_S^2 \mathrm{d}m + \int_{(m)} \dot{\vec{r}}_S(\vec{\omega}\times\vec{r})\mathrm{d}m + \\ &+\frac{1}{2}\int_{(m)} (\vec{\omega}\times\vec{r})^2\mathrm{d}m = \frac{1}{2}m\dot{\vec{r}}_S^2 + \dot{\vec{r}}_S.(\vec{\omega}\times\int_{(m)}\vec{r}\mathrm{d}m) + \\ &+\frac{1}{2}\int_{(m)} \boldsymbol{\omega}^T(\mathbf{r}\times(\boldsymbol{\omega}\times\mathbf{r}))\mathrm{d}m = \frac{1}{2}m\mathbf{v}_S^2 + \mathbf{v}_S.(\boldsymbol{\omega}\times\mathbf{0}) + \\ &+\frac{1}{2}\int_{(m)} \boldsymbol{\omega}^T(-\mathbf{R}\mathbf{R})\boldsymbol{\omega}\mathrm{d}m = \frac{1}{2}m\mathbf{v}_S^2 + \frac{1}{2}\boldsymbol{\omega}^T\int_{(m)} -\mathbf{R}\mathbf{R}\mathrm{d}m\boldsymbol{\omega} = \\ &= \frac{1}{2}m\mathbf{v}_S^T\mathbf{v}_S + \frac{1}{2}\boldsymbol{\omega}^T\mathbf{I}_S\boldsymbol{\omega} \end{aligned} \tag{9.34b}$$

where, according to (9.9), $\mathbf{I}_S$ denotes the tensor of inertia. This expression for $\mathbf{v}_S$ is most simply expressed in the coordinate system of the frame and $\boldsymbol{\omega}$ in the local coordinate system of the body.

We have described the potential energy for the most frequent cases by relation (8.24) in chapter 8.

The work W of force $\vec{F}$ acting on path $\vec{r}$ of the mass point is given by

$$W = \int_{\vec{r}_0}^{\vec{r}} \vec{F}.\mathrm{d}\vec{r} = \int_0^t \vec{F}.\frac{\mathrm{d}\vec{r}}{\mathrm{d}t}\mathrm{d}t \tag{9.35}$$

9.2.2.1 The Principle of Virtual Work

One of the most fundamental principles of analytical mechanics is the principle of virtual work. This is written for a system of bodies in the form

$$\sum_{i=1}^{p} \{(\mathbf{F}_i^W - m_i\mathbf{a}_i)^T\delta\mathbf{r}_i + (\mathbf{M}_i^W - (\mathbf{I}_i\dot{\boldsymbol{\omega}}_i + \boldsymbol{\omega}_i \times \mathbf{I}_i\boldsymbol{\omega}_i))^T\delta\boldsymbol{\varphi}_i\} = 0 \quad (9.36)$$

where p is the number of bodies, $\mathbf{F}_i^W$ and $\mathbf{M}_i^W$ are the resultant work effect of forces substituted in the centre of mass of each body, and $\delta\mathbf{r}_i$ and $\delta\boldsymbol{\varphi}_i$ are virtual translational and angular displacements of the body for basic decomposition with the point of reference in the centre of mass. Equation (9.36) is sometimes designated in literature as the central equation of dynamics, the d'Alembert-Lagrange equation or the analytical expression of d'Alembert's principle in Lagrangean formulation. It says that a system of bodies performs such a motion as to maintain the algebraic sum of virtual work of all work effects and inertial effects equal to zero. Working forces (as defined below) are all applied forces and, for constraints with passive resistances, also the working (friction) components of reactions. The virtual translational and angular displacements are the elementary changes of body position; these are in agreement with the geometric constraint conditions of the examined body, in which these conditions are considered not to be variable with respect to time. The principle (9.36) is applicable for systems with holonomic and linear nonholonomic constraints. Let us also note that in analytical mechanics, forces are divided into working forces and constraint forces. Working forces are those which perform work, and constraint forces are defined as the non-working components of constraint reactions. For ideal constraints there is no difference between reaction and constraint effects, and applied forces are identical to working forces. As the constraint forces perform no work, they are not considered from the beginning in several methods of analytical mechanics. If we wish to calculate them, we must reformulate the problem to have reaction forces taking the place of applied forces, or we must introduce them into the calculation in some other way, for example, by Lagrange multipliers.

9.2.2.2 Jourdain's Principle

Jourdain's principle in the form

$$\sum_{i=1}^{p} \{(\mathbf{F}_i^W - m_i\mathbf{a}_i)^T\delta\mathbf{v}_i + (\mathbf{M}_i^W - (\mathbf{I}_i\dot{\boldsymbol{\omega}}_i + \boldsymbol{\omega}_i \times \mathbf{I}_i\boldsymbol{\omega}_i))^T\delta\boldsymbol{\omega}_i\} = 0 \quad (9.37)$$

is particularly suitable for systems with nonholonomic constraints. Virtual velocities $\delta\mathbf{v}_i, \delta\boldsymbol{\omega}_i$ calculated without respect to time at a fixed position of the system, i.e. $\delta\boldsymbol{\varphi}_i = \mathbf{0}, \delta\mathbf{r}_i = \mathbf{0}$, take their place here; for this reason the principle is often called the principle of virtual power.

9.2.2.3 Gauss's Principle

The Gauss's principle, valid for a system of bodies in the form

$$\sum_{i=1}^{p} \{(\mathbf{F}_i^W - m_i\mathbf{a}_i)^T\delta\mathbf{a}_i + (\mathbf{M}_i^W - (\mathbf{I}_i\dot{\boldsymbol{\omega}}_i + \boldsymbol{\omega}_i \times \mathbf{I}_i\boldsymbol{\omega}_i))^T\delta\boldsymbol{\alpha}_i\} = 0 \quad (9.38)$$

is simple and generally applicable. Virtual accelerations $\delta\mathbf{a}_i, \delta\boldsymbol{\alpha}_i$ considered without respect to time at $\delta\mathbf{r}_i = \mathbf{0}, \delta\mathbf{v}_i = \mathbf{0}, \delta\boldsymbol{\varphi}_i = \mathbf{0}, \delta\boldsymbol{\omega}_i = \mathbf{0}$ are used here.

9.2.2.4 Hamilton's Principle

Hamilton's principle is one of the basic integral variation principles. Its usual form is

$$\int_{t_1}^{t_2} (\delta T + \delta^* W)\mathrm{d}t = 0 \quad (9.39)$$

or

$$\delta J = \delta \int_{t_1}^{t_2} T\mathrm{d}t + \int_{t_1}^{t_2} \delta^* W\mathrm{d}t = 0 \quad (9.40)$$

In the case of a system of p bodies, equations

$$\begin{aligned} (\delta\vec{r}_i)_{t=t_1} &= \vec{0} \quad (\delta\vec{r}_i)_{t=t_2} = \vec{0} \\ (\delta\vec{\varphi}_i)_{t=t_1} &= \vec{0} \quad (\delta\vec{\varphi}_i)_{t=t_2} = \vec{0} \end{aligned} \quad (9.41)$$

must be added. δT is the exact isochronous variation of function T. Symbol δ^*, used for the designation of virtual work $\delta^* W$, does not express the variation of function W, but expresses the virtual work where radius vectors $\vec{r}_i$ are varied but not forces.

Hamilton's principle may be formulated also, for example, as follows. Observing the motion of a system along arbitrary isochronously varied paths which pass in the end points of the time interval through the same points as for the real motion, the variation of integral J for the real motion equals zero.

The path sought is the extreme of functional J, and the equations of motion are identical with Euler equations of the corresponding variation problem.

The formulation of Hamilton's principle contains no details of the description of the position of the observed system. Obviously independent coordinates $\mathbf{q}$ in the number of i degrees of freedom may be used. In the same way, r coordinates $\mathbf{s}$ (where $r > i$) constrained for example by holonomic skleronomic conditions may be used, so that

$$f_k(\mathbf{s}) = 0, \qquad k = 1, 2, \ldots, m = r - i \tag{9.42}$$

The reason for using the latter method is the easier expression of kinetic energy and virtual work. Hamilton's principle maintains its validity; however, in addition to conditions (9.40) and (9.41), condition (9.42) must be satisfied.

In analytical mechanics, two principal functions of state are defined to describe the state of motion. These are the Lagrangean function L and the Hamiltonian function H. The former is defined as the difference between the kinetic and potential energies of the system

$$L(\mathbf{q}, \dot{\mathbf{q}}, t) = T - V \tag{9.43}$$

The latter is defined by formula

$$H(\mathbf{q}, \mathbf{p}, t) = \sum_{r=1}^{i} p_r \dot{q}_r - L(\mathbf{q}, \mathbf{p}, t) \tag{9.44}$$

where i is the number of degrees of freedom, p_r is the generalized momentum corresponding to the r-th coordinate

$$p_r = \frac{\partial L(\mathbf{q}, \dot{\mathbf{q}}, t)}{\partial \dot{q}_r} \tag{9.45}$$

and $L(\mathbf{q}, \mathbf{p}, t)$ is expressed with respect to generalized momentum (9.45) values. The generalized momentum is a linear function of generalized velocities $\dot{q}_r$. It can be shown that, for skleronomic conservative systems (Leech, 1958; Dobronravov, 1976), the Hamiltonian function equals the total energy of the mechanical system

$$H = T + V \tag{9.46}$$

It is possible, using the Lagrangean function L, to formulate Hamilton's principle in the form

$$\int_{t_1}^{t_2} \delta L \mathrm{d}t = 0 \tag{9.47}$$

9.2.2.5 Lagrange's Equations

Lagrange's equations are the basic method of assembling equations of motion of a mechanical system based on principles of analytical mechanics. If the system is described by generalized coordinates q_r, we can use Lagrange's equations of the second type (Leech, 1968; Dobronravov, 1976)

$$\frac{\mathrm{d}}{\mathrm{dt}}\frac{\partial T}{\partial \dot{q}_r} - \frac{\partial T}{\partial q_r} = Q_r, \qquad r = 1, \ldots, i \tag{9.48}$$

Apart from the symbols introduced above, the generalized force Q_r corresponding to the r-th coordinate appears in these equations. It is defined by the equality of virtual works

$$Q_r \delta q_r = \sum_l \mathbf{F}_l^T \delta \mathbf{r}_l + \sum_m \mathbf{M}_m^T \delta \boldsymbol{\varphi}_m \tag{9.49}$$

of generalized force and working forces $\mathbf{F}_l$ acting on coordinates $\mathbf{r}_l$ and working moment $\mathbf{M}_m$ acting on coordinates $\boldsymbol{\varphi}_m$. If these forces are conservative, i.e. their action can be described by the potential energy of the system,

$$Q_r = -\frac{\partial V}{\partial q_r} \tag{9.50}$$

then the Lagrange's equations of the second type can be described by the relation

$$\frac{\mathrm{d}}{\mathrm{dt}}\frac{\partial L}{\partial \dot{q}_r} - \frac{\partial L}{\partial q_r} = 0 \tag{9.51}$$

If the system is described by n coordinates s_j where n is greater than number of degrees of freedom and therefore bound by constraints (9.42) we assemble the equations of motion using Lagrange's equations of mixed type (LEMT). LEMT is described in more detail in Section 9.3.

9.2.2.6 Hamilton's Canonical Equations

The motion of a mechanical system can be described by Lagrange's equations. These are differential equations of the second order. An equivalent description of the first order of a symmetric form

$$\dot{q}_r = \frac{\partial H}{\partial p_r}, \qquad \dot{p}_r = -\frac{\partial H}{\partial q_r} \tag{9.52}$$

These are known as Hamilton's canonical equations.

For skleronomic holonomic conservative systems, the kinetic energy of the system (9.34) may be expressed in the form

$$T = \frac{1}{2}\dot{\mathbf{q}}^T \mathbf{M} \dot{\mathbf{q}} \tag{9.53}$$

where $\mathbf{M}$ is the generalized system mass matrix. Generalized momentum quantities are then of the form

$$\mathbf{p} = \frac{\partial T}{\partial \dot{\mathbf{q}}^T} = \mathbf{M}\dot{\mathbf{q}} \tag{9.54}$$

The Hamilton function after transformation (9.44) has the form

$$H = \frac{1}{2}\mathbf{p}^T\mathbf{M}^{-1}\mathbf{p} + V(\mathbf{q}) \tag{9.55}$$

and Hamilton's canonical equations (Shet and Uicker, 1972)

$$\begin{aligned} \dot{\mathbf{q}} &= \mathbf{M}^{-1}\mathbf{p} \\ \dot{\mathbf{p}} &= \frac{1}{2}\mathbf{p}^T\left(\mathbf{M}^{-1}\frac{\partial \mathbf{M}}{\partial \mathbf{q}^T}\mathbf{M}^{-1}\right)\mathbf{p} - \frac{\partial V}{\partial \mathbf{q}^T} \end{aligned} \tag{9.56}$$

They had originally been derived only for holonomic conservative systems. However, they can easily be modified for systems of another type. The modification always concerns the second group of equations (9.52), which can be written more generally in the form

$$\dot{p}_r = -\frac{\partial H}{\partial q_r} + \sum_k \lambda_k \frac{\partial f_k}{\partial q_r} + \sum_l \mu_l \frac{\partial g_l}{\partial q_r^{(h)}} + \bar{Q}_r - \frac{\partial F}{\partial \dot{q}_r} \tag{9.57}$$

where the second and third term, respectively, of the right-hand side contains the influence of the holonomic and nonholonomic constraints respectively, the fourth term expresses the contribution of applied forces not included in potential energy, and the last term expresses the effect of damping forces depending on velocities.

9.3 Lagrange's Equations of Mixed Type

Let us assume a MBS with i degrees of freedom. We describe its position by n physical coordinates

$$s_j, \quad j = 1, 2, \ldots, n, \quad n \geq i \tag{9.58}$$

the number of which is generally greater than the number of degrees of freedom. These coordinates are therefore in general not independent; they are subject to holonomic rheonomic constraints

$$f_k(s_j, t) = 0, \quad k = 1, 2, \ldots, \quad r = n - i \tag{9.59}$$

or nonholonomic rheonomic generally nonlinear constraints of order h

$$g_k(s_j, \dot{s}_j, ..., s_j^{(h)}, t) = 0, \quad k = 1, 2, ..., r \tag{9.60}$$

We are going to assume an arbitrary order $h \geq 0$, as the following equations of motion hold for any $h \geq 0$; this generality allows us to consider, for example, a MBS with a feedback control loop as a MBS with a nonholonomic constraint. We are going to describe constraints (9.59) as nonholonomic even though they may be integrable. If the total number of independent constraint conditions (9.59) and (9.60) is r, the MBS has $i = n - r$ degrees of freedom. From n coordinates (9.58) only i are independent, the rest being dependent. In the case of nonholonomic nonintegrable constraints, the relation between coordinates represents virtual changes from the given position.

We are going to base the dynamic MBS description on the expression of MBS kinetic energy $T = T(s_j, \dot{s}_j, t)$, for example as in Section 9.2.2.5. The Lagrange mixed type equations for holonomic constraint (9.59) are then expressed as follows (Dobronravov, 1976)

$$\frac{\mathrm{d}}{\mathrm{dt}}\frac{\partial T}{\partial \dot{s}_j} - \frac{\partial T}{\partial s_j} = Q_j + \sum_{k=1}^{r} \lambda_k \frac{\partial f_k}{\partial s_j}, \quad j = 1, 2, \ldots, n \tag{9.61}$$

where λ_k are Lagrange's multipliers corresponding to constraints (9.59) and Q_j is the generalized force corresponding to coordinate s_j, expressing the effect of all working forces P_l acting along coordinates $p_l = p_l(s_j, t)$. The generalized force Q_j is given by the relation

$$Q_j = \sum_l P_l \frac{\partial p_l}{\partial s_j} \tag{9.62}$$

If some work forces are potential, their effect may be substituted by expressing the corresponding potential energy $V = V(s_j, t)$. Similarly some dissipation forces proportional to velocities $P_l = -b_l \dot{s}_l$ may be substituted by the expression of the corresponding Rayleigh's dissipation function

$$D = D(\mathrm{s}, \dot{\mathrm{s}}, t) = \frac{1}{2} \sum_i \sum_j b_{ij} \dot{s}_i \dot{s}_j \tag{9.63}$$

Lagrange's equations (9.61) may be expressed in the form

$$\frac{\mathrm{d}}{\mathrm{dt}}\frac{\partial T}{\partial \dot{s}_j} - \frac{\partial T}{\partial s_j} = Q_j - \frac{\partial V}{\partial s_j} - \frac{\partial D}{\partial \dot{s}_j} + \sum_{k=1}^{r} \lambda_k \frac{\partial f_k}{\partial s_j}, \quad j = 1, 2, \ldots, n \tag{9.64}$$

If we consider nonholonomic constraints (9.60) instead of holonomic ones (9.59), the Lagrange equations of the mixed type (LEMT) have the form (Dobronravov, 1970; Golst et al., 1979)

$$\frac{\mathrm{d}}{\mathrm{dt}}\frac{\partial T}{\partial \dot{s}_j} - \frac{\partial T}{\partial s_j} = Q_j + \sum_{k=1}^{r} \lambda_k \frac{\partial g_k}{\partial s_j^{(h)}}, \qquad j = 1, 2, ..., n \tag{9.65}$$

where $s_j^{(h)}$ expresses the highest order of differentiating in the constraint g_k (9.60). The individual cases of holonomic or nonholonomic constraints, potential or dissipation function, may be mutually combined. It is obvious from relation (9.60) that holonomic constraints (9.59) are a special case of nonholonomic ones for $h = 0$. Analogously, equations (9.61) are a special case of equations (9.65) for $h = 0$. This makes it possible for further derivation to consider in general only constraints (9.60) and equations (9.65). The system of equations (9.65) and (9.60) contains $n + r$ equations for n unknown functions $s_j(t)$ and r Lagrange multipliers $\lambda_k(t)$, which express the reaction forces of the constraints.

For a great many MBS's we can use equations (9.61) with constraints (9.59) or nonholonomic constraints of the first order (9.3) or at most of the second order. The principal outline of the solution of that system of equations is as follows. We treat the equations (9.61)

$$\sum_{a=1}^{n} \frac{\partial^2 T}{\partial \dot{s}_a \partial \dot{s}_j} \ddot{s}_a + \sum_{a=1}^{n} \frac{\partial^2 T}{\partial s_a \partial \dot{s}_j} \dot{s}_a + \frac{\partial^2 T}{\partial t \partial \dot{s}_j} - \frac{\partial T}{\partial s_j} = Q_j + \sum_{k=1}^{r} \lambda_k \frac{\partial f_k}{\partial s_j} \tag{9.66}$$

We substitute the equations of constraints (9.59) by their second total derivative with respect to time

$$\frac{\partial^2 f_k}{\partial t^2} = \sum_{a=1}^{n} \frac{\partial f_k}{\partial s_a} \ddot{s}_a + \sum_{j=1}^{n} \sum_{a=1}^{n} \frac{\partial^2 f_k}{\partial s_j \partial s_a} \dot{s}_a \dot{s}_j + 2 \sum_{a=1}^{n} \frac{\partial^2 f_k}{\partial t \partial s_a} \dot{s}_a + \frac{\partial^2 f_k}{\partial t^2} = 0 \tag{9.67}$$

or we substitute the nonholonomic constraints of the first order (9.3) by the first total derivative with respect to time in a form similar to (9.67). The system of equations (9.66), (9.67) now depends linearly on $\ddot{s}_j$ and λ_k. It is thus possible to re-write it in matrix form

$$\begin{bmatrix} \mathbf{M} & \mathbf{\Phi}_s^T \\ \mathbf{\Phi}_s & \mathbf{0} \end{bmatrix} \begin{bmatrix} \ddot{\mathbf{s}} \\ -\boldsymbol{\lambda} \end{bmatrix} = \begin{bmatrix} \mathbf{p}_1 \\ \mathbf{p}_2 \end{bmatrix} \tag{9.68}$$

where $\ddot{\mathbf{s}} = [\ddot{s}_1, \ddot{s}_2, \ldots, \ddot{s}_n], \boldsymbol{\lambda} = [\lambda_1, \lambda_2, \ldots, \lambda_n], \mathbf{M}$ is a square regular symmetric (n, n) matrix, $\mathbf{\Phi}_s$ is an (r, n) matrix, $\mathbf{0}$ is an (r, r) null-matrix, $\mathbf{p}_1$ is

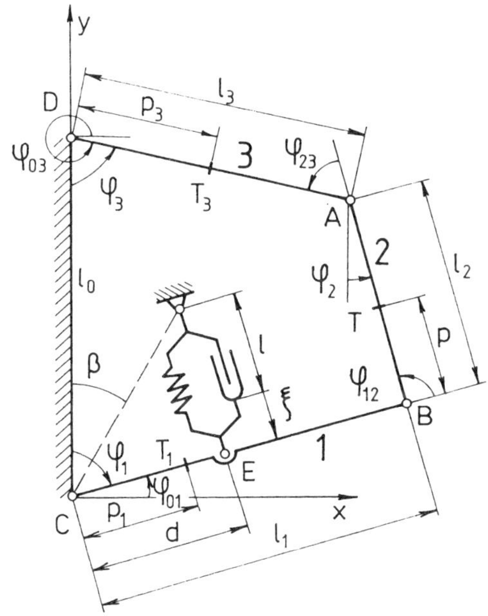

Figure 9.2. Front wheel suspension.

an n-dimensional vector and $\mathbf{p}_2$ is an r-dimensional vector. Then

$$
\begin{aligned}
(\mathbf{M})_{ij} &= \frac{\partial^2 T}{\partial \dot{s}_i \partial \dot{s}_j}, \qquad (\mathbf{\Phi}_s)_{ij} = \frac{\partial f_i}{\partial s_j}, \\
(\mathbf{p}_1)_j &= \frac{\partial T}{\partial s_j} - \frac{\partial^2 T}{\partial t \partial \dot{s}_j} - \sum_{a=1}^{n} \frac{\partial^2 T}{\partial s_a \partial \dot{s}_j} \dot{s}_a + Q_j \\
(\mathbf{p}_2)_k &= -\sum_{j=1}^{n}\sum_{a=1}^{n} \frac{\partial^2 f_k}{\partial s_j \partial s_a} \dot{s}_j \dot{s}_a - 2\sum_{a=1}^{n} \frac{\partial^2 f_k}{\partial t \partial s_a} \dot{s}_a - \frac{\partial^2 f_k}{\partial t^2}
\end{aligned} \tag{9.69}
$$

Obviously matrices $\mathbf{M}, \mathbf{\Phi}_s, \mathbf{p}_1, \mathbf{p}_2$ are functions only of $s_j, \dot{s}_j, t$. The numerical solution of this system of equations is considered in detail in Section 9.8. We present here only an outline principle of the solution. If in some time instant t_i values $s_j(t_i), \dot{s}_j(t_i),\ j = 1, \ldots, n$ are known, we determine from them the values of matrices $\mathbf{M}, \mathbf{\Phi}_s, \mathbf{p}_1, \mathbf{p}_2$ and calculate from (9.68) $\ddot{s}_j(t_i), \lambda_k(t_i)$. Integrating $\ddot{s}_j(t_i)$ numerically, we obtain the values of $\dot{s}_j(t_{i+1}), s_j(t_{i+1}),\ j = 1, \ldots, n$ for time $t_{i+1} = t_i + \Delta t_i$ and the procedure may be repeated. However, this simple straightforward scheme is for longer time numerically unstable. For reliable solution another schemes must be used (see Section 9.8).

Let us demonstrate the application of LEMT to a planar four-bar mechanism (Figure 9.2) representing a simplified model of an automobile front wheel trapezoidal suspension. The dimensions are l_0, l_1, l_2 and l_3, moments of inertia of elements 1 and 3 are I_1 and I_3, the mass of the connecting rod is m_2 and the moment of inertia of the rod with respect to the centre of mass T is $I_{2T}, BT = p$. The rocker arm is connected to the frame by a nonlinear spring and damper. The spring axis is, in the equilibrium position, perpendicular to the rocker arm; its free length is l and $CE = d$. We examine large (gross) oscillations of the mechanism around the equilibrium position.

For the mechanism description we choose angles φ_1, φ_2 and φ_3 as coordinates. We express the kinetic energy

$$T = \frac{1}{2}(I_1\dot{\varphi}_1^2 + I_{2T}\dot{\varphi}_2^2 + m_2(\dot{x}_2^2 + \dot{y}_2^2) + I_3\dot{\varphi}_3^2) \tag{a}$$

where

$$\begin{aligned} x_2 &= l_1 \mathrm{s}\varphi_1 - p\mathrm{s}\varphi_2 \\ y_2 &= l_1 \mathrm{c}\varphi_1 + p\mathrm{c}\varphi_2 \\ \dot{x}_2^2 + \dot{y}_2^2 &= l_1^2\dot{\varphi}_1^2 + p^2\dot{\varphi}_2^2 - 2l_1 p\mathrm{c}(\varphi_1 + \varphi_2)\dot{\varphi}_1\dot{\varphi}_2 \end{aligned} \tag{b}$$

are the coordinates and the square of the velocity of the centre of mass of the connecting rod. The coordinates φ_1, φ_2 and φ_3 are not independent, but are related by constraint conditions

$$\begin{aligned} f_1 &\equiv l_1\mathrm{s}\varphi_1 - l_2\mathrm{s}\varphi_2 - l_3\mathrm{s}\varphi_3 = 0 \\ f_2 &\equiv l_1\mathrm{c}\varphi_1 + l_2\mathrm{c}\varphi_2 + l_3\mathrm{c}\varphi_3 - l_0 = 0 \end{aligned} \tag{c}$$

The force of the spring and damper $F(\xi, \dot{\xi})$ acts on the mechanism along coordinate ξ. According to Figure 9.2

$$(l + \xi)^2 = l^2 + 2d^2 - 2d\sqrt{l^2 + d^2}\mathrm{c}(\varphi_1 - \beta) \tag{d}$$

Therefore

$$\xi = \sqrt{a - 2b\mathrm{c}(\varphi_1 - \beta)} - l$$

where $a = l^2 + 2d^2$ and $b = d\sqrt{l^2 + d^2}$. Substituting into (9.50) we obtain the following system of equations which we alter into the from (9.68)

$$\begin{bmatrix} I_1 + m_2 l_1^2, & -m_2 l_1 p\mathrm{c}(\varphi_1 + \varphi_2), & 0, & l_1\mathrm{c}\varphi_1, & -l_1\mathrm{s}\varphi_1 \\ & I_{2T} + m_2 p^2, & 0, & -l_2\mathrm{c}\varphi_2, & -l_2\mathrm{s}\varphi_2 \\ & & I_3, & -l_3\mathrm{c}\varphi_3, & -l_3\mathrm{s}\varphi_3 \\ & & & 0, & 0, \\ \text{sym.} & & & & 0, \end{bmatrix} \begin{bmatrix} \ddot{\varphi}_1 \\ \ddot{\varphi}_2 \\ \ddot{\varphi}_3 \\ -\lambda_1 \\ -\lambda_2 \end{bmatrix} =$$

$$= \begin{bmatrix} -m_2 l_1 p\dot{\varphi}_2^2 \mathrm{s}(\varphi_1 + \varphi_2) - F(\xi, \dot{\xi})(b/(l+\xi))\mathrm{s}(\varphi_1 - \beta) \\ -m_2 l_1 p\dot{\varphi}_1^2 \mathrm{s}(\varphi_1 + \varphi_2) \\ 0 \\ l_1\dot{\varphi}_1^2 \mathrm{s}\varphi_1 - l_2\dot{\varphi}_2^2 \mathrm{s}\varphi_2 - l_3\dot{\varphi}_3^2 \mathrm{s}\varphi_3 \\ l_1\dot{\varphi}_1^2 \mathrm{c}\varphi_1 + l_2\dot{\varphi}_2^2 \mathrm{c}\varphi_2 + l_3\dot{\varphi}_3^2 \mathrm{c}\varphi_3 \end{bmatrix} \tag{e}$$

9.3.1 Lagrange's Equations Written in Quasi-coordinates

Another important way of formulating MBS equations of motion is to use quasi-coordinates (Kane et al., 1983; Dobronravov, 1970; Neimark and Fufajev, 1972; Golst et al., 1979; Valášek and Stejskal, 1987). In a MBS described by coordinates (9.58) we introduce quantities

$$\dot{\pi}_{\varkappa} = \sum_{j=1}^{n} \alpha_{\varkappa j}(s_k)\dot{s}_j, \quad \varkappa = 1, 2, \dots, n \tag{9.70}$$

where $\alpha_{\varkappa j}(s_k)$ are functions of coordinates (9.58). For $\dot{\pi}_k$ to be a total differential, it is sufficient to establish that

$$\frac{\partial \alpha_{\mu j}}{\partial s_\nu} - \frac{\partial \alpha_{\mu\nu}}{\partial s_j} = 0, \quad \mu, j, \nu = 1, 2, \dots, n \tag{9.71}$$

If equations (9.71) are satisfied, we may consider $\pi_{\varkappa}$ to be ordinary coordinates. However, we are going to examine the general case of

$$\frac{\partial \alpha_{\mu j}}{\partial s_\nu} - \frac{\partial \alpha_{\mu\nu}}{\partial s_j} \neq 0 \tag{9.72}$$

Quantities (9.70) are then designated as quasi-velocities, and quantities

$$\pi_{\varkappa} = \int_0^t \sum_{m=1}^{n} \alpha_{\varkappa m}(s_j)\dot{s}_m \mathrm{d}t \tag{9.73}$$

as quasi-coordinates. Quasi-coordinates differ from ordinary coordinates in the dependence of their value on the integration path in (9.73).

We assume determinant $|\alpha_{\varkappa j}|$ to be non-zero; thus velocities $\dot{s}_m$ from (9.70) may be expressed from quasi-velocities $\dot{\pi}_k$

$$\dot{s}_m = \sum_{\varkappa=1}^{n} \beta_{m\varkappa}(s_j)\dot{\pi}_{\varkappa}, \quad m = 1, 2, ..., n \tag{9.74}$$

Using (9.74) we transform the kinetic energy

$$T'(s_j, \dot{\pi}_{\varkappa}, t) = T(s_j, \dot{s}_j, t) \tag{9.75}$$

and similarly also transform the constraints (9.60)

$$g'_k(s_j, \dot{\pi}_{\varkappa}, \ldots, \dot{\pi}_{\varkappa}^{(h)}, t) = g_k(s_j, \dot{s}_j, \ldots, s_j^{(h)}, t) \tag{9.76}$$

Further we introduce formally the operation

$$\frac{\partial}{\partial \pi_{\varkappa}} = \sum_{m=1}^{n} \beta_{m\varkappa} \frac{\partial}{\partial s_m} \tag{9.77}$$

Then, instead of equations (9.65) we have equations

$$\frac{\mathrm{d}}{\mathrm{dt}} \frac{\partial T'}{\partial \dot{\pi}_{\varkappa}} - \frac{\partial T'}{\partial \pi_{\varkappa}} + \sum_{\mu=1}^{n} \sum_{j=1}^{n} \gamma_{\varkappa\mu j} \frac{\partial T'}{\partial \dot{\pi}_m} \dot{\pi}_j = Q'_{\varkappa} + \sum_{k=1}^{r} \lambda_k \frac{\partial g'_k}{\partial \pi_{\varkappa}^{(h)}} \tag{9.78}$$

where

$$\gamma_{\varkappa\mu j} = \sum_{m=1}^{n} \sum_{\nu=1}^{n} \beta_{mj} \beta_{\nu\varkappa} \left(\frac{\partial \alpha_{\mu m}}{\partial s_{\nu}} - \frac{\partial \alpha_{\mu\nu}}{\partial s_m} \right) \tag{9.79}$$

$$Q'_{\varkappa} = \sum_{m=1}^{n} \beta_{m\varkappa -} Q_m \tag{9.80}$$

We call the equations (9.78) Lagrange equations of the mixed type written in quasi-coordinates (Dobronravov, 1970; Butenin, 1971; Valášek and Stejskal, 1987). To simplify their calculation we designate

$$A^{\mu}_{m\nu} = \frac{\partial \alpha_{\mu m}}{\partial s_{\nu}} - \frac{\partial \alpha_{\mu\nu}}{\partial s_m} \tag{9.81}$$

From definition (9.81),

$$\begin{aligned} A^{\mu}_{mm} &= 0 \\ A^{\mu}_{m\nu} &= -A^{\mu}_{\nu m} \\ \gamma_{\varkappa\mu j} &= \sum_{m=1}^{n} \sum_{\nu=1}^{n} \beta_{mj} \beta_{\nu\varkappa} A^{\mu}_{m\nu} \end{aligned} \tag{9.82}$$

Hence

$$\begin{aligned} \gamma_{j\mu j} &= 0 \\ \gamma_{\varkappa\mu j} &= \gamma_{j\mu\varkappa} \end{aligned} \tag{9.83}$$

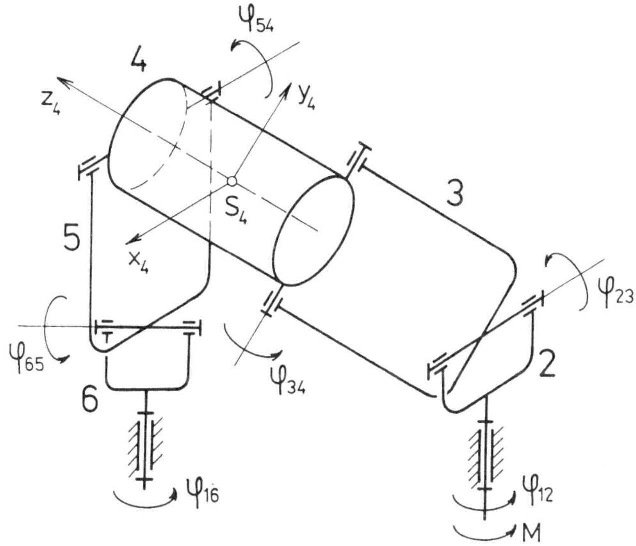

Figure 9.3. Dynamics of Turbula.

The above knowledge makes the calculation of $\gamma_{\varkappa\mu j}$ for equations (9.78) much easier. We first compute expressions $A^{\mu}_{m\nu}$, from which we obtain the expressions $\gamma_{\varkappa\mu j}$.

For example let us examine the motion of Turbula, the mechanism from Chapter 7 in which moment M acts on coordinate φ_{12} according to Figure 9.3. We first describe the MBS using coordinates $\varphi_{12}, \varphi_{23}, \varphi_{34}$ $\varphi_{54}, \varphi_{65}$ and φ_{16}. We then express the angular velocity of body 4. According to (7.56),

$$\mathbf{S}_{14} = \mathbf{S}_{Z6}(\varphi_{12})\mathbf{S}_{Z4}(\varphi_{23})\mathbf{S}_{Z5}(\varphi_{34}) \tag{a}$$

From (a)

$$\begin{aligned}\mathbf{\Omega}_{14} &= \mathbf{S}_{Z5}(-\varphi_{34})\mathbf{S}_{Z4}(-\varphi_{23})\mathbf{D}_{Z6}(\dot{\varphi}_{12})\mathbf{S}_{Z4}(\varphi_{23})\mathbf{S}_{Z5}(\varphi_{34}) + \\ &+\mathbf{S}_{Z5}(-\varphi_{34})\mathbf{D}(\dot{\varphi}_{23})\mathbf{S}_{Z5}(\varphi_{34}) + \mathbf{D}_{Z5}(\dot{\varphi}_{34})\end{aligned} \tag{b}$$

Thus, according to (4.69),

$$\begin{aligned}\omega_{4x} &= \dot{\varphi}_{12}c_{23}s_{34} - \dot{\varphi}_{23}c_{34} \\ \omega_{4y} &= -\dot{\varphi}_{12}s_{23} + \dot{\varphi}_{34} \\ \omega_{4z} &= -(\dot{\varphi}_{12}c_{23}c_{34} + \dot{\varphi}_{23}s_{34})\end{aligned} \tag{c}$$

where $c_{ij} = c\varphi_{ij}, s_{ij} = s\varphi_{ij}$. We introduce quasi-coordinates

$$\dot{\pi}_1 = \omega_{4x}, \qquad \dot{\pi}_2 = \omega_{4y}, \qquad \dot{\pi}_3 = \omega_{4z} \tag{d}$$

and

$$\dot{\pi}_4 = \dot{\varphi}_{54}, \qquad \dot{\pi}_5 = \dot{\varphi}_{65}, \qquad \dot{\pi}_6 = \dot{\varphi}_{16} \tag{e}$$

We express α_{kl} from (d) according to (9.70). For example

$$\alpha_{11} = \mathrm{c}_{23}\mathrm{s}_{34},\ \alpha_{12} = -\mathrm{c}_{34},\ \alpha_{22} = 0,\ \alpha_{23} = 1,\ \alpha_{44} = 1 \tag{f}$$

and according to (9.74) we obtain for example

$$\begin{aligned} \dot{\varphi}_{12} &= \frac{\dot{\pi}_1\mathrm{s}_{34} - \dot{\pi}_3\mathrm{c}_{34}}{\mathrm{c}_{23}} \\ \dot{\varphi}_{34} &= \dot{\pi}_1\mathrm{s}_{34}\tan\varphi_{23} + \dot{\pi}_2 - \dot{\pi}_3\mathrm{c}_{34}\tan\varphi_{23} \\ \dot{\varphi}_{54} &= \dot{\pi}_4 \\ \beta_{11} &= \frac{\mathrm{s}_{34}}{\mathrm{c}_{23}},\quad \beta_{12} = 0,\quad \beta_{32} = 1 \end{aligned} \tag{g}$$

It is important that, for the above mechanism, $\mathrm{c}_{23} \neq 0$ as follows from (7.52a) in Chapter 7. Treating this further according to (9.81) and (9.82) we obtain

$$A_{11}^1 = 0,\quad A_{12}^1 = -\mathrm{s}_{23}\mathrm{s}_{34},\quad A_{13}^1 = \mathrm{c}_{23}\mathrm{c}_{34}$$

$$\gamma_{111} = 0,\ \gamma_{213} = 1,\ \gamma_{231} = -(c_{34}^2 - s_{34}^2),\ \gamma_{233} = 2\mathrm{s}_{34}\mathrm{c}_{34} \tag{h}$$

We can now express the MBS kinetic energy

$$\begin{aligned} T \ = \ & (I_2\dot{\varphi}_{12}^2 + I_{3x}\omega_{3x}^2 + I_{3y}\omega_{3y}^2 + I_{3z}\omega_{3z}^2 + m_3 v_{S3}^2 + \\ & + I_{4x}\omega_{4x}^2 + I_{4y}\omega_{4y}^2 + I_{4z}\omega_{4z}^2 + m_4 v_{S4}^2 + \\ & + I_{5x}\omega_{5x}^2 + I_{5y}\omega_{5y}^2 + I_{5z}\omega_{5z}^2 + m_5 v_{S5}^2 + I_6\dot{\varphi}_{16}^2) \end{aligned} \tag{i}$$

where the I's represent the moments of inertia about the respective axes. For bodies 3, 4 and 5 we assume that the axes of coordinate systems according to Figure 9.3 are principal central axes of inertia of the bodies. Symbols m_3, m_4 and m_5 represent the respective masses and v_{S3}, v_{S4} and v_{S5} the respective velocities of the centers of mass of bodies 3, 4 and 5. We express for particular bodies, for example

$$\begin{aligned} \mathbf{T}_{15} &= \mathbf{T}_{Z2}(-b)\mathbf{T}_{Z6}(\varphi_{16})\mathbf{T}_{Z5}(\varphi_{65})\mathbf{T}_{Z3}(t_5) \\ \boldsymbol{\Omega}_{15} &= \mathbf{S}_{Z5}(-\varphi_{65})\mathbf{D}_{Z6}(\dot{\varphi}_{16})\mathbf{S}_{Z5}(\varphi_{65}) + \mathbf{D}_{Z5}(\dot{\varphi}_{65}) \\ \omega_{5x} &= -\mathrm{s}_{65}\dot{\varphi}_{16} \\ \mathbf{r}_{S5} &= [t_5\mathrm{c}_{16}\mathrm{s}_{65}, t_5\mathrm{s}_{16}\mathrm{s}_{65} - b, t_5\mathrm{c}_{65}, 1] \\ v_{S5}^2 &= t_5^2(\dot{\varphi}_{65}^2 + \mathrm{s}_{65}^2\dot{\varphi}_{16}^2) \\ \omega_{4x} &= \dot{\pi}_1, \quad \omega_{4y} = \dot{\pi}_2, \quad \omega_{4z} = \dot{\pi}_3 \end{aligned} \tag{j}$$

We express the kinetic energy from these relations and alter it into the form (9.75)

$$\begin{aligned} T &= \frac{1}{2}(\dot\varphi_{12}^2(I_2 + I_{3y}\mathrm{s}_{23}^2 + I_{3z}\mathrm{c}_{23}^2 + m_3t_3^2\mathrm{s}_{23}^2 + \ldots) + \\ &\quad + \dot\varphi_{23}^2(I_{3x} + m_3t_3^2 + \ldots) + \ldots + \\ &\quad + I_{4x}\omega_{4x}^2 + I_{4y}\omega_{4y}^2 + I_{4z}\omega_{4z}^2 + \ldots) \\ T' &= \frac{1}{2}\left(\left(\frac{\dot\pi_1\mathrm{s}_{34} - \dot\pi_3\mathrm{c}_{34}}{\mathrm{c}_{23}}\right)^2 (I_2 + I_{3y}\mathrm{s}_{23}^2 + I_{3z}\mathrm{c}_{23}^2 + \right. \\ &\quad + m_3t_3^2\mathrm{s}_{23}^2 + \ldots) + (\dot\pi_1\mathrm{c}_{34} + \dot\pi_3\mathrm{s}_{34})^2(I_{3x} + m_3t_3^2 + \ldots) + \\ &\quad \left. + I_{4x}\dot\pi_1^2 + I_{4y}\dot\pi_2^2 + I_{4z}\dot\pi_3^2 + \ldots\right) \end{aligned} \tag{k}$$

Finally we may, according to (9.78), assemble the equations of motion. For example, for $\varkappa = 2$ we obtain

$$\begin{aligned} &\frac{\mathrm{d}}{\mathrm{dt}}\frac{\partial T'}{\partial\dot\pi_2} - \frac{\partial T'}{\partial\varphi_{34}} + \frac{\partial T'}{\partial\dot\pi_1}\dot\pi_3 - (\mathrm{c}_{34}^2 - \mathrm{s}_{34}^2)\frac{\partial T'}{\partial\dot\pi_3}\dot\pi_1 + \\ &\qquad + 2\mathrm{s}_{34}\mathrm{c}_{34}\frac{\partial T'}{\partial\dot\pi_3}\dot\pi_3 = \sum_{k=1}^{5}\lambda_k\frac{\partial g_{r_k}}{\partial\varphi_{34}} \end{aligned} \tag{l}$$

where $g_{r_k}(r_k = 1, \ldots, 13)$ represent constraint equations (7.38) modified to give right-hand sides equal to zero, from which five independent constraints must be chosen. Let us present the calculation of the first term

$$\begin{aligned} \frac{\partial T'}{\partial\dot\pi_2} &= 2(\dot\pi_1\mathrm{s}_{34}\tan\varphi_{23} + \dot\pi_2 - \dot\pi_3\mathrm{c}_{34}\tan\varphi_{23})m_4t_4^2 + \ldots \\ \frac{\mathrm{d}}{\mathrm{dt}}\frac{\partial T'}{\partial\dot\pi_2} &= 2(\ddot\pi_1\mathrm{s}_{34}\tan\varphi_{23} + \ddot\pi_2 - \ddot\pi_3\mathrm{c}_{34}\tan\varphi_{23})m_4t_4^2 + \ldots \end{aligned}$$

In this way we obtain a system of differential equations for $\ddot\pi_i$, hence by numerical integration we determine $\dot\pi_i$, from which we express $\dot\varphi_{ij}$ according to (9.74), i.e. (g); by further integration we obtain φ_{ij}.

9.4 Equivalence between Vector and Analytical Approaches

We have seen several principles of mechanics (Section 9.2) which make possible the creation of a MBS dynamic description. In the next section we shall see a number of specific formalisms for assembling MBS equations of motion. All these approaches to MBS dynamics are in principle equivalent, as they describe the same physical model. Different formalisms for the solution of MBS dynamics lead however to equations of motion which differ in their structure and computational complexity. We are confronted with the problem of finding the appropriate transformations from one set of equations of motion to the other. These equations are structurally different. What we are seeking is the structural equivalence of these equations of motion, however obvious their physical equivalence as a matter of principle may be.

We are going to derive the structural equivalence for Newton-Euler's equations as representing the vector approach, and Lagrange's equations of the mixed type as representing the analytical approach.

Let the MBS in question be composed of n bodies. We connect to each body i a fixed system of coordinates with its origin in the centre of mass. Newton's equations of motion for the i-th body are of the form

$$m_i \dot{\mathbf{v}}_i = \mathbf{F}_i \tag{9.84}$$

Euler's equations can be written in the form

$$\mathbf{I}_i \dot{\boldsymbol{\omega}}_i + \boldsymbol{\omega}_i \times \mathbf{I}_i \boldsymbol{\omega}_i = \mathbf{M}_i \tag{9.85}$$

where m_i is the mass of body i, $\mathbf{r}_i$ and $\mathbf{v}_i$ are the position and velocity vector of its centre of mass expressed in the fixed system of coordinates connected with the basic frame, $\mathbf{F}_i$ is the resultant force acting in the centre of mass of body i and expressed in the system of coordinates of the frame, and $\mathbf{I}_i$ is the tensor of inertia of body i expressed in the coordinate system connected with body i and moving along with it. A symbol $\boldsymbol{\omega}_i$ is the angular velocity of body i and $\mathbf{M}_i$ is the resultant moment, both expressed in the same local system of coordinates.

For Lagrange's equations let us first determine the kinetic energy of the body i

$$T_i = \frac{1}{2} m_i \mathbf{v}_i^T \mathbf{v}_i + \frac{1}{2} \boldsymbol{\omega}_i^T \mathbf{I}_i \boldsymbol{\omega}_i \tag{9.86}$$

After performing the operations on the left-hand side of Lagrange's equations we obtain

$$\begin{aligned}
\frac{\partial T_i}{\partial \dot{s}_j} &= m_i \mathbf{v}_i^T \frac{\partial \mathbf{v}_i}{\partial \dot{s}_j} + \boldsymbol{\omega}_i^T \mathbf{I}_i \frac{\partial \boldsymbol{\omega}_i}{\partial \dot{s}_j} \\
\frac{\mathrm{d}}{\mathrm{d}t} \frac{\partial T_i}{\partial \dot{s}_j} &= m_i \dot{\mathbf{v}}_i^T \frac{\partial \mathbf{v}_i}{\partial \dot{s}_j} + m_i \mathbf{v}_i^T \frac{\mathrm{d}}{\mathrm{d}t} \left(\frac{\partial \mathbf{v}_i}{\partial \dot{s}_j} \right) + \\
&\quad + \dot{\boldsymbol{\omega}}_i^T \mathbf{I}_i \frac{\partial \boldsymbol{\omega}_i}{\partial \dot{s}_j} + \boldsymbol{\omega}_i^T \mathbf{I}_i \frac{\mathrm{d}}{\mathrm{d}t} \left(\frac{\partial \boldsymbol{\omega}_i}{\partial \dot{s}_j} \right)
\end{aligned}$$

$$\frac{\partial T_i}{\partial s_j} = m_i \mathbf{v}_i^T \frac{\partial \mathbf{v}_i}{\partial s_j} + \boldsymbol{\omega}_i^T \mathbf{I}_i \frac{\partial \boldsymbol{\omega}_i}{\partial s_j} \tag{9.87}$$

Before substituting them into the left-hand sides of Lagrange's equations for individual bodies, we use the relations (9.87). First

$$\mathbf{v}_i = \frac{\mathrm{d}}{\mathrm{d}t}\mathbf{r}_i(s_j, t) = \sum_j \frac{\partial \mathbf{r}_i}{\partial s_j}\dot{s}_j + \frac{\partial \mathbf{r}_i}{\partial t}$$

and hence

$$\frac{\partial \mathbf{v}_i}{\partial \dot{s}_j} = \frac{\partial \mathbf{r}_i}{\partial s_j} \tag{9.88}$$

Further

$$\begin{aligned}\frac{\mathrm{d}}{\mathrm{d}t}\frac{\partial \mathbf{r}_i}{\partial s_j} &= \sum_l \frac{\partial^2 \mathbf{r}_i}{\partial s_j \partial s_l}\dot{s}_l + \frac{\partial^2 \mathbf{r}_i}{\partial s_j \partial t} = \\ &= \frac{\partial}{\partial s_j}\left(\sum_l \frac{\partial \mathbf{r}_i}{\partial s_l}\dot{s}_l + \frac{\partial \mathbf{r}_i}{\partial t}\right) = \frac{\partial \mathbf{v}_i}{\partial s_j}\end{aligned}$$

and hence

$$\frac{\mathrm{d}}{\mathrm{d}t}\frac{\partial \mathbf{v}_i}{\partial \dot{s}_j} = \frac{\partial \mathbf{v}_i}{\partial s_j} \tag{9.89}$$

The respective relations for angular velocity $\boldsymbol{\omega}_i$ are more complicated as we have to deal with derivatives of a vector expressed in a moving space. Our starting point is the derivative using (4.63)

$$\frac{\mathrm{d}}{\mathrm{d}t}\left(\mathbf{S}_i \frac{\partial \boldsymbol{\omega}_i}{\partial \dot{s}_j}\right) = \mathbf{S}_i \boldsymbol{\Omega}_i \frac{\partial \boldsymbol{\omega}_i}{\partial \dot{s}_j} + \mathbf{S}_i \frac{\mathrm{d}}{\mathrm{d}t}\left(\frac{\partial \boldsymbol{\omega}_i}{\partial \dot{s}_j}\right) \tag{9.90}$$

where $\mathbf{S}_i$ is the transformation matrix (of directional cosines (4.29)) between the expression of vectors in the space of body i and the space of the frame and $\boldsymbol{\Omega}_i$ is the matrix of angular velocity (4.69) of body i (the expression of angular velocity $\boldsymbol{\omega}_i$ as a skew-symmetric matrix). However, the same differentiation also can be performed as follows

$$\frac{\mathrm{d}}{\mathrm{d}t}\left(\mathbf{S}_i \frac{\partial \boldsymbol{\omega}_i}{\partial \dot{s}_j}\right) = \frac{\mathrm{d}}{\mathrm{d}t}\frac{\partial}{\partial \dot{s}_j}(\mathbf{S}_i \boldsymbol{\omega}_i) = \frac{\partial}{\partial \dot{s}_j}\frac{\mathrm{d}}{\mathrm{d}t}(\mathbf{S}_i \boldsymbol{\omega}_i) \tag{9.91}$$

The interchangeability of derivatives in (9.91) follows from the fact that the vector $\mathbf{S}_i\boldsymbol{\omega}_i$ is linear in $\dot{s}_j$ and is expressed in the fixed system of coordinates of the frame. It is therefore not necessary to use (4.3). From (9.91), again using (4.63), it follows that

$$\frac{\partial}{\partial \dot{s}_j}\frac{\mathrm{d}}{\mathrm{d}t}(\mathbf{S}_i \boldsymbol{\omega}_i) = \frac{\partial}{\partial \dot{s}_j}(\mathbf{S}_i \boldsymbol{\Omega}_i \boldsymbol{\omega}_i + \mathbf{S}_i \dot{\boldsymbol{\omega}}_i) = \frac{\partial}{\partial \dot{s}_j}(\mathbf{S}_i \dot{\boldsymbol{\omega}}_i) =$$

$$= \mathbf{S}_i \frac{\partial \dot{\boldsymbol{\omega}}_i}{\partial \dot{s}_j} = \mathbf{S}_i \frac{\partial}{\partial \dot{s}_j} \left(\sum_l \frac{\partial \boldsymbol{\omega}_i}{\partial \dot{s}_l} \ddot{s}_l + \sum_l \frac{\partial \boldsymbol{\omega}_i}{\partial s_l} \dot{s}_l + \frac{\partial \boldsymbol{\omega}_i}{\partial t} \right) = \mathbf{S}_i \frac{\partial \boldsymbol{\omega}_i}{\partial s_j} \tag{9.92}$$

We use $\boldsymbol{\Omega}_i \boldsymbol{\omega}_i = \boldsymbol{\omega}_i \times \boldsymbol{\omega}_i = \mathbf{0}$ and $[\mathrm{d}\boldsymbol{\omega}_{ab}/\mathrm{d}t]_a = [\mathrm{d}\boldsymbol{\omega}_{ab}/\mathrm{d}t]_b$ from (4.3), expressing the independence of calculating the time derivative of angular velocity in the space of expression. Comparing (9.90) and (9.91), and using (9.92), we obtain

$$\mathbf{S}_i \boldsymbol{\Omega}_i \frac{\partial \boldsymbol{\omega}_i}{\partial \dot{s}_j} + \mathbf{S}_i \frac{\mathrm{d}}{\mathrm{d}t} \frac{\partial \boldsymbol{\omega}_i}{\partial \dot{s}_j} = \mathbf{S}_i \frac{\partial \boldsymbol{\omega}_i}{\partial s_j}$$

Hence

$$\frac{\mathrm{d}}{\mathrm{d}t} \frac{\partial \boldsymbol{\omega}_i}{\partial \dot{s}_j} = \frac{\partial \boldsymbol{\omega}_i}{\partial s_j} - \boldsymbol{\omega}_i \times \frac{\partial \boldsymbol{\omega}_i}{\partial \dot{s}_j} \tag{9.93}$$

Using (9.87), (9.88), (9.89) and (9.93), the left-hand side of Lagrange's equations for body i can be written in the from

$$\begin{aligned} \frac{\mathrm{d}}{\mathrm{d}t} \frac{\partial T_i}{\partial \dot{s}_j} - \frac{\partial T_i}{\partial s_j} &= m_i \dot{\mathbf{v}}_i^T \frac{\partial \mathbf{v}_i}{\partial \dot{s}_j} + m_i \mathbf{v}_i^T \frac{\partial \mathbf{v}_i}{\partial s_j} + \dot{\boldsymbol{\omega}}_i^T \mathbf{I}_i \frac{\partial \boldsymbol{\omega}_i}{\partial \dot{s}_j} + \\ &\quad + \boldsymbol{\omega}_i^T \mathbf{I}_i \left(\frac{\partial \boldsymbol{\omega}_i}{\partial s_j} - \boldsymbol{\omega}_i \times \frac{\partial \boldsymbol{\omega}_i}{\partial \dot{s}_j} \right) - m_i \mathbf{v}_i^T \frac{\partial \mathbf{v}_i}{\partial s_j} - \boldsymbol{\omega}_i^T \mathbf{I}_i \frac{\partial \boldsymbol{\omega}_i}{\partial s_j} = \\ &= (m_i \dot{\mathbf{v}}_i)^T \frac{\partial \mathbf{v}_i}{\partial \dot{s}_j} + \left[\dot{\boldsymbol{\omega}}_i^T \mathbf{I}_i \frac{\partial \boldsymbol{\omega}_i}{\partial \dot{s}_j} - \boldsymbol{\omega}_i^T \mathbf{I}_i \left(\boldsymbol{\omega}_i \times \frac{\partial \boldsymbol{\omega}_i}{\partial \dot{s}_j} \right) \right] \end{aligned} \tag{9.94}$$

With regard to the symmetry of the tensor of inertia $\mathbf{I}_i$ and the skew-symmetry of the angular velocity matrix $\boldsymbol{\Omega}_i$, we alter the last term in (9.94)

$$\begin{aligned} -\boldsymbol{\omega}_i^T \mathbf{I}_i \left(\boldsymbol{\omega}_i \times \frac{\partial \boldsymbol{\omega}_i}{\partial \dot{s}_j} \right) &= -\boldsymbol{\omega}_i^T \mathbf{I}_i \boldsymbol{\Omega}_i \frac{\partial \boldsymbol{\omega}_i}{\partial \dot{s}_j} = (-\boldsymbol{\Omega}_i^T \mathbf{I}_i^T \boldsymbol{\omega}_i)^T \frac{\partial \boldsymbol{\omega}_i}{\partial \dot{s}_j} = \\ &= (\boldsymbol{\Omega}_i \mathbf{I}_i \boldsymbol{\omega}_i)^T \frac{\partial \boldsymbol{\omega}_i}{\partial \dot{s}_j} = (\boldsymbol{\omega}_i \times \mathbf{I}_i \boldsymbol{\omega}_i)^T \frac{\partial \boldsymbol{\omega}_i}{\partial \dot{s}_j} \end{aligned} \tag{9.95}$$

Substituting (9.95) into (9.94) we finally obtain

$$\begin{aligned} \frac{d}{dt} \frac{\partial T_i}{\partial \dot{s}_j} - \frac{\partial T_i}{\partial s_j} &= (m_i \dot{\mathbf{v}}_i)^T \frac{\partial \mathbf{v}_i}{\partial \dot{s}_j} + \left[\dot{\boldsymbol{\omega}}_i^T \mathbf{I}_i^T \frac{\partial \boldsymbol{\omega}_i}{\partial \dot{s}_j} + (\boldsymbol{\omega}_i \times \mathbf{I}_i \boldsymbol{\omega}_i)^T \frac{\partial \boldsymbol{\omega}_i}{\partial \dot{s}_j} \right] = \\ &= (m_i \dot{\mathbf{v}}_i)^T \frac{\partial \mathbf{v}_i}{\partial \dot{s}_j} + (\mathbf{I}_i \dot{\boldsymbol{\omega}}_i + \boldsymbol{\omega}_i \times \mathbf{I}_i \boldsymbol{\omega}_i)^T \frac{\partial \boldsymbol{\omega}_i}{\partial \dot{s}_j} \end{aligned} \tag{9.96}$$

Comparing (9.96) and (9.84), (9.85) and (9.65), we obtain the agreement of the left-hand sides of Newton-Euler's and Lagrange's equations for body i by a linear combination given by vectors $\partial \mathbf{v}_i / \partial \dot{s}_j$, $\partial \boldsymbol{\omega}_i / \partial \dot{s}_j$ representing the projection of equations of motion into the space of chosen coordinates.

The left-hand sides of Lagrange's equations and the corresponding Newton-Euler's equations for the whole MBS (see the following section) are obtained simply by summing the expressions on the left-hand side and on the right-hand side of (9.96) over all bodies for individual coordinates j.

It remains to prove the agreement of the right-hand sides of the equations of motion. Let us start with body i. We substitute all forces and couples including those of reactions acting on body i by the resultant force $\mathbf{F}_i$ in its centre of mass and the resultant couple $\mathbf{M}_i$. We can then write according to the principle of virtual work for generalized force Q_j acting on coordinate s_j

$$Q_j \delta s_j = \mathbf{F}_i^T \delta \mathbf{r}_i + \mathbf{M}_i^T \delta \boldsymbol{\pi}_i \tag{9.97}$$

where $\delta \mathbf{r}_i$ is the virtual displacement of the centre of mass of body i expressed in the fixed system of coordinates of the frame, and $\delta \boldsymbol{\pi}_i$ is the virtual angular displacement of body i expressed in the coordinate system moving along with the i-th body. From (9.97)

$$Q_j = \mathbf{F}_i^T \frac{\partial \mathbf{r}_i}{\partial s_j} + \mathbf{M}_i^T \frac{\partial \boldsymbol{\pi}_i}{\partial s_j} \tag{9.98}$$

According to (9.88)

$$\mathbf{F}_i^T \frac{\partial \mathbf{r}_i}{\partial s_j} = \mathbf{F}_i^T \frac{\partial \mathbf{v}_i}{\partial \dot{s}_j} \tag{9.99}$$

Further

$$\delta \boldsymbol{\pi}_i = \boldsymbol{\omega}_i \delta t \tag{9.100}$$

and, as $\dot{\boldsymbol{\pi}}_i = \boldsymbol{\omega}_i$, we can write

$$\boldsymbol{\omega}_i = \sum_l \frac{\partial \boldsymbol{\pi}_i}{\partial s_l} \dot{s}_l + \frac{\partial \boldsymbol{\pi}_i}{\partial t} \tag{9.101}$$

Thus

$$\frac{\partial \boldsymbol{\omega}_i}{\partial \dot{s}_j} = \frac{\partial \boldsymbol{\pi}_i}{\partial s_j} \tag{9.102}$$

$$\mathbf{M}_i^T \frac{\partial \boldsymbol{\pi}_i}{\partial s_j} = \mathbf{M}_i^T \frac{\partial \boldsymbol{\omega}_i}{\partial \dot{s}_j} \tag{9.103}$$

From (9.88) to (9.103) we obtain

$$Q_j = \mathbf{F}_i^T \frac{\partial \mathbf{v}_i}{\partial \dot{s}_j} + \mathbf{M}_i^T \frac{\partial \boldsymbol{\omega}_i}{\partial \dot{s}_j} \tag{9.104}$$

which, compared with (9.96), (9.84), (9.85) and (9.65), proves the agreement of the right-hand sides of Lagrange's and Newton-Euler's equations of body i. We assume that coordinates s_j represent the generalized coordinates for each independently considered body i; thus for each body i these are Lagrange's equations of the second type. In the following, we are going to assume that the coordinates s_j are not independent for the whole MBS. For a body where that is the case, it would be necessary to use the following procedure for its equations of motion.

The equations of the whole MBS are obtained by summing equations (9.96) and (9.104) (their left-hand sides and their right-hand sides) over all bodies i for each coordinate s_j. The summation of the right-hand sides of the equations of motion given by relations (9.104) cancels the terms containing the reaction corresponding to disconnected (split) links (constraints) of MBS bodies, which are identically satisfied by the chosen coordinates. For the description using Newton-Euler's equations, this statement follows from the following. From the derivation of relations (9.104) from (9.97), it follows that relation (9.104) multiplied by δs_j expresses the virtual work of forces $\mathbf{F}_i$ and moments $\mathbf{M}_i$ on the coordinate s_j. The virtual work of forces is invariant to a change of expression of forces from one system of coordinates to another given by (8.8). Let $T_{1\alpha}$ and $T_{1\beta}$ be transformations from the frame space into spaces α and β. The virtual work of force $\mathbf{F}_\beta$ and moment $\mathbf{M}_\beta$ in space β is

$$\mathbf{F}_\beta^T \mathbf{S}_{1\beta}^{-1} \delta \mathbf{r}_{1\beta} + \mathbf{M}_\beta^T \delta \boldsymbol{\pi}_{1\beta} \tag{9.105}$$

where $\delta \mathbf{r}_{1\beta}$ is in space 1, $\mathbf{S}_{1\beta}^{-1} \delta \mathbf{r}_{1\beta}$ is in space β, and $\delta \boldsymbol{\pi}_{1\beta}$ is in space β. The virtual work of force $\mathbf{F}_\alpha$ and moment $\mathbf{M}_\alpha$ in space α is

$$\mathbf{F}_\alpha^T \mathbf{S}_{1\alpha}^{-1} \delta \mathbf{r}_{1\alpha} + \mathbf{M}_\alpha^T \delta \boldsymbol{\pi}_{1\alpha} \tag{9.106}$$

According to (8.9)

$$\mathbf{F}_\alpha = \mathbf{S}_{\alpha b} \mathbf{F}_\beta, \quad \mathbf{M}_\alpha = \mathbf{S}_{\alpha b} \mathbf{M}_\beta + \mathbf{R}_{\alpha b} \mathbf{S}_{\alpha b} \mathbf{F}_\beta \tag{9.107}$$

Because the virtual work does not depend on the coordinate system in which it is expressed, the expressions (9.105) and (9.106) are equal.

For the formulation of Newton-Euler's equations (9.84) and (9.85), we have substituted the reaction effects of kinematic pair constraints according to (9.107) in the origin of the system of coordinates connected with body i. We now sum the expression (9.104) expressing the virtual work of these reaction forces. This virtual work is independent of the system in which it is expressed. We express it again in the place where reaction forces act. For the reaction effect on the one body and on the other body the law of action and reaction holds. For example, between bodies i and k in place of contact β

$$\mathbf{F}_{\beta i} = -\mathbf{F}_{\beta k}, \quad \mathbf{M}_{\beta i} = -\mathbf{M}_{\beta k} \tag{9.108}$$

Summing (9.104) we obtain from (9.105)

$$\mathbf{F}_i^T \frac{\partial \mathbf{v}_i}{\partial \dot{s}_j} + \mathbf{M}_i^T \frac{\partial \boldsymbol{\omega}_i}{\partial \dot{s}_j} + \mathbf{F}_k^T \frac{\partial \mathbf{v}_k}{\partial \dot{s}_j} + \mathbf{M}_k^T \frac{\partial \boldsymbol{\omega}_k}{\partial \dot{s}_j} =$$

$$= \mathbf{F}_{\beta i}^T \mathbf{S}_{1i\beta}^{-1} \frac{\partial \mathbf{r}_{1i\beta}}{\partial s_j} + \mathbf{M}_{\beta i}^T \frac{\partial \boldsymbol{\pi}_{1i\beta}}{\partial s_j} + \mathbf{F}_{\beta k}^T \mathbf{S}_{1k\beta}^{-1} \frac{\partial \mathbf{r}_{1k\beta}}{\partial s_j} + \mathbf{M}_{\beta k}^T \frac{\partial \boldsymbol{\pi}_{1k\beta}}{\partial s_j} =$$

$$= \mathbf{F}_{\beta i}^T \left(\mathbf{S}_{1i\beta}^{-1} \frac{\partial \mathbf{r}_{1i\beta}}{\partial s_j} - \mathbf{S}_{1k\beta}^{-1} \frac{\partial \mathbf{r}_{1k\beta}}{\partial s_j} \right) + \mathbf{M}_{\beta i}^T \left(\frac{\partial \boldsymbol{\pi}_{1i\beta}}{\partial s_j} - \frac{\partial \boldsymbol{\pi}_{1k\beta}}{\partial s_j} \right) \quad (9.109)$$

where e.g. $\mathbf{r}_{1i\beta}$ denotes the expression of $\mathbf{r}_{1\beta}$ with respect to body i. Should the expressions in parentheses in (9.109) identically equal zero, the reaction $\mathbf{F}_{\beta i}$, $\mathbf{M}_{\beta i}$ would not appear in the resultant equations.

This is then the case when the expression of position $\mathbf{r}_{1\beta}$ and rotational position $\boldsymbol{\pi}_{1\beta}$ are identical expressing them with regard to both body i and body k which are in contact in constraints β. This is then the case when the choice of MBS description implicitly contains the connection of bodies i and k in location β.

Adding the left-hand sides of (9.104), we start by expressing the generalized forces Q_j according to (9.62) for bodies i and k. According to (9.108) these are reaction forces in general $\mathbf{R}_{\beta i}$, $\mathbf{R}_{\beta k}$ acting on coordinates $\mathbf{u}_{\beta i}$, $\mathbf{u}_{\beta k}$. Then, according to (9.62), we have the sum

$$\begin{aligned} \mathbf{Q}_j^i + \mathbf{Q}_j^k &= \sum_R \left(\mathbf{R}_{\beta i} \frac{\partial \mathbf{u}_{\beta i}}{\partial s_j} + \mathbf{R}_{\beta k} \frac{\partial \mathbf{u}_{\beta k}}{\partial s_j} \right) = \sum_R \mathbf{R}_{\beta i} \left(\frac{\partial \mathbf{u}_{\beta i}}{\partial s_j} - \frac{\partial \mathbf{u}_{\beta k}}{\partial s_j} \right) = \\ &= \sum_R \mathbf{R}_{\beta i} \frac{\partial (\mathbf{u}_{\beta i} - \mathbf{u}_{\beta k})}{\partial s_j} \end{aligned} \quad (9.110)$$

Again, should the expression $\mathbf{u}_{\beta i}(s_j, t) - \mathbf{u}_{\beta k}(s_i, t)$ be identically satisfied by the chosen MBS description using s_j, the sum (9.110) is identically equal to zero and does not occur in Lagrange's equations. In the contrary case we set $\boldsymbol{\lambda} = \mathbf{R}_{\beta i}$ and $\mathbf{f}(s_j, t) = \mathbf{u}_{\beta i}(s_j, t) - \mathbf{u}_{\beta k}(s_j, t)$, obtaining a term of the right-hand side of LEMT.

This finally proves the agreement of the right-hand sides of LEMT for the whole MBS and of the sum of right-hand sides of Newton-Euler's equations multiplied by terms $\partial \mathbf{v}_i / \partial \dot{s}_j$, $\partial \boldsymbol{\omega}_i / \partial \dot{s}_j$

$$Q_j + \sum_{k=1}^{r} \lambda_k \frac{\partial g_k}{\partial s_j^{(h)}} = \sum_i \left(\mathbf{F}_i^T \frac{\partial \mathbf{v}_i}{\partial \dot{s}_j} + \mathbf{M}_i^T \frac{\partial \boldsymbol{\omega}_i}{\partial \dot{s}_j} \right) \quad (9.111)$$

where Q_j contains only work forces and no reactions. The agreement of the left-hand sides is expressed by the relation

$$\frac{\mathrm{d}}{\mathrm{dt}} \frac{\partial T}{\partial \dot{s}_j} - \frac{\partial T}{\partial s_j} = \sum_i \left((m_i \dot{\mathbf{v}}_i)^T \frac{\partial \mathbf{v}_i}{\partial \dot{s}_j} + (\mathbf{I}_i \dot{\boldsymbol{\omega}}_i + \boldsymbol{\omega}_i \times \mathbf{I}_i \boldsymbol{\omega}_i)^T \frac{\partial \boldsymbol{\omega}_i}{\partial \dot{s}_j} \right) \quad (9.112)$$

where

$$T = \sum_i T_i$$

We now convert the equivalence relations obtained (9.111) and (9.112) into a form using quasi-coordinates from Section 9.3. We have obtained equations (9.78) from the sum of equation (9.65) multiplied by function β_{jk}. Using the same procedure we obtain from (9.112)

$$\frac{\mathrm{d}}{\mathrm{d}t}\frac{\partial T'}{\partial \dot{\pi}_{\varkappa}} - \frac{\partial T'}{\partial \pi_{\varkappa}} + \sum_{\mu=1}^{n}\sum_{j=1}^{n} \gamma_{\varkappa\mu j}\frac{\partial T'}{\partial \dot{\pi}_{\mu}}\dot{\pi}_j = \sum_i \left(\sum_{j=1}^{n} \beta_{j\varkappa}(m_i \dot{\mathbf{v}}_i)^T \frac{\partial \mathbf{v}_i}{\partial \dot{s}_j} + \right.$$

$$\left. + \sum_{j=1}^{n} \beta_{j\varkappa}(\mathbf{I}_i\dot{\boldsymbol{\omega}}_i + \boldsymbol{\omega}_i \times \mathbf{I}_i\boldsymbol{\omega}_i)^T \frac{\partial \boldsymbol{\omega}_i}{\partial \dot{s}_j} \right) = \sum_i \left((m_i\dot{\mathbf{v}}_i)^T \sum_{j=1}^{n} \beta_{j\varkappa}\frac{\partial \mathbf{v}_i}{\partial \dot{s}_j} + \right.$$

$$\left. + (\mathbf{I}_i\dot{\boldsymbol{\omega}}_i + \boldsymbol{\omega}_i \times \mathbf{I}_i\boldsymbol{\omega}_i)^T \sum_{j=1}^{n} \beta_{j\varkappa}\frac{\partial \boldsymbol{\omega}_i}{\partial \dot{s}_j} \right) \tag{9.113}$$

Further, from (9.74),

$$\begin{aligned} \mathbf{v}_i &= \frac{\mathrm{d}}{\mathrm{d}t}\mathbf{r}_i = \sum_{j=1}^{n} \frac{\partial \mathbf{r}_i}{\partial s_j}\dot{s}_j + \frac{\partial \mathbf{r}_i}{\partial t} = \sum_{j=1}^{n} \frac{\partial \mathbf{r}_i}{\partial s_j} \sum_{\varkappa=1}^{n} \beta_{j\varkappa}\dot{\pi}_{\varkappa} + \frac{\partial \mathbf{r}_i}{\partial t} = \\ &= \sum_{\varkappa=1}^{n} \left(\sum_{j=1}^{n} \frac{\partial \mathbf{r}_i}{\partial s_j}\beta_{j\varkappa} \right) \dot{\pi}_{\varkappa} + \frac{\partial \mathbf{r}_i}{\partial t} = \mathbf{v}_i' \end{aligned} \tag{9.114}$$

Here $\mathbf{v}_i'$ is the velocity of the centre of mass of body i expressed in quasi-coordinates. Hence

$$\begin{aligned} \frac{\partial \mathbf{v}_i}{\partial \dot{s}_j} &= \frac{\partial \mathbf{r}_i}{\partial s_j} \\ \sum_{j=1}^{n} \beta_{j\varkappa}\frac{\partial \mathbf{v}_i}{\partial \dot{s}_j} &= \sum_{j=1}^{n} \beta_{j\varkappa}\frac{\partial \mathbf{r}_i}{\partial s_j} = \frac{\partial \mathbf{v}_i'}{\partial \dot{\pi}_{\varkappa}} \end{aligned} \tag{9.115}$$

and similarly for $\boldsymbol{\omega}_i$. The derivative of momentum and moment of momentum are independent of the choice of coordinates. Hence, from (9.113)

$$\begin{aligned} \frac{\mathrm{d}}{\mathrm{d}t}\frac{\partial T'}{\partial \dot{\pi}_{\varkappa}} &- \frac{\partial T'}{\partial \pi_{\varkappa}} + \sum_{\mu=1}^{n}\sum_{k=1}^{n} \gamma_{\varkappa\mu k}\frac{\partial T'}{\partial \dot{\pi}_{\mu}}\dot{\pi}_k = \\ &= \sum_i \left((m_i\dot{\mathbf{v}}_i')^T \frac{\partial \mathbf{v'}_i}{\partial \dot{\pi}_{\varkappa}} + (\mathbf{I}_i\dot{\boldsymbol{\omega}}_i' + \boldsymbol{\omega}_i' \times \mathbf{I}_i\boldsymbol{\omega}_i')^T \frac{\partial \boldsymbol{\omega}_i'}{\partial \dot{\pi}_{\varkappa}} \right) \end{aligned} \tag{9.116}$$

Similarly from (9.111) and (9.115)

$$Q'_j + \sum_{k=1}^{r} \lambda_k \frac{\partial g'_k}{\partial \pi_{\varkappa}^{(h)}} = \sum_i \left(\mathbf{F}_i^T \sum_{j=1}^{n} \beta_{j\varkappa} \frac{\partial \mathbf{v}_i}{\partial \dot{s}_j} + \mathbf{M}_i^T \sum_{j=1}^{n} \beta_{j\varkappa} \frac{\partial \boldsymbol{\omega}_i}{\partial \dot{s}_j} \right)$$

$$Q'_j + \sum_{k=1}^{r} \lambda_k \frac{\partial g'_k}{\partial \pi_{\varkappa}^{(h)}} = \sum_i \left(\mathbf{F}_i^T \frac{\partial \mathbf{v}'_i}{\partial \dot{\pi}_{\varkappa}} + \mathbf{M}_i^T \frac{\partial \boldsymbol{\omega}'_i}{\partial \dot{\pi}_{\varkappa}} \right) \quad (9.117)$$

The left-hand sides of (9.116) to (9.117) represent LEMT expressed in quasi-coordinates for the whole MBS, and the right-hand sides represent the new form of Newton–Euler's equations for the whole MBS. These are the so-called Kane's equations generalized for MBS with constraints (Valášek and Stejskal, 1987; Wang and Huston, 1987) which is the subject of the following section.

9.5 Various Multibody Formalisms

There are two important groups of multibody formalisms – non-recursive and recursive. The notion of recursion is used with many meanings in multibody system theory. We shall use the adjective recursive in the following way.

The non-recursive multibody formalisms are formalisms which assemble the generalized mass matrix $\mathbf{M}$ and the vector $\mathbf{Q}$ of generalized and applied forces for the formulation of equations of motion into the form

$$\mathbf{M}\ddot{\mathbf{q}} = \mathbf{Q} \quad (9.118)$$

The system of differential equations of motion can be viewed as a system of n linear algebraic equations for the accelerations $\ddot{\mathbf{q}}$. These formalisms must always solve this system of linear algebraic equations (mostly by the Choleski algorithm) in order to integrate the accelerations, but this needs $O(n^3)$ operations every time and is not efficient in comparison with recursive formalisms at least for large number n of degrees of freedom.

The recursive multibody formalisms are formalisms which do not need to assemble and solve a system of linear algebraic equations (9.118) but solve the dynamic problem for the acceleration by a sequence of recursive statements for which the computational complexity is only $O(n)$. This is at least asymptotically very efficient.

In the following text we shall describe the individual formalisms using symbols which are in common use when dealing with those formalisms. These symbols are not fully consistent with the previous text.

9.5.1 Non-recursive Multibody Formalisms

We present here a survey of various non-recursive multibody formalisms for the assembly of MBS equations of motion from the viewpoint of structural equivalence of Newton-Euler's and Langrange's equations. All these various formalisms are obtained from the equivalences (9.112), (9.111) and (9.116), (9.117) by various choices of MBS descriptions, i.e. various choices of coordinates.

9.5.1.1 NEWEUL Approach

Let us choose generalized, i.e. independent, coordinates for the MBS description. The left-hand sides of equivalences (9.112), 9.111) lead to well known Lagrange's equations of the second type (9.48) as a special case of LEMT with the absence of constraints. The right-hand sides of equivalences (9.112), (9.111) lead to the Schiehlen-Kreuzer formalism (Schiehlen and Kreuzer, 1978; Kreuzer, 1979; Schiehlen, 1984) in the NEWEUL system. We assemble in this formalism Newton's and Euler's equations of motion for each freed body in the coordinate system of the frame. According to the notation in Section 9.4 we have

$$
\begin{gathered}
m_i \dot{\mathbf{v}}_i = \mathbf{F}_i^a + \mathbf{F}_i^r \\
\mathbf{S}_i \mathbf{I}_i \mathbf{S}_i^T (\mathbf{S}_i \dot{\boldsymbol{\omega}}_i) + (\mathbf{S}_i \boldsymbol{\omega}_i) \times (\mathbf{S}_i \mathbf{I}_i \mathbf{S}_i^T)(\mathbf{S}_i \boldsymbol{\omega}_i) = \mathbf{S}_i \mathbf{M}_i^a + \mathbf{S}_i \mathbf{M}_i^r
\end{gathered}
\tag{9.119}
$$

where superscript a denotes action effects, and superscript r reaction effect of MBS constraints. The second equation of (9.119) is the Euler equation (9.17) or (9.85) transformed into the system of coordinates of the frame. We express all quantities in the equations using the chosen generalized coordinates $\mathbf{q}$

$$
\begin{aligned}
\mathbf{r}_i &= \mathbf{r}_i(\mathbf{q}, t) \\
\mathbf{v}_i &= \dot{\mathbf{r}}_i = \frac{\partial \mathbf{r}_i}{\partial \mathbf{q}} \dot{\mathbf{q}} + \frac{\partial \mathbf{r}_i}{\partial t} = \mathbf{J}_{Ti} \dot{\mathbf{q}} + \frac{\partial \mathbf{r}_i}{\partial t} \\
\dot{\mathbf{v}}_i &= \mathbf{J}_{Ti} \ddot{\mathbf{q}} + \dot{\mathbf{q}}^T \frac{\partial \mathbf{J}_{Ti}}{\partial \mathbf{q}} \dot{\mathbf{q}} + \frac{\partial^2 \mathbf{r}_i}{\partial \mathbf{q} \partial t} \dot{\mathbf{q}} + \frac{\partial^2 \mathbf{r}_i}{\partial t^2} \\
\boldsymbol{\pi}_i &= \boldsymbol{\pi}_i(\mathbf{q}, t) \\
\mathbf{S}_i \boldsymbol{\omega}_i &= \dot{\boldsymbol{\pi}}_i = \frac{\partial \boldsymbol{\pi}_i}{\partial \mathbf{q}} \dot{\mathbf{q}} + \frac{\partial \boldsymbol{\pi}_i}{\partial t} = \mathbf{J}_{Ri} \dot{\mathbf{q}} + \frac{\partial \boldsymbol{\pi}_i}{\partial t} \\
\mathbf{S}_i \dot{\boldsymbol{\omega}}_i &= \mathbf{J}_{Ri} \ddot{\mathbf{q}} + \dot{\mathbf{q}}^T \frac{\partial \mathbf{J}_{Ri}}{\partial \mathbf{q}} \dot{\mathbf{q}} + \frac{\partial^2 \boldsymbol{\pi}_i}{\partial \mathbf{q} \partial t} \dot{\mathbf{q}} + \frac{\partial^2 \boldsymbol{\pi}_i}{\partial t^2}
\end{aligned}
\tag{9.120}
$$

We then add equations (9.119), obtaining

$$
\bar{\mathbf{M}}(\mathbf{q}, t) \mathbf{J}(\mathbf{q}, t) \ddot{\mathbf{q}} + \bar{\mathbf{q}}^C(\mathbf{q}, \dot{\mathbf{q}}, t) = \bar{\mathbf{q}}^a(t) + \bar{\mathbf{Q}} \mathbf{g}(t) \tag{9.121}
$$

where $\bar{\mathbf{M}}(\mathbf{q},t)$ is the matrix of mass of the whole MBS, $\mathbf{J}(\mathbf{q},t)$ is Jacobi's matrix of transmission indixes of the whole MBS, $\bar{\mathbf{q}}^C(\mathbf{q},\dot{\mathbf{q}},t)$ contains the remaining terms of the left-hand sides expressing Coriolis' and centrifugal forces, $\bar{\mathbf{q}}^a(t)$ expresses action forces acting on the MBS and $\bar{\mathbf{Q}}\mathbf{g}(t)$ expresses reaction forces of MBS constraints given by the distribution matrix $\bar{\mathbf{Q}}$ of generalized constraint forces $\mathbf{g}(t)$. The virtual work of the reaction forces of ideal MBS constraints according to d'Alembert's principle equals zero, i.e.

$$\delta W = \delta\mathbf{r}_i^T\mathbf{F}_i^R = \delta\mathbf{q}^T\mathbf{J}^T(\mathbf{q},t)\bar{\mathbf{Q}}\mathbf{g}(t) = 0 \tag{9.122}$$

Due to the independence of generalized coordinates $\mathbf{q}$ and generalized constraint forces $\mathbf{g}(t)$

$$\mathbf{J}^T(\mathbf{q},t)\bar{\mathbf{Q}} = \mathbf{0} \tag{9.123}$$

Therefore, premultiplying (9.121) by $\mathbf{J}^T(\mathbf{q},t)$, we finally obtain

$$\mathbf{J}^T(\mathbf{q},t)\bar{\mathbf{M}}(\mathbf{q},t)\mathbf{J}(\mathbf{q},t)\ddot{\mathbf{q}} + \mathbf{J}^T(\mathbf{q},t)\bar{\mathbf{q}}^C(\mathbf{q},\dot{\mathbf{q}},t) = \mathbf{J}^T(\mathbf{q},t)\bar{\mathbf{q}}^a(t) \tag{9.124}$$

Equations (9.124) follow directly from the right-hand sides of equations (9.112) and (9.111) after substituting the MBS description (9.120). Similarly, the analogical equations of motion written in quasi-coordinates can be derived by the application of Jourdain's principle instead of d'Alembert's principle. These equations are the same as Kane's equations.

Generalized constraint forces can also be expressed by premultiplying equations (9.121) by $\bar{\mathbf{Q}}^T\bar{\mathbf{M}}^{-1}(\mathbf{q},t)$. According to (9.123) we obtain

$$\bar{\mathbf{Q}}^T\bar{\mathbf{M}}^{-1}(\mathbf{q},t)\bar{\mathbf{q}}^C(\mathbf{q},\dot{\mathbf{q}},t) = \bar{\mathbf{Q}}^T\bar{\mathbf{M}}^{-1}(\mathbf{q},t)\bar{\mathbf{q}}^a(t) + \bar{\mathbf{Q}}^T\bar{\mathbf{M}}^{-1}(\mathbf{q},t)\bar{\mathbf{Q}}\mathbf{g}(t)$$

$$\mathbf{g}(t) = (\bar{\mathbf{Q}}^T\bar{\mathbf{M}}^{-1}(\mathbf{q},t)\bar{\mathbf{Q}})^{-1}\bar{\mathbf{Q}}^T\bar{\mathbf{M}}^{-1}(\mathbf{q},t)(\bar{\mathbf{q}}^C(\mathbf{q},\dot{\mathbf{q}},t) - \bar{\mathbf{q}}^a(t)) \tag{9.125}$$

The important idea in the NEWEUL approach (Kreuzer, 1979; Schmoll, 1988; Bestle, 1989) and the MESA VERDE approach (Wittenburg and Wolz, 1985 ab) to MBS with kinematic loops is the application of so-called help variables. For the description of all quantities in (9.120) we use some coordinates which consist of independent (generalized) $\mathbf{q}$ and dependent (help) $\mathbf{h}$ coordinates. The dependency of help coordinates $\mathbf{h}$ on the generalized coordinates $\mathbf{q}$ can generally be computed numerically only in the proper solution of MBS motion. Nevertheless, the equations of motion may be formulated using both coordinates $\mathbf{q}$ and $\mathbf{h}$ together with the formal symbolical expressions $\mathbf{h}(\mathbf{q})$, $\partial\mathbf{h}/\partial\mathbf{q}$ and $\partial^2\mathbf{h}/\partial\mathbf{q}^2$, for example, the equations (9.120) are

$$\begin{aligned} \mathbf{r}_i &= \mathbf{r}_i(\mathbf{q},\mathbf{h}(\mathbf{q}),t) \\ \mathbf{v}_i &= \mathbf{v}_i(\mathbf{q},\mathbf{h}(\mathbf{q}),\dot{\mathbf{q}},\partial\mathbf{h}/\partial\mathbf{q},t) \end{aligned}$$

$$\begin{aligned}
\dot{\mathbf{v}}_i &= \mathbf{a}_i(\mathbf{q},\mathbf{h}(\mathbf{q}),\dot{\mathbf{q}},\ddot{\mathbf{q}},\partial\mathbf{h}/\partial\mathbf{q},\partial^2\mathbf{h}/\partial\mathbf{q}^2,t)\\
\boldsymbol{\pi}_i &= \boldsymbol{\pi}_i(\mathbf{q},\mathbf{h}(\mathbf{q}),t)\\
\mathbf{S}_i\boldsymbol{\omega}_i &= \boldsymbol{\omega}_{1,i}(\mathbf{q},\mathbf{h}(\mathbf{q}),\dot{\mathbf{q}},\partial\mathbf{h}/\partial\mathbf{q},t)\\
\mathbf{S}_i\dot{\boldsymbol{\omega}}_i &= \boldsymbol{\alpha}_{1,i}(\mathbf{q},\mathbf{h}(\mathbf{q}),\dot{\mathbf{q}},\ddot{\mathbf{q}},\partial\mathbf{h}/\partial\mathbf{q},\partial^2\mathbf{h}/\partial\mathbf{q}^2,t)
\end{aligned} \tag{9.126}$$

and similarly the equations of motion (9.124). The formal expressions are computed numerically in the integration of equations of motion from the constraints and their time derivatives

$$\begin{aligned}
\mathbf{f}(\mathbf{q},\mathbf{h},t) &= \mathbf{0}\\
\frac{d\mathbf{f}}{dt} &= \mathbf{f}_v(\mathbf{q},\mathbf{h}(\mathbf{q}),\dot{\mathbf{q}},\frac{\partial\mathbf{h}}{\partial\mathbf{q}},t)=\mathbf{0}\\
\frac{d^2\mathbf{f}}{dt^2} &= \mathbf{f}_a\left(\mathbf{q},\mathbf{h}(\mathbf{q}),\dot{\mathbf{q}},\ddot{\mathbf{q}},\frac{\partial\mathbf{h}}{\partial\mathbf{q}},\frac{\partial^2\mathbf{h}}{\partial\mathbf{q}^2},t\right)=\mathbf{0}
\end{aligned} \tag{9.127}$$

The coordinates used can be divided into generalized and help coordinates before the proper numerical solution is performed. However, such division is often not valid during the whole motion of MBS (see also Chapter 7, Section 9.5.1.4, 9.5.1.5, and 9.8) and must be repeated. This makes it necessary to derive several symbolical variants of equations of motion according to different divisions of coordinates and to switch among these equations of motion during the numerical solution. A new and promising approach (Leister, 1990), is to consider all coordinates as help coordinates and to introduce the symbolical dependencies $\mathbf{h}(\mathbf{q})$, $\partial\mathbf{h}/\partial\mathbf{q}$ and $\partial\mathbf{h}/\partial\mathbf{q}^2$ quite formally in the corresponding theoretical number of generalized coordinates. Then not only the values of these dependencies but even the choice of independent coordinates in the form of some combination of introduced coordinates can be computed numerically during the numerical solution (see also Section 9.8). We than need just one variant of the equations of motion.

Let us now solve the example from Section 9.3 (Figure 9.2) by this formalism. We use all the coordinates φ_1, φ_2 and φ_3 as the help coordina- which are the functions of some generalized coordinate q. We express (9.120)

$$\begin{aligned}
\mathbf{r}_1 &= [p_1 s\varphi_1, p_1 c\varphi_1]\\
\mathbf{v}_1 &= [p_1 c\varphi_1\varphi_{1q}\dot{q} - p_1 s\varphi_1\varphi_{1q}\dot{q}]\\
\mathbf{a}_1 &= [p_1 c\varphi_1\varphi_{1q}\ddot{q} + (p_1 c\varphi_1\varphi_{1qq} - p_1 s\varphi_1\varphi_{1q}^2)\dot{q}^2, -p_1 s\varphi_1\varphi_{1q}\ddot{q} +\\
&\quad +(-p_1 s\varphi_1\varphi_{1qq} - p_1 c\varphi_1\varphi_{1q}^2)\dot{q}^2]\\
\mathbf{r}_2 &= [l_1 s\varphi_1 - p s\varphi_2, l_1 c\varphi_1 + p c\varphi_2]\\
\mathbf{v}_2 &= [l_1 c\varphi_1\varphi_{1q}\dot{q} - p c\varphi_2\varphi_{2q}\dot{q}, -l_1 s\varphi_1\varphi_{1q}\dot{q} - p s\varphi_2\varphi_{2q}\dot{q}]\\
\mathbf{a}_2 &= [(l_1 c\varphi_1\varphi_{1q} - p c\varphi_2\varphi_{2q})\ddot{q} + (l_1 c\varphi_1\varphi_{1qq} - l_1 s\varphi_1\varphi_{1q}^2 -
\end{aligned}$$

$$
\begin{aligned}
& -pc\varphi_2\varphi_{2qq} + ps\varphi_2\varphi_{2q}^2)\dot{q}^2, (-l_1 s\varphi_1\varphi_{1q} - ps\varphi_2\varphi_{2q})\ddot{q} + \\
& +(-l_1 s\varphi_1\varphi_{1qq} - l_1 c\varphi_1\varphi_{1q} - ps\varphi_2\varphi_{2qq} - pc\varphi_2\varphi_{2q}^2)\dot{q}^2] \\
\mathbf{r}_3 &= [p_3 s\varphi_3, l_0 - p_3 c\varphi_3] \\
\mathbf{v}_3 &= [p_3 c\varphi_3\varphi_{3q}\dot{q}, p_3 s\varphi_3\varphi_{3q}\dot{q}] \\
\mathbf{a}_3 &= [p_3 c\varphi_{3q}\ddot{q} + (p_3 c\varphi_3\varphi_{3qq} - p_3 s\varphi_3\varphi_{3q}^2)\dot{q}^2, p_3 s\varphi_3\varphi_{3q}\ddot{q} + \\
& +(p_3 s\varphi_3\varphi_{3qq} + p_3 c\varphi_3\varphi_{3q}^2)\dot{q}^2]
\end{aligned}
$$

$$
\begin{array}{lll}
\pi_1 = -\varphi_1 + \frac{\pi}{2}, & \omega_1 = -\varphi_{1q}\dot{q}, & \alpha_1 = -\varphi_{1q}\ddot{q} - \varphi_{1qq}\dot{q}^2, \\
\pi_2 = \varphi_2 - \frac{\pi}{2}, & \omega_2 = \varphi_{2q}\dot{q}, & \alpha_2 = \varphi_{2q}\ddot{q} + \varphi_{2qq}\dot{q}^2, \\
\pi_3 = \varphi_3 - \frac{\pi}{2}, & \omega_3 = \varphi_{3q}\dot{q}, & \alpha_3 = \varphi_{3q}\ddot{q} + \varphi_{3qq}\dot{q}^2
\end{array} \tag{a}
$$

Hence, for example, we have

$$
\begin{aligned}
\mathbf{J}_{T1} &= [p_1 c\varphi_1\varphi_{1q}, -p_1 s\varphi_1\varphi_{1q}] \\
\mathbf{J}_{R1} &= [-\varphi_{1q}]
\end{aligned} \tag{b}
$$

Further

$$
\begin{aligned}
\bar{\mathbf{M}} &= [m_1, m_1, m_2, m_2, m_3, m_3, I_{1T_1}, I_{2T}, I_{3T_3}] \\
\mathbf{J} &= [\mathbf{J}_{T1}, \mathbf{J}_{T2}, \mathbf{J}_{T3}, \mathbf{J}_{R1}, \mathbf{J}_{R2}, \mathbf{J}_{R3}]
\end{aligned} \tag{c}
$$

Finally we obtain

$$
\begin{aligned}
\mathbf{J}^T\mathbf{M}\mathbf{J} &= (m_1 p_1^2 + I_{1T_1})\varphi_{1q}^2 + (m_3 p_3^2 + I_{3T_3})\varphi_{3q}^2 + \\
& +m_2(l_1^2\varphi_{1q}^2 + p^2\varphi_{2q}^2 - 2l_1 pc(\varphi_1 + \varphi_2)\varphi_{1q}\varphi_{2q}) + I_{2T}\varphi_{2q}^2 \\
\mathbf{J}^T\bar{\mathbf{q}}^C &= ((p_1 c\varphi_1\varphi_{1q}(p_1 c\varphi_1\varphi_{1qq} - p_1 s\varphi_1\varphi_{1q}^2) + \\
& +p_1 s\varphi_1\varphi_{1q}(p_1 s\varphi_1\varphi_{1qq} + p_1 c\varphi_1\varphi_{1q}^2))m_1 + \\
& +((l_1 c\varphi_1\varphi_{1q} - pc\varphi_2\varphi_{2q})(l_1 c\varphi_1\varphi_{1qq} - l_1 s\varphi_1\varphi_{1q}^2 - \\
& -pc\varphi_2\varphi_{2qq} + ps\varphi_2\varphi_{2q}^2) + (l_1 s\varphi_1\varphi_{1q} + ps\varphi_2\varphi_{2q})(l_1 s\varphi_1\varphi_{1qq} + \\
& +l_1 c\varphi_1\varphi_{1q}^2 + ps\varphi_2\varphi_{2qq} + pc\varphi_2\varphi_{2q}^2))m_2 + \\
& +(p_3 c\varphi_3\varphi_{3q}(p_3 c\varphi_3\varphi_{3qq} - p_3 s\varphi_3\varphi_{3q}^2) + \\
& +p_3 s\varphi_3\varphi_{3q}(p_3 s\varphi_3\varphi_{3qq} + p_3 c\varphi_3\varphi_{3q}^2))m_3 + \varphi_{1q}\varphi_{1qq}I_{1T_1} + \\
& +\varphi_{2q}\varphi_{2qq}I_{2T} + \varphi_{3q}\varphi_{3qq}I_{3T_3})\dot{q}^2 \\
\mathbf{J}^T\bar{\mathbf{q}}^a &= [p_1 c\varphi_1\varphi_{1q}, -p_1 s\varphi_1\varphi_{1q}, \ldots, -\varphi_{1q}, \ldots] \\
& [Ff_x(\varphi_1), Ff_y(\varphi_1), 0, 0, 0, 0, Ff_\varphi(\varphi)_1, 0, 0]
\end{aligned} \tag{d}
$$

where $f_x(\varphi_1)$, $f_y(\varphi_1)$ and $f_\varphi(\varphi_1)$ express the action of a unique spring force on the body 1 resolved in the centre of mass T_1. The partial derivatives

φ_{iq} and φ_{iqq} are determined from the partial derivatives of the constraints (c) in Section 9.3., for example

$$\begin{aligned}\frac{\partial f_1}{\partial q} &= l_1 c\varphi_1\varphi_{1q} - l_2 c\varphi_2\varphi_{2q} - l_3 c\varphi_3\varphi_{3q} = 0\\ \frac{\partial^2 f_1}{\partial q^2} &= l_1 c\varphi_1\varphi_{1qq} - l_2 c\varphi_2\varphi_{2qq} - l_3 c\varphi_3\varphi_{3qq} - l_1 s\varphi_1\varphi_{1q}^2 + \\ &\quad + l_2 s\varphi_2\varphi_{2q}^2 + l_3 s\varphi_3\varphi_{3q}^2 = 0 \qquad \text{(e)}\end{aligned}$$

and from the numerical solution of the independent coordinate (see Chapter 7 and Section 9.8). This solution has, for example, the form

$$q = \varkappa_1\varphi_1 + \varkappa_2\varphi_2 + \varkappa_3\varphi_3 \qquad \text{(f)}$$

where $\varkappa_i$ are constants. Hence, the equations (e) are completed by

$$\begin{aligned}\varkappa_1\varphi_{1q} + \varkappa_2\varphi_{2q} + \varkappa_3\varphi_{3q} - 1 &= 0\\ \varkappa_1\varphi_{1qq} + \varkappa_2\varphi_{2qq} + \varkappa_3\varphi_{3qq} &= 0 \qquad \text{(g)}\end{aligned}$$

9.5.1.2 MESA VERDE Approach

A special choice of generalized coordinates yields from the right-hand sides of (9.112), (9.111) the Roberson-Wittenburg formalism (Wittenburg, 1977; Roberson and Schwertassek, 1987; Wittenburg and Wolz, 1985 ab). In this formalism we assemble Newton's and Euler's equations for each body and for the MBS in the form (9.84), (9.85) and use Jourdain's principle

$$\sum_{i=1}^{n}(\delta\dot{\mathbf{r}}_i^T(m_i\ddot{\mathbf{r}}_i - \mathbf{F}_i) + \delta\boldsymbol{\omega}_i^T(\mathbf{I}_i\dot{\boldsymbol{\omega}}_i + \boldsymbol{\omega}_i \times \mathbf{I}_i\boldsymbol{\omega}_i - \mathbf{M}_i)) - \delta P = 0 \qquad (9.128)$$

where δP is the virtual power of all MBS internal forces (e.g. springs, dampers and motors), and $\mathbf{F}_i$ and $\mathbf{M}_i$ are the resultant external force and moment acting on body i. We convert the equations (9.128) into matrix form:

$$\begin{aligned}\delta\dot{\mathbf{r}}^T(\mathbf{m}\ddot{\mathbf{r}} - \mathbf{F}) + \delta\boldsymbol{\omega}^T(\mathbf{I}\dot{\boldsymbol{\omega}} + \mathbf{U} - \mathbf{M}) - \delta P = 0\\ \mathbf{U} = \lceil\boldsymbol{\omega}_i \times \mathbf{I}_i\boldsymbol{\omega}_i\rceil \qquad (9.129)\end{aligned}$$

where $\mathbf{m}$ and $\mathbf{I}$ are diagonal matrices of masses and tensors of inertia, and the others are column matrices of vectors. This equation ends the dynamic part of the problem; what remains to be done is the kinematic expression of the respective quantities.

The basis for expressing the kinematic relations is the assignment of a reduced system with a tree structure to the whole MBS and the formulation of additional constraint conditions to satisfy MBS loop closures.

The individual MBS loops are disconnected either by removing a chosen kinematic pair in the loop and describing the connection condition of the pair, or by splitting an arbitrary body in the loop into two partial bodies with identical dimensions and half mass each, with each containing one of the two kinematic pairs of the loop. The constraint condition of the loop closure is given in the second case by coincidence of both partial bodies. The coordinates used for the MBS description are formed by the relative coordinates of single kinematic pairs of the reduced system. Let $\dot{\mathbf{q}}$ be the column matrix of all velocity coordinates of both generalized velocities and quasicomponents of angular velocities. We may express

$$\begin{array}{lll} \delta\dot{\mathbf{r}} = \boldsymbol{\alpha}\delta\dot{\mathbf{q}} & \delta\boldsymbol{\omega} = \boldsymbol{\beta}\delta\dot{\mathbf{q}} & \\ \ddot{\mathbf{r}} = \boldsymbol{\alpha}\ddot{\mathbf{q}} + \mathbf{u} & \dot{\boldsymbol{\omega}} = \boldsymbol{\beta}\ddot{\mathbf{q}} + \mathbf{y} & \delta P = \delta\dot{\mathbf{q}}^T\mathbf{Q} \end{array} \tag{9.130}$$

Substituting (9.130) into (9.129), we obtain

$$\delta\dot{\mathbf{q}}^T[\underbrace{(\boldsymbol{\alpha}^T\mathbf{m}\boldsymbol{\alpha} + \boldsymbol{\beta}^T\mathbf{I}\boldsymbol{\beta})}_{\mathbf{A}}\ddot{\mathbf{q}} + \underbrace{\boldsymbol{\alpha}^T(\mathbf{mu} - \mathbf{F}) + \boldsymbol{\beta}^T(\mathbf{Iy} + \mathbf{U} - \mathbf{M}) - \mathbf{Q}}_{-\mathbf{B}}] = 0 \tag{9.131}$$

Due to the independence of $\delta\dot{\mathbf{q}}$, the equations of motion for the reduced system follow from (9.131)

$$\mathbf{A}\ddot{\mathbf{q}} = \mathbf{B} \tag{9.132}$$

The MBS description is composed of the reduced system and constraint conditions. The holonomic constraint conditions can be written in the form

$$\mathbf{f}(\mathbf{q}) = \mathbf{0} \tag{9.133}$$

Hence, differentiating with respect to time

$$\begin{aligned} \mathbf{J}^*\dot{\mathbf{q}} &= \mathbf{0} \\ \mathbf{J}^*\ddot{\mathbf{q}} + \mathbf{H}^* &= \mathbf{0} \end{aligned} \tag{9.134}$$

Solving equations (9.134), we find dependent variables $\mathbf{q}^{(2)}$ and independent variables $\mathbf{q}^{(1)}$

$$\begin{aligned} \dot{\mathbf{q}}^{(2)} &= \mathbf{J}\dot{\mathbf{q}}^{(1)} \\ \ddot{\mathbf{q}}^{(2)} &= \mathbf{J}\ddot{\mathbf{q}}^{(1)} + \mathbf{H} \end{aligned} \tag{9.135}$$

Nonholonomic constraints are in most cases described straight away in the form (9.135). We supply identities to equations (9.135) and express them in the form

$$\dot{\mathbf{q}} = \mathbf{J}_1\dot{\mathbf{q}}^{(1)}, \qquad \delta\dot{\mathbf{q}} = \mathbf{J}_1\delta\dot{\mathbf{q}}^{(1)}, \qquad \ddot{\mathbf{q}} = \mathbf{J}_1\ddot{\mathbf{q}}^{(1)} + \mathbf{H}_1 \tag{9.136}$$

We substitute from (9.136) into (9.132), obtaining

$$\delta\dot{\mathbf{q}}^{(1)T}[\underbrace{\mathbf{J}_1^T\mathbf{A}\mathbf{J}_1\ddot{\mathbf{q}}^{(1)}}_{\mathbf{A}_1} - \underbrace{\mathbf{J}_1^T(\mathbf{B}-\mathbf{A}\mathbf{H}_1)}_{\mathbf{B}_1}] = 0 \tag{9.137}$$

Due to the independence of $\delta\dot{\mathbf{q}}^{(1)}$, the relation (9.137) gives equations of motion for the whole MBS with loops

$$\mathbf{A}_1\ddot{\mathbf{q}}^{(1)} = \mathbf{B}_1 \tag{9.138}$$

Equations (9.138) follow directly from the right-hand sides of equations (9.112) and (9.111) after substituting MBS descriptions; thus

$$\left[\frac{\partial \mathbf{v}_i}{\partial \dot{s}_j}\right] = \left[\frac{\partial \dot{\mathbf{r}}_i}{\partial \dot{q}_k}\frac{\partial \dot{q}_k}{\partial \dot{q}_j^{(1)}}\right] = \boldsymbol{\alpha}\mathbf{J}_1 \tag{9.139a}$$

and

$$\left[\frac{\partial \boldsymbol{\omega}_i}{\partial \dot{s}_j}\right] = \left[\frac{\partial \boldsymbol{\omega}_i}{\partial \dot{q}_k}\frac{\partial \dot{q}_k}{\partial \dot{q}_j^{(1)}}\right] = \boldsymbol{\beta}\mathbf{J}_1 \tag{9.139b}$$

Using LEMT the above description of MBS is used by (Maisser, 1988).

Let us again solve the example from Section 9.3 (Figure 9.2) by this formalism. The loop is cut in the revolute pair A. We use the relative coordinates φ_{01}, φ_{12} and φ_{03} for the reduced system. We express the radius vectors, velocities and accelerations of all bodies

$$\begin{aligned}
\mathbf{r}_1 &= [p_1 c\varphi_{01}, p_1 s\varphi_{01}] \\
\mathbf{v}_1 &= [-p_1 s\varphi_{01}\dot{\varphi}_{01}, p_1 c\varphi_{01}\dot{\varphi}_{01}] \\
\mathbf{a}_1 &= [-p_1 s\varphi_{01}\ddot{\varphi}_{01} - p_1 c\varphi_{01}\dot{\varphi}_{01}^2, p_1 c\varphi_{01}\ddot{\varphi}_{01} - p_1 s\varphi_{01}\dot{\varphi}_{01}^2] \\
\mathbf{r}_2 &= [l_1 c\varphi_{01} + p c(\varphi_{01}+\varphi_{12}), l_1 s\varphi_{01} + p s(\varphi_{01}+\varphi_{12})] \\
\mathbf{v}_2 &= [-l_1 s\varphi_{01}\dot{\varphi}_{01} - p s(\varphi_{01}+\varphi_{12})(\dot{\varphi}_{01}+\dot{\varphi}_{12}), l_1 c\varphi_{01}\dot{\varphi}_{01} + \\
&\quad + p c(\varphi_{01}+\varphi_{12})(\dot{\varphi}_{01}+\dot{\varphi}_{12})] \\
\mathbf{a}_2 &= [-l_1 s\varphi_{01}\ddot{\varphi}_{01} - p s(\varphi_{01}+\varphi_{12})(\ddot{\varphi}_{01}+\ddot{\varphi}_{12}) - l_1 c\varphi_{01}\dot{\varphi}_{01}^2 - \\
&\quad - p c(\varphi_{01}+\varphi_{12})(\dot{\varphi}_{01}+\dot{\varphi}_{12})^2, l_1 c\varphi_{01}\ddot{\varphi}_{01} + p c(\varphi_{01}+\varphi_{12}) . \\
&\quad .(\ddot{\varphi}_{01}+\ddot{\varphi}_{12}) - l_1 s\varphi_{01}\dot{\varphi}_{01}^2 - p s(\varphi_{01}+\varphi_{12})(\dot{\varphi}_{01}+\dot{\varphi}_{12})^2] \\
\mathbf{r}_3 &= [p_3 c\varphi_{03}, p_3 s\varphi_{03}] \\
\mathbf{v}_3 &= [-p_3 s\varphi_{03}\dot{\varphi}_{03}, p_3 c\varphi_{03}\dot{\varphi}_{03}] \\
\mathbf{a}_3 &= [-p_3 s\varphi_{03}\ddot{\varphi}_{03} - p_3 c\varphi_{03}\dot{\varphi}_{03}^2, p_3 c\varphi_{03}\ddot{\varphi}_{03} - p_3 s\varphi_{03}\dot{\varphi}_{03}^2]
\end{aligned}$$

$$\begin{array}{lll}
\pi_1 = \varphi_{01}, & \omega_1 = \dot{\varphi}_{01}, & \alpha_1 = \ddot{\varphi}_{01} \\
\pi_2 = \varphi_{01}+\varphi_{12}, & \omega_2 = \dot{\varphi}_{01}+\dot{\varphi}_{12}, & \alpha_2 = \ddot{\varphi}_{01}+\ddot{\varphi}_{12} \\
\pi_3 = \varphi_{03}, & \omega_3 = \dot{\varphi}_{03}, & \alpha_3 = \ddot{\varphi}_{03}
\end{array} \tag{a}$$

Hence

$$\boldsymbol{\alpha} = \begin{bmatrix} -p_1 c\varphi_{01}, & 0, & 0 \\ p_1 s\varphi_{01}, & 0, & 0 \\ -l_1 s\varphi_{01} - p s(\varphi_{01}+\varphi_{12}), & -p s(\varphi_{01}+\varphi_{12}), & 0 \\ l_1 c\varphi_{01} + p c(\varphi_{01}+\varphi_{12}), & p c(\varphi_{01}+\varphi_{12}), & 0 \\ 0, & 0, & -p_3 s\varphi_{03} \\ 0, & 0, & p_3 c\varphi_{03} \end{bmatrix}$$

$$\mathbf{u} = \lceil -p_1 c\varphi_{01}\dot\varphi_{01}^2, -p_1 s\varphi_{01}\dot\varphi_{01}^2, -l_1 c\varphi_{01}\dot\varphi_{01}^2 - p c(\varphi_{01}+\varphi_{12})(\dot\varphi_{01}+\dot\varphi_{12})^2,$$
$$-l_1 s\varphi_{01}\dot\varphi_{01}^2 - p s(\varphi_{01}+\varphi_{12})(\dot\varphi_{01}+\dot\varphi_{12})^2, -p_3 c\varphi_{03}\dot\varphi_{03}^2, -p_3 s\varphi_{03}\dot\varphi_{03}^2\rceil$$

$$\begin{aligned} \boldsymbol{\beta} &= \begin{bmatrix} 1 & 1 & 0 \\ 0 & 1 & 0 \\ 0 & 0 & 1 \end{bmatrix} \\ \mathbf{U} &= \lceil 0, 0, 0 \rceil \end{aligned} \tag{b}$$

and further from (9.131)

$$\begin{aligned} \mathbf{A} &= \boldsymbol{\alpha}^T \lceil m_1, m_1, m_2, m_2, m_3, m_3 \rfloor \boldsymbol{\alpha} + \boldsymbol{\beta}^T \lceil I_{1T_1}, I_{2T}, I_{3T_3} \rfloor \boldsymbol{\beta} = \\ &= \begin{bmatrix} I_{1T_1} + m_1 p_1^2 + I_{2T} + m_2(l_1^2 + p^2 + 2 l_1 p c\varphi_{12}), & I_{2T} + m_2(p^2 + l_1 p c\varphi_{12}), & 0 \\ I_{2T} + m_2(p^2 + l_1 p c\varphi_{12}), & I_{2T} + m_2 p^2, & 0 \\ 0, & 0, & I_{3T_3} + m_3 p_3^2 \end{bmatrix} \\ \mathbf{B} &= \lceil m_2 l_1 p s\varphi_{12}(2\dot\varphi_{01} + \dot\varphi_{12})\dot\varphi_{12} + (-p_1 s\varphi_{01}) F f_x(\varphi_1) + \\ &\quad + p_1 c\varphi_{01} F f_y(\varphi_1) + F f_\varphi(\varphi_1), -m_2 l_1 p s\varphi_{12}\dot\varphi_{01}^2, 0 \rceil \end{aligned} \tag{c}$$

The constraints (c) in Section 9.3 are now formulated in the relative coordinates. From their derivatives

$$\begin{aligned} -l_1 s\varphi_{01}\dot\varphi_{01} - l_2 s(\varphi_{01}+\varphi_{12})(\dot\varphi_{01}+\dot\varphi_{12}) - l_3 s\varphi_{03}\dot\varphi_{03} &= 0 \\ l_1 c\varphi_{01}\dot\varphi_{01} + l_2 c(\varphi_{01}+\varphi_{12})(\dot\varphi_{01}+\dot\varphi_{12}) + l_3 c\varphi_{03}\dot\varphi_{03} &= 0 \end{aligned} \tag{d}$$

we find the independent coordinate (see Chapter 7 and Section 9.8), for example φ_{01}, and express

$$\begin{aligned} \dot\varphi_{12} &= f_2(\varphi_{01}, \varphi_{12}, \varphi_{03})\dot\varphi_{01} \\ \dot\varphi_{03} &= f_3(\varphi_{01}, \varphi_{12}, \varphi_{03})\dot\varphi_{01} \end{aligned} \tag{e}$$

According to (9.136), we obtain

$$\begin{aligned} [\dot\varphi_{01}, \dot\varphi_{12}, \dot\varphi_{03}] &= [1, f_2, f_3]\dot\varphi_{01} \\ [\ddot\varphi_{01}, \ddot\varphi_{12}, \ddot\varphi_{03}] &= [1, f_2, f_3]\ddot\varphi_{01} + [0, \dot f_2\dot\varphi_{01}, \dot f_3\dot\varphi_{01}] \end{aligned} \tag{f}$$

and finally from (9.137) we obtain (9.138)

$$\begin{aligned} \mathbf{A}_1 &= [a_{11} + (a_{12} + a_{21})f_2 + a_{22}f_2^2 + a_{33}f_3^2] \\ \mathbf{B}_1 &= [b_1 - a_{12}\dot f_2\dot\varphi_{01} + f_2(b_2 - a_{22}\dot f_2\dot\varphi_{01}) + f_3(-a_{33}\dot f_3\dot\varphi_{01})] \end{aligned} \tag{g}$$

9.5.1.3 Kane's Approach

Another formalism based on MBS description using independent coordinates is based on Kane's equations of motion (Kane et al., 1983; Huston, 1990). They can be obtained from the right-hand sides of equivalence (9.116), (9.117). In Kane's formalism, the MBS position is described by generalized (independent) coordinates $q_1, \ldots, q_n$ where n is the number of degrees of freedom of the MBS. Further, for the MBS description n so-called generalized velocities $u_1, \ldots, u_n$ are to be chosen

$$u_r = \sum_{s=1}^{n} A_{rs}\dot q_s + A_r, \quad r = 1, \ldots, n \tag{9.140}$$

where A_{rs} and A_r are functions of q_j and time t. It may therefore be possible to express from (9.140) $\dot q_i (i = 1, \ldots, n)$ uniquely from $u_r (r = 1, \ldots, n)$

$$\dot q_i = \sum_{s=1}^{n} B_{is}u_s + B_i, \quad i = 1, \ldots, n \tag{9.141}$$

where B_{is} and B_i are functions of $q_1, \ldots, q_n$ and time t. The MBS is composed of N bodies. For each body A we express the translational velocity of the centre of mass $\mathbf{v}^A$ and the angular velocity $\boldsymbol{\omega}^A$ of the body in the form

$$\begin{aligned} \mathbf{v}^A &= \sum_{r=1}^{n} \mathbf{v}_r^A u_r + \mathbf{v}_t^A \\ \boldsymbol{\omega}^A &= \sum_{r=1}^{n} \boldsymbol{\omega}_r^A u_r + \boldsymbol{\omega}_t^A \end{aligned} \tag{9.142}$$

where $\mathbf{v}_r^A, \mathbf{v}_t^A, \boldsymbol{\omega}_r^A, \boldsymbol{\omega}_t^A$ are functions of q_j and t and are called the partial translational velocity of body A or the partial angular velocity of body

A. We may also find this expression by expressing the respective velocities of body A as functions of generalized coordinates and using (9.141). The simplicity of the resultant expression (9.142) shows the reasons for the above choice (9.140). Generalized velocities are often chosen equal to components of velocities of MBS significant bodies.

We must further express the translational acceleration $\mathbf{a}^A$ of the centre of mass of body A and the angular acceleration $\boldsymbol{\alpha}^A$ of body A as functions of $q_1, \ldots, q_n, u_1, \ldots, u_n, \dot{u}_1, \ldots, \dot{u}_n$. These are easy to obtain by differentiating (9.142) with respect to time and substituting from (9.141).

Several forces obtained after freeing body A act on the MBS body A. They are the reaction forces of constraints, forces of gravitation and other external work forces. Denoting the resultant force $\mathbf{R}_i$ acting in the centre of mass of body A_i and $\mathbf{T}_i$ the resultant moment acting on body A_i, we have

$$K_r = \sum_{i=1}^{n} (\mathbf{v}_r^{A_i}.\mathbf{R}_i + \boldsymbol{\omega}_r^{A_i}.\mathbf{T}_i) \qquad (9.143)$$

where "." expresses the dot (scalar) product of vectors. Also acting on the MBS body A_i are the forces of inertia $\mathbf{R}_i^*$ and $\mathbf{T}_i^*$

$$\begin{aligned} \mathbf{R}_i^* &= -m_{A_i}\mathbf{a}^{A_i} \\ \mathbf{T}_i^* &= -\mathbf{I}^{A_i}\boldsymbol{\alpha}^{A_i} - \boldsymbol{\omega}^{A_i} \times \mathbf{I}^{A_i}\boldsymbol{\omega}^{A_i} \end{aligned} \qquad (9.144)$$

where m^{A_i} is the mass, and $\mathbf{I}^{A_i}$ the tensor of inertia of the body A_i. We then define

$$K_r^* = \sum_{i=1}^{N} (\mathbf{v}_r^{A_i}.\mathbf{R}_i^* + \boldsymbol{\omega}_r^{A_i}.\mathbf{T}_i^*) \qquad (9.145)$$

as the r-th generalized force of inertia acting on the MBS.

The equations of motion describing MBS behaviour are of the form

$$K_r^* + K_r = 0, \quad r = 1, \ldots, n \qquad (9.146)$$

Relations (9.146) are called Kane's equations of motion. They have the form of equations of motion in generalized velocities u_r. This could be an advantage.

Kane's equations of motion are identical to the original Newton-Euler equations with Jourdain's principle as they are for example used in the NEWEUL or MESA VERDE approaches; they are equivalent to Lagrange's equations written in quasi-coordinates. The most important idea of Kane's equations of motion is the possibility of obtaining simple and decoupled equations of motion (for example, in the ideal case the diagonal mass matrix of the system) by the use of quasi-coordinates instead of classical coordinates. The common example of this possibility is Euler's dynamic equations

(9.15) of a single body in the space (the elements of angular velocity as the quasi-velocities) instead of Lagrange's equations for this body with Cardan's angles or Euler parameters. However, until now, a way has not been derived for finding out in a computationally easy way such quasi-coordinates for a general MBS that significantly simplifies its equations of motion. In the general case of a MBS there is only the possibility (Maisser, 1990) of starting from classical coordinates or other quasi-coordinates and by some method of vector orthogonalization finding particular quasimotion. However, such an orthogonalization process is symbolically very demanding, or it must be numerically realized at each step of the integration of MBS motion, and therefore it is also very computationally demanding. Another serious problem is that the quasi-coordinates must be independent and this is often not fulfilled (see also the example in Section 9.3.1 where this is fulfilled). Therefore Kane's equations of motion are mostly used only with classical coordinates.

Let us solve again the example from Section 9.3 (Figure 9.2). We choose φ_1 as the generalized (independent) coordinate and $\dot{\varphi}_1, \dot{\varphi}_2$ and $\dot{\varphi}_3$ as the generalized velocities u_1, u_2 and u_3 which however are not independent;

$$u_1 = \dot{\varphi}_1, \quad u_2 = \dot{\varphi}_2, \quad u_3 = \dot{\varphi}_3 \tag{a}$$

From the time derivatives of constraints (c) from Section 9.3

$$\begin{aligned} l_1 c\varphi_1 u_1 - l_2 c\varphi_2 u_2 - l_3 c\varphi_3 u_3 &= 0 \\ -l_1 s\varphi_1 u_1 - l_2 s\varphi_2 u_2 - l_3 s\varphi_3 u_3 &= 0 \end{aligned} \tag{b}$$

we express

$$\begin{aligned} u_2 &= \frac{l_1 s(\varphi_1 + \varphi_3)}{l_2 s(\varphi_2 - \varphi_3)} u_1 = Z_1 u_1 \\ u_3 &= -\frac{l_1 s(\varphi_1 + \varphi_2)}{l_3 s(\varphi_2 - \varphi_3)} u_1 = Z_2 u_1 \end{aligned} \tag{c}$$

where the Z symbols are used as a means of simplifying the expressions. Each Z is used in subsequent computations in place of the particular expression being substituted in order to eliminate unnecessary repetition of operations. From (a) in Section 9.5.1.1 we obtain the expressions (9.142)

$$\begin{aligned} v^1 &= [Z_3 u_1, Z_4 u_1] \\ \omega^1 &= -u_1 \\ v^2 &= [Z_5 u_1 + Z_6 u_2, Z_7 u_1 + Z_8 u_2] = [Z_9 u_1, Z_{10} u_1] \\ \omega^2 &= u_2 = Z_1 u_1 \\ v^3 &= [Z_{11} u_3, Z_{12} u_3] = [Z_{13} u_1, Z_{14} u_1] \\ \omega^3 &= u_3 = Z_2 u_1 \end{aligned} \tag{d}$$

where, for example,

$$Z_3 = p_1 c\varphi_1, \qquad Z_5 = l_1 c\varphi_1, \qquad Z_6 = -pc\varphi_2,$$

$$Z_9 = Z_5 + Z_6 Z_1, \qquad Z_{13} = Z_{11} Z_2 \tag{e}$$

Hence

$$\begin{array}{llll} \mathbf{v}_1^1 = [Z_3, Z_4], & \omega_1^1 = -1, & \mathbf{v}_t^1 = \mathbf{0}, & \omega_t^1 = 0 \\ \mathbf{v}_1^2 = [Z_9, Z_{10}], & \omega_1^2 = Z_1, & \mathbf{v}_t^2 = \mathbf{0}, & \omega_t^2 = 0 \\ \mathbf{v}_1^3 = [Z_{13}, Z_{14}], & \omega_1^3 = Z_2, & \mathbf{v}_t^3 = \mathbf{0}, & \omega_t^3 = 0 \end{array} \tag{f}$$

Further, we express the accelerations

$$\begin{array}{ll} \mathbf{a}^1 = [Z_3\dot{u}_1 + \dot{Z}_3 u_1, Z_4\dot{u}_1 + \dot{Z}_4 u_1], & \alpha^1 = -\dot{u}_1 \\ \mathbf{a}^2 = [Z_9\dot{u}_1 + \dot{Z}_9 u_1, Z_{10}\dot{u}_1 + \dot{Z}_{10} u_1], & \alpha^2 = Z_1\dot{u}_1 + \dot{Z}_1 u_1 \\ \mathbf{a}^3 = [Z_{13}\dot{u}_1 + \dot{Z}_{13} u_1, Z_{14}\dot{u}_1 + \dot{Z}_{14} u_1], & \alpha^3 = Z_2\dot{u}_1 + \dot{Z}_2 u_1 \end{array} \tag{g}$$

where, for example,

$$\dot{Z}_3 = -p_1 s\varphi_1\dot{\varphi}_1 = Z_4 u_1, \quad \dot{Z}_4 = -p_1 c\varphi_1\dot{\varphi}_1 = -Z_3 u_1 \tag{h}$$

According to the expression of the force in Section 9.5.1.1, we obtain from (9.143)

$$K_1 = Z_3 F f_x(\varphi_1) + Z_4 F f_y(\varphi_1) + (-1) F f_\varphi(\varphi_1) \tag{i}$$

Then the forces of inertia are

$$R_i^* = -m_i a^i, \qquad T_i^* = -I_{iT_i}\alpha^i, \qquad i = 1, 2, 3 \tag{j}$$

and finally the generalized force of inertia K_1^*

$$\begin{aligned} K_1^* \;=\;& Z_3(-m_1)(Z_3\dot{u}_1 + \dot{Z}_3 u_1) + Z_4(-m_1)(Z_4\dot{u}_1 + \dot{Z}_4 u_1) + \\ &+(-1)(-I_{1T_1})(-\dot{u}_1) + Z_9(-m_2)(Z_9\dot{u}_1 + \dot{Z}_9 u_1) + \\ &+Z_{10}(-m_2)(Z_{10}\dot{u}_1 + \dot{Z}_{10} u_1) + Z_1(-I_{2T})(Z_1\dot{u}_1 + \dot{Z}_1 u_1) + \\ &+Z_{13}(-m_3)(Z_{13}\dot{u}_1 + \dot{Z}_{13} u_1) + Z_{14}(-m_3)(Z_{14}\dot{u}_1 + \dot{Z}_{14} u_1) + \\ &+Z_2(-I_{3T_3})(Z_2\dot{u}_1 + \dot{Z}_2 u_1) \end{aligned} \tag{k}$$

9.5.1.4 ADAMS Approach

There are other formalisms which are based on the use of dependent coordinates. The left-hand sides of equivalences (9.112), (9.111) lead to Lagrange's equations of mixed type. Their particular form depends on the choice of coordinates used. Let us present three formalisms based on a systematic choice of coordinates for the MBS description.

The first is the basis of system ADAMS (Automatic Dynamic Analysis of Mechanical Systems) (Orlandea et al., 1976, Ryan, 1990). The position of each MBS body is determined by three Cartesian coordinates of its mass centre and three Euler angles for its orientation. The total number of coordinates for a MBS with n bodies is $6n$. Using LEMT the kinematic and dynamic equations of motion for unconstrained bodies have been derived a priori and pre-coded similarly as in FEM. The algebraic (or differential) constraint equation for each kinematic joint are added according to (5.125-5.129) together corresponding Lagrange multipliers.

The resulting DAE system is solved numerically usually by DAE solver described in Section 9.8.

Let us solve the example from Section 9.3. (Figure 9.2) once more. We introduce the physical coordinates $x_1, y_1, \varphi_1, x_2, y_2, \varphi_2, x_3, y_3$ and φ_3, where x_i, y_i $(i = 1, 2, 3)$ are the cartesian coordinates of the centres of mass of bodies 1, 2 and 3 and φ_i $(i = 1, 2, 3)$ are as shown in Figure 9.2. If we do not include the redundant constraints due to planar MBS treated as a special case of spatial MBS, the coordinates must fulfill the following constraint equations

$$\begin{aligned}
x_1 - p_1 \mathrm{s}\varphi_1 &= 0 \\
y_1 - p_1 \mathrm{c}\varphi &= 0 \\
x_1 + (l_1 - p_1)\mathrm{s}\varphi_1 - (x_2 + p\mathrm{s}\varphi_2) &= 0 \\
y_1 + (l_1 - p_1)\mathrm{c}\varphi_1 - (y_2 - p\mathrm{c}\varphi_2) &= 0 \\
x_2 - (l_2 - p)\mathrm{s}\varphi_2 - (x_3 - (l_3 - p_3)\mathrm{s}\varphi_3) &= 0 \\
y_2 + (l_2 - p)\mathrm{c}\varphi_2 - (y_3 + (l_3 - p_3)\mathrm{c}\varphi_3) &= 0 \\
x_3 - p_3 \mathrm{s}\varphi_3 &= 0 \\
y_3 - l_0 + p_3 \mathrm{c}\varphi_3 &= 0 \qquad \text{(a)}
\end{aligned}$$

Then the equations of motion (9.149) are

$$\begin{aligned}
m_1 \ddot{x}_1 &= \lambda_1 + \lambda_3 \\
m_1 \ddot{y}_1 &= \lambda_2 + \lambda_4 \\
I_{1T_1} \ddot{\varphi}_1 &= -\lambda_1 p_1 \mathrm{c}\varphi_1 + \lambda_2 p_1 \mathrm{s}\varphi_1 + \lambda_3 (l_1 - p_1)\mathrm{c}\varphi_1 - \lambda_4 (l_1 - p_1)\mathrm{s}\varphi_1 \\
m_2 \ddot{x}_2 &= \lambda_5 - \lambda_3 \\
m_2 \ddot{y}_2 &= \lambda_6 - \lambda_4 \\
I_{2T} \ddot{\varphi}_2 &= -\lambda_3 p \mathrm{c}\varphi_2 - \lambda_4 p \mathrm{s}\varphi_2 - \lambda_5 (l_2 - p)\mathrm{c}\varphi_2 - \lambda_6 (l_2 - p)\mathrm{s}\varphi_2 \\
m_3 \ddot{x}_3 &= \lambda_7 - \lambda_5 \\
m_3 \ddot{y}_3 &= \lambda_8 - \lambda_6 \\
I_{3T_3} \ddot{\varphi}_3 &= \lambda_5 (l_3 - p_3)\mathrm{c}\varphi_3 - \lambda_6 (l_3 - p_3)\mathrm{s}\varphi_3 + \lambda_7 p_3 \mathrm{c}\varphi_3 - \lambda_8 p_3 \mathrm{s}\varphi_3 \qquad \text{(b)}
\end{aligned}$$

where Lagrange's multipliers correspond to the constraints (a).

9.5.1.5 DADS Approach

The second is the basis of system DADS (Dynamic Analysis and Design System) (Haug et al., 1982; Haug, 1989; Nikravesh, 1988). The position of each MBS body in space is determined by three coordinates of its centre of mass and four Euler parameters giving its orientation in space. For an MBS composed of n bodies, the total number of coordinates equals $7n$. In planar MBS each body is described by two Cartesian coordinates of its mass centre and one angle. Thus in plane n bodies have $3n$ coordinates. There are m independent constraint equations of the type

$$\mathbf{\Phi}(\mathrm{s}, t) = 0 \tag{9.147}$$

where s are all MBS coordinates. These constraint equations are described by relations (5.125–5.129) for individual kinematic pairs. The equations of motion are derived from LEMT. The kinetic energy of the i-the body is

$$T_i = \frac{1}{2}(\dot{\mathbf{r}}_i^T \mathbf{J}_i \dot{\mathbf{r}}_i + \boldsymbol{\omega}_{ei}^T \mathbf{I}_i \boldsymbol{\omega}_{ei}) \tag{9.148}$$

where $\mathbf{r}_i = \lceil x_i, y_i, z_i \rceil$ are coordinates of the centre of mass, $\mathbf{J}_i = \lceil m_i \rfloor$ is the diagonal mass matrix, $\mathbf{I}_i$ the tensor of inertia, and $\boldsymbol{\omega}_{ei}$ is the angular velocity expressed by Euler's parameters $\mathbf{e}_i$. Equations of motion for body i are

$$\mathbf{M}_i \ddot{\mathrm{s}}_i = \mathrm{g}_i - \mathbf{\Phi}_{s_i}^T \boldsymbol{\lambda}_i \tag{9.149}$$

where $\mathbf{M}_i$ is the generalized mass matrix, $\mathrm{s}_i = \lceil \mathbf{r}_i, \mathrm{e}_i \rceil$, $\mathbf{\Phi}_{s_i}$ is the matrix of the system of constraint equations expressed in velocities, $\boldsymbol{\lambda}_i$ are Lagrange's multipliers and g_i contains the remaining terms. For the whole MBS

$$\mathbf{M}(\mathrm{s})\ddot{\mathrm{s}} = \mathrm{g}(\mathrm{s}, \dot{\mathrm{s}}, t) - \mathbf{\Phi}_s^T(\mathrm{s}, t)\boldsymbol{\lambda}_i \tag{9.150}$$

Equation (9.150), together with constraint equations (9.147), represent differential-algebraic equations describing MBS dynamics.

From the viewpoint of numerical computational efficiency, it is beneficial to reduce the system in such a way as to leave only a minimum necessary number of coordinates equal to the number of degrees of freedom. A unique approach (the so-called coordinate partitioning) is to be found in the DADS system (Wehage and Haug, 1982). We differentiate the constraint equations with respect to time

$$\mathbf{\Phi}_s \dot{\mathrm{s}} + \mathbf{\Phi}_t = \mathbf{0} \tag{9.151}$$

Assuming independency of constraint equations, the matrix $\mathbf{\Phi}_s$ has its full row rank, and the equation (9.151) can be written in the form

$$\mathbf{\Phi}_s \dot{\mathbf{s}} = [\mathbf{\Phi}_u, \mathbf{\Phi}_v] \begin{bmatrix} \dot{\mathbf{u}} \\ \dot{\mathbf{v}} \end{bmatrix} = -\mathbf{\Phi}_t \tag{9.152}$$

where $\mathbf{\Phi}_u$ is the non-singular submatrix of $\mathbf{\Phi}_s$ corresponding to the dependent velocities $\dot{\mathbf{u}}$, and submatrix $\mathbf{\Phi}_v$ corresponds to independent velocities $\dot{\mathbf{v}}$. For dependent velocities and accelerations

$$\dot{\mathbf{u}} = -\mathbf{\Phi}_u^{-1}\mathbf{\Phi}_v \dot{\mathbf{v}} - \mathbf{\Phi}_u^{-1}\mathbf{\Phi}_t \tag{9.153a}$$
$$\ddot{\mathbf{u}} = -\mathbf{\Phi}_u^{-1}\mathbf{\Phi}_v \ddot{\mathbf{v}} - \mathbf{\Phi}_u^{-1}((\mathbf{\Phi}_s \dot{\mathbf{s}})_s \dot{\mathbf{s}} + 2\mathbf{\Phi}_{ts}\dot{\mathbf{s}} + \mathbf{\Phi}_{tt}) \tag{9.153b}$$

We use the same decomposition for the equations of motion

$$\begin{aligned} \mathbf{M}^{uu}\ddot{\mathbf{u}} + \mathbf{M}^{uv}\ddot{\mathbf{v}} &= \mathbf{g}^u - \mathbf{\Phi}_u^T \boldsymbol{\lambda} \\ \mathbf{M}^{vu}\ddot{\mathbf{u}} + \mathbf{M}^{vv}\ddot{\mathbf{v}} &= \mathbf{g}^v - \mathbf{\Phi}_v^T \boldsymbol{\lambda} \end{aligned} \tag{9.154}$$

We determine $\boldsymbol{\lambda}$ from the former equation and substitute it into the latter

$$(\mathbf{M}^{vv} - (\mathbf{\Phi}_u^{-1}\mathbf{\Phi}_v)^T \mathbf{M}^{uv})\ddot{\mathbf{v}} + (\mathbf{M}^{vu} - (\mathbf{\Phi}_u^{-1}\mathbf{\Phi}_v)^T \mathbf{M}^{uu})\ddot{\mathbf{u}} =$$

$$= \mathbf{g}^v - (\mathbf{\Phi}_u^{-1}\mathbf{\Phi}_v)^T \mathbf{g}^u \tag{9.155}$$

Finally, substituting (9.153b), we obtain a system of differential equations of second order for $\ddot{\mathbf{v}}$. To complete the calculation, the dependent velocities $\dot{\mathbf{u}}$ and coordinates $\mathbf{u}$ can be computed from equations (9.153a).

The equation of motion for the example from Section 9.3 are the same as by ADAMS approach, but they are solved numerically by the algorithm of coordinate partitioning, see Section 9.8.

9.5.1.6 Natural Coordinates Approach

Yet another formalism using dependent coordinates is the approach of Garcia de Jalón (Garcia de Jalón et al., 1986, 1987; Garcia de Jalón and Bayo, 1994) using natural coordinates. The MBS is described by natural coordinates as in Section 5.11.4. The virtual power output of inertial forces acting on a body described by natural coordinates of two points A and B and two unit vectors $\mathbf{u}, \mathbf{v}$ according to Figure 5.23. and Figure 9.4. is

$$\delta W_i = -\int_{V_i} (\delta \dot{\mathbf{x}}^P)^T \ddot{\mathbf{x}}^P \mathrm{d}m_i \tag{9.156}$$

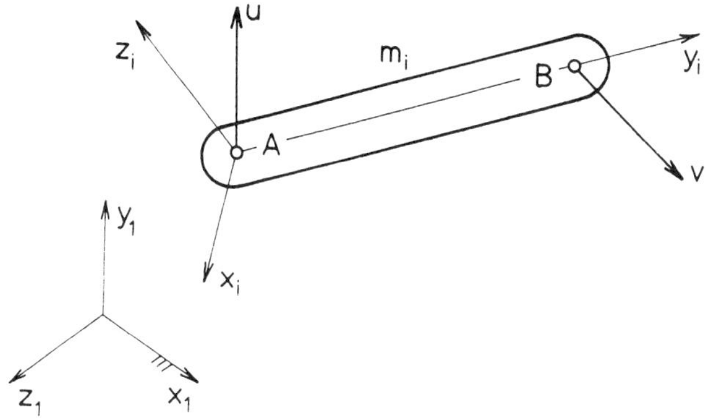

Figure 9.4. Dynamics of a body by natural coordinates.

We introduce matrix $\mathbf{S}_{1i}$ of directional cosines of the transformation between frame 1 and body i. Then, according to (4.28)

$$\begin{aligned} \mathbf{x}^A - \mathbf{x}^B &= \mathbf{S}_{1i}(\mathbf{x}_i^A - \mathbf{x}_i^B) \\ \mathbf{u} &= \mathbf{S}_{1i}\mathbf{u}_i \\ \mathbf{v} &= \mathbf{S}_{1i}\mathbf{v}_i \end{aligned} \tag{9.157}$$

We introduce matrices

$$\mathbf{X} = \begin{bmatrix} \mathbf{x}^A - \mathbf{x}^B \\ \mathbf{u} \\ \mathbf{v} \end{bmatrix}, \qquad \mathbf{X}_i = \begin{bmatrix} \mathbf{x}_i^A - \mathbf{x}_i^B \\ \mathbf{u}_i \\ \mathbf{v}_i \end{bmatrix}$$

so that

$$\mathbf{X} = \mathbf{S}_{1i}\mathbf{X}_i, \qquad \dot{\mathbf{X}} = \dot{\mathbf{S}}_{1i}\mathbf{X}_i \qquad \dot{\mathbf{S}}_{1i} = \dot{\mathbf{X}}\mathbf{X}_i^{-1} \qquad \ddot{\mathbf{S}}_{1i} = \ddot{\mathbf{X}}\mathbf{X}_i^{-1} \tag{9.158}$$

Further,

$$\begin{aligned} \mathbf{x}^P &= \mathbf{x}^A + \mathbf{S}_{1i}\mathbf{x}_i^P \\ \dot{\mathbf{x}}^P &= \dot{\mathbf{x}}^A + \dot{\mathbf{S}}_{1i}\mathbf{x}_i^P \\ \ddot{\mathbf{x}}^P &= \ddot{\mathbf{x}}^A + \ddot{\mathbf{S}}_{1i}\mathbf{x}_i^P \end{aligned} \tag{9.159}$$

Substituting into (9.156) and altering, we obtain

$$\delta W_i = [\delta\dot{\mathbf{x}}^A, \delta\dot{\mathbf{x}}^B, \delta\dot{\mathbf{u}}, \delta\dot{\mathbf{v}}]^T \mathbf{M}_i [\ddot{\mathbf{x}}^A, \ddot{\mathbf{x}}^B, \ddot{\mathbf{u}}, \ddot{\mathbf{v}}] \tag{9.160}$$

$\mathbf{M}_i$ is the constant generalized mass matrix of element i

$$\mathbf{M}_i = \begin{bmatrix} (a_{11}+2b_1+m_i)\mathbf{E}_3, & -(a_{11}+b_1)\mathbf{E}_3, & -(a_{12}+b_2)\mathbf{E}_3, & -(a_{13}+b_3)\mathbf{E}_3 \\ & a_{11}\mathbf{E}_3 & -a_{12}\mathbf{E}_3 & -a_{13}\mathbf{E}_3 \\ \text{sym.} & & a_{22}\mathbf{E}_3, & a_{23}\mathbf{E}_3 \\ & & & a_{33}\mathbf{E}_3 \end{bmatrix} \tag{9.161}$$

where $\mathbf{E}_3$ is the unit (3,3) matrix,

$$\begin{aligned} a_{kl} &= \left[\mathbf{X}_i^{-1}\right]_k^T \mathbf{I}_i \left[\mathbf{X}_i^{-1}\right]_l \\ \mathbf{I}_i &= \int_{V_i} \mathbf{x}_i^P (\mathbf{x}_i^P)^T \mathrm{d}m_i \\ b_l &= m_i \mathbf{g}_i^T \left[\mathbf{X}_i^{-1}\right]_l \end{aligned} \tag{9.162}$$

$[\mathbf{X}_i^{-1}]_k$ is the k-th row of matrix $\mathbf{X}_i^{-1}$, and $\mathbf{g}_i$ is the centre of mass vector of body i. Then, from the principle of virtual power for the whole MBS,

$$-\delta\dot{\mathbf{s}}^T\mathbf{M}\ddot{\mathbf{s}} + \delta\dot{\mathbf{s}}^T\mathbf{f} = 0 \tag{9.163}$$

where $\mathbf{f}$ is the vector of generalized forces. In this relation, coordinates $\mathbf{s}$ are, however, not independent. For the MBS description we choose independent coordinates $\mathbf{q}$ (see Section 9.8). Then

$$\dot{\mathbf{s}} = \mathbf{R}\dot{\mathbf{q}} \tag{9.164}$$

where $\mathbf{R}$ is the transmission matrix. Expressing $\delta\dot{\mathbf{s}} = \mathbf{R}\delta\dot{\mathbf{q}}$ and substituting into (9.163) we obtain from the independence of the equations of motion

$$\mathbf{R}^T\mathbf{M}\ddot{\mathbf{s}} = \mathbf{R}^T\mathbf{f} \tag{9.165}$$

which, together with constraint conditions (5.133–5.138), form LEMT. Expressing from (9.164)

$$\ddot{\mathbf{s}} = \mathbf{R}\ddot{\mathbf{q}} + \dot{\mathbf{R}}\dot{\mathbf{q}} \tag{9.166}$$

and substituting into (9.165), we obtain equations of motion

$$\mathbf{R}^T\mathbf{M}\mathbf{R}\ddot{\mathbf{q}} = \mathbf{R}^T\mathbf{f} - \mathbf{R}^T\mathbf{M}\dot{\mathbf{R}}\dot{\mathbf{q}} \tag{9.167}$$

in the form of Lagrange's equations of the second type.

Let us again solve the example from Section 9.3. (Figure 9.2). We introduce the natural coordinates $x_1, y_1, x_2, y_2, x_3, y_3, x_4$ and y_4 as the cartesian coordinates of centres of revolute pairs C, B, A and D, respectively. Let us first study the body 2. The coordinates of its centre of mass are

$$\begin{aligned} x &= \frac{x_3 p + x_2(l_2 - p)}{l_2} \\ y &= \frac{y_3 p + y_2(l_2 - p)}{l_2} \end{aligned} \tag{a}$$

For the angle $\varphi = \varphi_2 - \pi/2$ between the axis x and the bar 2 we have

$$\mathrm{s}\varphi = \frac{y_3 - y_2}{l_2}, \qquad \mathrm{c}\varphi = \frac{x_3 - x_2}{l_2} \tag{b}$$

Hence

$$\begin{aligned} \mathrm{c}\varphi\delta\varphi &= \frac{\delta y_3 - \delta y_2}{l_2}, & -\mathrm{s}\varphi\delta\varphi &= \frac{x_3 - x_2}{l_2} \\ \mathrm{c}\varphi\ddot{\varphi} - \mathrm{s}\varphi\dot{\varphi}^2 &= \frac{\ddot{y}_3 - \ddot{y}_2}{l_2}, & -\mathrm{s}\varphi\ddot{\varphi} - \mathrm{c}\varphi\dot{\varphi}^2 &= \frac{\ddot{x}_3 - \ddot{x}_2}{l_2} \end{aligned} \tag{c}$$

The principle of virtual power gives

$$\delta x m_2 \ddot{x} + \delta y m_2 \ddot{y} + \delta\varphi I_{2T}\ddot{\varphi} = \frac{\delta x_3 p + \delta x_2(l_2 - p)}{l_2} m_2 \cdot$$

$$\begin{aligned} &\cdot \frac{\ddot{x}_3 p + \ddot{x}_2(l_2 - p)}{l_2} + \frac{\delta y_3 p + \delta y_2(l_2 - p)}{l_2} m_2 \frac{\ddot{y}_3 p + \ddot{y}_2(l_2 - p)}{l_2} + \\ &+((\mathrm{c}\varphi\delta\varphi)(\mathrm{c}\varphi\ddot{\varphi} - \mathrm{s}\varphi\dot{\varphi}^2) + (-\mathrm{s}\varphi\delta\varphi)(-\mathrm{s}\varphi\ddot{\varphi} - \mathrm{c}\varphi\dot{\varphi}^2))I_{2T} = \\ = \; &\delta x_2 \left(\left(\frac{m_2(l_2 - p)^2}{l_2^2} + \frac{I_{2T}}{l_2^2} \right) \ddot{x}_2 + \left(\frac{p(l_2 - p)}{l_2^2} m_2 - \frac{I_{2T}}{l_2^2} \right) \ddot{x}_3 \right) + \\ &+\delta y_2 \left(\left(\frac{m_2(l_2 - p)^2}{l_2^2} + \frac{I_{2T}}{l_2^2} \right) \ddot{y}_2 + \left(\frac{p(l_2 - p)}{l_2^2} m_2 - \frac{I_{2T}}{l_2^2} \right) \ddot{y}_3 \right) + \\ &+\delta x_3 \left(\left(\frac{p(l_2 - p)}{l_2^2} m_2 - \frac{I_{2T}}{l_2^2} \right) \ddot{x}_2 + \left(m_2 \frac{p^2}{l_2^2} + \frac{I_{2T}}{l_2^2} \right) \ddot{x}_3 \right) + \\ &+\delta y_3 \left(\left(\frac{p(l_2 - p)}{l_2^2} m_2 - \frac{I_{2T}}{l_2^2} \right) \ddot{y}_2 + \left(m_2 \frac{p^2}{l_2^2} + \frac{I_{2T}}{l_2^2} \right) \ddot{y}_3 \right) \end{aligned} \tag{d}$$

After repeating this derivation for all bodies we obtain the mass matrix of the whole MBS

$$\mathbf{M} = \left[\begin{array}{ccccc} A_{11} & 0 & A_{12} & 0 & 0 \\ 0 & A_{11} & 0 & A_{12} & 0 \\ A_{12} & 0 & A_{13} + A_{21} & 0 & A_{22} \\ 0 & A_{12} & 0 & A_{13} + A_{21} & 0 \\ 0 & 0 & A_{22} & 0 & A_{22} + A_{31} \\ 0 & 0 & 0 & A_{22} & 0 \\ 0 & 0 & 0 & 0 & A_{32} \\ 0 & 0 & 0 & 0 & 0 \end{array} \right. \quad \cdots$$

$$\cdots \quad \begin{bmatrix} 0 & 0 & 0 \\ 0 & 0 & 0 \\ 0 & 0 & 0 \\ A_{22} & 0 & 0 \\ 0 & A_{32} & 0 \\ A_{22}+A_{31} & 0 & A_{32} \\ 0 & A_{33} & 0 \\ A_{32} & 0 & A_{33} \end{bmatrix} \qquad \text{(e)}$$

where

$$\begin{aligned} A_{i1} &= \frac{I_{iT_i} + m_i(p_i - l_i)^2}{l_i^2} \\ A_{i2} &= \frac{m_i p_i (l_i - p_i) - I_{iT_i}}{l_i^2} \\ A_{i3} &= \frac{I_{iT_i} + m_i p_i^2}{l_i^2} \end{aligned} \qquad \text{(f)}$$

The natural coordinates must fulfil the constraints

$$\begin{aligned} x_1 &= 0 \\ y_1 &= 0 \\ (x_2 - x_1)^2 + (y_2 - y_1)^2 &= l_1^2 \\ (x_3 - x_2)^2 + (y_3 - y_2)^2 &= l_2^2 \\ (x_4 - x_3)^2 + (y_4 - y_3)^2 &= l_3^2 \\ x_4 &= 0 \\ y_4 &= l_0 \end{aligned} \qquad \text{(g)}$$

The virtual power of the spring force is

$$\delta x F f_x(\varphi_1) + \delta y F f_y(\varphi_1) + \delta\varphi F f_\varphi(\varphi_1) =$$

$$\begin{aligned} &= \delta x_1 \left(\frac{(l_1 - p_1)}{l_1} F f_x(\varphi_1) + \frac{(y_2 - y_1)}{l_1^2} F f_\varphi(\varphi_1) \right) + \\ &+ \delta y_1 \left(\frac{(l_1 - p_1)}{l_1} F f_y(\varphi_1) - \frac{(x_2 - x_1)}{l_1^2} F f_\varphi(\varphi_1) \right) + \\ &+ \delta x_2 \left(\frac{p_1}{l_1} F f_x(\varphi_1) - \frac{(y_2 - y_1)}{l_1^2} F f_\varphi(\varphi_1) \right) + \\ &+ \delta y_2 \left(\frac{p_1}{l_1} F f_y(\varphi_1) + \frac{(x_2 - x_1)}{l_1^2} F f_\varphi(\varphi_1) \right) = \\ &= F_1 \delta x_1 + F_2 \delta y_1 + F_3 \delta x_2 + F_4 \delta y_2 \end{aligned} \qquad \text{(h)}$$

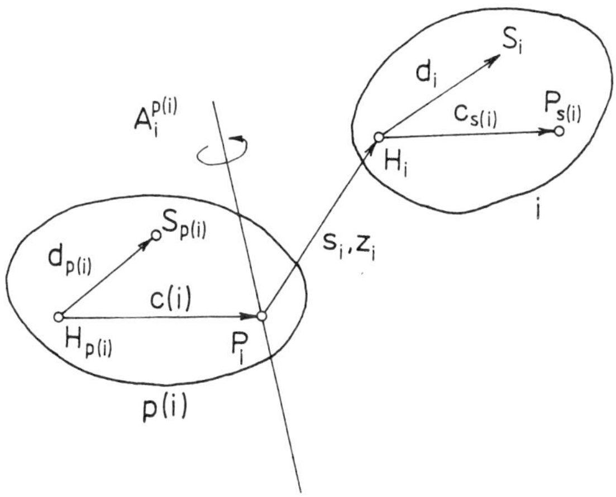

Figure 9.5. Kinematics of relative motion.

Hence the equations of motion

$$\mathbf{M}[\ddot{x}_1, \ddot{y}_1, \ddot{x}_2, \ddot{y}_2, \ddot{x}_3, \ddot{y}_3, \ddot{x}_4, \ddot{y}_4] = [F_1, F_2, F_3, F_4, 0, 0, 0, 0] + \mathbf{\Phi}_s \boldsymbol{\lambda} \quad \text{(i)}$$

where $\mathbf{\Phi}_s$ is the Jacobi matrix of constraints (g) and $\boldsymbol{\lambda}$ are corresponding Lagrange's multipliers. Futher reduction of these equations of motion (i) is realized numerically (see Section 9.8).

The great advantage of these equations of motion is that the mass matrix of the system is constant, there are no Coriolis or centrifugal forces, and the Jacobi matrix $\mathbf{\Phi}_s$ is only a linear function of coordinates. The disadvantages are that the size of the system of equations of motion is often large and the interpretation of Lagrange's multipliers as real reaction forces is not straightforward.

9.5.1.7 Composite Rigid-Body Approach

The use of relative coordinates enables us to derive very efficient equations of motion based on the notion of composite rigid-body inertia from Section 9.2.1.1. Let us have a kinematic chain with n bodies and with only revolute or prismatic joints. Let us have two successive bodies named i and $p(i)$ (p as previous body) or $s(i)$ and i (s as subsequent body). On each of the two bodies several fixed points are defined. These are the so-called hinge points P_i and H_i on the kinematic joint and the two bodies (Figure 9.5), and the mass centre S_i of the particular body. To each body there are attached fixed coordinate systems with centre in point S_i or H_i with parallel coordinate axes. The inertia tensor of each body is expressed in these coordinate systems. The path from the coordinate system in point $H_{p(i)}$ or $S_{p(i)}$ of the body $p(i)$ to the corresponding coordinate system in point H_i or S_i of body i consists of translation to the point P_i, revolution around the axes going through the point P_i and translation to the point H_i or then further to the point S_i. Several different vectors and matrices are defined for the description of relative kinematics:

$\mathbf{d}_i = \mathbf{d}_{i,i}$	vector from H_i to S_i resolved in frame i
$\mathbf{z}_i = \mathbf{z}_{p(i),i}$	vector from P_i to H_i resolved in frame $p(i)$
$\mathbf{s}_i = \mathbf{s}_{i,i}$	vector from P_i to H_i resolved in frame i
$\mathbf{c}_i = \mathbf{c}_{p(i),i}$	vector from $H_{p(i)}$ to P_i resolved in frame $p(i)$
$\mathbf{r}_i = \mathbf{r}_{p(i),i} = \mathbf{c}_i + \mathbf{z}_i$	vector from $H_{p(i)}$ to H_i resolved in frame $p(i)$
$\mathbf{A}_i = \mathbf{A}_i^{p(i)} = \mathbf{S}_{p(i)i}^{-1}$	direction cosine matrix transforms vectors from frame $p(i)$ into frame i, $\mathbf{S}_{p(i)i}$ transforms vectors from frame i into frame $p(i)$, i.e. $\mathbf{a}_i = \mathbf{A}_i\mathbf{a}_{p(i)} = \mathbf{A}_i^{p(i)}\mathbf{a}_{p(i)} = \mathbf{S}_{p(i)i}^{-1}\mathbf{a}_{p(i)}$
$\mathbf{r}_{1S}$ (or $\mathbf{r}_{iS}$)	radius vector of centre of mass of body i resolved in the basic frame 1 (or in frame i)
$\mathbf{r}_{1,iH}$ (or $\mathbf{r}_{i,iH}$)	radius vector of hinge point H of body i resolved in the basic frame 1 (or in frame i)
$\mathbf{v}_{1S}$ (or $\mathbf{v}_{iS}$)	velocity vector of centre of mass of body i resolved in the basic frame 1 (or in frame i)
$\mathbf{v}_{1,iH}$ (or $\mathbf{v}_{i,iH}$)	velocity vector of hinge point H of body i resolved in the basic frame 1 (or in frame i)
$\mathbf{a}_{1S}$ (or $\mathbf{a}_{iS}$)	acceleration vector of centre of mass of body i resolved in the basic frame 1 (or in frame i)
$\mathbf{a}_{1,iH}$ (or $\mathbf{a}_{i,iH}$)	acceleration vector of hinge point H of body i resolved in the basic frame 1 (or in frame i)
$\boldsymbol{\omega}_{1,i}$ (or $\boldsymbol{\omega}_{i,i}$)	angular velocity vector of body i resolved in the basic frame 1 (or in frame i)
$\boldsymbol{\alpha}_{1,i}$ (or $\boldsymbol{\alpha}_{i,i}$)	angular acceleration vector of body i resolved in the basic frame 1 (or in frame i)
s_i	coordinate of the joint between body $p(i)$ and i.

Using matrix transformations for the kinematic chain from $H_{p(i)}$ to H_i and H_i to S_i we obtain

$$\begin{bmatrix} \mathbf{S}_i & \mathbf{r}_{1,iH} \\ 0 & 1 \end{bmatrix} = \begin{bmatrix} \mathbf{S}_{p(i)} & \mathbf{r}_{1,p(i)H} \\ 0 & 1 \end{bmatrix} \begin{bmatrix} \mathbf{E} & \mathbf{c}_i \\ 0 & 1 \end{bmatrix} \begin{bmatrix} \mathbf{S}_{p(i)i} & \mathbf{0} \\ 0 & 1 \end{bmatrix} \begin{bmatrix} \mathbf{E} & \mathbf{s}_i \\ 0 & 1 \end{bmatrix}$$

$$\begin{bmatrix} \mathbf{S}_i & \mathbf{r}_{1,iS} \\ 0 & 1 \end{bmatrix} = \begin{bmatrix} \mathbf{S}_i & \mathbf{r}_{1,iH} \\ 0 & 1 \end{bmatrix} \begin{bmatrix} \mathbf{E} & \mathbf{d}_i \\ 0 & 1 \end{bmatrix} \tag{9.168}$$

For the description of kinematic parameters we use the Denavit-Hartenberg notation (see Section 5.11.1). The great advantage is that in general between two successive bodies we need only two elementary rotations $\mathbf{S}_{p(i)i} = \mathbf{T}_{Z4}(\alpha_i)\mathbf{T}_{Z6}(\vartheta_i)$, and the vector $\mathbf{c}_i$ has generally only two non-zero elements. According to Denavit-Hartenberg's notation the axis of the kinematic pair is always the z-axis. Then from (9.168) we obtain

$$\mathbf{a}_i^H = \mathbf{C}_i \mathbf{a}_{p(i)}^H + \mathbf{\Phi}_i \ddot{s}_i + \boldsymbol{\eta}_i \tag{9.169}$$

where

$$\mathbf{C}_i = \begin{bmatrix} \mathbf{A}_i^{p(i)} & \mathbf{A}_i^{p(i)}(\tilde{\mathbf{c}}_i + \tilde{\mathbf{z}}_i)^T \\ 0 & \mathbf{A}_i^{p(i)} \end{bmatrix} \tag{9.170}$$

$$\boldsymbol{\eta}_i = \begin{bmatrix} \mathbf{A}_i^{p(i)}\left(\tilde{\boldsymbol{\omega}}_{p(i),p(i)}^2(\tilde{\mathbf{c}}_i + \tilde{\mathbf{z}}_i)\right) + \left(\mathbf{A}_i^{p(i)}\boldsymbol{\omega}_{p(i),p(i)}\right) \times 2\dot{\mathbf{s}}_i \\ \left(\mathbf{A}_i^{p(i)}\boldsymbol{\omega}_{p(i),p(i)}\right) \times \boldsymbol{\omega}_{i,p(i)i} \end{bmatrix} \tag{9.171}$$

and for the revolute kinematic pair

$$\mathbf{\Phi}_i = \lceil 0,0,0,0,0,1 \rceil \tag{9.172a}$$

and for the prismatic kinematic pair

$$\mathbf{\Phi}_i = \lceil 0,0,1,0,0,0 \rceil \tag{9.172b}$$

The matrix $\mathbf{C}_i = \mathbf{C}_i^{p(i)}$ enables us to describe the general transformation of accelerations, forces and composite rigid-body inertias

$$\begin{aligned} \mathbf{a}_j^H &= \mathbf{C}_j^i \mathbf{a}_i^H \\ \mathbf{F}_i^H &= \mathbf{C}_i^{jT} \mathbf{F}_i^H \\ \mathbf{I}_j^H &= \mathbf{C}_i^{jT} \mathbf{I}_i^H \mathbf{C}_i^j \end{aligned} \tag{9.173}$$

where

$$\mathbf{C}_i^j = \begin{bmatrix} \mathbf{A}_i^j & \mathbf{A}_i^j \tilde{\mathbf{r}}_{i,ij}^T \\ 0 & \mathbf{A}_i^j \end{bmatrix} \tag{9.174}$$

The direction cosine matrix $\mathbf{A}_j^i$ expresses the transformation from the coordinate system i to j, and the vector $\mathbf{r}_{i,ij}$ is the radius vector from the origin of the coordinate system i to j resolved in the coordinate system i.

Because of (9.88) and (9.102),

$$\frac{\partial \mathbf{v}_i^H}{\partial \dot{s}_j} = \frac{\partial \mathbf{a}_i^H}{\partial \ddot{s}_j} \tag{9.175}$$

and we obtain from (9.169)

$$\begin{aligned} \frac{\partial \mathbf{v}_i^H}{\partial \dot{s}_j} &= \mathbf{C}_i \frac{\partial \mathbf{v}_{p(i)}^H}{\partial \dot{s}_j}, \qquad i \neq j \\ \frac{\partial \mathbf{v}_i^H}{\partial \dot{s}_i} &= \mathbf{\Phi}_i \end{aligned} \tag{9.176}$$

We may now express

$$\mathbf{a}_i^H = \sum_k \frac{\partial \mathbf{v}_i^H}{\partial \dot{s}_k} \ddot{s}_k + \underline{\mathbf{a}}_i^H \tag{9.177}$$

where $\underline{\mathbf{a}}_i^H$ is the part of vector $\mathbf{a}_i^H$ which does not contain the second time derivatives $\ddot{s}_i$ of joint coordinates. It is a very important concept of the so-called acceleration remainder $\underline{\mathbf{a}}$ of vector $\mathbf{a}$. After time differentiations and modifications we obtain from (9.168)

$$\begin{aligned} \boldsymbol{\omega}_{i,i} &= \mathbf{S}_{p(i)i}^{-1} \boldsymbol{\omega}_{p(i),p(i)} + \boldsymbol{\omega}_{i,p(i)i} \\ \underline{\boldsymbol{\alpha}}_{i,i} &= \mathbf{S}_{p(i)i}^{-1} \underline{\boldsymbol{\alpha}}_{p(i),p(i)} + (\mathbf{S}_{p(i)i}^{-1} \boldsymbol{\omega}_{p(i),p(i)}) \times \boldsymbol{\omega}_{i,p(i)i} \\ \underline{\mathbf{a}}_{i,iH} &= \mathbf{S}_{p(i)i}^{-1} \big(\underline{\mathbf{a}}_{p(i),p(i)H} + (\underline{\tilde{\boldsymbol{\alpha}}}_{p(i),p(i)} + \tilde{\boldsymbol{\omega}}_{p(i),p(i)}^2)\mathbf{c}_i\big) + \\ &\quad + (\underline{\tilde{\boldsymbol{\alpha}}}_{i,i} + \tilde{\boldsymbol{\omega}}_{i,i}^2)\mathbf{s}_i + (\mathbf{S}_{p(i)i}^{-1} \boldsymbol{\omega}_{p(i),p(i)}) \times 2\dot{s}_1 \\ \underline{\mathbf{a}}_{i,iS} &= \underline{\mathbf{a}}_{i,iH} + (\underline{\tilde{\boldsymbol{\alpha}}}_{i,i} + \tilde{\boldsymbol{\omega}}_{i,i}^2)\mathbf{d}_i \end{aligned} \tag{9.178}$$

From (9.112), (9.111), (9.27), (9.176) and (9.177) we derive

$$\begin{aligned} M_{jk} &= \sum_i \left(\left(\frac{\partial \mathbf{v}_i^H}{\partial \dot{s}_j} \right)^T \mathbf{I}_i^H \left(\frac{\partial \mathbf{v}_i^H}{\partial \dot{s}_k} \right) \right) \\ Q_j &= \sum_i \left(\left(\frac{\partial \mathbf{v}_i^H}{\partial \dot{s}_j} \right) (-\mathbf{I}_i^H \underline{\mathbf{a}}_i^H + \mathbf{F}_i^H + \boldsymbol{\beta}_i^H) \right) \end{aligned} \tag{9.179}$$

Further, the equations of motion (9.27) may be written

$$\mathbf{I}_i^H \mathbf{a}_i^H = \mathbf{F}_i - \mathbf{C}_{s(i)}^T \mathbf{F}_{s(i)} + \mathbf{F}_{Ei}^H + \boldsymbol{\beta}_i^H \tag{9.180}$$

where $\mathbf{F}_i$ is the reaction force in the joint i acting on the body i, and $\mathbf{F}_{Ei}^H$ is the vector of the resulting applied forces acting on the body i resolved in point H_i. In the joint i there may also act the applied joint force (for example, an actuator) according to the degree of freedom of the joint χ_i. The nature of the joints will be described by the derived matrices $\mathbf{\Phi}_i$ (9.172). The structure of reaction force $\mathbf{F}_i$ of joint i is complementary to the structure of applied force of joint i (Brandl et al., 1988). For the revolute pair it is (M torque, F force)

$$\mathbf{F}_i = [F_{ix}, F_{iy}, F_{iz}, M_{ix}, M_{iy}, 0] \tag{9.181a}$$

and for the prismatic pair it is

$$\mathbf{F}_i = [F_{ix}, F_{iy}, 0, M_{ix}, M_{iy}, M_{iz}] \tag{9.181b}$$

Then the applied force $\mathbf{F}_{Ei}^H$ may be expressed as

$$\mathbf{F}_{Ei}^H = \mathbf{F}_{Ai}^H + \mathbf{\Phi}_i \chi_i - \mathbf{C}_{s(i)}^T \mathbf{\Phi}_{s(i)} \chi_{s(i)} \tag{9.182}$$

where $\mathbf{F}_{Ai}^H$ is the applied force without gravitational forces and without applied joint forces. From (9.182) it can be shown (Valášek, 1990a) or from (9.62) it follows that each element Q_j of the vector $\mathbf{Q}$ includes for a pure kinematic chain only χ_j

$$Q_j = \ldots + \chi_j \tag{9.183}$$

and the applied joint forces are easily computed.

It is very efficient to compute the effects of gravitational forces outside the applied force $\mathbf{F}_{Ai}^H$. The gravitational forces are easily computed as the influence of a non-inertial coordinate system in terms of acceleration remainders by substitution

$$\underline{\mathbf{a}}_{0,0}^H = \underline{\mathbf{a}}_{p(1),p(1)}^H = -\mathbf{g} \tag{9.184}$$

and then by successive application of (9.178). Hence, we obtain the ways of computing of the effects of gravitational forces and applied joint forces in (9.179).

The equations of motion (9.179) are the so-called direct equations of motion (terminology according to (Nielan, 1986) because of direct computation according to basic mechanical principles of virtual work (9.36)). They express in another form the structure of equations of the above described formalisms which can use the relative coordinates The structure of these equations of motion demonstrates the computational complexity of these formalisms. There are $\sim n^2$ elements of the matrix $\mathbf{M}$ and the computation of each element according to (9.179) needs $\sim n$ operations. There are $\sim n$ elements of the vector $\mathbf{Q}$ and the computation of each element according to (9.179) needs $\sim n$ operations. Thus the complexity of the computation of only the generalized mass matrix $\mathbf{M}$ is $O(n^3)$ and of the vector $\mathbf{Q}$ of generalized and applied forces is $O(n^2)$.

However, the computation of the matrices $\mathbf{M}$ and $\mathbf{Q}$ may be achieved in a more efficient way. If we substitute (9.176) into (9.179), we obtain for the first left upper terms of the generalized mass matrix $\mathbf{M}$ which is a symmetric matrix

$$\begin{aligned}
M_{n,n} &= \mathbf{\Phi}_n^T \mathbf{I}_n^H \mathbf{\Phi}_n \\
M_{n,n-1} &= \mathbf{\Phi}_n^T \mathbf{I}_n^H \mathbf{C}_n \mathbf{\Phi}_{n-1}
\end{aligned}$$

$$
\begin{aligned}
M_{n,n-2} &= \mathbf{\Phi}_n^T \mathbf{I}_n^H \mathbf{C}_n \mathbf{C}_{n-1} \mathbf{\Phi}_{n-2} \\
M_{n-1n} &= \mathbf{\Phi}_{n-1}^T \mathbf{C}_n^T \mathbf{I}_n^H \mathbf{\Phi}_n \\
M_{n-1n-1} &= \mathbf{\Phi}_{n-1}^T \mathbf{I}_{n-1}^H \mathbf{\Phi}_{n-1} + \mathbf{\Phi}_{n-1}^T \mathbf{C}_n^T \mathbf{I}_n^H \mathbf{C}_n \mathbf{\Phi}_{n-1} = \\
&= \mathbf{\Phi}_{n-1}^T (\mathbf{I}_{n-1}^H + \mathbf{C}_n^T \mathbf{I}_n^H \mathbf{C}_n) \mathbf{\Phi}_{n-1} \\
M_{n-1n-2} &= \mathbf{\Phi}_{n-1}^T \mathbf{I}_{n-1}^H \mathbf{C}_{n-1} \mathbf{\Phi}_{n-2} + \mathbf{\Phi}_{n-1}^T \mathbf{C}_n^T \mathbf{I}_n^H \mathbf{C}_n \mathbf{C}_{n-1} \mathbf{\Phi}_{n-2} = \\
&= \mathbf{\Phi}_{n-1}^T (\mathbf{I}_{n-1}^H + \mathbf{C}_n^T \mathbf{I}_n^H \mathbf{C}_n) \mathbf{C}_{n-1} \mathbf{\Phi}_{n-2} \\
M_{n-2n} &= \mathbf{\Phi}_{n-2}^T \mathbf{C}_{n-1}^T \mathbf{C}_n^T \mathbf{I}_n^H \mathbf{\Phi}_n \\
M_{n-2n-1} &= \mathbf{\Phi}_{n-2}^T \mathbf{C}_{n-1}^T \mathbf{I}_{n-1}^H \mathbf{\Phi}_{n-1} + \mathbf{\Phi}_{n-2}^T \mathbf{C}_{n-1}^T \mathbf{C}_n^T \mathbf{I}_n^H \mathbf{C}_n \mathbf{\Phi}_{n-1} = \\
&= \mathbf{\Phi}_{n-2}^T \mathbf{C}_{n-1}^T (\mathbf{I}_{n-1}^H + \mathbf{C}_n^T \mathbf{I}_n^H \mathbf{C}_n) \mathbf{\Phi}_{n-1} \\
M_{n-2n-2} &= \mathbf{\Phi}_{n-2}^T \mathbf{I}_{n-2}^H \mathbf{\Phi}_{n-2} + \mathbf{\Phi}_{n-2}^T \mathbf{C}_{n-1}^T \mathbf{I}_{n-1}^H \mathbf{C}_{n-1} \mathbf{\Phi}_{n-2} + \\
&\quad + \mathbf{\Phi}_{n-2}^T \mathbf{C}_{n-1}^T \mathbf{C}_n^T \mathbf{I}_n^H \mathbf{C}_n \mathbf{C}_{n-1} \mathbf{\Phi}_{n-2} = \\
&= \mathbf{\Phi}_{n-2}^T (\mathbf{I}_{n-2}^H + \mathbf{C}_{n-1}^T \mathbf{I}_{n-1}^H \mathbf{C}_{n-1} + \mathbf{C}_{n-1}^T \mathbf{C}_n^T \mathbf{I}_n^H \mathbf{C}_n \mathbf{C}_{n-1}) \mathbf{\Phi}_{n-2}
\end{aligned}
\tag{9.185}
$$

and similarly for the first elements of the vector of generalized forces

$$
\begin{aligned}
Q_n &= \mathbf{\Phi}_n^T (-\mathbf{I}_n^H \underline{\mathbf{a}}_n^H + \mathbf{F}_n^H + \beta_n^H) \\
Q_{n-1} &= \mathbf{\Phi}_{n-1}^T \mathbf{C}_n^T (-\mathbf{I}_n^H \underline{\mathbf{a}}_n^H + \mathbf{F}_n^H + \beta_n^H) + \mathbf{\Phi}_{n-1}^T (-\mathbf{I}_{n-1}^H \underline{\mathbf{a}}_{n-1}^H + \mathbf{F}_{n-1}^H + \\
&\quad + \beta_{n-1}^H) = \mathbf{\Phi}_{n-1}^T (\mathbf{C}_n^T (-\mathbf{I}_n^H \underline{\mathbf{a}}_n^H + \beta_n^H + \beta_n^H) + (-\mathbf{I}_{n-1}^H \underline{\mathbf{a}}_{n-1}^H + \\
&\quad + \mathbf{F}_{n-1}^H + \beta_{n-1}^H)) \\
Q_{n-2} &= \mathbf{\Phi}_{n-2}^T \mathbf{C}_{n-1}^T \mathbf{C}_n^T (-\mathbf{I}_n^H \underline{\mathbf{a}}_n^H + \mathbf{F}_n^H + \beta_n^H) + \mathbf{\Phi}_{n-2}^T \mathbf{C}_{n-1}^T (-\mathbf{I}_{n-1}^H \underline{\mathbf{a}}_{n-1}^H + \\
&\quad + \mathbf{F}_{n-1}^H + \beta_{n-1}^H) + \mathbf{\Phi}_{n-2}^T (-\mathbf{I}_{n-2}^H \underline{\mathbf{a}}_{n-2}^H + \mathbf{F}_{n-2}^H + \beta_{n-2}^H) = \\
&= \mathbf{\Phi}_{n-2}^T (\mathbf{C}_{n-1}^T \mathbf{C}_n^T (-\mathbf{I}_n^H \underline{\mathbf{a}}_n^H + \mathbf{F}_n^H + \beta_n^H) + \mathbf{C}_{n-1}^T (-\mathbf{I}_{n-1}^H \underline{\mathbf{a}}_{n-1}^H + \\
&\quad + \mathbf{F}_{n-1}^H + \beta_{n-1}^H) + (-\mathbf{I}_{n-2}^H \underline{\mathbf{a}}_{n-2}^H + \mathbf{F}_{n-2}^H + \beta_{n-2}^H))
\end{aligned}
\tag{9.186}
$$

The derived structure of equations (9.185) and (9.186) is the basis of the so-called composite-rigid-body method (or Walker–Orin method as method 3 in (Walker and Orin, 1982)), see also (Nielan, 1986), (Featherstone, 1987), (Bae et al., 1988). The generalized mass matrix is computed in such a way that we first compute the composite rigid-body inertias

$$
\begin{aligned}
\mathbf{I}_n &= \mathbf{I}_n^H \\
\mathbf{I}_{j-1} &= \mathbf{I}_{j-1}^H + \mathbf{C}_j^T \mathbf{I}_j \mathbf{C}_j, \qquad j = n, \ldots, 2
\end{aligned}
\tag{9.187}
$$

They are the basis of the elements on the diagonal. The particular elements of generalized mass matrix are then computed by multiplication of the composite rigid-body inertias from the right-hand side by $\mathbf{\Phi}_j$ and from the left-hand side successively by $\mathbf{\Phi}_j^T, \mathbf{\Phi}_{j-1}^T, \mathbf{C}_j^T, \mathbf{\Phi}_{j-2}^T \mathbf{C}_{j-1}^T \mathbf{C}_j^T, \ldots, \mathbf{\Phi}_1^T \mathbf{C}_2^T \ldots \mathbf{C}_{j-1}^T$. $.\mathbf{C}_j^T$. We obtain

$$
\begin{array}{rcll}
\mathbf{J}_j^j & = & \mathbf{I}_j \mathbf{\Phi}_j, & j = n, \ldots, 1 \\
\mathbf{J}_j^{k-1} & = & \mathbf{C}_k^T \mathbf{J}_j^k, & k = j, \ldots, 2 \\
M_{jk} & = & \mathbf{\Phi}_k^T \mathbf{J}_j^k, & k = j, \ldots, 1
\end{array}
\qquad (9.188)
$$

Similarly the elements of the vector of generalized forces are computed

$$
\left.
\begin{array}{rcl}
\mathbf{F}_j & = & -\mathbf{I}_j^H \underline{\mathbf{a}}_j^H + \mathbf{F}_j^H + \beta_j^H \\
\mathbf{K}_n & = & \mathbf{F}_n \\
\mathbf{K}_{j-1} & = & \mathbf{F}_{j-1} + \mathbf{C}_j^T \mathbf{K}_j \\
Q_j & = & \mathbf{\Phi}_j^T \mathbf{K}_j
\end{array}
\right\} j = n, \ldots, 2 \qquad (9.189)
$$

where the acceleration remainder $\underline{\mathbf{a}}_j^H$ is computed according to (9.178).

This scheme (9.187) – (9.189) of computation of elements of matrices in equations of motion (9.118) has the great advantage that the complexity of computation of a generalized mass matrix is only $O(n^2)$ and of the vector of generalized forces only $O(n)$. We compute the diagonal elements of the generalized mass matrix by successive transformation of generalized inertias of previous bodies and by successive addition to the next body; the elements outside the diagonal are then computed by some transformation from the corresponding element on the diagonal. The vector of generalized forces is computed as a successive addition of dynamic force acting on a body due to acceleration remainders and the transformed generalized force from the previous body. However, this scheme can be computed in different ways. These different ways can be explained from a mechanical point of view by a different interpretation but their common differences are the different coordinate systems in which the particular quantities are resolved (Valášek, 1990a). There are at least four quite different coordinate systems, because we can change the direction of coordinate axes and the origin of the coordinate system. The directions of coordinate axes can be either local, i.e. attached to a particular body, or global, i.e. the directions of axes of inertial (basic) coordinate system. Similarly the origin of the coordinate system can lie in the particular body or in the origin of the inertial (basic) coordinate system. We have these possibilities:

1. The local variant. The coordinate system is the local coordinate system of a particular body whose origin is in some point of the body (centre of mass or hinge points) and which is firmly attached to the body, i.e. axes of this system move with the body.

2. The global variant. The coordinate system is the inertial coordinate system, i.e. origin and axes of the basic frame.

3. The variant of local axes and global origin. The coordinate system is the coordinate system whose origin is in the origin of the inertial (basic) frame but whose axes have the direction of the local coordinate system of a particular body.

4. The variant of global axes and local origin. The coordinate system is the coordinate system whose origin is in some point of the body (centre of mass or hinge points) but whose coordinate axes have the directions of coordinate axes of the inertial (basic) coordinate system.

Further, we can also use different coordinate systems for the computation of the generalized mass matrix and for the computation of the generalized force vector. The other possibilities are given by the possibility of expressing the composite rigid-body inertias and the acceleration remainders in the centre of mass or in the hinge point of the particular body. All these possibilities have been investigated (Valášek, 1990a). These results have shown that the most efficient possibility for a small number ($n < 9$) of bodies in the kinematic chain is the first variant and asymptotically for a large number of bodies the second variant. The best way of computation is to express the composite rigid-body inertias in the hinge points and the acceleration remainders in the centres of mass (Valášek, 1990a). The reasons for this could be the following. The expressions for generalized mass matrix are transformed in any case to the hinge point so the best way is to realize this immediately. The computation of forces due to acceleration remainders needs the computation of many additional terms, the computational complexity of which is greater than the transformation of the resultant forces. Therefore the forces $\mathbf{F}_j$ in (9.189) are computed in the following way:

$$\mathbf{F}_j = \begin{bmatrix} \mathbf{E} & \mathbf{0} \\ \tilde{\mathbf{d}}_j & \mathbf{E} \end{bmatrix} (-\mathbf{I}_j^S \underline{\mathbf{a}}_j^S + \mathbf{F}_j^S + \boldsymbol{\beta}_j^S) \tag{9.190}$$

Let us again solve the example from Section 9.3 (Figure 9.2). The composite rigid-body approach increases the efficiency of equations of motion for kinematic chains. The equations of motion for a MBS with loops are then assembled using Lagrange's multipliers or help variables as in Sections 9.5.1.1–9.5.1.3. We demonstrate the assembling of equations of motion for the chain consisting of bodies 1 and 2. We use the relative coordinates φ_{01} and φ_{12}. The composite rigid-body inertia of body 2 is

$$\mathbf{I}_2 = \mathbf{I}_2^H = \begin{bmatrix} m_2 & 0 & 0 & 0 & 0 & 0 \\ 0 & m_2 & 0 & 0 & 0 & m_2 p \\ 0 & 0 & m_2 & 0 & -m_2 p & 0 \\ 0 & 0 & 0 & 0 & 0 & 0 \\ 0 & 0 & -m_2 p & 0 & m_2 p^2 & 0 \\ 0 & m_2 p & 0 & 0 & 0 & I_{2T} + m_2 p^2 \end{bmatrix} \tag{a}$$

According to (9.188) we obtain

$$\begin{aligned} \mathbf{J}_2^2 &= \mathbf{I}_2\mathbf{\Phi}_2 = [0, m_2p, 0, 0, 0, I_{2T} + m_2p^2] \\ M_{22} &= \mathbf{\Phi}_2^T\mathbf{J}_2^2 = I_{2T} + m_2p^2 \end{aligned} \tag{b}$$

The transformation matrix $\mathbf{C}_2$ is

$$\mathbf{C}_2 = \begin{bmatrix} \mathrm{c}\varphi_{12} & \mathrm{s}\varphi_{12} & 0 & 0 & 0 & l_1\mathrm{s}\varphi_{12} \\ -\mathrm{s}\varphi_{12} & \mathrm{c}\varphi_{12} & 0 & 0 & 0 & l_1\mathrm{c}\varphi_{12} \\ 0 & 0 & 1 & 0 & -l_1 & 0 \\ 0 & 0 & 0 & \mathrm{c}\varphi_{12} & \mathrm{s}\varphi_{12} & 0 \\ 0 & 0 & 0 & -\mathrm{s}\varphi_{12} & \mathrm{c}\varphi_{12} & 0 \\ 0 & 0 & 0 & 0 & 0 & 1 \end{bmatrix} \tag{c}$$

Then from (9.188),

$$\mathbf{J}_2^1 = \mathbf{C}_2^T\mathbf{J}_2^2 = [-m_2p\mathrm{s}\varphi_{12}, m_2p\mathrm{c}\varphi_{12}, 0, 0, 0, I_{2T} + m_2p^2 + m_2pl_1\mathrm{c}\varphi_{12}] \tag{d}$$

and finally

$$M_{21} = \mathbf{\Phi}_1^T\mathbf{J}_2^1 = I_{2T} + m_2p^2 + m_2pl_1\mathrm{c}\varphi_{12} \tag{e}$$

Further, according to (9.187)

$$\mathbf{C}_2^T\mathbf{I}_2^H\mathbf{C}_2 =$$

$$\begin{bmatrix} m_2 & 0 & 0 & 0 & 0 & -m_2p\mathrm{s}\varphi_{12} \\ 0 & m_2 & 0 & 0 & 0 & m_2A \\ 0 & 0 & m_2 & m_2p\mathrm{s}\varphi_{12} & -m_2A & 0 \\ 0 & 0 & m_2p\mathrm{s}\varphi_{12} & m_2p^2s^2\varphi_{12} & -m_2p\mathrm{s}\varphi_{12}A & 0 \\ 0 & 0 & -m_2A & -m_2p\mathrm{s}\varphi_{12}A & m_2A^2 & 0 \\ -m_2p\mathrm{s}\varphi_{12} & m_2A & 0 & 0 & 0 & I_{2T} + m_2B \end{bmatrix}$$

$$\begin{aligned} A &= p\mathrm{c}\varphi_{12} + l_1 \\ B &= l_1^2 + p^2 + 2pl_1\mathrm{c}\varphi_{12} \end{aligned} \tag{f}$$

and hence

$$\begin{aligned} \mathbf{J}_1^1 &= \mathbf{I}_1\mathbf{\Phi}_1 = (\mathbf{I}_1^H + \mathbf{C}_2^T\mathbf{I}_2^H\mathbf{C}_2)\mathbf{\Phi}_1 \\ M_{11} &= \mathbf{\Phi}_1^T\mathbf{J}_1^1 = I_{1T_1} + m_1p_1^2 + I_{2T} + m_2B \end{aligned} \tag{g}$$

Similarly from (9.189) and (9.190)

$$\mathbf{F}_2 = \begin{bmatrix} 1 & 0 & 0 & 0 & 0 & 0 \\ 0 & 1 & 0 & 0 & 0 & 0 \\ 0 & 0 & 1 & 0 & 0 & 0 \\ 0 & 0 & 0 & 1 & 0 & 0 \\ 0 & 0 & -p & 0 & 1 & 0 \\ 0 & p & 0 & 0 & 0 & 1 \end{bmatrix} \begin{bmatrix} -m_1a_{2Sx} \\ -m_2a_{2Sy} \\ 0 \\ 0 \\ 0 \\ -I_{2T}\alpha_{2z} \end{bmatrix} \tag{h}$$

where from (9.178)

$$\begin{aligned} \underline{\mathbf{a}}_{2S} &= \underline{\mathbf{a}}_{2,2S} = [-l_1 c\varphi_{12}\dot{\varphi}_{01}^2 - p(\dot{\varphi}_{01} + \dot{\varphi}_{12})^2, l_1 s\varphi_{12}\dot{\varphi}_{01}^2, 0] \\ \underline{\boldsymbol{\alpha}}_2 &= \underline{\boldsymbol{\alpha}}_{2,2} = [0, 0, 0] \end{aligned} \tag{i}$$

Then

$$Q_2 = \mathbf{\Phi}_2^T \mathbf{K}_2 = -m_2 p l_1 s\varphi_{12}\dot{\varphi}_{01}^2 \tag{j}$$

Further, from (9.189) we obtain

$$\begin{aligned} \mathbf{K}_1 &= \mathbf{F}_1 + \mathbf{C}_2^T \mathbf{K}_2 \\ Q_1 &= \mathbf{\Phi}_1^T \mathbf{K}_1 = m_2 l_1 p s\varphi_{12}(2\dot{\varphi}_{01} + \dot{\varphi}_{12})\dot{\varphi}_{12} \end{aligned} \tag{k}$$

The efficiency of these approaches is much more obvious for longer kinematic chains (see Section 9.5.3).

9.5.1.8 Inverse Multibody Formalism

Until now we have studied multibody formalisms only for the solution of the direct (forward or simulation) dynamic problem of MBS. The composite rigid-body approach enable us also to derive an efficient multibody formalism for the solution of inverse (backward or control) dynamic problems of MBS, especially of robots modelled by a kinematic chain. One method of solution is described in Section 8.1 by equations (8.20). The other method of solution follows from (9.183). According to this we may write

$$\mathbf{M}\ddot{\mathbf{q}} = \mathbf{Q} + \boldsymbol{\chi} \tag{9.191}$$

where the vector $\mathbf{Q}$ does not include the applied joint forces χ_i from the vector $\boldsymbol{\chi}$. From (9.191) together with (9.179) we derive

$$\boldsymbol{\chi} = (\mathbf{M}\ddot{\mathbf{q}} - \mathbf{Q}) = -\left[\sum_i \left(\left(\frac{\partial \mathbf{v}_i^H}{\partial \dot{s}_j}\right)^T (-\mathbf{I}_i^H \mathbf{a}_i^H + \mathbf{F}_i^H + \boldsymbol{\beta}_i^H)\right)\right] \tag{9.192}$$

The equation (9.192) can be efficiently computed from (9.189) and (9.190) by the substitution of the whole acceleration $\mathbf{a}_i^S$ for the acceleration remainder $\underline{\mathbf{a}}_i^S$ in (9.190). The procedure is that we compute the accelerations (9.169) using (9.178), we substitute them into (9.189) and then $\chi_j = Q_j$.

9.5.1.9 Residual Algorithm

A new and very promising approach which is based on the results of the last two sections is so-called residual algorithm (Eichberger et al., 1993, Andrzejewski et al., 1993). It uses the fact that the values of the mass matrix

$\mathbf{M}$ and its inverse $\mathbf{M}^{-1}$ change significantly only after several integration steps. It also uses the approach of modern DAE solvers (see Section 9.8) which require to compute the error — residuum of equations (9.118)

$$\mathbf{r} = \mathbf{M}\ddot{\mathbf{q}} - \mathbf{Q} \tag{9.193}$$

The direct evaluation of the residuum according (9.193) is very inefficient. However, if we apply the procedure from the previous section, we can compute the residuum in a very efficient way (see Section 9.5.3).

The MBS is described by relative coordinates. The solution of equations of motion has following steps

– in each integration step the residuum $\mathbf{r}$ (9.193) is being computed according to the Section 9.5.1.8

– in α-fraction of the integration steps the matrices $\mathbf{M}, \mathbf{Q}$ are computed according to (9.188)-(9.189) and the Choleski decomposition $\mathbf{M}^{-1} = (\mathbf{LDL}^T)^{-1}$ of the mass matrix is being computed, in other integration steps we use the previous decomposition

– in each integration step the new values of accelerations are computed

$$\begin{aligned} \Delta\ddot{\mathbf{q}} &= -\mathbf{M}^{-1}\mathbf{r} = -(\mathbf{LDL}^T)^{-1}\mathbf{r} && (9.194) \\ \ddot{\mathbf{q}}_{\text{new}} &= \ddot{\mathbf{q}}_{\text{old}} + \Delta\ddot{\mathbf{q}} && (9.195) \end{aligned}$$

- in each integration step $\dot{\mathbf{q}}$ and $\mathbf{q}$ are integrated from $\ddot{\mathbf{q}}$ and $\dot{\mathbf{q}}$.

This simple algorithm is valid only for MBS without loops. For MBS with loops we must apply the DAE solvers and compute also the Lagrange multipliers $\boldsymbol{\lambda}$ and their differences $\Delta\boldsymbol{\lambda}$. In the case of MBS with loops the above described procedure for the computation of residui (9.193) is directly applicable, only the forces $\mathbf{F}_i^H$ include the Lagrange multipliers

The value of α can be 0.05-0.1. The advantage of this approach is that the most computationally expensive step (evaluation of $\mathbf{M}$ and $\mathbf{M}^{-1}$) is computed only in the α-fraction of the total integration time.

9.5.2 Recursive Multibody Formalisms

The recursive multibody formalism is the method which computes the accelerations $\ddot{\mathbf{q}}$ in (9.118) by a sequence of recursive formulae without assembling and solving the system of linear algebraic equations (9.118) for the accelerations $\ddot{\mathbf{q}}$. In this way it reaches the computational complexity of only $O(n)$ operations. This formalism has been successively developed by (Armstrong, 1979), (Featherstone, 1983), (Brandl et al., 1988), (Bae and Haug, 1987) and others. We shall use the modified notation from (Brandl et al., 1988).

9.5.2.1 Open-Loop Systems

The simplest way of deriving a recursive multibody formalism is to use the equations of motion written for hinge points. We shall use the equations of motion (9.180). Let us have a kinematic chain which contains only revolute or prismatic pairs.

The equation of motion for the last body $i = n$ in the kinematic chain is

$$\mathbf{I}_i^H \mathbf{a}_i^H = \mathbf{F}_i + \mathbf{F}_{Ei}^H + \boldsymbol{\beta}_i^H \tag{9.196}$$

Let us multiply this equation from the left side by $\boldsymbol{\Phi}_i^T$.

$$\boldsymbol{\Phi}_i^T \mathbf{I}_i^H \mathbf{a}_i^H = \boldsymbol{\Phi}_i^T \mathbf{F}_i + \boldsymbol{\Phi}_i^T (\mathbf{F}_{Ei}^H + \boldsymbol{\beta}_i^H) \tag{9.197}$$

Because of the structure (9.181) of reaction force $\mathbf{F}_i$ corresponding to the structure of the matrix $\boldsymbol{\Phi}_i$ (9.172), the multiplication $\boldsymbol{\Phi}_i^T \mathbf{F}_i$ is always zero. It is another form of the orthogonality principle in mechanics ((Brandl et al., 1988), (Schiehlen, 1990)). We substitute (9.169) into (9.197)

$$\boldsymbol{\Phi}_i^T \mathbf{I}_i^H \boldsymbol{\Phi}_i \ddot{s}_i + \boldsymbol{\Phi}_i^T \mathbf{I}_i^H \mathbf{C}_i \mathbf{a}_{p(i)}^H + \boldsymbol{\Phi}_i^T \mathbf{I}_i^H \boldsymbol{\eta}_i = \boldsymbol{\Phi}_i^T (\mathbf{F}_{Ei}^H + \boldsymbol{\beta}_i^H) \tag{9.198}$$

We designate

$$\mathbf{M}_i = \boldsymbol{\Phi}_i^T \mathbf{I}_i^H \boldsymbol{\Phi}_i \tag{9.199}$$

and express $\ddot{s}_i$ from (9.198)

$$\ddot{s}_i = \mathbf{M}_i^{-1} \boldsymbol{\Phi}_i^T (\mathbf{F}_{Ei}^H + \boldsymbol{\beta}_i^H - \mathbf{I}_i^H (\mathbf{C}_i \mathbf{a}_{p(i)}^H + \boldsymbol{\eta}_i)) \tag{9.200}$$

Now we substitute (9.200) into (9.169) and then (9.169) into (9.196)

$$\mathbf{I}_i^H (\mathbf{C}_i \mathbf{a}_{p(i)}^H + \boldsymbol{\Phi}_i \mathbf{M}_i^{-1} \boldsymbol{\Phi}_i^T (\mathbf{F}_{Ei}^H + \boldsymbol{\beta}_i^H - \mathbf{I}_i^H (\mathbf{C}_i \mathbf{a}_{p(i)}^H + \boldsymbol{\eta}_i)) + \boldsymbol{\eta}_i) =$$

$$= \mathbf{F}_i + \mathbf{F}_{Ei}^H + \boldsymbol{\beta}_i^H \tag{9.201}$$

After modification

$$(\mathbf{E} - \mathbf{I}_i^H \boldsymbol{\Phi}_i \mathbf{M}_i^{-1} \boldsymbol{\Phi}_i^T) \mathbf{I}_i^H (\mathbf{C}_i \mathbf{a}_{p(i)}^H + \boldsymbol{\eta}_i) - (\mathbf{E} - \mathbf{I}_i^H \boldsymbol{\Phi}_i \mathbf{M}_i^{-1} \boldsymbol{\Phi}_i^T)(\mathbf{F}_{Ei}^H + \boldsymbol{\beta}_i^H) = \mathbf{F}_i$$

and designation

$$\begin{aligned} \mathbf{P}_i &= \mathbf{E} - \mathbf{I}_i^H \boldsymbol{\Phi}_i \mathbf{M}_i^{-1} \boldsymbol{\Phi}_i^T \\ \mathbf{N}_i &= (\mathbf{E} - \mathbf{I}_i^H \boldsymbol{\Phi}_i \mathbf{M}_i^{-1} \boldsymbol{\Phi}_i^T) \mathbf{I}_i^H = \mathbf{P}_i \mathbf{I}_i^H \\ \gamma_i &= \mathbf{N}_i \boldsymbol{\eta}_i - \mathbf{P}_i (\mathbf{F}_{Ei}^H + \boldsymbol{\beta}_i^H) \end{aligned} \tag{9.202}$$

we express the reaction force $\mathbf{F}_i$

$$\mathbf{F}_i = \mathbf{N}_i \mathbf{C}_i \mathbf{a}_{p(i)}^H + \gamma_i \tag{9.203}$$

The equation of motion for the predecessory body $p(i) = p(n)$ is

$$\mathbf{I}_{p(i)}^H \mathbf{a}_{p(i)}^H = \mathbf{F}_{p(i)} - \mathbf{C}_i^T \mathbf{F}_i + \mathbf{F}_{Ep(i)}^H + \boldsymbol{\beta}_{p(i)}^H \tag{9.204}$$

We substitute (9.203) and modify

$$(\mathbf{I}_{p(i)}^H + \mathbf{C}_i^T \mathbf{N}_i \mathbf{C}_i)\mathbf{a}_{p(i)}^H = \mathbf{F}_{p(i)} + \mathbf{F}_{Ep(i)}^H + \boldsymbol{\beta}_{p(i)}^H - \mathbf{C}_i^T \boldsymbol{\gamma}_i \tag{9.205}$$

If we now designate

$$\begin{aligned} \mathbf{I}_{p(i)}^{H^*} &= \mathbf{I}_{p(i)}^H + \mathbf{C}_i^T \mathbf{N}_i \mathbf{C}_i \\ \boldsymbol{\beta}_{p(i)}^{H^*} &= \boldsymbol{\beta}_{p(i)}^H - \mathbf{C}_i^T \gamma_i \end{aligned} \tag{9.206}$$

we obtain from (9.205)

$$\mathbf{I}_{p(i)}^{H^*} \mathbf{a}_{p(i)}^H = \mathbf{F}_{p(i)} + \mathbf{F}_{Ep(i)}^H + \boldsymbol{\beta}_{p(i)}^{H^*} \tag{9.207}$$

the equation of motion for the body $p(i) = p(n)$ which has the same structure as the equation of motion for the previous body i, i.e. $\mathbf{F}_{s(i)} = \mathbf{0}$, and this process can continue until $i = 1$. Then in the equation (9.200) the acceleration $\mathbf{a}_{p(1)}^H$ is known from (9.184) and the equation (9.200) can be solved for $\ddot{s}_1$. From the known $\ddot{s}_1$ we can recursively compute from (9.169) and (9.200) all the accelerations $\mathbf{a}_i^H$ and $\ddot{s}_i$ and if necessary from (9.203) all the reaction forces $\mathbf{F}_i$. The whole recursive multibody algorithm therefore follows. For all bodies along the kinematic chain $(i = 1, \ldots, n)$ we compute the quantities $\mathbf{c}_i, \boldsymbol{\Phi}_i, \boldsymbol{\eta}_i, \mathbf{F}_{Ei}^H, \boldsymbol{\beta}_i^H$ (first forward iteration). For the last body n we let $\mathbf{I}_n^{H*}$ and $\boldsymbol{\beta}_n^{H*}$ equal $\mathbf{I}_n^H$ and $\boldsymbol{\beta}_n^H$. Then we compute from the end to the beginning of the kinematic chain $(i = n, \ldots, 1)$ the quantities $\mathbf{M}_i, \mathbf{P}_i, \mathbf{N}_i, \boldsymbol{\gamma}_i, \mathbf{I}_{p(i)}^{H^*}$ and $\boldsymbol{\beta}_{p(i)}^{H^*}$ according to (9.199), (9.202) and (9.206) where the composite rigid-body inertias $\mathbf{I}_i^H$ are substituted by the articulated-body inertias $\mathbf{I}_i^{H^*}$ just computed (backward iteration). We then use (9.184), and for all bodies along the kinematic chain $(i = 1, \ldots, n)$ we compute the accelerations $\ddot{s}_i$ and $\mathbf{a}_i^H$ according to (9.200) and (9.169) (second forward iteration). Simultaneously we may also compute the reaction forces $\mathbf{F}_i$ from (9.203).

In this algorithm the only difficult step from the point of view of computational complexity is the computation of $\mathbf{C}_i^T \mathbf{N}_i \mathbf{C}_i$. In this computation we need to compute several times $\mathbf{A}_i^T \mathbf{N}_{ij} \mathbf{A}_i$ for some matrix $\mathbf{N}_{ij}$ of the type (3,3). For the computation of $\mathbf{A}_i^T \mathbf{N}_{ij} \mathbf{A}_i$, it is very important that the direction cosine matrix $\mathbf{A}_i$ can always be expressed as the multiplication of two basic direction cosine matrices for basic rotation around the z and x axes. We can then use the property of an invariant part of matrix $\mathbf{N}_{ij}$

with respect to the planar rotation (Brandl et al., 1988) for very efficient computation of the transformation $\mathbf{A}_i^T \mathbf{N}_{ij} \mathbf{A}_i$. We have

$$\mathbf{A}_i^T \mathbf{N}_{ij} \mathbf{A}_i = \mathbf{S}_{Z4} \mathbf{S}_{Z6} \mathbf{N}_{ij} \mathbf{S}_{Z6}^T \mathbf{S}_{Z4}^T = \mathbf{S}_{Z4} (\mathbf{S}_{Z6} \mathbf{N}_{ij} \mathbf{S}_{Z6}^T) \mathbf{S}_{Z4}^T \tag{9.208}$$

The computation of

$$\mathbf{R}^T \mathbf{B} \mathbf{R} = \begin{bmatrix} c\vartheta & -s\vartheta \\ s\vartheta & c\vartheta \end{bmatrix} \begin{bmatrix} a & b \\ c & d \end{bmatrix} \begin{bmatrix} c\vartheta & s\vartheta \\ -s\vartheta & c\vartheta \end{bmatrix} \tag{9.209}$$

is then realized as

$$\mathbf{R}^T \mathbf{B} \mathbf{R} = \mathbf{B}_1 + \mathbf{R}^T \mathbf{B}_2 \mathbf{R} \tag{9.210}$$

$$\mathbf{B} = \mathbf{B}_1 + \mathbf{B}_2$$

$$\mathbf{B}_1 = \begin{bmatrix} (a+d)/2 & (b-c)/2 \\ -(b-c)/2 & (a+d)/2 \end{bmatrix}, \quad \mathbf{B}_2 = \begin{bmatrix} (a-d)/2 & (b+c)/2 \\ (b+c)/2 & -(a-d)/2 \end{bmatrix} \tag{9.211}$$

and

$$\mathbf{R}^T \mathbf{B}_1 \mathbf{R} = \mathbf{B}_1 \tag{9.212}$$

is a valid relation.

These formulae are the basis for efficient computation of the transformation of articulated body inertias $\mathbf{N}_i$. The last part of this algorithm will also be used for the efficient computation of the transformation of the inertia tensor in the previous section.

The matrix $\mathbf{N}_i$ is the so-called articulated-body inertia (Featherstone, 1983, 1987) which is capable of expressing the influence of the motion of bodies $s(i)$ on the motion of body i. The basic idea of the formalism is therefore recursively to transform the dynamic (inertia) characteristics of bodies into previous bodies in the kinematic chain. But this transformation of complex inertia characteristics is the most expensive part of the algorithm from the computational point of view. The transformation of composite rigid-body inertias $\mathbf{I}_j$ in the previous section is very similar to the transformation of articulated-body inertias but the great difference is the symmetry of the structure of composite rigid-body inertias and articulated-body inertias. The articulated-body inertias are generally only symmetric (6,6) matrices and therefore they are described by 21 values. On the other hand, the composite rigid-body inertias include a symmetric (3,3) matrix of classical inertia tensor $\mathbf{I}_{jH}$, skew-symmetric (3,3) left lower and right upper matrices describing the same vector $m_j \mathbf{d}_j$ and a diagonal (3,3) matrix $m_j \mathbf{E}$. The composite rigid-body inertias are therefore described by only 10 values. This difference in values which must be transformed in order to transform a particular inertia quantity is the fundamental reason for the different computational costs for kinematic chains with few bodies (see

Section 9.5.3). This formalism can also be expressed in different coordinate systems as in Section 9.5.1.6. The most efficient formulation is in the local system (Valášek, 1990a).

9.5.2.2 Closed-Loop Systems

Recursive formalisms have also been developed for MBS's with loops. There are two approaches. One (Brandl et al., 1987), (Bae and Haug, 1987) uses Lagrange multipliers for the cutting of loops and has the computational complexity $O(n)+O(l^3)$ where l is the number of loops. The other (Valášek, 1990b), (Valášek and Stejskal, 1990) uses the kinematic solution of MBS with loops from Section 5.11.6 and could decrease the computational complexity for MBS's even more down to only $O(n)$.

In the first approach we cut all loops in the MBS. The equation of motion of the last body i of the kinematic chain of a cut loop is described not by (9.196) but by the equation (9.180) where, however, the reaction force $\mathbf{F}_{s(i)}$ contains the Lagrange multipliers $\boldsymbol{\lambda}$ corresponding to the constraints of the cut loop. The acceleration of the hinge point k of the cut kinematic pair may be described from the last body $i = p(k)$ of one kinematic chain

$$\mathbf{a}_k = \mathbf{C}_{s(i)}\mathbf{a}_i^H + \boldsymbol{\Phi}_{s(i)}\ddot{s}_{s(i)} + \boldsymbol{\eta}_{s(i)} \tag{9.213}$$

and from the last body j of another kinematic chain which completes the loop

$$\mathbf{a}_k = \mathbf{C}_{s(j)}\mathbf{a}_j^H + \boldsymbol{\eta}_{s(j)} \tag{9.214}$$

where

$$\mathbf{C}_{s(j)} = \begin{bmatrix} \mathbf{E} & \tilde{\mathbf{r}}_{s(j)}^T \\ \mathbf{0} & \mathbf{E} \end{bmatrix} \tag{9.215}$$

$$\boldsymbol{\eta}_{s(j)} = \begin{bmatrix} \boldsymbol{\omega}_{j,j} \times (\boldsymbol{\omega}_{j,j} \times \mathbf{c}_{s(j)}) \\ \mathbf{0} \end{bmatrix} \tag{9.216}$$

The constraint conditions are obtained from the equality of (9.213) and (9.214) which is pre-multiplied by the matrix $\boldsymbol{\Psi}_{s(i)}^T$

$$\boldsymbol{\Psi}_{s(i)}^T\left(\mathbf{C}_{s(i)}\mathbf{a}_i^H + \boldsymbol{\Phi}_{s(i)}\ddot{s}_{s(i)} + \boldsymbol{\eta}_{s(i)}\right) = \boldsymbol{\Psi}_{s(i)}^T\left(\mathbf{C}_{s(j)}\mathbf{a}_j^H + \boldsymbol{\eta}_{s(j)}\right) \tag{9.217}$$

where, for the revolute pair,

$$\boldsymbol{\Psi}_{s(i)} = \begin{bmatrix} 1 & 0 & 0 & 0 & 0 \\ 0 & 1 & 0 & 0 & 0 \\ 0 & 0 & 1 & 0 & 0 \\ 0 & 0 & 0 & 1 & 0 \\ 0 & 0 & 0 & 0 & 1 \\ 0 & 0 & 0 & 0 & 0 \end{bmatrix} \tag{9.218a}$$

and, for the prismatic pair,

$$\mathbf{\Psi}_{s(i)} = \begin{bmatrix} 1 & 0 & 0 & 0 & 0 \\ 0 & 1 & 0 & 0 & 0 \\ 0 & 0 & 0 & 0 & 0 \\ 0 & 0 & 1 & 0 & 0 \\ 0 & 0 & 0 & 1 & 0 \\ 0 & 0 & 0 & 0 & 1 \end{bmatrix} \tag{9.218b}$$

It is obvious that $\mathbf{\Psi}^T_{s(i)}\mathbf{\Phi}_{s(i)} = \mathbf{0}$ and from (9.217) we obtain the constraint condition

$$\mathbf{\Psi}^T_{s(i)}\mathbf{C}_{s(i)}\mathbf{a}^H_i - \mathbf{\Psi}^T_{s(i)}\mathbf{C}_{s(j)}\mathbf{a}^H_j = \mathbf{\Psi}^T_{s(i)}(\boldsymbol{\eta}_{s(j)} - \boldsymbol{\eta}_{s(i)}) \tag{9.219}$$

The recursive solution of both kinematic chains follows as in the case of an open-loop system. However, the solutions (9.200) and (9.203) contain the Lagrange multipliers $\boldsymbol{\lambda}$ corresponding to the constraints (9.219). We substitute these solutions into (9.169) and obtain

$$\begin{aligned} \mathbf{a}^H_i &= \mathbf{D}_i\mathbf{a}^H_0 + \mathbf{B}_i\boldsymbol{\lambda} \\ \mathbf{a}^H_j &= \mathbf{D}_j\mathbf{a}^H_0 + \mathbf{B}_j\boldsymbol{\lambda} \end{aligned} \tag{9.220}$$

with known matrices $\mathbf{D}$ and $\mathbf{B}$. After substitution of (9.220) into (9.219) we obtain the following system of linear algebraic equations for the Lagrange multipliers $\boldsymbol{\lambda}$

$$\begin{aligned} \boldsymbol{\lambda} = \left(\mathbf{\Psi}^T_{s(i)}\left(\mathbf{C}_{s(i)}\mathbf{B}_i - \mathbf{C}_{s(j)}\mathbf{B}_j\right)\right)^{-1} \mathbf{\Psi}^T_{s(i)}\left(\left(\mathbf{C}_{s(j)}\mathbf{B}_j - \right.\right. \\ \left.\left. -\mathbf{C}_{s(i)}\mathbf{B}_i\right)\mathbf{a}^H_0 + \boldsymbol{\eta}_{s(j)} - \boldsymbol{\eta}_{s(i)}\right) \end{aligned} \tag{9.221}$$

All variables can then be completely computed.

In the second approach we use the kinematic solution of the MBS with loops. The equations of motion of MBS, especially in the recursive formalism, can be derived by the synthetic method or by the analytic method. In the derivation by the synthetic method, all reaction forces and their elimination are considered in detail. In the derivation by the analytic method the reaction forces and their elimination are not considered because the reaction forces are according to d'Alembert's principle non-working forces and they do not enter the final equations of motion.

Let us again consider the simple open kinematic chain. The equations of motion (9.199)–(9.207) can be derived from Gauss's principle (as in (Bae and Haug, 1987))

$$\sum_i \partial(\mathbf{a}^H_i)^T(\mathbf{I}^H_i\mathbf{a}^H_i - \boldsymbol{\beta}^H_i) = 0 \tag{9.222}$$

We rearrange (9.222)

$$\sum_{i=1}^{n-1} \partial(\mathbf{a}_i^H)^T(\mathbf{I}_i^H\mathbf{a}_i^H - \boldsymbol{\beta}_i^H) + \partial(\mathbf{a}_n^H)(\mathbf{I}_n^H\mathbf{a}_n^H - \boldsymbol{\beta}_n^H) = 0 \tag{9.223}$$

From (9.169) we have

$$\partial\mathbf{a}_i^H = \mathbf{C}_i\partial\mathbf{a}_{p(i)}^H + \boldsymbol{\Phi}_i\partial\ddot{s}_i \tag{9.224}$$

and substitute into (9.223)

$$\sum_{i=1}^{n-1} \partial(\mathbf{a}_i^H)^T(\mathbf{I}_i^H\mathbf{a}_i^H - \boldsymbol{\beta}_i^H) + \partial(\mathbf{a}_{n-1}^H)^T\mathbf{C}_n^T(\mathbf{I}_n^H\mathbf{a}_n^H - \boldsymbol{\beta}_n^H) +$$

$$+ \partial\ddot{s}_n\boldsymbol{\Phi}_n^T(\mathbf{I}_n^H\mathbf{a}_n^H - \boldsymbol{\beta}_n^H) = 0 \tag{9.225}$$

Because $\partial\ddot{s}_n$ is independent we obtain from (9.225) the equations of motion

$$\sum_{i=1}^{n-1} \partial(\mathbf{a}_i^H)^T(\mathbf{I}_i^H\mathbf{a}_i^H - \boldsymbol{\beta}_i^H) + \partial(\mathbf{a}_{n-1}^H)^T\mathbf{C}_n^T(\mathbf{I}_n^H\mathbf{a}_n^H - \boldsymbol{\beta}_n^H) = 0 \tag{9.226}$$

$$\boldsymbol{\Phi}_n^T(\mathbf{I}_n^H\mathbf{a}_n^H - \boldsymbol{\beta}_n^H) = 0 \tag{9.227}$$

We substitute from (9.169) into (9.227)

$$\boldsymbol{\Phi}_n^T(\mathbf{I}_n^H(\mathbf{C}_n\mathbf{a}_{n-1}^H + \boldsymbol{\Phi}_n\ddot{s}_n + \boldsymbol{\eta}_n) - \boldsymbol{\beta}_n^H) = 0 \tag{9.228}$$

From (9.228) we express $\ddot{s}_n$ as in (9.200). Now we substitute (9.200) into (9.169), (9.169) into (9.226) and modify

$$\mathbf{a}_n^H = (\mathbf{E} - \boldsymbol{\Phi}_n\mathbf{M}_n^{-1}\boldsymbol{\Phi}_n^T\mathbf{I}_n^H)(\mathbf{C}_n\mathbf{a}_{n-1}^H + \boldsymbol{\eta}_n) + \boldsymbol{\Phi}_n\mathbf{M}_n^{-1}\boldsymbol{\Phi}_n^T\boldsymbol{\beta}_n^H \tag{9.229}$$

$$\sum_{i=1}^{n-2} \partial(\mathbf{a}_i^H)^T(\mathbf{I}_i^H\mathbf{a}_i^H - \boldsymbol{\beta}_i^H) + \partial(\mathbf{a}_{n-1}^H)^T((\mathbf{I}_{n-1}^H + \mathbf{C}_n^T\mathbf{I}_n^H\mathbf{P}_n\mathbf{C}_n)\mathbf{a}_{n-1}^H -$$

$$- (\boldsymbol{\beta}_{n-1}^H - \mathbf{C}_n^T g_n)) = 0 \tag{9.230a}$$

$$\sum_{i=1}^{n-2} \partial(\mathbf{a}_i^H)^T(\mathbf{I}_i^H\mathbf{a}_i^H - \boldsymbol{\beta}_i^H) + \partial(\mathbf{a}_{n-1}^H)^T(\mathbf{I}_{n-1}^{H^*}\mathbf{a}_{n-1}^H - \boldsymbol{\beta}_{n-1}^{H^*}) = 0 \tag{9.230b}$$

We have obtained the recursive equations of motion for the last body (9.200) and the transformation of the whole kinematic chain into the kinematic chain without the last body. Hence we can recursively continue until we reach the first body $n = 1$ in the chain. Then the equation (9.200) is

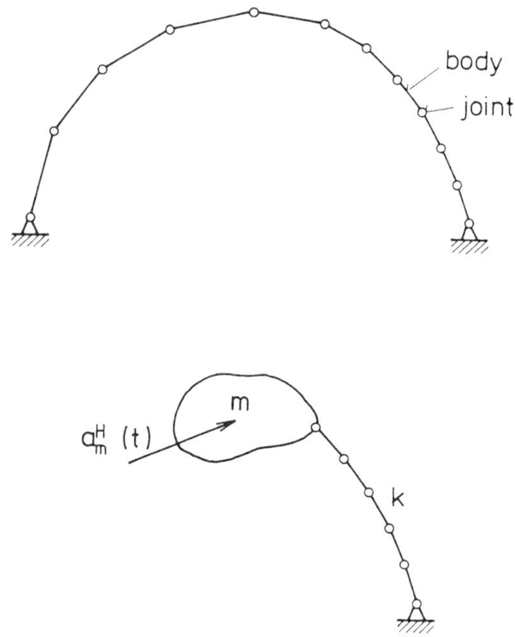

Figure 9.6. MBS with external loop.

solvable because the acceleration $\mathbf{a}_{p(1)}^{H}$ is given. Hence we can recursively compute accelerations by (9.169) and (9.200) from the first to the last body in the kinematic chain.

Let us apply this method of derivation of equations of motion by the recursive multibody formalism to the MBS's with kinematic loops. We suppose that kinematic joints have one DOF. We must distinguish two kinds of loops:

– external loops, which are closed through the frame (for example Figure 9.6)

– internal loops, which are closed by moving bodies (for example Figure 9.7)

In the case of external loops (Figure 9.6), we determine the kinematic solution from which the accelerations of the last 5 bodies k are expressed as functions of the acceleration of the previous body m. We use redundant coordinates s_k which are the relative coordinates of kinematic joints. The recursive relation (9.169) is valid for all bodies k. The kinematic solution,

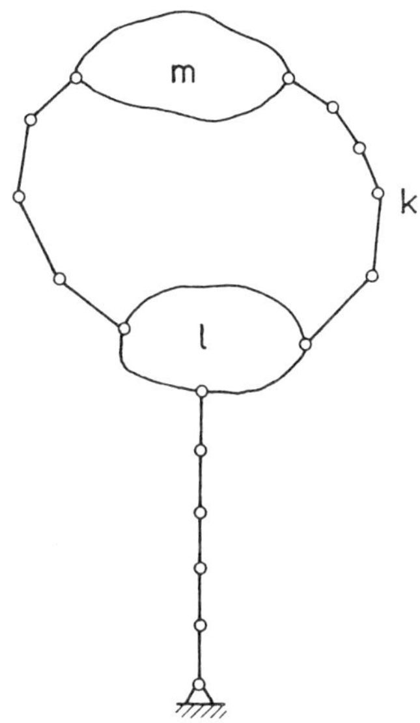

Figure 9.7. MBS with internal loop.

which is for example achieved by the methods from Section 5.11.6., gives

$$\ddot{s}_k = \mathbf{D}_k \mathbf{a}_m^H + b_k \tag{9.231}$$

with distribution matrix $\mathbf{D}_k$ and scalar b_k. After substitution of (9.231) into (9.169) we obtain

$$\mathbf{a}_k^H = \mathbf{C}_{km}\mathbf{a}_m^H + \boldsymbol{\eta}_{km} \tag{9.232}$$
$$\partial\mathbf{a}_k^H = \mathbf{C}_{km}\partial\mathbf{a}_m^H \tag{9.233}$$

The equations of motion of such MBS's are also formulated by (9.222). After modification we have

$$\sum_{i=1}^{m-1} \partial(\mathbf{a}_i^H)^T(\mathbf{I}_i^H\mathbf{a}_i^H - \boldsymbol{\beta}_i^H) + \partial(\mathbf{a}_m^H)^T(\mathbf{I}_m^H\mathbf{a}_m^H - \boldsymbol{\beta}_m^H) +$$
$$+ \sum_k \partial(\mathbf{a}_k^H)^T(\mathbf{I}_k^H\mathbf{a}_k^H - \boldsymbol{\beta}_k^H) = 0 \tag{9.234}$$

The difference between (9.222) and (9.234) is that the virtual accelerations $\partial\mathbf{a}_k^H$ in (9.234) are not independent. Therefore we cannot apply the method of derivation of equations of motion directly. We substitute (9.233) into (9.234) and modify

$$\sum_{i=1}^{m-1} \partial(\mathbf{a}_i^H)^T(\mathbf{I}_i^H \mathbf{a}_i^H - \boldsymbol{\beta}_i^H) + \partial(\mathbf{a}_m^H)^T(\mathbf{I}_m^H \mathbf{a}_m^H - \boldsymbol{\beta}_m^H) +$$

$$+ \sum_k \partial(\mathbf{a}_m^H)^T \mathbf{C}_{km}^T (\mathbf{I}_k^H (\mathbf{C}_{km}\mathbf{a}_m^H + \boldsymbol{\eta}_{km}) - \boldsymbol{\beta}_k^H) = 0$$

$$\sum_{i=1}^{m-1} \partial(\mathbf{a}_i^H)^T(\mathbf{I}_i \mathbf{a}_i^H - \boldsymbol{\beta}_i^H) + \partial(\mathbf{a}_m^H)^T \left((\mathbf{I}_m^H + \sum_k \mathbf{C}_{km}^T \mathbf{I}_k^H \mathbf{C}_{km})\mathbf{a}_m^H - \right.$$

$$\left. -(\boldsymbol{\beta}_m^H + \sum_k \mathbf{C}_{km}^T(-\mathbf{I}_k^H \boldsymbol{\eta}_{km} + \boldsymbol{\beta}_k^H) \right) = 0$$

$$\sum_{i=1}^{m-1} \partial(\mathbf{a}_i^H)^T(\mathbf{I}_i \mathbf{a}_i^H - \boldsymbol{\beta}_i^H) + \partial(\mathbf{a}_m^H)^T(\mathbf{I}_m^{H^*} \mathbf{a}_m^H - \boldsymbol{\beta}_m^{H^*}) = 0 \tag{9.235}$$

Now all the virtual accelerations $\partial \mathbf{a}_i^H, \partial \mathbf{a}_m^H$ are independent and thus we can apply the derivation of equations of motion from (9.222) to (9.230) in the same way. By this procedure the equations of motion of a loop are transformed into the equations of motion of an open kinematic chain.

In the case of internal loops (Figure 9.7) the derivation of equations of motion is more difficult. We use redundant coordinates s_k which are the relative coordinates of kinematic joints. We determine the kinematic solution from which the accelerations of 5 bodies k are expressed as functions of the accelerations of the bodies m and l. The recursive relation (9.169) is still valid for all bodies k. The kinematic solution, which is for example achieved by the methods from Section 5.11.6, gives

$$\ddot{s}_k = \mathbf{D}_k \mathbf{a}_m^H + \mathbf{D}_l \mathbf{a}_l^H + b_k \tag{9.236}$$

with some distribution matrices $\mathbf{D}_k, \mathbf{D}_l$ and scalar b_k. After substitution of (9.236) into (9.169) we obtain

$$\mathbf{a}_k^H = \mathbf{C}_{km}\mathbf{a}_m^H + \mathbf{C}_{kl}\mathbf{a}_l^H + \boldsymbol{\eta}_{km} \tag{9.237}$$

$$\partial \mathbf{a}_k^H = \mathbf{C}_{km}\partial \mathbf{a}_m^H + \mathbf{C}_{kl}\partial \mathbf{a}_l^H \tag{9.238}$$

The equations of motion of such MBS's are also formulated by (9.222). After modification we have

$$\sum_{i=1}^{l} \partial(\mathbf{a}_i^H)^T(\mathbf{I}_i^H \mathbf{a}_i^H - \boldsymbol{\beta}_i^H) + \sum_{i=l+1}^{m} \partial(\mathbf{a}_i^H)^T(\mathbf{I}_i^H \mathbf{a}_i^H - \boldsymbol{\beta}_i^H) +$$

$$+\sum_k \partial(\mathbf{a}_k^H)^T(\mathbf{I}_k^H\mathbf{a}_k^H - \boldsymbol{\beta}_k^H) = 0 \tag{9.239}$$

The virtual accelerations $\partial\mathbf{a}_k^H$ in (9.239) are again not independent as in (9.234). We must again first substitute (9.237) and (9.238) into (9.239) and modify

$$\sum_{i=1}^{l}\partial(\mathbf{a}_i^H)^T(\mathbf{I}_i^H\mathbf{a}_i^H - \boldsymbol{\beta}_i^H) + \sum_{i=l+1}^{m}\partial(\mathbf{a}_i^H)^T(\mathbf{I}_i^H\mathbf{a}_i^H - \boldsymbol{\beta}_i^H) +$$

$$+\sum_k(\mathbf{C}_{km}\partial\mathbf{a}_m^H + \mathbf{C}_{kl}\partial\mathbf{a}_l^H)^T(\mathbf{I}_k^H(\mathbf{C}_{km}\mathbf{a}_m^H + \mathbf{C}_{kl}\mathbf{a}_l^H + \boldsymbol{\eta}_{km}) - \boldsymbol{\beta}_k^H) = 0$$

$$\sum_{i=1}^{l}\partial(\mathbf{a}_i^H)^T(\mathbf{I}_i^H\mathbf{a}_i^H - \boldsymbol{\beta}_i^H) + \sum_{i=l+1}^{m-1}\partial(\mathbf{a}_i^H)^T(\mathbf{I}_i^H\mathbf{a}_i^H - \boldsymbol{\beta}_i^H) +$$
$$+\partial(\mathbf{a}_m^H)^T(\mathbf{I}_m^{H^*}\mathbf{a}_m^H + \mathbf{I}_{ml}^H\mathbf{a}_l^H - \boldsymbol{\beta}_m^{H^*}) +$$
$$+\partial(\mathbf{a}_l^H)^T(\mathbf{I}_{lm}^H\mathbf{a}_m^H + \mathbf{I}_l^{H^*}\mathbf{a}_l^H - \boldsymbol{\beta}_l^{H^*}) = 0 \tag{9.240}$$

where

$$\boldsymbol{\beta}_m^{H^*} = \boldsymbol{\beta}_m^H + \sum_k \mathbf{C}_{km}^T\boldsymbol{\beta}_k^H - \sum_k \mathbf{C}_{km}^T\mathbf{I}_k^H\boldsymbol{\eta}_{km}$$
$${}^m\boldsymbol{\beta}_l^{H^*} = \boldsymbol{\beta}_l^H + \sum_k \mathbf{C}_{kl}^T\boldsymbol{\beta}_k^H - \sum_k \mathbf{C}_{kl}^T\mathbf{I}_k^H\boldsymbol{\eta}_{kl}$$

$$\mathbf{I}_m^{H^*} = \mathbf{I}_m^H + \sum_k \mathbf{C}_{km}^T\mathbf{I}_k^H\mathbf{C}_{km}, \quad \mathbf{I}_{ml}^H = \sum_k \mathbf{C}_{km}^T\mathbf{I}_k^H\mathbf{C}_{kl}$$

$$\mathbf{I}_{lm}^H = \sum_k \mathbf{C}_{kl}^T\mathbf{I}_k^H\mathbf{C}_{km}, \qquad {}^m\mathbf{I}_l^{H^*} = \sum_k \mathbf{C}_{kl}^T\mathbf{I}_k^H\mathbf{C}_{kl} \tag{9.241}$$

From (9.169) for body m we obtain

$$\mathbf{a}_m^H = \mathbf{C}_m\mathbf{a}_{p(m)}^H + \boldsymbol{\Phi}_m\ddot{s}_m + \boldsymbol{\eta}_m \tag{9.242}$$

$$\partial\mathbf{a}_m^H = \mathbf{C}_m\partial\mathbf{a}_{p(m)}^H + \boldsymbol{\Phi}_m\partial\ddot{s}_m \tag{9.243}$$

We substitute (9.242) and (9.243) into (9.240)

$$\sum_{i=1}^{l}\partial(\mathbf{a}_i^H)^T(\mathbf{I}_i^H\mathbf{a}_i^H - \boldsymbol{\beta}_i^H) + \sum_{i=l+1}^{m-1}\partial(\mathbf{a}_i^H)^T(\mathbf{I}_i^H\mathbf{a}_i^H - \boldsymbol{\beta}_i^H) +$$

$$+(\mathbf{C}_m\partial\mathbf{a}_{p(m)}^H + \boldsymbol{\Phi}_m\partial\ddot{s}_m)^T\left(\mathbf{I}_m^{H^*}(\mathbf{C}_m\mathbf{a}_{p(m)}^H + \boldsymbol{\Phi}_m\ddot{s}_m + \boldsymbol{\eta}_m) +\right.$$

$$+\mathbf{I}_{ml}^{H}\mathbf{a}_{l}^{H} - \boldsymbol{\beta}_{m}^{H^{*}}\Big) + \partial(\mathbf{a}_{l}^{H})^{T}(\mathbf{I}_{lm}^{H}\mathbf{a}_{m}^{H} + {}^{m}\mathbf{I}_{l}^{H^{*}}\mathbf{a}_{l}^{H} - {}^{m}\boldsymbol{\beta}_{l}^{H^{*}}) = 0 \qquad (9.244)$$

Because $\partial\ddot{s}_m$ is independent we obtain from (9.244) the equations of motion

$$\sum_{i=1}^{l}\partial(\mathbf{a}_{i}^{H})^{T}(\mathbf{I}_{i}^{H}\mathbf{a}_{i}^{H} - \boldsymbol{\beta}_{i}^{H}) + \sum_{i=l+1}^{m-1}\partial(\mathbf{a}_{i}^{H})^{T}(\mathbf{I}_{i}^{H}\mathbf{a}_{i}^{H} - \boldsymbol{\beta}_{i}^{H}) + (\mathbf{C}_{m}\partial\mathbf{a}_{p(m)}^{H})^{T}\cdot$$

$$\cdot(\mathbf{I}_{m}^{H*}(\mathbf{C}_{m}\mathbf{a}_{p(m)}^{H} + \boldsymbol{\Phi}_{m}\ddot{s}_{m} + \boldsymbol{\eta}_{m}) + \mathbf{I}_{ml}^{H}\mathbf{a}_{l}^{H} - \boldsymbol{\beta}_{m}^{H^{*}})+$$

$$+\,\partial(\mathbf{a}_{l}^{H})^{T}(\mathbf{I}_{lm}^{H}\mathbf{a}_{m}^{H} + {}^{m}\mathbf{I}_{l}^{H^{*}}\mathbf{a}_{l}^{H} - {}^{m}\boldsymbol{\beta}_{l}^{H^{*}}) = 0 \qquad (9.245)$$

$$\boldsymbol{\Phi}_{m}^{T}(\mathbf{I}_{m}^{H^{*}}(\mathbf{C}_{m}\mathbf{a}_{p(m)}^{H} + \boldsymbol{\Phi}_{m}\ddot{s}_{m} + \boldsymbol{\eta}_{m}) + \mathbf{I}_{ml}^{H}\mathbf{a}_{l}^{H} - \boldsymbol{\beta}_{m}^{H*}) = 0 \qquad (9.246)$$

Then from (9.246) we express $\ddot{s}_m$, substitute it into (9.242) and these new expressions for $\ddot{s}_m$ and $\mathbf{a}_m^H$ are substituted back into the equations of motion (9.245) and after modification we obtain

$$\sum_{i=1}^{l}\partial(\mathbf{a}_{i}^{H})^{T}(\mathbf{I}_{i}^{H}\mathbf{a}_{i}^{H} - \boldsymbol{\beta}_{i}^{H}) + \sum_{i=l+1}^{m-2}\partial(\mathbf{a}_{i}^{H})^{T}(\mathbf{I}_{i}^{H}\mathbf{a}_{i}^{H} - \boldsymbol{\beta}_{i}^{H})+$$

$$+\partial(\mathbf{a}_{p(m)}^{H})^{T}(\mathbf{I}_{p(m)}^{H^{*}}\mathbf{a}_{p(m)}^{H} + \mathbf{I}_{p(m)l}^{H}\mathbf{a}_{l}^{H} - \boldsymbol{\beta}_{p(m)}^{H^{*}})+$$

$$+\,\partial(\mathbf{a}_{l}^{H})^{T}(\mathbf{I}_{lp(m)}^{H}\mathbf{a}_{p(m)}^{H} + {}^{p(m)}\mathbf{I}_{l}^{H^{*}}\mathbf{a}_{l}^{H} - {}^{p(m)}\boldsymbol{\beta}_{l}^{H^{*}}) = 0 \qquad (9.247)$$

The equation of motion (9.247) has the same structure as the equation of motion (9.240) but it has been recursively reduced by one body m. This recursion can continue until $p(m) = l$. Then instead of equation (9.247) we obtain the equation

$$\sum_{i=1}^{l-1}\partial(\mathbf{a}_{i}^{H})^{T}(\mathbf{I}_{i}^{H}\mathbf{a}_{i}^{H} - \boldsymbol{\beta}_{i}^{H}) + \partial(\mathbf{a}_{l}^{H})^{T}(\mathbf{I}_{l}^{H^{*}}\mathbf{a}_{l}^{H} - \boldsymbol{\beta}_{l}^{H^{*}}) = 0 \qquad (9.248)$$

where

$$\begin{aligned}\mathbf{I}_{l}^{H^{*}} &= \mathbf{I}_{p(m)}^{H^{*}} + \mathbf{I}_{p(m)l}^{H} + \mathbf{I}_{lp(m)}^{H} +^{p(m)}\mathbf{I}_{l}^{H^{*}}\\ \boldsymbol{\beta}_{l}^{H^{*}} &= \boldsymbol{\beta}_{p(m)}^{H^{*}} +^{p(m)}\boldsymbol{\beta}_{l}^{H^{*}}\end{aligned} \qquad (9.249)$$

This equation of motion is exactly the same as the equation of motion (9.223) for the simple open kinematic chain and the further derivation is the same as above.

If the computational complexity of kinematic solution is only $O(n)$ and the number of accelerations on which the dependent acceleration $\mathbf{a}_k^H$ depends (for example in equation (9.237) two) is limited to some fixed finite number, then the computational complexity of the dynamic solution is also only $O(n)$. This algorithm rapidly decreases the computational complexity of the solution of the dynamic problem of MBS. However, the proof of their linear computational complexity for general MBS's is still an open question. Further investigation is also necessary in order to study the application of these algorithms to singular cases in the mechanics of MBS's.

Let us solve the example from Section 9.3. (Figure 9.2) once more. We shall use the first method from this section. We cut the loop in the revolute pair A. The Lagrange multipliers corresponding to the constraints of this pair create the reaction force $\mathbf{F}_{s(2)}$, where

$$\mathbf{F}_{s(2)} = \lceil \lambda_1, \lambda_2, 0, 0, 0, 0 \rceil \tag{a}$$

The articulated-body inertia $\mathbf{I}_2^H$ is the same as the composite rigid-body inertia in Section 9.5.1.7. According to (9.199) and (9.202) we obtain

$$\mathbf{M}_2 = \mathbf{\Phi}_2^T \mathbf{I}_2^H \mathbf{\Phi}_2 = [M_2] = \left[I_{2T} + m_2 p^2\right] \tag{b}$$

$$\mathbf{N}_2 = \begin{bmatrix} m_2 & 0 & 0 & 0 & 0 & 0 \\ 0 & \frac{I_{2T} m_2}{M_2} & 0 & 0 & 0 & 0 \\ 0 & 0 & m_2 & 0 & -m_2 p & 0 \\ 0 & 0 & 0 & 0 & 0 & 0 \\ 0 & 0 & -m_2 p & 0 & m_2 p^2 & 0 \\ 0 & 0 & 0 & 0 & 0 & 0 \end{bmatrix}$$

$$\begin{aligned}
\boldsymbol{\eta}_1 &= \lceil 0, 0, 0, 0, 0, 0 \rceil \\
\boldsymbol{\beta}_1 &= \lceil m_1 p_1 \dot{\varphi}_{01}^2, 0, 0, 0, 0, 0 \rceil \\
\boldsymbol{\eta}_2 &= \lceil -l_1 c\varphi_{12} \dot{\varphi}_{01}^2, l_1 s\varphi_{12} \dot{\varphi}_{01}^2, 0, 0, 0, 0 \rceil \\
\boldsymbol{\beta}_2 &= \lceil m_2 p (\dot{\varphi}_{01} + \dot{\varphi}_{12})^2, 0, 0, 0, 0, 0 \rceil \\
\boldsymbol{\gamma}_2 &= \lceil -m_2 (l_1 c\varphi_{12} \dot{\varphi}_{01}^2 + p(\dot{\varphi}_{01} + \dot{\varphi}_{12})^2) - (\lambda_1 c\varphi_{23} - \lambda_2 s\varphi_{23}), \\
&\quad -(\lambda_1 s\varphi_{23} + \lambda_2 c\varphi_{23}) + m_2 (p l_2 (\lambda_1 s\varphi_{23} + \lambda_2 c\varphi_{23}) + \\
&\quad + I_{2T} l_1 s\varphi_{12} \dot{\varphi}_{01}^2)/M_2, 0, 0, 0, 0 \rceil
\end{aligned} \tag{d}$$

Then according to (9.206) we obtain the articulated-body inertia $\mathbf{I}_1^{H^*}$. As a first step we need $\mathbf{C}_2^T \mathbf{N}_2 \mathbf{C}_2$

$$\begin{aligned}
A &= 1 - m_2 p^2 c^2 \varphi_{12}/M_2, \quad B = m_2 p^2 s\varphi_{12}/M_2 \\
C &= B c\varphi_{12}, \quad D = p c\varphi_{12} + l_1, \quad E = p s\varphi_{12}
\end{aligned}$$

$$\mathbf{C}_2^T \mathbf{N}_2 \mathbf{C}_2 =$$

$$= \begin{bmatrix} m_2(1 - Bs\varphi_{12}) & & & & sym. & \\ m_2C & m_2A & & & & \\ 0 & 0 & m_2 & & & \\ 0 & 0 & m_2E & m_2E^2 & & \\ 0 & 0 & -m_2D & -m_2DE & m_2D^2 & \\ m_2l_1C & m_2l_1A & 0 & 0 & 0 & m_2l_1^2A \end{bmatrix} \quad \text{(e)}$$

Further

$$\begin{aligned} \beta_1^{H^*} &= [m_1p_1\dot{\varphi}_{01}^2 - (\gamma_2)_1c\varphi_{12} + (\gamma_2)_2s\varphi_{12}, \\ &\quad -((\gamma_2)_1s\varphi_{12} + (\gamma_2)_2c\varphi_{12}), 0, 0, 0, \\ &\quad -l_1((\gamma_2)_1s\varphi_{12} + (\gamma_2)_2c\varphi_{12})] \end{aligned} \quad \text{(f)}$$

From (9.199) we compute

$$\begin{aligned} \mathbf{M}_1 &= \mathbf{\Phi}_1^T\mathbf{I}_1^{H^*}\mathbf{\Phi}_1 = \mathbf{\Phi}_1^T(\mathbf{I}_1^H + \mathbf{C}_2^T\mathbf{N}_2\mathbf{C}_2)\mathbf{\Phi}_1 = [M_1] = \\ &= \left[I_{1T_1} + m_1p_1^2 + m_2l_1^2\left(1 - \frac{m_2p^2c^2\varphi_{12}}{M_2}\right)\right] \end{aligned} \quad \text{(g)}$$

and then from (9.200)

$$\ddot{s}_1 = \ddot{\varphi}_{01} = \frac{1}{M_1}(\mathbf{f}_{E1}^H + \beta_1^{H^*})_6 \quad \text{(h)}$$

Finally from (9.169) and (9.200)

$$\begin{aligned} \ddot{s}_2 = \ddot{\varphi}_{12} = -\ddot{\varphi}_{01} + \frac{1}{M_2}(-pm_2(c\varphi_{12}l_1\ddot{\varphi}_{01} + s\varphi_{12}l_1\dot{\varphi}_{01}^2) + \\ + l_2(\lambda_1s\varphi_{23} + \lambda_2c\varphi_{23})) \end{aligned} \quad \text{(i)}$$

Both relative accelerations are functions of Lagrange multipliers λ_1, and λ_2. Hence, the acceleration $\mathbf{a}_A$ in (9.213)–(9.214) of the hinge point of the revolute pair A is also the function of these multipliers. Similarly we express the acceleration $\mathbf{a}_A$ from the motion of body 3. Then the condition (9.217) gives five equations. However, three of them are identically zero because of the planar problem. From the remaining two equations we determine the Lagrange multipliers λ_1 and λ_2, and then all the accelerations.

9.5.3 Computational Complexity of Multibody Formalisms

The most important criterion for the choice of one of the multibody formalisms as the most suitable is the resultant computational complexity.

Research into this subject is, however, not yet complete. It is a fairly difficult subject, as we demonstrate in Section 9.8. The resultant solution is in general performed by the evaluation of symbolically assembled expressions, the numerically performed treatment of equations of motion and their actual integration. For each part of these solutions, there are more alternatives available and they have different properties for various kinds of MBS.

The computational complexity of the solution of inverse dynamic problems has been investigated for a long time. Several methods have been developed. The solution of inverse dynamic problems involves the calculation of joint actuation forces or torques, given the current kinematic state of the robot (position, velocity, acceleration of each joint), and it is performed only by evaluation of assembled expressions. The methods went through a stormy development procedure as described in (Hollerbach, 1980; Toogood and Warkentin, 1986; Toogood, 1987; Neuman and Murray, 1985, 1987ab). The number of operations ranged from n^4-dependence (n is the number of elements of a robot simple open chain, (Lagrange's equations (Uicker, 1985)), through n-dependence (Lagrange's equations, $(412n - 277)$ multiplications (Hollerbach, 1980), Newton-Euler equations, $(150n - 48)$ multiplications (Luh et al., 1980; Hollerbach, 1980), $(130n - 68)$ multiplications and $(101n - 56)$ additions (Featherstone, 1987), the composite rigid-body approach $(97n - 112)$ multiplications and $(85n - 105)$ additions (Valášek, 1990a)) (see Section 9.5.1.8) to individually assembled or treated (customized) equations. For example, in the case of the Stanford arm ($n = 6$) we obtain from Kane's equations 646 multiplications (Kane and Levinson, 1983), from Lagrange's equations 298 multiplications (Toogood, 1987) and from the composite rigid-body approach 171 multiplications (Valášek, 1990a).

Not so much is known about the solution of the general MBS direct dynamic problem. A great variety of MBS structures and possible choices of coordinates and numerical integration methods are used. The general comparison of multibody formalisms is therefore very difficult. It is shown (Unda et al., 1986) that the natural coordinates method is more efficient than the DADS method. Several facts are known about the computational complexity of composite rigid-body and recursive articulated-body formalisms for a simple open chain containing n quite general revolute pairs and described by relative coordinates. The composite rigid-body formalism needs $(3n^2 + 186n - 36)$ multiplications and $(2.5n^2 + 105.5n - 78)$ additions for assembling the equations and $1/6(n^3 + 9n^2 - 4n)$ multiplications and $1/6(n^3 + 6n^2 - 7n)$ additions for the Choleski decomposition using the global variant from Section 9.5.1.7 (Bae et al., 1988), and $(10n^2 + 112n - 174)$ multiplications and $(6n^2 + 111n - 188)$ additions for assembling the equations

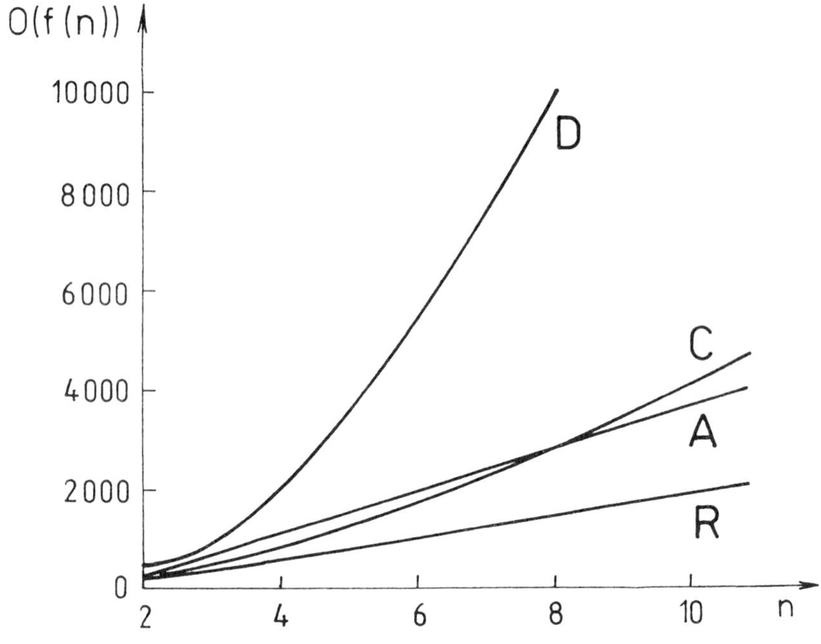

Figure 9.8. Computational complexity of direct equations of motion (D), composite rigid-body (C), recursive articulated-body inertia (A) and residual (R) formalisms.

using the local variant from Section 9.5.1.7 (Valášek, 1990a). For $n < 12$ the local variant is much better than the global variant, but both variants are worse than the recursive articulated-body formalism for $n > 8$. The recursive articulated-body formalism needs $(250n - 222)$ multiplications and $(220n - 198)$ additions (Brandl et al., 1986) and $(226n - 343)$ multiplications and $(206n - 345)$ additions (Valášek, 1990a). Comparisons have also been made (Valášek, 1990a) between the methods of direct equations of motion, composite rigid-body inertias and recursive articulated-body inertias (Figure 9.8). The method of direct equations of motion is worse than the others for all numbers of revolute pairs. The composite rigid-body (local variant) is the best when up to eight revolute pairs are involved and the recursive articulated-body is then asymptotically the best (both implementations from (Valášek, 1990a)). The newest residual algorithm requires only $\alpha((10n^2+112n-174+1/6(n^3+9n^2-4n))+(1-\alpha)(97n-112)$ multiplications and $\alpha((6n^2 + 111n - 188 + 1/6(n^3 + 6n^2 - 7n)) + (1 - \alpha)(85n - 105)$ additions. For $\alpha = 0.1$ see Figure 9.8. This method is better than any other method for $n < 61$ bodies and for greater number of bodies it is worse than articulated-body formalism. This is the partial conclusion which is at present available.

The computational complexity of various multibody formalisms also strongly depends on efficient implementation. It is based particularly on the technique of layering, i.e. the substitution of intermediate expressions by new symbols as in the example from Section 9.5.1.3 (Valášek, 1990a).

9.6 Physical Interpretation of Lagrange Multipliers

Most formalisms for the assembly of equations of motion usually eliminate the unknown reaction forces from the solved equations. In many cases, however, it is necessary to know the magnitudes of the reaction forces; this may be subsequently necessary, for example for the dimensioning of mechanism elements, or the reaction forces may be required for expressing other forces, for example Coloumb friction forces. In such a case, the reaction forces must be introduced into the MBS solution. For the Newton-Euler approach, the reaction forces of individual kinematic pairs in equations of motion are introduced in a natural way according to the natural interpretation – see Chapter 8. For the Lagrange approach, the reaction forces are expressed by Lagrange's multipliers. However, their physical meaning depends to a great extent on the choice of MBS description. We are now going to derive the conditions required for the Lagrange multipliers in LEMT in order to have the physical interpretation of reaction forces according to Chapter 8.

Let us again examine a general spatial MBS. From the viewpoint of structure, this is a mixed chain formed by several open and closed chains. Let us cut the MBS into two parts which we shall denote as the left-hand and right-hand parts. We describe the MBS by such coordinates s_j that the cut is of no concern to any coordinate, i.e. none of the coordinates describes the mutual (relative) position of the left-hand and right-hand MBS parts. In that case, the coordinates s_j can be divided into coordinates $s_j^L(s_j^R)$ describing the position of the left-hand (right-hand) MBS part. We express the conditions for connecting the left-hand and right-hand parts in the real MBS by constraint conditions, which are in general nonholonomic

$$g_l(s_j^L, \dot{s}_j^L, \ldots, s_j^{(h)L}, s_j^R, \dot{s}_j^R, \ldots, s_j^{(h)R}, t) = 0 \tag{9.250}$$

Note that we use Lagrange's principle of constraint removal (Dobronravov, 1976). We substitute the constraint conditions (9.250) in the disconnected MBS by the equivalent action of components of reaction forces. In the left-hand MBS part we may conceive them as actually interpretable forces R_l^L acting in coordinates u_l^L from the right-hand part on the left-hand part and analogously for the right-hand part. The number of components equals the number of degrees of freedom removed from the relative motion of both MBS parts by re-connecting the cut.

Due to the complete separation of the left-hand and right-hand MBS parts (i.e. for the left-hand part coordinates s_j^L, expression of energy $T^L = T^L(s_j^L, \dot{s}_j^L, t)$, of generalized force Q_j^L and possibly constraint conditions g_k inside the left-hand part for coordinates s_j) equations of motion e.g. for the left-hand MBS part can be written

$$\frac{\mathrm{d}}{\mathrm{d}t}\frac{\partial T^L}{\partial \dot{s}_j^L} - \frac{\partial T^L}{\partial s_j^L} = Q_j^L + \sum_k \lambda_k^L \frac{\partial g_k^L}{\partial s_j^{(h)L}} + \sum_l R_l^L \frac{\partial u_l^L}{\partial s_j^L} \tag{9.251}$$

A similar expression can be written for the right-hand MBS part. We now assemble the equations of motion for the whole MBS. The total kinetic energy is given by the sum

$$T(s_j, \dot{s}_j, t) = T^L(s_j^L, \dot{s}_j^L, t) + T^R(s_j^R, \dot{s}_j^R, t) \tag{9.252}$$

General forces acting on the whole MBS are to be obtained by joining forces Q_j^L and Q_j^R. The constraint condition for the MBS without disconnection are also to be obtained by combining conditions g_k^L containing coordinates s_j^L and conditions q_k^R written in coordinates s_j^R. We must further supply the constraint conditions (9.250) describing the connection of the left-hand and right-hand MBS parts. As we shall see shortly these conditions can always be expressed as functions of coordinates u_l^L, u_l^R on which equivalent force effects act

$$\begin{aligned} & g_l(s_j^L, \dot{s}_j^L, \ldots, s_j^{(h)L}, s_j^R, \dot{s}_j^R, \ldots, s_j^{(h)R}, t) = \\ & = G_l(u_l^L(s_j^L, t), \ldots, u_l^{(h)L}(s_j^L, t), u_l^R(s_j^R, t), \ldots, u_l^{(h)R}(s_j^R, t), t) = 0 \end{aligned} \tag{9.253}$$

Because the coordinates s_j^L and s_j^R are always related only to the left-hand and right-hand MBS parts respectively, the constraint conditions (9.253) can always be expressed as the difference of boundary coordinates

$$\begin{aligned} g_l(s_j^L, \ldots, s_j^R, \ldots, t) &= G_l(u_l^L, \ldots, u_l^R, \ldots, t) = \\ &= G_l^L(u_l^L, \ldots, t) - G_l^R(u_l^R, \ldots, t) = 0 \end{aligned} \tag{9.254}$$

For the equations of motion of the whole MBS written in coordinates s_j^L,

$$\frac{\mathrm{d}}{\mathrm{d}t}\frac{\partial T}{\partial \dot{s}_j^L} - \frac{\partial T}{\partial s_j^L} = Q_j^L + \sum_k \lambda_k^L \frac{\partial g_k^L}{\partial s_j^{(h)L}} + \sum_l \lambda_l \frac{\partial g_l}{\partial s_j^{(h)L}} \tag{9.255}$$

As the forces Q_j^R do not act on coordinates s_j^L, the terms g_k^R do not contain coordinates s_j^L, and

$$\frac{\partial g_l^R(s_j^R,\ldots,t)}{\partial s_j^{(h)L}} = 0$$

Further, from (9.252), it follows that

$$\frac{\mathrm{d}}{\mathrm{d}t}\frac{\partial T}{\partial \dot{s}_j^L} - \frac{\partial T}{\partial s_j^L} = \frac{\mathrm{d}}{\mathrm{d}t}\frac{\partial T^L}{\partial \dot{s}_j^L} - \frac{\partial T^L}{\partial s_j^L} \tag{9.256}$$

Substituting (9.256) into (9.255) and comparing with (9.251), we obtain

$$\sum_l R_l^L \frac{\partial u_l^L}{\partial s_j^L} = \sum_m \lambda_m \frac{\partial g_m}{\partial s_j^{(h)L}} \tag{9.257}$$

From equations (9.253), however

$$\frac{\partial g_m}{\partial s_j^{(h)L}} = \sum_l \frac{\partial G_m}{\partial u_l^{(h)L}} \frac{\partial u_l^{(h)L}}{\partial s_j^{(h)L}} = \sum_l \frac{\partial G_m}{\partial u_l^{(h)L}} \frac{\partial u_l^L}{\partial s_j^L} \tag{9.258}$$

as, analogously to (9.88),

$$\frac{\partial u_l^{(h)L}}{\partial s_j^{(h)L}} = \frac{\partial u_l^L}{\partial s_j^L}$$

We substitute (9.258) into (9.257) and alter, to give

$$\begin{aligned}
\sum_l R_l^L \frac{\partial u_l^L}{\partial s_j^L} &= \sum_m \lambda_m \sum_l \frac{\partial G_m}{\partial u_l^{(h)L}} \frac{\partial u_l^L}{\partial s_j^L} = \\
&= \sum_l \sum_m \lambda_m \frac{\partial G_m}{\partial s_j^{(h)L}} \frac{\partial u_l^L}{\partial s_j^L}
\end{aligned}$$

$$\sum_l \left(R_l^L - \sum_m \lambda_m \frac{\partial G_m}{\partial u_l^{(h)L}} \right) \frac{\partial u_l^L}{\partial s_j^L} = 0 \tag{9.259}$$

Due to the independence and therefore the arbitrariness of $\partial u_l^L / \partial s_j^L$, it follows from (9.259) that

$$R_l^L = \sum_m \lambda_m \frac{\partial G_m}{\partial u_l^{(h)L}} \tag{9.260}$$

Relations (9.260) yield a generally valid interpretation of Lagrange multipliers as actually interpretable forces. For example, in a MBS with a holonomic constraint $G_l = u_l^L - u_l^R$ in the disconnected kinematic pair, $R_l^L = \lambda_l$ and $R_l^R = -\lambda_l$. It is obvious from the derivation why it is necessary to maintain discipline in describing the left-hand MBS part (kinetic energy, generalized forces, constraint conditions) using coordinates s_j^L only and similarly for the right-hand part. Lagrange multipliers always have the meaning of force effects; however in MBS descriptions where the above principle is not kept, this convenient interpretation does not hold. For example, λ_1 and λ_2 from example in Section 9.3. (Figure 9.2) have the interpretation of x and y components of reaction in joint A, as the MBS description is divided in point A, i.e. the kinetic energy of element 2 is expressed using φ_1 and φ_2. Should we require λ_1 and λ_2 to have the meaning of x, y components of reaction in joint B, we would have to express the kinetic energy of element 2 using φ_3 and φ_2 in the form

$$
\begin{aligned}
T_2 &= \frac{1}{2} m_2(\dot{x}_2^2 + \dot{y}_2^2) + \frac{1}{2} I_{2T} \dot{\varphi}_2^2 \\
x_2 &= l_3 s\varphi_3 + (l_2 - p) s\varphi_2 \\
y_2 &= l_0 - l_3 c\varphi_3 - (l_2 - p) c\varphi_2
\end{aligned}
$$

In that case, the expression of constraint conditions is the same for both cases.

It is not possible using this MBS description to ensure that λ_1 and λ_2 have the meaning of x, y components of reaction in joint C or D. For example, to express reactions in joint C, the angle φ_1 would have to express the angle CBA.

When, on dividing the MBS description by a cut the above principle is not maintained, the easily interpretable meaning of Lagrange multipliers does not apply. In the above case, this would correspond, for example, to the following expression of kinetic energy of element 2

$$
\begin{aligned}
T_2 &= \frac{1}{2} m_2(\dot{x}_2^2 + \dot{y}_2^2) + \frac{1}{2} I_{2T} \dot{\varphi}_2^2 \\
x_2 &= l_1 s\varphi_1 - p s\varphi_2 \\
y_2 &= l_0 - l_3 c\varphi_3 - (l_2 - p) c\varphi_2
\end{aligned}
$$

However, to determine the real reaction forces directly from (9.260) is not simple. They may be more readily determined using the structural equivalence (9.111) derived in Section 9.4. We now derive for individual kinematic pairs the constraint conditions and the corresponding interpretation of Lagrange multipliers for the determination of reaction forces in the respective kinematic pair.

Let us have a MBS with a certain kinematic pair, the reaction of which we must determine. We describe the MBS by coordinates s_j in such a way as to be able to divide it into left-hand and right-hand parts maintaining the above principles. We describe the left-hand and right-hand MBS parts using the respective coordinates s_j^L, s_j^R. We set additional constraint conditions (9.253), (9.254) presented in Table 9.1 and Table 9.2 for the MBS disconnected in the manner described. Let us assume that the kinematic pair connects bodies i and j. We connect fixed coordinate systems with the origin in the centre of mass of the body and coordinate systems with the origin in the kinematic pair consisting of bodies i and j. In the case of a spherical pair, the origin is in the pair centre A and the axes are oriented in agreement with the coordinate system with the origin in the centre of mass. In the case of a revolute pair, the coordinate system origin B is on the axis of rotation and its axis z lies in the axis of rotation; origins B for bodies i, j coincide. With prismatic pairs, the origin P of the coordinate system is again on the shift axis, axis z is in the shift axis; axes x and y have an arbitrary orientation, which is however the same for both bodies i and j. For a cylindrical pair the coordinate system origin V is on the pair axis, and axis z lies in it. For a flat pair, the origin L of the coordinate system is in the pair plane, and axis z is perpendicular to it.

The left-hand sides of the constraint conditions corresponding to body i of the kinematic pair, i.e. expressions G_l^L in (9.254) are presented in column 2 of Tables 9.1 and 9.2 (Valášek and Stejskal, 1986). For body j, the other side is either structurally identical or this non-symmetric side is also given for body i as a second variant. Column three contains the number of constraint conditions and respective Lagrange multipliers, and the entry in parentheses gives the number of independent conditions. Columns 4 and 5 give relations for the action of reaction forces corresponding to the respective constraint conditions expressed in the form of generalized force (9.104). Column four gives the interpretation as the force acting on the centre of mass of body i expressed in the coordinate system of the frame. Column five gives the interpretation as the moment acting on body i expressed in the coordinate system with the centre in the centre of mass of body i. Column six shows interpretation of Lagrange multipliers as reaction effects traditionally related to individual kinematic pairs according to Chapter 8. Column seven gives the place of action and the system of coordinates of the expression of this interpretation.

The reaction effects of a spherical pair are represented by a force $\mathbf{R}_A$ acting in its centre. The reaction effects of a revolute pair are represented by a force $\mathbf{R}_B$ acting in its point B and the moment perpendicular to its axis with components M_x, M_y in the revolute pair coordinate system. The reaction effects of a prismatic pair are represented by a force perpendicular to the shift axis acting in its point P with components R_x, R_y in the coordinate system of the prismatic pair and moment $\mathbf{M}_P$. Reaction effects

Table 9.1. Interpretation of reaction forces for holonomic constraints

Kinematic pair	Holonomic constraint G_l^L $h = 0$	Number of λ	Interpretation Newton $()^T \partial \mathbf{v}_{1i} / \partial \dot{s}_j^L$
spherical	$\mathbf{r}_{1i} + \mathbf{S}_{1i}\mathbf{r}_{iA}$	3	λ
revolute	$\mathbf{r}_{1i} + \mathbf{S}_{1i}\mathbf{r}_{iB}$	3	λ
	$\mathbf{S}_{1i}\mathbf{z}_{i,iB}^0$	3(2)	0
prismatic	$(\mathbf{r}_{1i} + \mathbf{S}_{1i}\mathbf{r}_{iP}) \times (\mathbf{S}_{1i}\mathbf{z}_{i,iP}^0) = = -\mathbf{Z}_{1,iP}^0(\mathbf{r}_{1i} + \mathbf{S}_{1i}\mathbf{r}_{iP})$	3(2)	$\mathbf{z}_{1,iP}^0 \times \lambda$
	$\mathbf{S}_{1i}\mathbf{S}_{iP}$	3	0
cylindrical	$(\mathbf{r}_{1i} + \mathbf{S}_{1i}\mathbf{r}_{iV}) \times (\mathbf{s}_{1i}\mathbf{z}_{i,iV}^0) = = -\mathbf{Z}_{i,iV}^0(\mathbf{r}_{1i} + \mathbf{S}_{1i}\mathbf{r}_{iV})$	3(2)	$\mathbf{z}_{1,iV}^0 \times \lambda$
	$\mathbf{S}_{1i}\mathbf{z}_{i,iV}^0$	3(2)	0
flat	$\mathbf{r}_{1Li}(\mathbf{S}_{1i}\mathbf{z}_{i,iL}^0)^T$	1	$\mathbf{S}_{1i}\lambda\mathbf{z}_{i,iL}^0$
	$\mathbf{S}_{1i}\mathbf{z}_{i,iL}^0$	3(2)	0

Interpretation Euler $()^T \partial \boldsymbol{\omega}_{1i}/\partial \dot{s}_j^L$	Interpretation as reaction force	Point of action frame
$\mathbf{r}_{iA} \times \mathbf{S}_{1i}^{-1}\lambda$	$\lambda = \mathbf{R}_{1,A}$	A, frame
$\mathbf{r}_{iB} \times \mathbf{S}_{1i}^{-1}\lambda$	$\lambda = \mathbf{R}_{1,B}$	B, frame
$\mathbf{S}_{1i}^{-1}(\mathbf{z}_{1,iB}^0 \times \lambda) =$ $= \mathbf{z}_{i,iB} \times \mathbf{S}_{1i}^{-1}\lambda$	$\mathbf{z}_{1,iB}^0 \times \lambda =$ $= \mathbf{S}_{1i}\mathbf{S}_{iB}[M_{x,B}, M_{y,B}0]^T$	0, frame
$\mathbf{r}_{iP} \times \mathbf{S}_{1i}^{-1}(\mathbf{z}_{1,iP}^0 \times \lambda)+$ $+\mathbf{z}_{i,iP}^0 \times \mathbf{S}_{1i}^{-1}(\mathbf{r}_{1Pi} \times \lambda)$	$\mathbf{z}_{1,iP}^0 \times \lambda = \mathbf{R}_{1,P}$ and a part of moment $M_{P,P}$	P, frame
$\mathbf{S}_{1i}^{-1}(\mathbf{x}_{1,iP}^0 \times \lambda)$ $\mathbf{S}_{1i}^{-1}(\mathbf{y}_{1,iP}^0 \times \lambda)$ $\mathbf{S}_{1i}^{-1}(\mathbf{z}_{1,iP}^0 \times \lambda)$	$\mathbf{x}_{1,iP}^0 \times \lambda = \mathbf{S}_{1i}\mathbf{S}_{iP}[0, M_{y,P}, M_{z,P}]^T$ $\mathbf{y}_{1,iP}^0 \times \lambda = \mathbf{S}_{1i}\mathbf{S}_{iP}[M_{x,P}, 0, M_{z,P}]^T$ $\mathbf{z}_{1,iP}^0 \times \lambda = \mathbf{S}_{1i}\mathbf{S}_{iP}[M_{x,P}, M_{y,P}, 0]^T$	0, frame
$\mathbf{r}_{iV} \times \mathbf{S}_{1i}^{-1}(\mathbf{z}_{1,iV}^0 \times \lambda)+$ $+\mathbf{z}_{i,iV}^0 \times \mathbf{S}_{1i}(\mathbf{r}_{1,Vi} \times \lambda)$	$\mathbf{z}_{1,iV}^0 \times \lambda - \mathbf{R}_{1,V}$ and a part of moment $M_{V,V}$	V, frame
$\mathbf{S}_{1i}^{-1}(\mathbf{z}_{1,iV}^0 \times \lambda)$	$\mathbf{z}_{1,iV}^0 \times \lambda = \mathbf{S}_{1i}\mathbf{S}_{iV}[M_{x,V}, M_{y,V}, 0]^T$	0, frame
$-\mathbf{S}_{1i}^{-1}\mathbf{r}_{1i} \times \lambda\mathbf{z}_{i,iL}^0$	$\lambda = R_{z,L}$	origin of frame, flat KP
$\mathbf{S}_{1i}^{-1}(\mathbf{z}_{1,iL}^0 \times \lambda)$	$\mathbf{z}_{1,iL}^0 \times \lambda = \mathbf{S}_{1i}\mathbf{S}_{iL}[M_{x,L}, M_{y,L}, 0]^T$	0, frame

Table 9.2. Interpretation of reaction forces for nonholonomic constraints

Kinematic pair	Nonholonomic constraint G_i^L $h=1$	Number of λ	Interpretation Newton $()^T \partial \mathbf{v}_{1i} / \partial \dot{s}_j^L$
spherical	$\mathbf{v}_{1i} + \mathbf{S}_{1i}\mathbf{\Omega}_{1i}\mathbf{r}_{iA}$	3	λ
revolute	$\mathbf{v}_{1i} + \mathbf{S}_{1i}\mathbf{\Omega}_{1i}\mathbf{r}_{iB}$	3	λ
	$\boldsymbol{\omega}_{1Bi\vert x,y}$ $(\mathbf{S}_{1j}\mathbf{S}_{jB})^{-1}(\mathbf{S}_{1i}\mathbf{S}_{iB})\boldsymbol{\omega}_{1Bi\vert x,y}$	2 2	0 0
prismatic	$\mathbf{S}_{1Pi}^{-1}\mathbf{v}_{1i} - \mathbf{S}_{iP}^{-1}\mathbf{\Omega}_{1i}\mathbf{S}_{1i}^{-1}\mathbf{r}_{1i,x,y}$	2	$\mathbf{S}_{1Pi}[\lambda_x, \lambda_y, 0]^T$
	$\mathbf{S}_{iP}^{-1}\boldsymbol{\omega}_{1i}$	3	0
cylindrical	$\mathbf{S}_{1Vi}^{-1}(\mathbf{v}_{1i} + \mathbf{S}_{1i}\mathbf{\Omega}_{1i}\mathbf{r}_{iV})\vert_{x,y}$ $\mathbf{S}_{1Vj}^{-1}(\mathbf{v}_{1Vi} + \mathbf{S}_{1Vi}\mathbf{\Omega}_{11Vi}\cdot$ $\cdot\mathbf{S}_{1Vj}^{-1}(\mathbf{r}_{1Vi} - \mathbf{r}_{1Vi}))\vert_{x,y}$	2 2	$\mathbf{S}_{1Vi}[\lambda_x, \lambda_y, 0]^T$ $\mathbf{S}_{1Vj}[\lambda_x, \lambda_y, 0]^T$
	$\boldsymbol{\omega}_{1Vi\vert x,y}$ $(\mathbf{S}_{1j}\mathbf{S}_{jV})^{-1}(\mathbf{S}_{1i}\mathbf{S}_{iV})\boldsymbol{\omega}_{1Vi\vert x,y}$	2 2	0 0
flat	$\mathbf{S}_{1Lj}^{-1}\mathbf{v}_{1Li\vert z}$ $\mathbf{S}_{1Lj}^{-1}(\mathbf{v}_{1Li} + \mathbf{S}_{1Li}\mathbf{\Omega}_{1Li}\mathbf{S}_{1Li}^{-1}\cdot$ $\cdot(\mathbf{r}_{1Lj} - \mathbf{r}_{1Li}))\vert_z$	1 1	$\mathbf{S}_{1Li}[0,0,\lambda]^T$ $\mathbf{S}_{1Lj}[0,0,\lambda]^T$
	$\boldsymbol{\omega}_{1Li\vert x,y}$ $\mathbf{S}_{1Lj}^{-1}\mathbf{S}_{1Li}\boldsymbol{\omega}_{1Li\vert x,y}$	2 2	0

Interpretation Euler $()^T \partial \boldsymbol{\omega}_{1i} / \partial \dot{s}_j^L$	Interpretation as reaction force	Point of action, coordinate system
$\mathbf{r}_{iA} \times \mathbf{S}_{1i}^{-1} \lambda$	$\lambda = \mathbf{R}_{1,A}$	A, frame
$\mathbf{r}_{iB} \times \mathbf{S}_{1i}^{-1} \lambda$	$\lambda = \mathbf{R}_{1,B}$	B, frame
$\mathbf{S}_{iB}[\lambda_x, \lambda_y, 0]^T$ $\mathbf{S}_{1i}^{-1} \mathbf{S}_{1j} \mathbf{S}_{jB}[\lambda_z, \lambda_y, 0]^T$	$\lambda_x = M_{x,Bi}$ $\lambda_y = M_{y,Bi}$ $\lambda_x = M_{x,Bj}$ $\lambda_y = M_{y,Bj}$	0, revolute KP, B_i 0, revolute KP, B_j
$-\mathbf{S}_{1i}^{-1}(\mathbf{r}_{1i} \times \mathbf{S}_{1Pi}[\lambda_x, \lambda_y, 0]^T)$	$\lambda_x = R_{x,P}$ $\lambda_y = R_{y,P}$	origin of frame, prismatic KP
$\mathbf{S}_{iP} \lambda$	$\lambda = M_{P,P}$	0, prismatic KP
$\mathbf{r}_{iV} \times \mathbf{S}_{iV}[\lambda_x, \lambda_y, 0]^T$ $\mathbf{r}_{iV} \times \mathbf{S}_{1i}^{-T} \mathbf{S}_{1Vj}[\lambda_z, \lambda_y, 0]^T +$ $+\mathbf{S}_{1i}^{-1}((\mathbf{r}_{1Vj} - \mathbf{r}_{1Vi}) \cdot$ $\cdot \mathbf{S}_{1Vj}[\lambda_z, \lambda_y, 0]^T)$	$\lambda_x = R_{z,Vi}$ $\lambda_y = R_{y,Vi}$ $\lambda_x = R_{x,Vj}$ $\lambda_y = R_{y,Vj}$	V_i, cylindrical KP, V_i V_j, cylindrical KP, V_j
$\mathbf{S}_{iV}[\lambda_x, \lambda_y, 0]^T$ $\mathbf{S}_{1i}^{-T} \mathbf{S}_{1j} \mathbf{S}_{jV}[\lambda_x, \lambda_y, 0]^T$	$\lambda_x = M_{x,Vi}$ $\lambda_y = M_{y,Vi}$ $\lambda_x = M_{x,Vj}$ $\lambda_y = M_{y,Vj}$	0, cylindrical KP, V_i 0, cylindrical KP, V_j
$\mathbf{r}_{iL} \times \mathbf{S}_{iL}[0, 0, \lambda]^T$ $\mathbf{r}_{iL} \times \mathbf{S}_{1i}^{-1} \mathbf{S}_{1Lj}[0, 0, \lambda]^T +$ $+\mathbf{S}_{1i}^{-1}((\mathbf{r}_{1Lj} - \mathbf{r}_{1Li}) \cdot$ $\cdot \mathbf{S}_{1Lj}[0, 0, \lambda]^T)$	$\lambda = R_{z,Li}$ $\lambda = R_{z,Lj}$	L_i, flat KP, L_i L_j, flat KP, L_j
$\mathbf{S}_{1Lj}[\lambda_x, \lambda_y, 0]^T$ $\mathbf{S}_{1i}^{-1} \mathbf{S}_{1Lj}[\lambda_x, \lambda_y, 0]^T$	$\lambda_x = M_{x,Li}$ $\lambda_y = M_{y,Li}$ $\lambda_x = M_{x,Lj}$ $\lambda_y = M_{y,Lj}$	0, flat KP, L_i 0, flat KP, L_j

of a cylindrical pair are the force and moment perpendicular to the pair axis acting in its point V with components x, y in the coordinate system of the pair. The reaction effects of a flat pair are formed by the normal force acting in point L perpendicularly to the pair plane and the moment with components M_x, M_y in the flat pair system of coordinates. The coordinate system in which the reaction effects are expressed is given in the subscript after the comma. In Table 9.1 the constraint conditions are expressed as holonomic conditions; Table 9.2 gives similar nonholonomic conditions of order 1. Nonholonomic conditions are not integrable; however they originate from integrable nonholonomic conditions by dividing by a common factor. The constraint conditions are to be found in the relation between the transformation matrix $\mathbf{T}_{1i}$ of the motion of body i with respect to 1 and $\mathbf{T}_{1j}$ of body j motion with regard to 1 given by a kinematic pair. For spherical, revolute, prismatic, cylindrical and flat pair, the relations are, respectively

$$\begin{aligned}
\mathbf{T}_{1i}\mathbf{T}_{iA_i}\mathbf{T}_{Z4}(\varphi)\mathbf{T}_{Z5}(\vartheta)\mathbf{T}_{Z6}(\nu) &= \mathbf{T}_{1j}\mathbf{T}_{jA_j} \\
\mathbf{T}_{1i}\mathbf{T}_{iB_i}\mathbf{T}_{Z6}(\varphi) &= \mathbf{T}_{1j}\mathbf{T}_{jB_j} \\
\mathbf{T}_{1i}\mathbf{T}_{iP_i}\mathbf{T}_{Z3}(p) &= \mathbf{T}_{1j}\mathbf{T}_{jP_j} \\
\mathbf{T}_{1i}\mathbf{T}_{iV_i}\mathbf{T}_{Z6}(\varphi)\mathbf{T}_{Z3}(p) &= \mathbf{T}_{1j}\mathbf{T}_{jV_j} \\
\mathbf{T}_{1i}\mathbf{T}_{iL_i}\mathbf{T}_{Z6}(\varphi)\mathbf{T}_{Z1}(p_1)\mathbf{T}_{Z2}(p_2) &= \mathbf{T}_{1j}\mathbf{T}_{jL_j} \qquad (9.261)
\end{aligned}$$

In these relations, for example, $\mathbf{T}_{iBi}$ denotes the transformation matrix between body i revolute pair coordinate system motion and body i coordinate system with its centre in the centre of mass, and $\varphi, \theta, \nu, p, p_1, p_2$ denote kinematic pair coordinates. A higher kinematic pair transformation matrix equation (6.19) may also be written in form (9.261).

We obtain the holonomic constraint conditions by expanding and altering (9.261) into relations not containing coordinates of the kinematic pair. In the same way we obtain the nonholonomic constraint conditions from the time derivative (9.261). We find the interpretation in columns four and five of Table 9.1 and Table 9.2 by altering the expression for the virtual work of Lagrange multipliers $\boldsymbol{\lambda}$ corresponding to constraint condition $\mathbf{G}_l$ in coordinate s_j^L into the form (9.104)

$$\begin{aligned}
\boldsymbol{\lambda}^T \frac{\partial \mathbf{G}_l}{\partial s_j^L}\delta s_j^L &= \boldsymbol{\lambda}^T \frac{\partial(\mathbf{G}_l^L - \mathbf{G}_l^R)}{\partial s_j^L}\delta s_j^L = \boldsymbol{\lambda}^T \frac{\partial \mathbf{G}_l^L}{\partial s_j^L}\delta s_j^L = \\
&= \left(\mathbf{F}_i^T \frac{\partial \mathbf{v}_{1i}}{\partial \dot{s}_j^L} + \mathbf{M}_i^T \frac{\partial \boldsymbol{\omega}_{1i}}{\partial \dot{s}_j^L}\right)\delta s_j^L \qquad (9.262)
\end{aligned}$$

The interpretations in columns six and seven of Tables 9.1 and 9.2. are to be found by altering expressions for the virtual work of traditional reaction effects of kinematic pairs according to Chapter 8 into the form (9.104) and their comparison with (9.262). Altering (9.262) we use the following two typical methods of treatment (Valášek and Stejskal, 1986). The first one is

$$\boldsymbol{\lambda}^T \frac{\partial}{\partial \dot{s}_j^L}(\mathbf{r}_{1i} + \mathbf{S}_{1i}\mathbf{r}_{ix}) = \boldsymbol{\lambda}^T \frac{\partial \mathbf{r}_{1i}}{\partial \dot{s}_j^L} + \boldsymbol{\lambda}^T \mathbf{S}_{1i} \frac{\partial \boldsymbol{\Omega}_{1i}}{\partial \dot{s}_j^L}\mathbf{r}_{ix} =$$

$$= \boldsymbol{\lambda}^T \frac{\partial \mathbf{v}_{1i}}{\partial \dot{s}_j^L} + \boldsymbol{\lambda}^T \mathbf{S}_{1i}\mathbf{R}_{ix}^T \frac{\partial \boldsymbol{\omega}_{1i}}{\partial \dot{s}_j^L} = \boldsymbol{\lambda}^T \frac{\partial \mathbf{v}_{1i}}{\partial \dot{s}_j^L} + (\mathbf{R}_{ix}\mathbf{S}_{1i}^T\boldsymbol{\lambda})^T \frac{\partial \boldsymbol{\omega}_{1i}}{\partial \dot{s}_j^L} =$$

$$= \boldsymbol{\lambda}^T \frac{\partial \mathbf{v}_{1i}}{\partial \dot{s}_j^L} + (\mathbf{r}_{ix} \times \mathbf{S}_{1i}^{-1}\boldsymbol{\lambda})^T \frac{\partial \boldsymbol{\omega}_{1i}}{\partial \dot{s}_j^L} \tag{9.263}$$

where $\mathbf{r}_{ix}$ is a constant vector with respect to $\dot{s}_j^L$, $\mathbf{R}_{ix}$ is its skew-symmetric expression. The second one is

$$\boldsymbol{\lambda}^T \frac{\partial}{\partial \dot{s}_j^L}(\mathbf{S}_1\mathbf{v}_{1i} + \mathbf{S}_2^{-1}\boldsymbol{\Omega}_{1i}\mathbf{S}_2\mathbf{r}_{ix}) =$$

$$= \boldsymbol{\lambda}^T \mathbf{S}_1 \frac{\partial \mathbf{v}_{1i}}{\partial \dot{s}_j^L} + \boldsymbol{\lambda}^T \frac{\partial}{\partial \dot{s}_j^L}(\mathbf{R}_{ix}^T\mathbf{S}_2\boldsymbol{\omega}_{1i}) =$$

$$= (\mathbf{S}_1^T\boldsymbol{\lambda})^T \frac{\partial \mathbf{v}_{1i}}{\partial \dot{s}_j^L} + \boldsymbol{\lambda}^T \mathbf{R}_{ix}^T\mathbf{S}_2 \frac{\partial \boldsymbol{\omega}_{1i}}{\partial \dot{s}_j^L} =$$

$$= (\mathbf{S}_1^{-1}\boldsymbol{\lambda})^T \frac{\partial \mathbf{v}_{1i}}{\partial \dot{s}_j^L} + (\mathbf{S}_2^T\mathbf{R}_{ix}\boldsymbol{\lambda})^T \frac{\partial \boldsymbol{\omega}_{1i}}{\partial \dot{s}_j^L} =$$

$$= (\mathbf{S}_1^{-1}\boldsymbol{\lambda})^T \frac{\partial \mathbf{v}_{1i}}{\partial \dot{s}_j^L} + (\mathbf{S}_2^{-1}(\mathbf{r}_{ix} \times \boldsymbol{\lambda}))^T \frac{\partial \boldsymbol{\omega}_{1i}}{\partial \dot{s}_j^L} \tag{9.264}$$

Based on the above results, we derive further knowledge of the solution of MBS kinetostatic analysis. The solution of the kinetostatic problem is a special case of the indirect dynamic problem solution, where, for the prescribed MBS motion given by functions $s_j = s_j(t)$, we seek the action and reaction forces required for its realization. Knowing $s_j(t)$ we know the velocities and accelerations, and we may alter the first equation of (9.68) in order to have unknown values on the left-hand side of the equation

$$\boldsymbol{\Phi}_s^T\boldsymbol{\lambda} = -\mathbf{p}_1 + \mathbf{M}\ddot{\mathbf{s}} \tag{9.265}$$

The second equation in (9.68) is identically satisfied through the knowledge of $s_j(t)$. We include all unknown values of the kinetostatic problem under $\boldsymbol{\lambda}$. Looking for the action forces P_h acting in coordinates $p_h = p_h(s_j, t)$, we may also consider them as λ_h corresponding to the constraint

$$p_h(s_j, t) - p_h(s_j(t), t) = 0 \tag{9.266}$$

Thus the kinetostatic problem and the indirect dynamic problem may be formulated by equation (9.265). Hence it follows that the matrix of the system of the problem is the transposed matrix of the Jacobian of kinematic constraints and the transposed matrix of the system of solution of the position, velocities and acceleration problem respectively. This property holds, however, only when the description is adhered to. Solving the MBS kinetostatics using equation (9.265) we always obtain from multipliers $\boldsymbol{\lambda}$ so many reactions in MBS loops that the closed sequences of unknown reactions in the loops are interrupted. Thus the kinetostatics problem of MBS with loops is transformed into kinetostatic problem of open chains. It is solved recursively from (8.20a). This algorithm requires less computational effort than (8.16) and is probably the most efficient one.

Because the matrix of the system of equations is the same as in kinematics the similar problems of singular cases as in Chapter 7 can occur (Stejskal et al., 1990).

9.7 The Compartment Method in Dynamics

The compartment method applied in Section 5.11.5 for assembling kinematic equations of MBS's is a special case of the general method of compartments for assembling dynamic equations of MBS's.

The method of compartment composition follows the method of diakoptics (Kron, 1963) which has been developed for the modelling of large and complex technical systems; in the first place these were electrical networks. The principle of the method consists in the division of the whole system into single parts which we shall call compartments. These single parts are described by proper (independent) equations and their interactions with surroundings are described by their so-called clamp variables. The model of the whole complex system originates with the composition (connection) of the proper equations of each single compartment together with the constraint equations for clamp variables of compartments which are determined by the structure (topology) of part connections through their clamps.

The diakoptic method for system modelling and simulation can be more computationally efficient than traditional ways of modelling the whole system. If the execution time T_E of solving a model with dimension N is proportional to N^α, where $\alpha > 1$, then the execution time of solving m smaller models with dimensions N/m is shorter than the solution of the original model with dimension N. The drawback of the diakoptic method is that we must use and determine additional variables for the description of interactions among compartments. This corresponds to the free body method with MBS's which gives the greatest number of unknowns from different methods of solution.

This method has originated and is applied for electrical networks where clamps and clamp variables have obvious meaning. They are currents and voltages, and the constraint equations of compartment interconnections are Kirchhoff's laws. Later this method has also been applied to other physical systems such as mechanical translational, mechanical rotational, hydraulic, pneumatic, etc. (Andrews and Kesavan, 1978; Mann, 1986). The current and voltage have been generalized into "through" and "across" variables. Kirchhoff's laws have different particular analogies and are formulated as different system principles derived empirically.

Further, we analytically derive the universal analogy (Valášek, 1987c) between a system of arbitrary physical nature (even of arbitrary algebraic differential equations) and an electrical network. This analogy enables us to use the method of diakoptics for the simulation of such systems.

9.7.1 Compartment Model of an Arbitrary Physical System

Arbitrary physical phenomena or even arbitrary systems of algebraic differential equations may be described by variational principles (Berdichevskij, 1983). Let us have a physical system consisting of mutually connected elements. These elements will be modelled as compartments and the whole system will be modelled by compartment composition.

Let the proper equations of compartments be formulated in generalized coordinates q_i by Lagrange's function L as the universal variational formulation. Firstly the coordinates q_i are related only to one compartment and they are not relative coordinates describing the mutual constraints between compartments. Later we shall remove this assumption. The equations describing one compartment may be derived from Lagrange's function L by variational principles.

Let us have two compartments A and B which interact with each other. The mutual coupling (constraint) between them is expressed by the functional constraint for their variables and

$$f_k(q_i^A, q_j^B, t) = 0 \tag{9.267}$$

This case we shall call binary interaction. All physical fields may be expressed by binary interactions of compartments. However, it is not generally valid for arbitrary systems of equations or for cases such as systems described by relative coordinates, chemical reactions controlled by a catalyzer etc. In these cases we speak about n-ary interaction, if in the constraint equation (9.267) there are coordinates from n compartments. Later we shall modify the compartment method even for n-ary interactions.

We consider first a binary interaction (9.267). Because q_i^A and q_i^B are related only to one compartment in the spaces V_A and V_B respectively, and the constraint (9.267) describes the connection between these compartments on their boundary,

$$f_k(q_i^A, q_j^B, t) = f_{kA}(q_i^A, t) - f_{kB}(q_j^B, t) = 0 \tag{9.268}$$

where f_{kA} and f_{kB} express the boundary physical property of compartment and the equation (9.268) describes their equality on the common boundary. Lagrange's function L_{AB} for the description of the system of both interacted compartments A and B is determined by the sum of Lagrange's functions of both compartments and of the constraint conditions described by equations (9.267) or (9.268) by means of Lagrange's multipliers

$$L_{AB} = L_A(q_i^A, \dot{q}_i^A, \ldots, t) + L_B(q_j^B, \dot{q}_j^B, \ldots, t) + \sum_k \lambda_k f_k \tag{9.269}$$

From (9.268) we write

$$\lambda_k f_k = \lambda_k(f_{kA} - f_{kB}) = \lambda_{kA} f_{kA} + \lambda_{kB} f_{kB} \tag{9.270}$$

where

$$\lambda_k = \lambda_{kA} = -\lambda_{kB} \tag{9.271}$$

Hence, Lagrange's function

$$\begin{aligned} L_{AB} &= L_A + L_B + \sum_k \lambda_{kA} f_{kA} + \sum_k \lambda_{kB} f_{kB} = \\ &= \left(L_A(q_i^A, \dot{q}_i^A, \ldots, t) + \sum_k \lambda_{kA} f_{kA}(q_i^A, t) \right) + \\ &\quad + \left(L_B(q_j^B, \dot{q}_j^B, \ldots, t) + \sum_k \lambda_{kB} f_{kB}(q_j^B, t) \right) \end{aligned} \tag{9.272}$$

The equations describing the interaction of both compartments A, B may be derived from necessary extreme conditions of the functional given by the time integration of Lagrange's function (9.272) (principle of minimal action). Hence it follows that these equations are created by the proper equations describing particular compartments under the action of generalized forces

$$Q_{kA_i} = \lambda_{kA} \frac{\partial f_{kA}}{\partial q_i^A}, \qquad Q_{kB_j} = \lambda_{kB} \frac{\partial f_{kB}}{\partial q_j^B} \tag{9.273}$$

and by the constraints (9.268) on compartment boundaries together with the definition (9.271). The expressions

$$Q_k \delta q = \lambda_k \frac{\partial f_k}{\partial q} \delta q \quad \text{and} \quad Q_k \dot{q} = \lambda_k \frac{\partial f_k}{\partial q} \dot{q} \tag{9.274}$$

express the virtual work of generalized forces and the power outputs respectively.

In the case of n-ary interaction the constraint (9.267) has the form

$$f_k(q_i^A, q_{j_1}^{B_1}, q_{j_2}^{B_2}, \ldots, q_{j_{n-1}}^{B_{n-1}}, t) = 0 \tag{9.275}$$

where $q_{jl}^{B_l}\,(l = 1, \ldots, n-1)$ is the j_l-th variable of B_l-th compartment. The above derivation must be modified. Let us introduce the following new variables

$$q_i^A + \sum_{p=1}^{L-1} |j_p| + j_L, \qquad L = 1, \ldots, n-1,$$

$|\ |$ is the number of indices, for all compartments $B_l\,(l = 1, \ldots, n-1)$ in the equation (9.275)). We than add these to the description of the compartment A. The constraint (9.275) may then be described by the equation

$$f_k(q_i^A, q_{|i|+j_1}^A, q_{|i|+|j_1|+j_2}^A, \ldots, q_{|i|+m}^A, t) = 0$$

$$m = \sum_{p=1}^{n-2} |j_p| + j_{n-1} \tag{9.276}$$

and the interaction by the equations

$$q_{|i|+m}^A - q_{j_L}^{B_L} = 0 \qquad m = \sum_{p=1}^{L-1} |j_p| + j_L \tag{9.277}$$

which have the form of binary interaction. The compartment A is now described by a new Lagrange's function $L_{A'}$ containing all constraints (9.276) which act only inside the compartment

$$\begin{aligned} L_{A'} &= L_A + \sum_k \lambda_k f_k(q_i^A, q_{|i|+j_1}^A, \ldots, q_{|i|+m}^A, t) \\ m &= \sum_{p=1}^{n-2} |j_p| + j_{n-1} \end{aligned} \tag{9.278}$$

The interaction between compartment A and compartment B_l is then described by equations (9.277) that have the form (9.268) of a binary interaction. After these modifications, further derivation is the same as above, including the interpretation (9.273)–(9.274). The relation between Lagrange multipliers to constraints (9.276) and (9.277) can be derived from the equality of virtual work of original and new multipliers as in Section 9.6.

The case of n-ary interaction is important for a MBS described by relative coordinates. For a relative coordinate q_i^{AB} between compartments A and B we introduce new coordinates $q_{i,A}^{AB}$ and $q_{i,B}^{AB}$ and add them to the description of compartments A and B, and we use the new constraint $q_{i,A}^{AB} - q_{i,B}^{AB} = 0$. This procedure is also important for application within symbolic systems for modelling electrical networks (e.g. Dymola (Elmqvist, 1994)). The symbolic treatment of network modelling enables us to eliminate linearly dependent variables before the proper simulation starts. By this way we can apply all the algorithms of multibody formalisms in such modelling. The recursive quantities used in composite rigid body formalism or articulated body inertia are introduced as new variables similarly as in the case of n-ary interaction. The symbolic exclusion of these variables is exactly the recursive computation according to (9.169), (9.187-9.189), (9.206).

Hence a system or arbitrary physical nature or of arbitrary algebraic differential equations can be described by compartments; these enable us to assemble the system model from proper equations of particular compartments together with the constraints of equality of their dual variables λ_{kA}/f_{kA} on their boundaries (9.271), (9.268) or (9.277).

It is obvious that the compartments in kinematics (see Section 5.11.5) are the above compartments with the constraints (9.267). The kinetostatic or static problems of MBS can, according to equation (9.265) in the previous section, be described by the compartment method by means of constraints (9.267) and corresponding dual variables λ_{kA} without proper dynamic equations of compartments.

9.7.2 General Transformation of a Compartment into a Macromodel Satisfying Kirchhoff's Laws

A compartment A, where the variables q_{Ai} are subject to n constraints f_k from the surroundings described by

$$\begin{aligned} f_{kA} &= f_{kB} \\ \lambda_{kA} + \lambda_{kB} &= 0 \end{aligned} \tag{9.279}$$

can be described as a n-multipole (Figure 9.9a). Each pole has two variables λ_{kA}, f_{kA} attached to it. λ_{kA} is the "through" variable (current analogy) and f_{kA} is the "across" variable (voltage analogy). These variables do not generally fulfill Kirchhoff's laws

$$\begin{aligned} \sum_{k=1}^{n} \lambda_{kA} &= 0 \\ f_{kA} &= f_{k-i,A} + f_{i,A} \end{aligned} \tag{9.280}$$

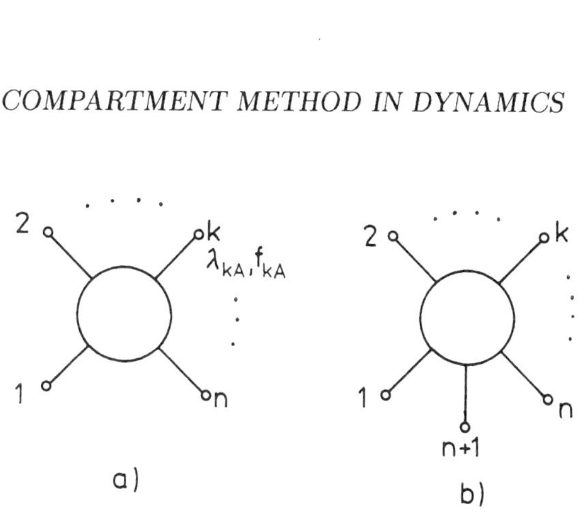

Figure 9.9. Transformation of compartment to macromodel.

We add the $(n+1)$-th pole and we obtain the $(n+1)$-multipole (Figure 9.9b). We define

$$\begin{aligned} \lambda_{n+1,A} &= -\sum_{k=1}^{n} \lambda_{kA} \\ f_{n+1,A} &= 0 \end{aligned} \tag{9.281}$$

and instead of variables f_{kA} we introduce variables $u^A_{k,n+1}$

$$\begin{aligned} u^A_{k,n+1} &= f_{kA} - f_{n+1,A} = f_{kA} \\ u^A_{k,l} &= f_{kA} - f_{lA} = u^A_{k,n+1} - u^A_{l,n+1} \end{aligned} \tag{9.282}$$

Hence, from (9.281)–(9.282),

$$\begin{aligned} &\sum_{k=1}^{n+1} \lambda_{kA} = 0 \\ u^A_{kl} = u^A_{kp} + u^A_{pl} &= u^A_{k,n+1} - u^A_{l,n+1} = u^A_{kp} - u^A_{lp} \end{aligned} \tag{9.283}$$

Thus the system of compartments is transformed into a network of $(n+1)$-multipoles interconnected according to particular constraints f_k satisfying Kirchhoff's laws.

Let the proper equations of compartment A have the form of state equations

$$\dot{q}^A_i = \dot{q}^A_i(q^A_l, ..., t) \tag{9.284}$$

then the macromodel of the compartment can be realized by the derived analogy with an electrical network in Figure 9.10a. The $(n+1)$-th pole of all compartments is connected to ground.

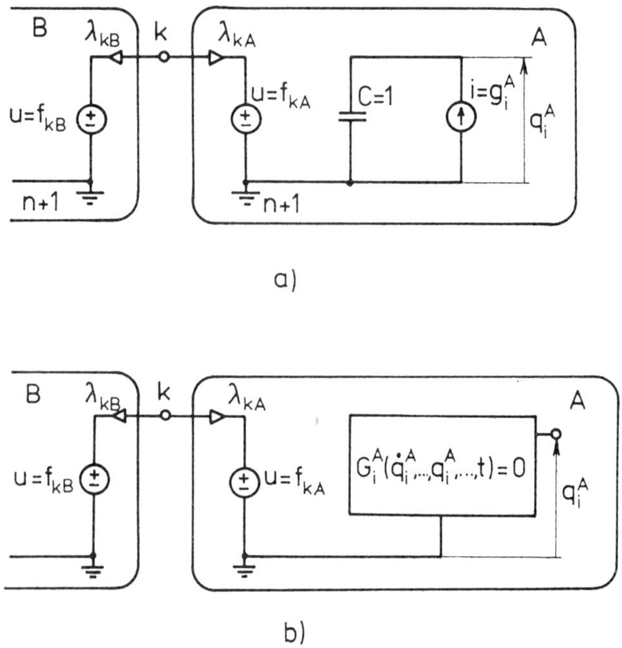

Figure 9.10. Electrical network model of a general compartment.

If the proper equations of compartment A can be described only in the form

$$G_i^A(\dot{q}_i^A, ..., q_i^A, t) = 0 \tag{9.285}$$

then the realization of the corresponding macromodel is as shown in Figure 9.10b. This completes the derivation of the universal analogy between an arbitrary physical system and an electrical network.

This universal analogy enables us to use the compartment method for modelling technical systems in a particular physical domain by simulation systems of electrical networks. The user describes the technical system in his language of particular domain (particular compartments, their parameters, structure of their connections), and inputs this data to a simulation system for electrical networks (e.g. system SADYS (Mann, 1986) in Section 10.3) or to a neutral simulation system for networks (e.g. Dymola (Elmqvist, 1994)). The simulation system automatically assembles particular equations according to the compartment method, solves them and returns the results of dual variables behaviour to the user. The user is able to interpret these results in the language of his particular domain by the concept of compartments without any knowledge of the analogy.

This approach is very important for modelling of technical system consisting of subsystems of different physical nature (mechanical, electrical, hydraulic, etc.) which are typical for mechatronic.

Theses algorithms also prove the principle of automatic transformation of a descriptive model to the corresponding operational model from Section 2.4.3. The application to planar MBSs is described in Section 10.3 and similarly also in (Andrews and Kesavan, 1978).

A simulation system for electrical networks assembles the particular equations of the compartment system from proper equations of compartments and from generalized Kirchhoff's laws described by means of topological matrices (3.12) (Andrews, 1977)

$$\begin{aligned} \vec{\mathbf{B}}\mathbf{f} &= \left[\mathbf{E}, \vec{\mathbf{B}}_v\right][f_{kA}] = \mathbf{0} \\ \vec{\mathbf{C}}\boldsymbol{\lambda} &= \left[\vec{\mathbf{C}}_t, \mathbf{E}\right][\lambda_{kA}] = \mathbf{0} \end{aligned} \tag{9.286}$$

These equations can be combined in different ways and excluded (Mann, 1984) according to different methods of equation formulation for electrical networks. The comparison of these methods and the assembling of traditional MBS equations can be fruitful areas for future research.

9.8 Numerical Solution

As the general solution of strongly nonlinear equations of MBS motion together with nonlinear constraint equations cannot be considered even in the simplest of cases, the only possibility is the numerical solution of the simulation of MBS motion. We resign on finding functional relations of kinematic and force quantities describing the dynamic state of the examined MBS which are continuous with respect to time. We are then satisfied with finding discontinuous values of the quantities in discontinuous instants, and all we require is that the discontinuous values satisfy the motion and constraint equations. In this section we deal with the possibilities which exist for the solution of the above problem, and we evaluate them briefly.

The most fundamental problem of the solution of equations of motion of MBS is the combination of differential equations with algebraic equations. However, there are two possibilities. The equations of motion which we obtain from the multibody formalism can be formulated in minimum (independent) coordinates or in redundant coordinates. The equations of motion formulated in minimum coordinates can generally be solved by the direct application of standard methods for the solution of ordinary differential equations designated as ODE, although it may be necessary to solve some algebraic constraints simultaneously.

For the actual integration, well-proven ODE integrators can be used, e.g. Gear's procedure DIFSUB (Gear, 1971) or Shampine-Gordon's procedure DE (Shampine and Gordon, 1975). Traditional formulae of Runge-Kutta, Hamming, Milne, etc. can also be used, but according to (Hall and Watt, 1976) the methods based on Adams predictor-corrector formulae are more efficient than, for example, Runge-Kutta methods in cases when it is necessary to carry out more than 25 arithmetic operations for one right-hand evaluation. MBS dynamic models are usually such cases. On the other hand, Runge-Kutta methods are convenient for systems with impacts, feedback control, etc. and are further developed with respect to DAE solution (Hairer et al., 1989, Simeon and Rentrop, 1993). Other possibility are the extrapolation methods (Lubich, 1989). The Shampine-Gordon's procedure DE is especially very efficient (Leister, 1990).

The solution of equations of motion formulated in redundant coordinates is generally not possible by the direct application of ODE methods. From the viewpoint of mathematics this formulation is a system of nonlinear ordinary differential equations of the second order and a system of nonlinear algebraic equations in literature often briefly designated as DAE (Gear, 1984). The minimum coordinate formulation can easily be obtained for MBSs without loops. Formalisms for the minimum coordinate formulation of a MBS with loops are under development (see Section 9.5). The relative success of the methods for DAE solution enables us to use redundant coordinate formulation generally for all MBSs. The solution of DAE is generally realized either by some special methods (DAE standard solvers, minimization of overdetermined system of all constraints and integration predictor) or by some transformation to ODE methods (stabilization, numerical transformation to minimum coordinates).

Because we do not study in this book the methods of solution of differential equations themselves we limit ourselves to the type of equations of motion yielded by the LEMT method, i.e. equations (9.68). It is the general case of DAE. We shall study the transformation of the DAE problem to the ODE which is realized by methods based on mechanics.

An important aspect of the choice of method for the solution of such a system is its dimension, i.e. the number of equations in the simplest cases of MBS (Cardan's joint, universal joint, etc.); this is a system formed by three mobile elements. Describing it by physical coordinates gives 18 differential equations of second order and (as these are systems with one degree of freedom) 17 constraint equations must be supplied. Although these equations are in principle solvable in the derived form by some of the methods for the direct numerical integration of second order systems of differential equations, this approach is not usual. The most common way is to convert the second order equations into twice the number of first order equations followed by integration. We then obtain for the simplest of cases 36 + 17 equations. For not very complex systems the number of equations is greater than 100.

Another characteristic feature of this problem is the low density of the system matrix which is the consequence of weak dependencies between the equations. For the solution of such systems, algorithms for sparse matrices can be recommended.

The principal problem of the numerical solution of DAE with differentiated algebraic equations is the problem of the numerical stability of the solution. This problem arises as a result of combining not the constraint equations (9.59) but the twice-differentiated equations (9.67) with the Lagrange mixed type equations. This is convenient from the viewpoint of uniting the two equations (9.65) and (9.59) for an acceleration in which both equations are linear. From the viewpoint of numerical solution this causes difficulties as the integration of equation (9.67) is unstable, so that after a short integration time the validity of equations (9.59) fails.

The difficulties of the solution of DAE are characterized by so-called index of DAE (Gear, 1991). There are several definitions, the useful are differential and perturbation ones. The differential index d says how many times increased by one the original DAE must be differentiated in order its Jacobian has full rank for the highest derivatives. The perturbation index p says that the roundoff errors of size ε in computation cause errors in the solution of size εh^{-p} where h is the step size used. This means that decrease of step size may not lead to decrease of the solution error. The perturbation index is equal to d or $d-1$.

Let us demonstrate in more detail this problem for the case of only one constraint equation. Integrating numerically the equation $\ddot{f} = \varepsilon_a$ obtained with error ε_a with vanishing initial conditions, we obtain after several computational steps, as a result of solution errors the result $\dot{f} = \varepsilon_v$, $f = \varepsilon_p$, instead of the correct result $\dot{f} = 0, f = 0$. Even if further steps in the solution are totally exact, we would obtain

$$f = 1/2\varepsilon_a t^2 + \varepsilon_v t + \varepsilon_p$$

so that the constraint equation has an instability with respect to time.

The methods of numerical solution of MBS dynamics depend substantially on the coordinates used. We may use physical coordinates or generalized (independent) coordinates, the number of which equals the number of degrees of freedom. We describe both alternatives in more detail. The independent coordinates may be obtained directly from the multibody formalism or by the numerical transformation of redundant coordinates to minimum coordinates. From this point of view we shall divide our further description into the methods which integrate the physical (redundant) coordinates and the methods which integrate the independent (generalized, minimum) coordinates.

9.8.1 Solution in Physical Coordinates

We deal with the stable numerical solution of equation (9.68) whereby we distinguish three alternatives
- stabilization technique,
- reformulation of equations (9.66) and (9.59),
- proper solution of (9.66) and (9.59) as DAE.

9.8.1.1 Baumgarte Stabilization

The procedure proposed by Baumgarte (Baumgarte, 1972) transforms the DAE problem to the ODE problem. The condition $\ddot{f} = 0$ is to be substituted by an analytically equivalent one

$$\ddot{f} + 2\alpha\dot{f} + \beta^2 f = 0, \qquad \alpha > 0 \tag{9.287}$$

The solution stability can be accomplished by a suitable choice of constants α, β. For example, for $\alpha = \beta = 1$ we obtain

$$f = e^{-t}(C_1 + C_2 t) \tag{9.288}$$

In the case of equations with more constraints, the same conclusions hold. Instead of (9.287), we then use the relation

$$\ddot{\mathbf{f}} + 2\alpha\dot{\mathbf{f}} + \beta^2\mathbf{f} = \mathbf{0} \tag{9.289}$$

where $\alpha > 0, \beta$ are constants. We thus obtain the equation

$$\boldsymbol{\Phi}_s\ddot{\mathbf{s}} + \dot{\boldsymbol{\Phi}}_s\dot{\mathbf{s}} + 2\alpha\boldsymbol{\Phi}_s\dot{\mathbf{s}} + \beta^2\mathbf{f} = \mathbf{0} \tag{9.290}$$

and combining it with LEMT (9.50) we obtain a modification of equation (9.68)

$$\begin{bmatrix} \mathbf{M} & \boldsymbol{\Phi}_s^T \\ \boldsymbol{\Phi}_s & \mathbf{0} \end{bmatrix} \begin{bmatrix} \ddot{\mathbf{s}} \\ -\boldsymbol{\lambda} \end{bmatrix} = \begin{bmatrix} \mathbf{p}_1 \\ -\dot{\boldsymbol{\Phi}}_s\dot{\mathbf{s}} - 2\alpha\boldsymbol{\Phi}_s\dot{\mathbf{s}} - \beta^2\mathbf{f} \end{bmatrix} \tag{9.291}$$

For the sake of conciseness we introduce the notation

$$\mathbf{p}_2 = -\dot{\boldsymbol{\Phi}}_s\dot{\mathbf{s}} - 2\alpha\boldsymbol{\Phi}_s\dot{\mathbf{s}} - \beta^2\mathbf{f} \tag{9.292}$$

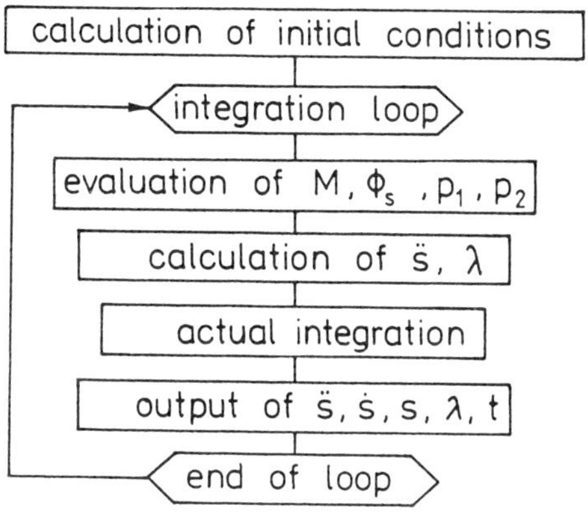

Figure 9.11. Flow chart of LEMT traditional solution.

We solve a DAE system (9.291) applying the proceduree from the flowchart in Figure 9.11.

Before beginning the integration loop, the initial conditions, i.e. the state of motion of MBS at the time $t = 0$ given by values $\mathbf{s}(0), \dot{\mathbf{s}}(0)$, must be known. Their calculation is in general not simple. Only m coordinates and m velocities can be chosen; the rest must be subsequently calculated from equations (9.59), (5.151) and (5.169). It is sufficient to perform this calculation once only at the beginning of the integration.

Before starting with the integration it is useful to convert the system of second order differential equations to twice the number of first order equations. The main reason for this approach is the availability of quality programs for integrating first order equations. The conversion can be done, for example, as follows.

We introduce vectors

$$\mathbf{y}_1 = \mathbf{s}, \quad \mathbf{y}_2 = \dot{\mathbf{y}}_1 = \dot{\mathbf{s}}, \qquad \dot{\mathbf{y}}_2 = \ddot{\mathbf{s}} \tag{9.293}$$

and equation (9.291) becomes

$$\begin{bmatrix} \mathbf{E} & 0 & 0 \\ 0 & \mathbf{M} & \mathbf{\Phi}_s^T \\ 0 & \mathbf{\Phi}_s & 0 \end{bmatrix} \begin{bmatrix} \dot{\mathbf{y}}_1 \\ \dot{\mathbf{y}}_2 \\ -\boldsymbol{\lambda} \end{bmatrix} = \begin{bmatrix} \mathbf{y}_2 \\ \mathbf{p}_1 \\ \mathbf{p}_2 \end{bmatrix} \tag{9.294}$$

whereby the symmetry of the system matrix is maintained.

The computation of $\dot{\mathbf{y}}_1, \dot{\mathbf{y}}_2$ and $\boldsymbol{\lambda}$ from (9.294) can be performed either by eliminating or by factoring the system matrix. The Choleski decomposition or the procedure GELS from (IBM SSP) are suitable for eliminating, while LU decomposition is suitable for the factoring method, for which the procedure MFGR from (IBM SSP) may be used. Other advantageous possibilities are SVD decomposition (see Chapter 7) or QR decomposition. For larger systems the use of special algorithms for symmetric sparse matrices is convenient.

Baumgarte's method may be interpreted in several ways (Ostermayer, 1991). The most useful interpretation is that the additional terms are the PD controller of the system $\ddot{f} = u$. This can be further developed. Based on Bode diagram consideration of this controller the better choice of parameters α and β is $\beta = \alpha\sqrt{2}$ or $\beta = \sqrt{2\alpha}$. The concepts from control theory can be followed further. In order to guarantee the asymptotic zero value of the constraints the PID controller with integration term is introduced instead of (9.289)

$$\ddot{\mathbf{f}} + \frac{1}{2T}(4\dot{\mathbf{f}} + \frac{1}{T}\mathbf{f} + \frac{1}{16T^2}\int_{t_0}^{t} \mathbf{f}\,\mathrm{d}\tau) = \mathbf{0} \tag{9.295}$$

with parameter T which can be even set adaptively. The larger values of α or $1/T$ are keeping the constraint violations smaller. However, their upper limit is given by the stability condition of the integration procedure (Bae and Yang, 1991) which also depends on the step size h. Hence the adaptive choice of the parameters α or $1/T$ as a function of the step size h follows. They are taken to be the greatest value that satisfy the stability condition.

Baumgarte's method is a very simple and computationally cheap method to take care of DAE. We are not forced to bound the types of integration codes.

9.8.1.2 Different Variational Formulation

Instead of traditional formulation of equations of motion of MBS by (9.47), (9.60)–(9.61), (9.68) there are other formulations available. Let us mention two approaches.

Borri's Approach

The functional (9.47) is modified to (Borri, 1990)

$$\int_{t_1}^{t_2} (\delta L + \mathbf{f}^T\delta\mathbf{s})\mathrm{d}t = \mathbf{f}^T\delta\mathbf{s})|_{t_1}^{t_2} \tag{9.296}$$

Instead of classical momenta $\mathbf{p}$ (9.45) the extended generalized momenta are introduced (Borri,1991)

$$\mathbf{p}^* = \mathbf{p} + \mathbf{\Phi}_s^T \boldsymbol{\mu} \tag{9.297}$$

where $\boldsymbol{\mu}$ are new different type of Lagrange's multipliers which are the time integrals of classical multipliers $\boldsymbol{\lambda}$

$$\dot{\boldsymbol{\mu}} = -\boldsymbol{\lambda} \tag{9.298}$$

They correspond to the velocity (holonomic or nonholonomic) constraints

$$\boldsymbol{\Phi}_s \dot{\mathbf{s}} + \mathbf{a} = \mathbf{0} \tag{9.299}$$

Using (9.54) and (9.299) we obtain

$$\begin{bmatrix} \mathbf{M} & \boldsymbol{\Phi}_s^T \\ \boldsymbol{\Phi}_s & 0 \end{bmatrix} \begin{bmatrix} \dot{\mathbf{s}} \\ \boldsymbol{\mu} \end{bmatrix} = \begin{bmatrix} \mathbf{p}^* \\ -\mathbf{a} \end{bmatrix} \tag{9.300}$$

and

$$\dot{\mathbf{p}}^* = \frac{\partial L}{\partial \mathbf{s}^T} + \mathbf{Q} + \dot{\boldsymbol{\Phi}}_s^T \boldsymbol{\mu} \tag{9.301}$$

The solution of these equations starts with some $\mathbf{s}, \dot{\mathbf{s}}$, $\mathbf{p}^* = \mathbf{p}$, $\boldsymbol{\mu} = \mathbf{0}$ and computes from (9.300)–(9.301) $\dot{\mathbf{p}}^*, \dot{\mathbf{s}}$ which is integrated into new $\mathbf{p}^*, \mathbf{s}$. The new Lagrange's multipliers $\boldsymbol{\mu}$ have increasing values which correspond to the implicitely integrated equation (9.298). After some time interval of integration their values are very large and they must be again set to zero by $\mathbf{p}^* = \mathbf{p}$.

This procedure satisfies exactly the constraints on the level of accelerations and velocities. The violations on the level of position are very small and grows very slowly.

Geradin's Approach

The functional (9.47) is augmented into (Geradin, 1993)

$$\int_{t_1}^{t_2} \delta(L - k\boldsymbol{\lambda}^T\mathbf{f} - p\mathbf{f}^T\mathbf{f})dt = 0 \tag{9.302}$$

where k is a scaling and p a penalty factor. Instead of (9.68) we obtain

$$\begin{aligned} \mathbf{M}\ddot{\mathbf{s}} + \boldsymbol{\Phi}_s^T(k\boldsymbol{\lambda} + p\mathbf{f}) &= \mathbf{p}_1 \\ k\mathbf{f} &= \mathbf{0} \end{aligned} \tag{9.303}$$

These equations are discretized, solved for $\Delta\mathbf{s}$, $\Delta\dot{\mathbf{s}}$, $\Delta\ddot{\mathbf{s}}$, $\Delta\boldsymbol{\lambda}$ and then integrated by Newmark's integration scheme.

This approach is a part of a new approach of joining the multibody and finite element concepts.

9.8.1.3 DAE Solvers

The integration scheme which starts from the equations (9.66), (9.59) are called DAE solvers.

The actual numerical solution of DAE is however the most serious problem. A general solution algorithm was described in 1971 by Gear (1984). His idea is based on expanding algebraic equations into Taylor series and using approximate difference formulae for differential equations. We thus obtain a system of linear algebraic equations for the increments of the desired variables. In some cases, however, this system of linear equations is singular. This is always the case for equations of MBS motion. A general solution (Gear, 1984) is based on substituting some equations from the DAE system by their time derivative. The order of the derivative represents the index of complexity of the DAE problem. Equations of MBS motion are problems of index 3. In the above text we used this method directly. Another possible solution is the development of special algorithms for the numerical solution of DAE (Griepentrog and März, 1986; Gear and Petzold, 1984; Gear et al., 1985; Lubich, 1989; Führer, 1988; Andrzejewski et al., 1993; Simeon and Rentrop, 1993).

The common approach is to apply ODE integration schemes but to correct the differencies $\Delta\mathbf{s}$, $\Delta\dot{\mathbf{s}}$, $\Delta\ddot{\mathbf{s}}$, $\Delta\boldsymbol{\lambda}$ in such a way that the constraints (9.59) and their time derivatives are satisfied.

The straightforward realization of this procedure is within ADAMS approach (Ryan, 1990). They are many new variants and developments of this approach (Andrzejewski et al., 1993). Equations (9.66) and (9.59) can be generally written

$$\begin{aligned} \mathbf{M}(\mathbf{s},\mathbf{u},\dot{\mathbf{u}},\boldsymbol{\lambda},t) &= \mathbf{0} \\ \mathbf{u}-\dot{\mathbf{s}} &= \mathbf{0} \\ \mathbf{f}(\mathbf{s},\boldsymbol{\lambda},t) &= \mathbf{0} \end{aligned} \tag{9.304}$$

or with state $\mathbf{y} = \lceil \mathbf{s}, \mathbf{u}, \boldsymbol{\lambda} \rceil$ as

$$\mathbf{G}(\mathbf{y},\dot{\mathbf{y}},t) = \mathbf{0} \tag{9.305}$$

If we knew the j-th iteration $\mathbf{y}_j, \dot{\mathbf{y}}_j$ we expand the equation (9.305) into Taylor series

$$\mathbf{G}_j + \frac{\partial \mathbf{G}}{\partial \mathbf{y}^T}\Delta\mathbf{y}_j + \frac{\partial \mathbf{G}}{\partial \dot{\mathbf{y}}^T}\Delta\dot{\mathbf{y}}_j + \ldots = \mathbf{0} \tag{9.306}$$

Using the implicit integration scheme (here multistep implicit) we have

$$\mathbf{y}_{j+1} = \sum_{i=1}^{k} \alpha_i \mathbf{y}_{j-i+1} - h\beta_0 \dot{\mathbf{y}}_{j+1} \tag{9.307}$$

Hence

$$\Delta\dot{\mathbf{y}}_j = -\frac{1}{h\beta_0}\Delta\mathbf{y}_j \tag{9.308}$$

and using this in (9.306) we get

$$\left[\frac{\partial\mathbf{G}}{\partial\mathbf{y}^T} - \frac{1}{h\beta_0}\frac{\partial\mathbf{G}}{\partial\dot{\mathbf{y}}^T}\right]\Delta\mathbf{y}_j = -\mathbf{G}_j \tag{9.309}$$

Then the steps of solution are following:

1. Predict values $\mathbf{y}_j, \dot{\mathbf{y}}_j$.
2. Evaluate $\mathbf{G}$. If it is within error bound then goto 5.
3. Solve (9.309) for $\Delta\mathbf{y}_j$ and compute $\mathbf{y}_{j+1}, \dot{\mathbf{y}}_{j+1}$.
4. Repeat steps 2 through 3.
5. Control the order and integration step size.

The similar idea is in many modifications and improvements applied with different ODE integration schemes (Andrzejewski et al., 1993) or with overdetermined constraints (Führer, 1989) with zero, first and second derivatives of (9.59) treated altogether.

This approach is under rapid development aiming into computationally efficient, accurate and reliable automatic integration procedure of DAE.

9.8.2 Solution in Independent Coordinates

This method is based on the elimination of dependent coordinates as well as Lagrange's multipliers $\boldsymbol{\lambda}$ and thus finding the proper equations of motion. Further described procedures are different variants of coordinate partitioning algorithm from Section 9.5. Its purpose is the reduction of computational complexity, which we accomplish by integrating the smallest possible number of differential equations in the given case. Whether or not the saving due to integrating a smaller number of equations is canceled by the complexity of the operations required to find system of equations needs to be considered, however.

The position of a MBS with m degrees of freedom is determined by m independent coordinates $\mathbf{q}$. Let the independent coordinates be related with all physical coordinates $\mathbf{s}$ by the relation

$$\mathbf{q} = \mathbf{B}\mathbf{s} \tag{9.310}$$

where $\mathbf{B}$ is a constant (m, n)-matrix. Differentiating the skleronomic constraint conditions with respect to time, we calculate

$$\mathbf{\Phi}_s\dot{\mathbf{s}} = \mathbf{0} \tag{9.311}$$

Combining this equation with the relation

$$\dot{\mathbf{q}} = \mathbf{B}\dot{\mathbf{s}} \tag{9.312}$$

we obtain

$$\begin{bmatrix} \mathbf{\Phi}_s \\ \mathbf{B} \end{bmatrix} \dot{\mathbf{s}} = \begin{bmatrix} \mathbf{0} \\ \dot{\mathbf{q}} \end{bmatrix} \tag{9.313}$$

and, assuming the existence of the inverse matrix,

$$\dot{\mathbf{s}} = \begin{bmatrix} \mathbf{\Phi}_s \\ \mathbf{B} \end{bmatrix}^{-1} \begin{bmatrix} \mathbf{0} \\ \dot{\mathbf{q}} \end{bmatrix} = [\mathbf{R}^*, \mathbf{R}] \begin{bmatrix} \mathbf{0} \\ \dot{\mathbf{q}} \end{bmatrix} = \mathbf{R}\dot{\mathbf{q}} \tag{9.314}$$

From equation (9.311), it is apparent that vector $\dot{\mathbf{s}}$ of velocities must be from the null space of matrix $\mathbf{\Phi}_s$, the equation (9.314) may be understood in such a way that the matrix $\mathbf{R}$ is the basis of the null space of matrix $\mathbf{\Phi}_s$ and the coefficients of the linear combination are independent velocities $\dot{\mathbf{q}}$. Then of course

$$\mathbf{\Phi}_s\mathbf{R} = \mathbf{R}^T\mathbf{\Phi}_s^T = \mathbf{0} \tag{9.315}$$

Pre-multiplying the first of equations (9.68), i.e.

$$\mathbf{M}\ddot{\mathbf{s}} - \mathbf{\Phi}_s^T\boldsymbol{\lambda} = \mathbf{p}_1 \tag{9.316}$$

by the matrix $\mathbf{R}^T$ yields

$$\mathbf{R}^T\mathbf{M}\ddot{\mathbf{s}} - \mathbf{R}^T\mathbf{\Phi}_s^T\boldsymbol{\lambda} = \mathbf{R}^T\mathbf{p}_1 \tag{9.317}$$

and

$$\mathbf{R}^T\mathbf{M}\ddot{\mathbf{s}} = \mathbf{R}^T\mathbf{p}_1 \tag{9.318}$$

respectively, as the second term of the left-hand side of (9.317) vanishes because of (9.315).

The computation itself assumes the knowledge of transformation (9.314) between physical and independent coordinates. Differentiating (9.314) with respect to time, we compute

$$\ddot{\mathbf{s}} = \mathbf{R}\ddot{\mathbf{q}} + \dot{\mathbf{R}}\dot{\mathbf{q}} \tag{9.319}$$

and substituting (9.319) into (9.318) we obtain the required minimum system of differential equations in the form

$$\mathbf{R}^T\mathbf{M}\mathbf{R}\ddot{\mathbf{q}} = \mathbf{R}^T\mathbf{p}_1 - \mathbf{R}^T\mathbf{M}\dot{\mathbf{R}}\dot{\mathbf{q}} \tag{9.320}$$

written in independent coordinates $\mathbf{q}$.

Matrix $\mathbf{R}$ plays the key part in deriving system (9.320), and to solve the system we further need to know the term $\dot{\mathbf{R}}\dot{\mathbf{q}}$. How this can be computed depends on the method chosen for the computation of matrix **R**. The possibilities of computing that matrix are dealt with in a separate section.

The ways of determining matrix **R** which has a close connection to matrix **B** from equation (9.310) may be divided into subjective and objective. Let us start with the subjective methods.

When solving simple MBS's formed by a small number of elements we may by estimation say which of the physical coordinates are independent. There may be several possible combinations of m independent coordinates. For each estimate we actually select a matrix **B** in the form of a Boolean matrix. The relation between physical and independent velocities is then characterized by matrix **R** defined by equation (9.314). This is the transmission matrix. We may find it according to definition (9.314) or in an elementary way by subsequently choosing $\dot{q}_i = 1,\ \dot{q}_j = 0,\ j \neq i,\ i, j = 1, 2, \ldots, n$, and we solve the MBS velocities for each choice. We thus obtain the elements of matrix **R** column-wise as vectors $\dot{\mathbf{s}}$.

The objective methods of determining the independent coordinates may be divided into projection methods and orthogonalizing methods.

9.8.2.1 Projection Methods of Determining R

We call the methods dealt with in this section projection methods because of the concept used. We imagine the vector of dependent velocities projected into the space of independent velocities, and we take the columns of matrix **R** for base vectors of that space.

We start from the following consideration. If the constraint equations (9.59) are independent, the matrix $\mathbf{\Phi}_s$ of dimensions (r, n) has its full row rank. Therefore at least one non-singular submatrix $\mathbf{\Phi}_z$ exists for which

$$[\mathbf{\Phi}_z \mathbf{\Phi}_q] \begin{bmatrix} \dot{\mathbf{z}} \\ \dot{\mathbf{q}} \end{bmatrix} = \mathbf{\Phi}_z \dot{\mathbf{z}} + \mathbf{\Phi}_q \dot{\mathbf{q}} = \mathbf{0} \tag{9.321}$$

The splitting of the original matrix $\mathbf{\Phi}_s$ into submatrices also defines at the same time the splitting of the original n physical velocities $\dot{\mathbf{s}}$ into r dependent velocities $\dot{\mathbf{z}}$ and m independent velocities $\dot{\mathbf{q}}$. The same holds for coordinates. As $\mathbf{\Phi}_z$ is regular, we may write

$$\dot{\mathbf{z}} = -\mathbf{\Phi}_z^{-1} \mathbf{\Phi}_q \dot{\mathbf{q}} = \mathbf{P} \dot{\mathbf{q}} \tag{9.322}$$

where **P** is the transmission matrix between dependent and independent velocities of the type (r, m). Combining (9.322) with the identity $\dot{\mathbf{q}} = \dot{\mathbf{q}}$, we obtain

$$\begin{bmatrix} \dot{\mathbf{z}} \\ \dot{\mathbf{q}} \end{bmatrix} = \dot{\mathbf{s}} = \begin{bmatrix} -\mathbf{\Phi}_z^{-1} \mathbf{\Phi}_q \\ \mathbf{E} \end{bmatrix} \dot{\mathbf{q}} \tag{9.323}$$

and, comparing with (9.314), we find

$$\mathbf{R} = \begin{bmatrix} -\mathbf{\Phi}_z^{-1}\mathbf{\Phi}_q \\ \mathbf{E} \end{bmatrix} \tag{9.324}$$

Matrix $\mathbf{R}$ of dimension (n, m) is now the objectively determined transmission (conversion) matrix between physical and independent velocities.

It is obvious from the computational method that in this case also the matrix $\mathbf{B}$ from equation (9.310) is a Boolean matrix; this is because we have determined the independent coordinates again by selection from the physical coordinates. This is a definite limitation. To enlarge the possibility of choice we allow according to (Wehage and Haug, 1982) independent coordinates in the form of a linear combination of physical coordinates. On one hand such coordinates lose their physical meaning; on the other hand, better numerical properties of the solution can be expected from them.

To find the new n coordinates which we designate $\mathbf{c}$ and call composite coordinates, we use the singular value decomposition (SVD) of matrix $\mathbf{\Phi}_s$ in the form

$$\mathbf{\Phi}_s = \mathbf{U}^T\mathbf{D}\mathbf{V} \tag{9.325}$$

We define the composite coordinates by the transformation

$$\mathbf{c} = \mathbf{V}\mathbf{s} \tag{9.326}$$

Assuming the (n, n) matrix $\mathbf{V}$ constant, we also have

$$\dot{\mathbf{c}} = \mathbf{V}\dot{\mathbf{s}} \quad , \quad \ddot{\mathbf{c}} = \mathbf{V}\ddot{\mathbf{s}} \tag{9.327}$$

As matrix $\mathbf{V}$ is orthogonal, we obtain by substituting (9.326) into (9.311) the equation

$$\mathbf{\Phi}_s\mathbf{V}^T\dot{\mathbf{c}} = \mathbf{0} \tag{9.328}$$

which, after substituting for $\mathbf{\Phi}_s$ from (9.325), yields

$$\mathbf{U}^T\mathbf{D}\dot{\mathbf{c}} = \mathbf{0} \tag{9.329}$$

Pre-multiplying the last equation by matrix $\mathbf{U}$ we obtain the equation

$$\mathbf{D}\dot{\mathbf{c}} = \mathbf{0} \tag{9.330}$$

Due to the structure and magnitude of matrix $\mathbf{D}$ elements constraints (assuming the existence on m non-vanishing singular numbers) only the first r composite velocities $\dot{c}_1, \dot{c}_2, \ldots, \dot{c}_r$ can be calculated. The remaining composite velocities can be calculated only from the equations of motion and the constraint equations must be satisfied for all of their values.

The equation (9.326) gives an objective division of composite coordinates $\mathbf{c}$ into dependent composite coordinates

$$\mathbf{c}_z = \lceil c_1, c_2, \ldots, c_r \rceil \tag{9.331}$$

and independent composite coordinates

$$\mathbf{c}_q = \lceil c_{r+1}, c_{r+2}, \ldots, c_n \rceil \tag{9.332}$$

Neither $\mathbf{c}_z$ nor $\mathbf{c}_q$ are chosen from the set of physical coordinates $\mathbf{s}$, but they are their linear combinations.

It is obvious from equation (9.326) written in expanded matrices

$$\mathbf{c} = \begin{bmatrix} \mathbf{c}_z \\ \mathbf{c}_q \end{bmatrix} = \begin{bmatrix} \mathbf{V}_z \\ \mathbf{V}_q \end{bmatrix} \mathbf{s} \tag{9.333}$$

that the independent composite coordinates are connected with the physical coordinates by the relation

$$\mathbf{c}_q = \mathbf{V}_q \mathbf{s} \tag{9.334}$$

where the (m, n)-matrix $\mathbf{V}_q$ is formed by the last m rows of matrix $\mathbf{V}$. Differentiating equation (9.334) with respect to time and comparing with (9.314) yields

$$\mathbf{R} = \mathbf{V}_q^T \tag{9.335}$$

Further,

$$\mathbf{B} = \mathbf{V}_q \tag{9.336}$$

Matrix $\mathbf{B}$ defined by equation (9.310) is no longer Boolean; it is a general and full matrix.

Let us return to equation (9.313). In all the above methods of determining matrix $\mathbf{B}$ we assumed the matrix to be constant. This assumption is essential for the computation of the term $\dot{\mathbf{R}}\dot{\mathbf{q}}$ from equation (9.319). Differentiating (9.313) with respect to time we find

$$\begin{bmatrix} \mathbf{\Phi}_s \\ \mathbf{B} \end{bmatrix} \ddot{\mathbf{s}} = \begin{bmatrix} 0 \\ \ddot{\mathbf{q}} \end{bmatrix} - \begin{bmatrix} \dot{\mathbf{\Phi}}_s \\ 0 \end{bmatrix} \dot{\mathbf{s}} \tag{9.337}$$

and hence

$$\ddot{\mathbf{s}} = \begin{bmatrix} \mathbf{\Phi}_s \\ \mathbf{B} \end{bmatrix}^{-1} \left(\begin{bmatrix} 0 \\ \ddot{\mathbf{q}} \end{bmatrix} - \begin{bmatrix} \dot{\mathbf{\Phi}}_s \\ 0 \end{bmatrix} \cdots \right) \tag{9.338}$$

Comparing (9.338) and (9.319), we come to the conclusion that the term $\dot{\mathbf{R}}\dot{\mathbf{q}}$ expressing accelerations depending on independent velocities can be computed from (9.338) as the physical accelerations $\ddot{\mathbf{s}}$ for vanishing independent accelerations $\ddot{\mathbf{q}} = \mathbf{0}$.

9.8.2.2 Orthogonalizing Methods of Determining R

We designate as orthogonalizing the methods which construct the matrix $\mathbf{R}$ directly from the orthogonality condition with matrix $\mathbf{\Phi}_s$ formulated by relation (9.315). One of the possibilities is the Gram-Schmidt's orthogonalizing process. It is used in the procedure (Liang and Lance, 1987) which we briefly describe.

We first construct the matrix $\mathbf{Z}$ with dimensions (n, n) using the formula

$$\mathbf{Z} = [\mathbf{\Phi}_s^T, \mathbf{B}^T] \tag{9.339}$$

where $\mathbf{B}$ is arbitrary although $\mathbf{Z}$ must be regular. Matrix $\mathbf{B}$ may be chosen subjectively or determined objectively by any procedure from the above section. Using Gram-Schmidt's process, we transform matrix $\mathbf{Z}$ into matrix

$$\mathbf{V} = [\mathbf{V}_z, \mathbf{V}_q] \tag{9.340}$$

where matrix $\mathbf{V}_z$ has dimensions (n, r) and matrix $\mathbf{V}_q$ is an (n, m)-matrix. It can be proved that matrix $\mathbf{V}_q$ defines a null space of matrix $\mathbf{\Phi}_s$, i.e.

$$\mathbf{\Phi}_s \mathbf{V}_q = \mathbf{0} \tag{9.341}$$

We now use the matrix $\mathbf{V}$ to determine the matrix $\mathbf{R}$. We define new composite velocities $\dot{\mathbf{g}}$ by relation

$$\dot{\mathbf{g}} = \mathbf{V}^T \dot{\mathbf{s}} \tag{9.342}$$

Also, due to the orthogonality,

$$\dot{\mathbf{s}} = \mathbf{V}\dot{\mathbf{g}} \tag{9.343}$$

The expansion of (9.343) yields

$$\dot{\mathbf{s}} = \begin{bmatrix} \dot{\mathbf{z}} \\ \dot{\mathbf{q}} \end{bmatrix} = [\mathbf{V}_z, \mathbf{V}_q] \begin{bmatrix} \dot{\mathbf{g}}_z \\ \dot{\mathbf{g}}_q \end{bmatrix} = \mathbf{V}_z \dot{\mathbf{g}}_z + \mathbf{V}_g \dot{\mathbf{g}}_q \tag{9.344}$$

By substituting (9.344) into (9.311) and altering, it can be proved that $\dot{\mathbf{g}}_z = \mathbf{0}$ and hence

$$\dot{\mathbf{s}} = \mathbf{V}_q \dot{\mathbf{g}}_q \tag{9.345}$$

Comparing with (9.314)

$$\mathbf{R} = \mathbf{V}_q \tag{9.346}$$

It may be further shown that $\ddot{\mathbf{g}}_z = \mathbf{0}$ also. Differentiating further the equation (9.345) with respect to time, we obtain

$$\ddot{\mathbf{s}} = \mathbf{V}_q \ddot{\mathbf{g}}_q + \dot{\mathbf{V}}_q \dot{\mathbf{g}}_q \tag{9.347}$$

The equation of motion in coordinates $\mathbf{g}_q$ can now be obtained by substituting (9.347) into (9.316)

$$\mathbf{M}(\mathbf{V}_q\ddot{\mathbf{g}}_q + \dot{\mathbf{V}}_q\dot{\mathbf{g}}_q) - \mathbf{\Phi}_s^T\boldsymbol{\lambda} = \mathbf{p}_1 \tag{9.348}$$

Pre-multiplying the last equation by matrix $\mathbf{V}_q^T$, we obtain

$$\mathbf{V}_q^T\mathbf{M}\mathbf{V}_q\ddot{\mathbf{g}}_q = \mathbf{V}_q^T\mathbf{p}_1 - \mathbf{V}_q^T\mathbf{M}\dot{\mathbf{V}}_q\dot{\mathbf{g}}_q \tag{9.349}$$

as $\mathbf{V}_q^T\mathbf{\Phi}_s^T$ vanishes due to (9.341).

Equation (9.349) enables the solution of motion in coordinates $\mathbf{g}_q$ to be achieved, assuming we can compute the expression $\dot{\mathbf{V}}_q\dot{\mathbf{g}}_q$. Liang and Lance suggest the direct differentiation of matrix $\mathbf{V}_q$ with respect to time. The time requirements of this approach are more demanding than the indirect calculation presented in the previous section.

The main advantage of orthogonalizing methods is the better stability of numerical integration which does not require subsequent stabilizing measures. The time requirements are a disadvantage of this method. One of the possibilities for reducing this disadvantage is the so-called *M*-orthogonalization proposed in (Garcia de Jalon et al., 1987). Another possibility is to construct a genuine orthonormal and differentiable basis of tangent space for constrained system based on Riemannian geometry (Blajer, 1993). Further development is expected along these lines.

In conclusion, it is necessary to stress the fact that matrix $\mathbf{R}$ depends on the position of the solved MBS. For reasons of economy it is not always possible to compute the matrix again from the beginning in each integration step, whether by projection or orthogonalizing methods. Various criteria can be set for the determining whether the computed matrix $\mathbf{R}$ is still satisfactory. For practical reasons the most suitable method is to test the magnitude of the pivot by factoring the equation (9.313). Should the pivot value drop below the given limit, a new selection of independent coordinates must be made and the integration process must be started again.

However, we have not considered the problems of singular cases in the choice of independent coordinates. The methods of solution of singularities from Chapter 7 must be included. Besides kinematic singularities there are also dynamic singularities caused by the neglecting of body masses (Stejskal et al., 1990).

The accuracy of numerical integration can be checked out using expressions for mechanical energy in the case of conservative systems.

9.9 Change of MBS Structure in the Course of Motion

By change of MBS structure we mean a change in the number of degrees of freedom of MBS. The cases when such a MBS structure change occurs are relatively frequent and they are usually characterized by adding or removing a constraint or constraints (Haug et al., 1986). The reasons for this may be in the kinematics or in the forces. The whole large group of mechanisms with interrupted motion of individual elements (e.g. Geneva mechanism) may be assigned to the former class. We put into the latter class systems with friction where, due to convenient force relations, kinematically mobile bodies would unite (so-called stiction = static friction); we would also include cases where the additional mobility is the consequence of damage to the integrity of bodies or kinematic pairs caused by excessive force loading. In this section we deal with the consequences of the changes presented for the numerical solution of dynamics. We assume for that purpose that the description of system is in the form (9.68).

Let us begin with the simplest case where, at the time instant t^d, the equation

$$f^d(\mathrm{s}, t) = 0 \qquad , \qquad t > t^d \tag{9.350}$$

was deleted from the original system of constraint equations. This causes no difficulty for the computation. Up to the time instant t^d, we integrate the system (9.68) as usual. At the specified time t^d, we terminate the original integration and start a new one with a system of equations where the condition (9.350), differentiated twice with respect to time, has been left out and where the initial conditions are given by the state of motion of the system at the time t^d. At the same time the unknown λ^d corresponding to constraint (9.350) is eliminated.

The case of adding a constraint is more complicated. Let us assume that, at the specified time t^a, a new constraint equation

$$f^a(\mathrm{s}, t) = 0 \tag{9.351}$$

is added to the original system of equations. It is essential that both the condition

$$f^a\left(\mathrm{s}(t^a), t^a\right) = 0 \tag{9.352}$$

and the condition

$$\dot{f}^a\left(\mathrm{s}(t^a), t^a\right) = 0 \tag{9.353}$$

must be satisfied to avoid discontinuity in velocities. For the computation in the time interval $0 < t < t^a$, we integrate the original equations (9.68). At the time instant t^a, we interrupt the integration; we check the validity of (9.352) and (9.353) and, if the outcome of the checks is positive, we start the integration of the system of equations with the additional constraint condition with a new unknown λ^a. The state of motion of the system at the instant t^a is now used as the initial conditions.

In the kinematic case, the instants t^a of adding or t^d of deleting a constraint equation are given by the position of the system being examined. In the force case they depend on the satisfaction of additional conditions containing the magnitude of internal forces in the cuts of individual bodies or the magnitude of reactions. These forces are usually non-linear functions of Lagrange's multipliers and accelerations. The determination of time instants t^a, t^d is therefore more complicated in the force case. We present the solution procedure for MBS with friction.

Friction is a non-conservative force, the magnitude of which is usually expressed in relation to the magnitude of reaction in the place of origin of the friction. The most common representation of a friction force is

$$F_f = \mu N \tag{9.354}$$

depending on the magnitude of the normal reaction N acting between the surface of friction or the moment of pin friction in the pin of radius r_p

$$M_f = \mu_\alpha r_p F_p \tag{9.355}$$

which depends on the resultant reaction F_p transmitted by the pin. The given friction depends however on a series of additional circumstances such as the material of the bodies in contact, the surface quality, the absence or presence of lubricant, the magnitude of the relative velocity in the place of contact, etc. In general this behaviour would have to be described as a complex dynamic process by a system of partial differential equations. Because of the complexity of the dependence on these influences, they would in common technical applications be summarized in a constant coefficient μ, μ_α of friction whose value is to be determined experimentally.

Designating the friction force in general by F_f and the coordinate on which the force acts by $g(s,t)$, we include it into the computation using the equivalence of virtual work

$$F_j = \sum_l F_{fl} \frac{\partial g_l}{\partial s_j} = F_j(\ddot{\mathbf{s}}, \dot{\mathbf{s}}, \mathbf{s}, \boldsymbol{\lambda}, t) \tag{9.356}$$

To be able to distinguish the case without friction we now present the equations (9.68) in the form

$$\begin{bmatrix} \mathbf{M} & \boldsymbol{\Phi}_s^T \\ \boldsymbol{\Phi}_s & 0 \end{bmatrix} \begin{bmatrix} \ddot{\mathbf{s}} \\ -\boldsymbol{\lambda} \end{bmatrix} = \begin{bmatrix} \mathbf{p}_1 \\ \mathbf{p}_2 \end{bmatrix} + \begin{bmatrix} \mathbf{f}_f \\ 0 \end{bmatrix} \tag{9.357}$$

where the subvector $\mathbf{f}_f$ of dimension n including all forces of friction corresponding to coordinates $\mathbf{s}$ contains elements dependent on $\boldsymbol{\lambda}$, $\mathbf{s}$, $\dot{\mathbf{s}}$ and $\ddot{\mathbf{s}}$. The solution of equation (9.357) is therefore of substantially higher complexity than the solution of equation (9.68).

The relation of (9.356) with respect to $\boldsymbol{\lambda}$ and $\ddot{\mathbf{s}}$ may be linear; in that case it is possible to convert (9.357) to (9.68). If it is non-linear, we must in each integration step solve a nonlinear system (9.357) for $\boldsymbol{\lambda}$ and $\ddot{\mathbf{s}}$. If we apply Newton's method we obtain the iterative formula

$$\begin{bmatrix} \mathbf{M} - \dfrac{\partial \mathbf{f}_f}{\partial \ddot{\mathbf{s}}^T} & , & \boldsymbol{\Phi}_s^T - \dfrac{\partial \mathbf{f}_f}{\partial \boldsymbol{\lambda}^T} \\ \boldsymbol{\Phi}_s & , & 0 \end{bmatrix} \begin{bmatrix} \Delta \ddot{\mathbf{s}} \\ -\Delta \boldsymbol{\lambda} \end{bmatrix} =$$

$$= -\begin{bmatrix} \mathbf{M} & \boldsymbol{\Phi}_s^T \\ \boldsymbol{\Phi}_s & 0 \end{bmatrix} \begin{bmatrix} \ddot{\mathbf{s}} \\ -\boldsymbol{\lambda} \end{bmatrix} + \begin{bmatrix} \mathbf{p}_1 \\ \mathbf{p}_2 \end{bmatrix} + \begin{bmatrix} \mathbf{f}_f \\ 0 \end{bmatrix} \tag{9.358}$$

Another way of solving (9.357) is as follows. We solve (9.357) similarly as in Chapter 8 using vector $\mathbf{f}_f$ of friction forces known from the previous integration step or equal to zero. Using $\ddot{\mathbf{s}}$ and $\boldsymbol{\lambda}$ we compute new values of friction forces and we repeat the iteration until a tolerance is achieved.

However, these methods are then only applicable when the relative velocities $\mathbf{v}_r$ do not vanish in all the places where friction is considered. These quantities must be monitored at the time of computation. If at some places the relative velocities drop to zero, the structure of the solved system would change. We speak of a lock-up or stiction. MBS loses one or more degrees of freedom. It is therefore necessary to leave out the constraint conditions which have been characterizing motion under constraint and substitute for them new conditions describing the relative rest. To these new Lagrange's multipliers correspond from which the magnitude of force at the contact of bodies can be determined. This force, however, ought to be smaller than that given by formulae (9.354) and (9.355). Should the magnitude of force in the constraint exceed the given limit, an opposite change of structure would occur and the MBS gains one or more degrees of freedom. Changes in number of degrees of freedom need changes in structure of equations (9.357).

In the whole computation the orientation of the friction force is important. It must always be opposite to the orientation of the relative velocity. At the change from rest to motion (stiction $\rightarrow$ friction) the presented criterion fails as the relative velocity is zero. The orientation of force acting in the constraint at rest can be used and maintained in the subsequent computation.

Let us examine the motion of a double pendulum according to Figure 9.12a. Let us consider the effect of pin friction in its revolute pairs. At the beginning there is relative motion in both revolute pairs A, B. From

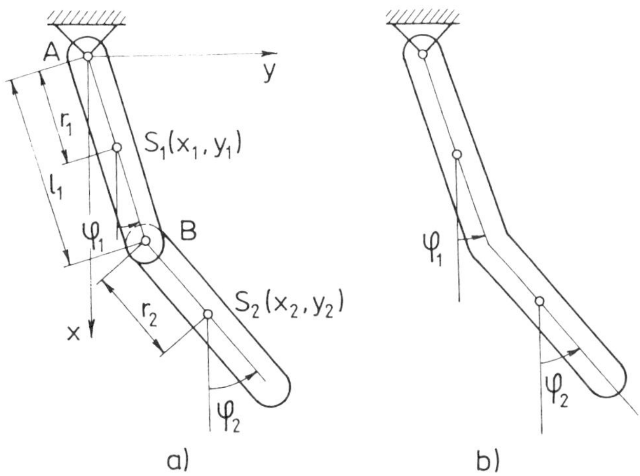

Figure 9.12. Change of MBS structure by Coulomb friction.

a certain instant, the magnitude of friction may cause the relative motion in pair B due to friction to cease to exist. From that instant, the MBS structure and the structure of its dynamic equations change as we now have a motion of one body according to Figure 9.12b. At some other instant, the friction at rest in pair B is overcome and the MBS structure changes again from that in Figure 9.12b to that in Figure 9.12a.

To describe the MBS we use coordinates x_i, y_i, φ_i $(i = 1, 2)$ and LEMT, which, according to (9.68) for the structure as in (9.347), have the form

$$\begin{aligned}
\ddot{\mathbf{s}} &= \lceil \ddot{x}_1, \ddot{y}_1, \ddot{x}_2, \ddot{y}_2, \ddot{\varphi}_1, \ddot{\varphi}_2 \rceil \\
\mathbf{M} &= \lceil m_1, m_1, m_2, m_2, I_{1S_1}, I_{2S_2} \rfloor \\
\mathbf{\Phi}_s &= \begin{bmatrix} 1 & 0 & 0 & 0 & r_1 s\varphi_1 & 0 \\ 0 & 1 & 0 & 0 & -r_1 c\varphi_1 & 0 \\ -1 & 0 & 1 & 0 & (l_1 - r_1)s\varphi_1 & r_2 s\varphi_2 \\ 0 & -1 & 0 & 1 & -(l_1 - r_1)c\varphi_1 & -r_2 c\varphi_2 \end{bmatrix}
\end{aligned} \qquad \text{(a)}$$

for constraints

$$\begin{aligned}
x_1 - r_1 c\varphi_1 &= 0 \\
y_1 - r_1 s\varphi_1 &= 0 \\
x_2 - r_2 c\varphi_2 - x_1 - (l_1 - r_1)c\varphi_1 &= 0 \\
y_2 - r_2 s\varphi_2 - y_1 - (l_1 - r_1)s\varphi_1 &= 0
\end{aligned} \qquad \text{(b)}$$

and

$$\begin{aligned}
\mathbf{p}_1 &= [m_1 g, 0, m_2 g, 0, 0, 0] \\
\mathbf{p}_2 &= [-r_1 c\varphi_1 \dot{\varphi}_1^2, -r_1 s\varphi_1 \dot{\varphi}_1^2, -(l_1 - r_1) c\varphi_1 \dot{\varphi}_1^2 - r_2 c\varphi_2 \dot{\varphi}_2^2, \\
&\quad -(l_1 - r_1) s\varphi_1 \dot{\varphi}_1^2 - r_2 s\varphi_2 \dot{\varphi}_2^2] \\
\mathbf{f}_f &= [0, 0, 0, 0, M_{f1} - M_{f2}, M_{f2}] \\
M_{f1} &= -\mu_{\alpha 1} r_{p1} \sqrt{\lambda_1^2 + \lambda_2^2}\ \operatorname{sgn}(\dot{\varphi}_1) \\
M_{f2} &= -\mu_{\alpha 2} r_{p2} \sqrt{\lambda_3^2 + \lambda_4^2}\ \operatorname{sgn}(\dot{\varphi}_2 - \dot{\varphi}_1)
\end{aligned} \tag{c}$$

For the structure as in Figure 9.12b we have an additional constraint $\dot{\varphi}_2 - \dot{\varphi}_1 = 0$. The Lagrange multiplier λ_5 corresponding to this new constraint expresses the reaction torque due to static friction. The friction moment M_2 is deleted from the from the equations. The new structure of LEMT has matrices $\mathbf{\Phi}_s$ and $\mathbf{f}$ of the form

$$\mathbf{\Phi}_s = \begin{bmatrix} 1 & 0 & 0 & 0 & r_1 s\varphi_1 & 0 \\ 0 & 1 & 0 & 0 & -r_1 c\varphi_1 & 0 \\ -1 & 0 & 1 & 0 & (l_1 - r_1) s\varphi_1 & r_2 s\varphi_2 \\ 0 & -1 & 0 & 1 & -(l_1 - r_1) c\varphi_1 & -r_2 c\varphi_2 \\ 0 & 0 & 0 & 0 & -1 & 1 \end{bmatrix}$$

$$\mathbf{f}_f = [0, 0, 0, 0, M_{f1}, 0] \tag{d}$$

These equations of motion are valid if the condition $|\lambda_5| < |M_{f2}|$ is fulfilled.

The MBS motion is often accompanied by impacts. From the view point of constraints, this corresponds to an additional constraint of the type (9.352) without satisfying condition (9.353). From the viewpont of physics, this means the appearance of an additional contact of bodies moving with different velocities. In some cases impacts are produced on purpose and are inherent in the function of the device being considered (e.g. mechanical hammers, power picks, shock vibration dampers, etc.); in other cases impacts are not deliberate and are caused, for example, by backlash in constraints. This is a complex phenomenon for the exact solution of which the MBS elements must be considered to be a viscous elastic continuum; contact deformations at the point of contact and elastic waves propagation in bodies and their reflections must be solved, and dispersion properties, etc. must be considered. It most technical cases concerning mechanisms, the elements of the system are relatively rigid and their dimensions small with respect to the elastic stress wavelength; in these cases it is usually sufficient to use Newton's theory of impact based on the assumption that an impact is an instantaneous process where, without change of position

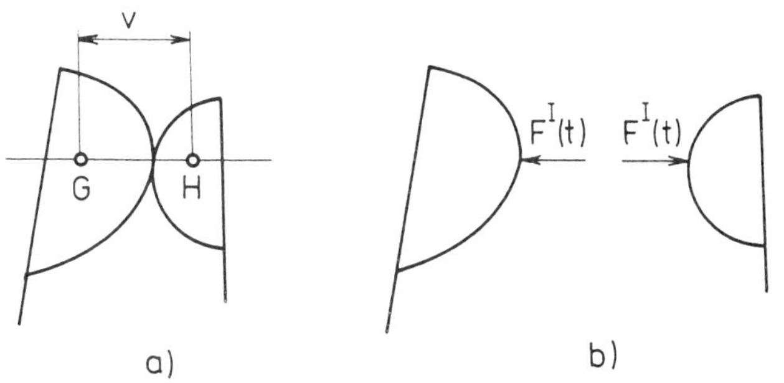

Figure 9.13. Impact of two bodies.

of the system, a finite change of element velocities takes place and loss of energy may expressed by the restitution coefficient ε. In some cases the impact time τ is considered a finite but very short time interval. The behaviour of impact forces in the interval is estimated. We use this approach and show the necessary modification of equations (LEMT) for the case of impact and the numerical solution procedure. We assume:

- the impact occurs in one place and is direct,
- the state of motion of the system just before impact is known,
- the restitution coefficient ε is known.

Our goal is the determination of the magnitudes of system velocities just after the impact, impulses of forces in the point of impact and the impulses of the reaction impact forces.

Modelling the whole process as an elastic impact we must abandon the assumption of ideal rigidity of the bodies and allow their deformability. Let us denote $\mathbf{v}(\mathbf{s}(t))$ as the distance of two chosen fixed points G, H lying in the common normal of the colliding bodies, and let the two points be at a sufficient distance from the place of impact, Figure 9.13. We denote $\mathbf{F}^I$ as the force of impact. We compute the vector of generalized $\mathbf{Q}^I$ forces of impact using the principle of virtual work. We have

$$\delta\mathbf{s}^T\mathbf{Q}^I = \delta\mathbf{v}^T\mathbf{F}^I = \delta\mathbf{s}^T\left(\frac{\partial\mathbf{v}}{\partial\mathbf{s}^T}\right)^T\mathbf{F}^I \tag{9.359}$$

and hence

$$\mathbf{Q}^I = \mathbf{F}^I\frac{\partial\mathbf{v}}{\partial\mathbf{s}^T} \tag{9.360}$$

For motion at impact, the equation of motion is

$$\mathbf{M}\ddot{\mathbf{s}} = \mathbf{\Phi}_s\boldsymbol{\lambda} + \mathbf{p}_1 + \mathbf{Q}^I \tag{9.361}$$

We integrate this between limits $0, \tau$ and in accordance with Newton's theory of impact we assume $\tau \to 0$. The individual terms of the equation then have the following meaning

$$\lim_{\tau \to 0} \int_0^\tau \mathbf{M}\ddot{\mathbf{s}}dt = \lim_{\tau \to 0} \left[\mathbf{M}\dot{\mathbf{s}} \Big|_0^\tau - \int_0^\tau \dot{\mathbf{M}}\dot{\mathbf{s}}dt \right] = \mathbf{M}(\dot{\mathbf{s}} - \dot{\mathbf{s}}_0) = \mathbf{M}\Delta\dot{\mathbf{s}} \quad (9.362)$$

$$\lim_{\tau \to 0} \int_0^\tau \mathbf{\Phi}_s^T \boldsymbol{\lambda} dt = \mathbf{\Phi}_s \lim_{\tau \to 0} \int_0^\tau \boldsymbol{\lambda} dt = \mathbf{\Phi}_s^T \boldsymbol{\lambda}^I \quad (9.363)$$

where $\boldsymbol{\lambda}^I$ is the vector expressing impulses of reaction impact forces,

$$\lim_{\tau \to 0} \int_0^\tau \mathbf{p}_1 dt = \mathbf{0} \quad (9.364)$$

$$\begin{aligned} \lim_{\tau \to 0} \int_0^\tau \mathbf{Q}^I dt &= \lim_{\tau \to 0} \int_0^\tau \mathbf{F}^I \frac{\partial \mathbf{v}}{\partial \mathbf{s}^T} dt = \left(\frac{\partial \mathbf{v}}{\partial \mathbf{s}^T} \right)^T \lim_{\tau \to 0} \int_0^\tau \mathbf{F}^I dt = \\ &= \left(\frac{d\mathbf{v}}{d\mathbf{s}^T} \right)^T \mathbf{I}^I \end{aligned} \quad (9.365)$$

where $\mathbf{I}^I$ is the impulse of impact force. Using (9.362), (9.363), (9.364) and (9.365), we obtain instead of (9.361) for the impact instant n equations

$$\mathbf{M}\Delta\ \dot{\mathbf{s}} = \mathbf{\Phi}_s^T \boldsymbol{\lambda}^I + \left(\frac{\partial \mathbf{v}}{\partial \mathbf{s}^T} \right)^T \mathbf{I}^I \quad (9.366)$$

containing $(n + m + 1)$ unknowns $\Delta\dot{\mathbf{s}}, \boldsymbol{\lambda}^I; \mathbf{I}^I$. It is therefore necessary to add further conditions. The first expresses the fact that velocity changes $\Delta\dot{\mathbf{s}}$ must satisfy m constraint conditions

$$\mathbf{\Phi}_s \Delta\dot{\mathbf{s}} = \mathbf{0} \quad (9.367)$$

The second condition expresses the impact character using the restitution coefficient ε. At the point of impact according to definition

$$\dot{\mathbf{v}} = -\varepsilon \dot{\mathbf{v}}_0 \quad (9.368)$$

where $\dot{\mathbf{v}}$ is the relative velocity just after the impact and $\dot{\mathbf{v}}_0$ the relative velocity just before the impact. Using the relation

$$\dot{\mathbf{v}} = \frac{\partial \mathbf{v}}{\partial \mathbf{s}^T} \dot{\mathbf{s}} \quad (9.369)$$

to substitute into (9.368) we obtain

$$\frac{\partial \mathbf{v}}{\partial \mathbf{s}^T}\dot{\mathbf{s}} = -\varepsilon \frac{\partial \mathbf{v}}{\partial \mathbf{s}^T}\dot{\mathbf{s}}_0 \tag{9.370}$$

where $\dot{\mathbf{s}}$ and $\dot{\mathbf{s}}_0$ denote the vectors of physical velocities just after and just before the impact. Subtracting the expression $\partial \mathbf{v}/\partial \mathbf{s}^T \cdot \dot{\mathbf{s}}_0$ from both sides of equation (9.370), we finally obtain the condition for the change $\Delta\dot{\mathbf{s}}$ of physical velocities caused by the impact in the form

$$\frac{\partial \mathbf{v}}{\partial \mathbf{s}^T}\Delta\dot{\mathbf{s}} = -(1+\varepsilon)\frac{\partial \mathbf{v}}{\partial \mathbf{s}^T}\dot{\mathbf{s}}_0 \tag{9.371}$$

Writing equations (9.366), (9.367) and (9.371) together in matrix form yields a system of $(n+m+1)$ linear equations

$$\begin{bmatrix} \mathbf{M} & \mathbf{\Phi}_s^T & \left(\frac{\partial \mathbf{v}}{\partial \mathbf{s}^T}\right)^T \\ \mathbf{\Phi}_s & 0 & 0 \\ \frac{\partial \mathbf{v}}{\partial \mathbf{s}^T} & 0 & 0 \end{bmatrix} \begin{bmatrix} \Delta\dot{\mathbf{s}} \\ -\boldsymbol{\lambda}^I \\ -\mathbf{I}^I \end{bmatrix} = \begin{bmatrix} 0 \\ 0 \\ -(1+\varepsilon)\frac{\partial \mathbf{v}}{\partial \mathbf{s}^T}\dot{\mathbf{s}}_0 \end{bmatrix} \tag{9.372}$$

for n changes $\Delta\dot{\mathbf{s}}$ of velocities, for m quantities $\boldsymbol{\lambda}^I$ expressing impulses of reaction impact forces, and for the impulse $\mathbf{I}^I$ of impact force in the point of impact.

To perform the calculation of MBS motion with impacts, we integrate equations (9.68) in the usual way, monitoring the collision conditions of the type (9.352). Having done this we interrupt the integration and calculate from system (9.372) the velocity changes $\Delta\dot{\mathbf{s}}$. We add these to the velocities obtaining at the instant of satisfying (9.352), and we use them, together with the values of coordinates at satisfaction of (9.352), as initial conditions for the start of further integration.

9.10 MBS with Flexible Elements

We have assumed, unless otherwise stated as an exception, that MBS elements are rigid bodies. This approach is satisfactory in partial cases where the highest frequencies of the spectrum of applied forces which we consider is sufficiently small with respect to the first natural frequency of the body (Brepta, 1986). With the development of machines, the requirements on their output and lower material consumption increase. The consequences of this are more yielding and more loadable light-weight structures. The requirement of exact function does not allow us to ignore the flexibility of individual elements, i.e. we must drop the analytical methods based on the

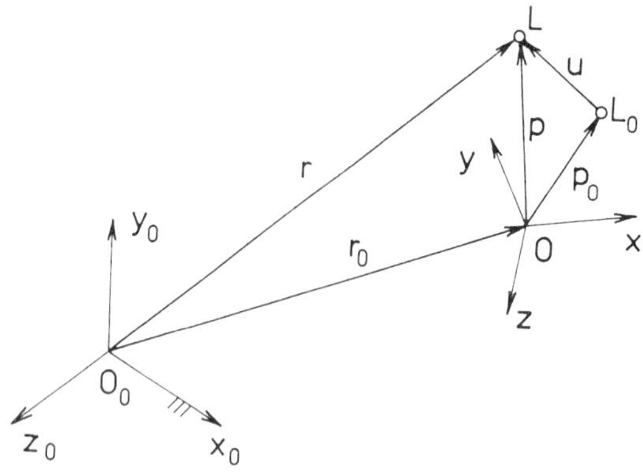

Figure 9.14. Flexible body dynamic description.

assumption of ideally rigid bodies and we must consider individual elements of the system as being of the elastic continuum type. The dynamics of such systems is however a problem of a higher order of complexity. In principle this is a gross motion given by the kinematic pairs used with superposed oscillating motion of flexible elements caused by their loading and by their motion which has a feedback influence on the gross motion. The exact dynamic description of such systems requires the use of partial differential equations. However the completely general solution is only required in exceptional cases. An approximate solution would therefore be chosen. It is possible if elastic deformations are small enough with respect to the body dimensions. Its type depends to a great extent on the purpose of the calculation. Less exact procedures suffice for the calculation of dynamic loading of MBS elements, whereas more perfect methods must be used for analysis of dynamic precision. Such methods usually involve approaches based on the finite element method (FEM) (Bathe, 1982) and have been used since the early 1980's. We present a brief outline of such an approach.

We consider every flexible MBS body to be a MBS substructure. We choose in it a local system of coordinates $Oxyz$, Figure 9.14. This is going to move along with the corresponding body considered as not deformed. The position of system $Oxyz$ in the basic fixed global system $O_0x_0y_0z_0$ is determined by radius vector $\mathbf{r}_0$ of origin O and, for example, by three Cardan angles (or four Euler parameters). We divide the substructure being considered into finite elements in such a way that we choose the position of their nodes. We enter the node coordinates of each element into vector $\mathbf{V}^e$. At the same time we choose the type of finite elements, the deformation properties of which are characterized by matrix $\mathbf{N}$ of shape functions.

We describe the position of general point L of a flexible body by the vector

$$\mathbf{p} = \mathbf{p}_0 + \mathbf{u} \tag{9.373}$$

where the coordinates of vector $\mathbf{p}_0$ correspond to the position of point L_0 before the deformation of the body and are expressed in system $Oxyz$; vector $\mathbf{u} \equiv \overrightarrow{L_0 L}$ determines the magnitude of elastic deformation of the body in the respective place again in system $Oxyz$.

Using elastic displacements $\mathbf{v}^e$ of the nodes of the e-th element, the vector $\mathbf{u}$ of an internal point L of this element can be expressed in the form

$$\mathbf{u} = \mathbf{N}\mathbf{v}^e \tag{9.374}$$

The matrix $\mathbf{N}$ can be used to express the coordinates of vector $\mathbf{p}_0$ using coordinates $\mathbf{V}^e$ of nodes of the e-th element

$$\mathbf{p}_0 = \mathbf{N}\mathbf{V}^e \tag{9.375}$$

The absolute position of point L is then expressed by coordinates of vector $\mathbf{r}$ in the following way

$$\mathbf{r} = \mathbf{r}_0 + \mathbf{S}\mathbf{p} = \mathbf{r}_0 + \mathbf{S}(\mathbf{p}_0 + \mathbf{u}) = \mathbf{r}_0 + \mathbf{S}\mathbf{N}\mathbf{V}^e + \mathbf{S}\mathbf{N}\mathbf{v}^e \tag{9.376}$$

where $\mathbf{S}$ is the matrix of directional cosines (4.29); we use this to transform the components of vector quantities from space $Oxyz$ into space $O_0x_0y_0z_0$. The equation (9.376) is of the type (4.25), and expresses the resultant motion of point L as the sum of motion of point L_0 expressed by translation motion $\mathbf{r}_0$ and relative spherical motion $\mathbf{S}\mathbf{N}\mathbf{V}^e$ of the rigid body, and oscillating motion LL_0 by $\mathbf{S}\mathbf{N}\mathbf{v}^e$ expressing the motion of the flexible body with respect to the undeformed state. All motions are expressed by components in the basic system of co ordinates $Ox_0y_0z_0$.

Vector $\mathbf{r}$ is a function of body position (expressed by three coordinates of vector $\mathbf{r}_0$ and, for example, three Cardan angles in $\mathbf{S}$), the position $\mathbf{V}^e$ of element e in the body (given, for example, by 24 coordinates of its nodes for an eight-node element), the position $\mathbf{p}_0$ of the point in the element (given by three coordinates) and finally of elastic dis placements $\mathbf{v}^e$ of nodes (24 in the given case). The coordinates determining the position of the body and the elastic displacement of nodes are functions of time.

For the description of the state of motion of a body, it is necessary to know the absolute velocities of all its points. We obtain these by differentiating (9.376) with respect to time. Assuming the elements of matrix $\mathbf{N}$ are not functions of time, we obtain

$$\begin{aligned}\dot{\mathbf{r}} &= \dot{\mathbf{r}}_0 + \dot{\mathbf{S}}\mathbf{N}(\mathbf{V}^e + \mathbf{v}^e) + \mathbf{S}\mathbf{N}\dot{\mathbf{v}}^e = \\ &= \dot{\mathbf{r}}_0 + \mathbf{S}\mathbf{\Omega}\mathbf{N}(\mathbf{V}^e + \mathbf{v}^e) + \mathbf{S}\mathbf{N}\dot{\mathbf{v}}^e\end{aligned} \tag{9.377}$$

Assuming that deformations $\mathbf{v}^e$ are small, we may neglect $\mathbf{v}^e$ with respect to $\mathbf{V}^e$ and the second term of the right-hand side of (9.377) using (4.39) and (4.64) e.g. in the form (5.37) is

$$\mathbf{S}\boldsymbol{\Omega}\mathbf{N}\mathbf{V}^e = \mathbf{S}\boldsymbol{\Omega}\mathbf{z} = -\mathbf{S}\mathbf{Z}\boldsymbol{\omega} = -\mathbf{S}\mathbf{Z}\mathbf{D}\dot{\boldsymbol{\varphi}} = \mathbf{B}^e\dot{\boldsymbol{\varphi}}$$

$$\mathbf{z} = \mathbf{N}\mathbf{V}^e, \qquad \boldsymbol{\omega} = \mathbf{D}\dot{\boldsymbol{\varphi}}, \qquad \mathbf{B}_1 = -\mathbf{S}\mathbf{Z},$$

$$\mathbf{S}\boldsymbol{\Omega}\mathbf{N}\mathbf{V}^e = \mathbf{B}_1\boldsymbol{\omega} = \mathbf{B}^e\dot{\boldsymbol{\varphi}} \tag{9.378}$$

where vector

$$\dot{\boldsymbol{\varphi}} = \lceil \dot{\varphi}_x, \dot{\varphi}_y, \dot{\varphi}_z \rceil \tag{9.379}$$

Relation (9.377) can then be written in the matrix form

$$\dot{\mathbf{r}} = [\mathbf{E}, \mathbf{B}^e, \mathbf{S}\mathbf{N}] \begin{bmatrix} \dot{\mathbf{r}}_0 \\ \dot{\boldsymbol{\varphi}} \\ \dot{\mathbf{v}}^e \end{bmatrix} \tag{9.380}$$

The kinetic energy of the considered element is given by the relation

$$T^e = \frac{1}{2}\int_{(m^e)} \dot{\mathbf{r}}^T\dot{\mathbf{r}}dm = \frac{1}{2}\dot{\mathbf{q}}^{e^T}\mathbf{M}^e\dot{\mathbf{q}}^e \tag{9.381}$$

where

$$\dot{\mathbf{q}}^e = \lceil \dot{\mathbf{r}}_0, \dot{\varphi}, \mathbf{v}^e \rceil \tag{9.382}$$

and $\mathbf{M}^e$ is the mass matrix of element e

$$\mathbf{M}^e = \int_{(m^e)} \begin{bmatrix} \mathbf{E} & \mathbf{B}^e & \mathbf{S}\mathbf{N} \\ & \mathbf{B}^{e^T}\mathbf{B}^e & \mathbf{B}^{e^T}\mathbf{S}\mathbf{N} \\ \text{sym} & & \mathbf{N}^T\mathbf{N} \end{bmatrix} dm \tag{9.383}$$

the dimension of which is $(6 + v^e, 6 + v^e)$.

The kinetic energy of the whole body is obtained as the sum of kinetic energy of all its elements; it can be written in the same form as (9.53), namely

$$T_b = \sum_e T^e = \frac{1}{2}\dot{\mathbf{q}}_b^T\mathbf{M}_b\dot{\mathbf{q}}_b \tag{9.384}$$

where the body mass matrix is

$$\mathbf{M}_b = \begin{bmatrix} m_b\mathbf{E} & \bar{\mathbf{B}} & \mathbf{S}\mathbf{N} \\ & \hat{\mathbf{B}} & \tilde{\mathbf{B}} \\ \text{sym} & & \mathbf{M}_v \end{bmatrix} \tag{9.385}$$

and the velocity vector

$$\dot{\mathbf{q}}_b = \lceil \dot{\mathbf{r}}_0, \dot{\varphi}, \dot{\mathbf{v}} \rceil \tag{9.386}$$

Vector $\mathbf{v}$ includes the node displacements of all elements forming the body, matrix $\hat{\mathbf{B}}$ expresses the inertial properties of the body in rotation, matrix $\mathbf{M}_v$ is the mass matrix of the flexible body currently used by FEM analysis, matrices $\bar{\mathbf{B}}, \bar{\mathbf{N}}, \tilde{\mathbf{B}}$ are constraint matrices, and m_b is the body mass.

The potential energy of an element disregarding the potential field of gravity is given by its deformation energy. Using the stiffness matrix $\mathbf{K}^e$ of an element it can be expressed by the relation

$$\Pi^e = \frac{1}{2}\mathbf{v}^{e^T}\mathbf{K}^e\mathbf{v}^e \tag{9.387}$$

The potential energy of the whole body is the sum of the potential energies of the individual elements

$$\Pi_b = \sum_e \Pi^e = \frac{1}{2}\mathbf{q}_b^T\mathbf{K}_b\mathbf{q}_b \tag{9.388}$$

The whole body stiffness matrix $\mathbf{K}_b$ has the structure

$$\mathbf{K}_b = \begin{bmatrix} 0 & 0 & 0 \\ & 0 & 0 \\ \text{sym} & & \mathbf{K}_v \end{bmatrix} \tag{9.389}$$

where $\mathbf{K}_v$ is the stiffness matrix commonly used in FEM.

We divide the forces acting on the body into forces from spring-damper elements, drives and external applied forces. We express their effect on the motion of the body by respective generalized forces, which we calculate using the principle of virtual work. Let force $\mathbf{F}_C^e$ act in point C of element e. Its virtual work is

$$\begin{aligned} \delta W &= \mathbf{F}_C^{e^T}\delta\mathbf{r}_C = \mathbf{F}_C^{e^T}(\delta\mathbf{r}_0 + \mathbf{B}^e\delta\boldsymbol{\varphi} + \mathbf{SN}\delta\mathbf{v}^e) = \\ &= [\mathbf{F}_C^{e^T}, \mathbf{F}_C^{e^T}\mathbf{B}^e, \mathbf{F}_C^{e^T}\mathbf{SN}]\begin{bmatrix} \delta\mathbf{r}_0 \\ \delta\boldsymbol{\varphi} \\ \delta\mathbf{v}^e \end{bmatrix} \end{aligned} \tag{9.390}$$

The virtual work of all forces acting on element e is to be obtained as a sum, hence the generalized force $\mathbf{Q}^e$ acting on element e is

$$\begin{aligned} \mathbf{Q}^e &= [\sum_C \mathbf{F}_C^{e^T}, \sum_C \mathbf{F}_C^{e^T}\mathbf{B}^e, \sum_C \mathbf{F}_C^{e^T}\mathbf{SN}] = \\ &= [\mathbf{Q}_r^e, \mathbf{Q}_\varphi^e, \mathbf{Q}_{v^e}^e] \end{aligned} \tag{9.391}$$

Analogously the generalized force $\mathbf{Q}_b$ for the whole body is

$$\mathbf{Q}_b = [\mathbf{Q}_r, \mathbf{Q}_\varphi, \mathbf{Q}_v] \tag{9.392}$$

where

$$\mathbf{Q}_r = \sum_e \mathbf{Q}_r^e, \qquad \mathbf{Q}_\varphi = \sum_e \mathbf{Q}_\varphi^e, \qquad \mathbf{Q}_v = \sum_e \mathbf{Q}_{v^e}^e \tag{9.393}$$

The last sum must be performed with regard to the order deformations $\mathbf{v}^e$ in vector $\mathbf{v}$.

Using the derived relations and applying Lagrange's second type of equations, it is possible to formulate the equations of motion of the whole body. Their form is

$$\mathbf{M}_b \ddot{\mathbf{q}}_b + \dot{\mathbf{M}}_b \dot{\mathbf{q}}_b + \mathbf{K}_b \mathbf{q}_b + \mathbf{c}_b = \mathbf{Q}_b \tag{9.394}$$

which, apart from obvious symbols, contains the vector

$$\mathbf{c}_b = \lceil \mathbf{0}, \mathbf{c}_{b\varphi}, \mathbf{c}_{bv} \rceil \tag{9.395}$$

This comprises all terms created by differentiation according to Lagrange's equations, the result of which does not depend linearly on $\dot{\mathbf{q}}_b$ and $\mathbf{q}_b$. The sub-vector $\mathbf{c}_{b\varphi}$ has three elements, and sub-vector $\mathbf{c}_{bv}$ as many elements as all finite elements of the body have degrees of freedom.

It is convenient for the numerical solution to re-write (9.394) in the form

$$\mathbf{M}_b \ddot{\mathbf{q}}_b = \mathbf{Q}_b + \mathbf{d}_b \tag{9.396a}$$

$$\mathbf{d}_b = -\dot{\mathbf{M}}_b \dot{\mathbf{q}}_b - \mathbf{K}_b \mathbf{q}_b - \mathbf{c}_b \tag{9.396b}$$

Analogous equations hold for all flexible MBS bodies. Writing them into a common formula, we obtain the equations of motion of MBS in the form

$$\mathbf{M}\ddot{\mathbf{q}} = \mathbf{Q} + \mathbf{d} \tag{9.397}$$

where vector $\mathbf{q}$ comprises all the coordinates used, matrix $\mathbf{M}$ is block wise diagonal and is composed of matrices $\mathbf{M}_b$, just as vectors $\mathbf{Q}, \mathbf{d}$ are formed as a union of vectors $\mathbf{Q}_b, \mathbf{d}_b$.

As we are dealing with a constrained mechanical system, not all coordinates in $\mathbf{q}$ are independent. They must satisfy the constraint equations. For individual kinematic pairs, these are of the same type as presented in Chapter 5 for systems of rigid bodies with the difference that some vectors are dependent on the deformation of bodies and are therefore time variables. For example, for connecting bodies i, j by a spherical pair according to Table 9.1,

$$\mathbf{r}_{1i} + \mathbf{S}_{1i}\mathbf{r}_{iA} - \mathbf{r}_{1j} - \mathbf{S}_{1j}\mathbf{r}_{jA} = \mathbf{0} \tag{9.398}$$

where $\mathbf{r}_{iA}, \mathbf{r}_{jA}$ depend on the deformation of both bodies and must be evaluated again in each integration step. Using (9.373), (9.374) and (9.375), condition (9.378) can be written in the form

$$\mathbf{r}_{1i} + \mathbf{S}_{1i}\mathbf{N}(\mathbf{V}^e + \mathbf{v}^e)_{A_i} - \mathbf{r}_{1j} - \mathbf{S}_{1j}\mathbf{N}(\mathbf{V}^e + \mathbf{v}^e)_{A_j} = \mathbf{0} \tag{9.399}$$

where expressions $\mathbf{N}(\mathbf{V}^e + \mathbf{v}^e)_{A_{i,j}}$ represent the radius vector of point A in system i and j expressing the position of centre A of the spherical pair in system i and j respectively at the deformation of both bodies.

In a similar way, the constraint conditions can be introduced into the calculation of all kinematic pairs. First, after adding all necessary constraint equations of the type (9.399) to the system of equations (9.397), the resultant system of equations is uniquely solvable. For its numerical solution, the procedures presented in Section 9.8 or other special methods (Bathe, 1982) can be used.

The disadvantage of the procedure described is the great dimension of vector $\mathbf{v}$ containing the node displacements of all finite elements whose number may run into hundreds; the computer memory and the number of operations required are tremendous. This disadvantage can be substantially reduced by expressing the deformation form of flexible bodies by chosen functions which we call deformation modes. This approach is in literature known as the "component mode synthesis" (Celiqüeta and Garcia de Jalón, 1987). These deformation modes can also be determined for bodies with simple shape by the analytical solution of deformation or by its approximation (Shabana, 1989). For bodies with complex shape FEM solution is necessary.

Thus instead of (9.374) we describe the deformation of a point by

$$\mathbf{u} = \mathbf{\Psi}\boldsymbol{\xi} \tag{9.400}$$

where $\mathbf{\Psi}$ is the modal matrix whose columns are the deformation modes and $\boldsymbol{\xi}$ is a vector of new unknown modal coordinates (coefficients of the linear combination) (Hurty, 1991).

As deformation modes, we can choose a combination of static deformation modes $\boldsymbol{\psi}^S$ in the number equal to the number of boundary degrees of freedom and dynamic modes $\boldsymbol{\psi}^D$, which are the vibration modes computed for free vibration of the bodies with all the boundaries perfectly fixed. The static deformation modes can be calculated with the choice of one boundary displacement equal to one and the others equal to zero. Their number equals the number of boundary degrees of freedom. The vibration modes are chosen to correspond to the lowest eigenfrequencies in the number given by the required range of validity of the solution frequency. The problem of suitable choice of deformation modes is treated in detail in (Yoo and Haug, 1985). Another possibility is to use the deformation modes determined for a great variety of different boundary conditions.

Using the method of expansion with respect to deformation modes, we described the deformed body position not by vector $\mathbf{q}_b$ but by a new vector in which coordinates $\mathbf{v}$ are substituted by new coordinates $\boldsymbol{\xi}$. They may be divided firstly into coordinates $\boldsymbol{\xi}^S$ of static deformation modes in the number of boundary degrees of freedom and coordinates ξ^D of dynamic deformation modes, the number of which we choose.

The above described derivation of equations of motion (9.397), (9.399) is completely valid for our case of deformation modes. However, these equations of motion correspond to the direct equations of motion from Section 9.5.

The question is whether and how can be applied the new efficient MBS formalisms (composite rigid-body, articulated-body inertia) to flexible MBS.

The straightforward extension of articulated-body inertia has been done by (Kim and Haug, 1988). They applied the recursive computation of articulated-body inertia described in global coordinates to each flexible body of MBS. This means that the mass matrix $\mathbf{M}_i$ of each body i must be inverted according to (9.200) and this leads to the computational complexity $\sum_i O((N_i + \sigma_i)^3)$ where N_i is the number of modal coordinates of body i and σ_i is the number of joint coordinates between body i and the previous body $p(i)$.

This can be substantially improved. In fact we can obtain almost complete extension of articulated-body inertia and composite rigid-body formalisms from rigid MBS to flexible one (Valášek et al., 1995). For the flexible articulated-body inertia formalism the resulting computational complexity is $\sum_i O(N_i^2 + \sigma_i)$ and if the change of inertia tensor of a body due to small flexible deformation is neglected the complexity is only $\sum_i O(N_i + \sigma_i)$, i.e. it corresponds to recursion applied to each DOF both rigid and flexible. Further, the local coordinate system is used and it improves the computational complexity similarly as for rigid MBS.

The key point of this approach is the modification of the set of deformation modes in order to be orthonormalized and not interacting with the deformation of hinge points. It is achieved as follows. For each hinge point of a body we add a sixtuple of additional deformation modes with zero values everywhere except for the hinge point. We substitute zero values for all elements of other modes which describe the deformation of hinge points. The ability of the set of deformation modes to describe the deformation of the body is not changed. This set of deformation modes is then orthonormalized. By this way the nutual interaction of modes is minimized and maintained during the recursion. The recursive derivation of equations of motion similarly as after the equation (9.222) follows.

Also the composite rigid-body formalism can be completely extended for flexible bodies (Valášek et al., 1995). The recursive relation among the elements of mass matrix and generalized forces is again achieved, however, not so simply as for rigid MBS (9.187)–(9.189). This enables to use the promising residual algorithm also for flexible MBS.

9.11 MBS with Feedback Control

MBS's often contain control systems which, with the help of feed-back, determine the required resultant behaviour of the MBS. To include control systems in the MBS dynamic model is not difficult in principle, but there are some peculiarities to be dealt with.

MBS control is only possible through action forces. These can however be functions not only of the position and velocities of the MBS elements but also of accelerations or even higher order derivatives and of the quantities of state of the control system (e.g. hydraulic or electric). By a detailed description of control system action on MBS, it is possible to measure on the MBS only the position, velocities and acting forces; only on those quantities may the controlling action forces depend. We then obtain the overall model of the controlled MBS simply by combining the description of state of the control system with the actual equations of motion of the MBS.

In an idealized description of the control system (given, for example, by neglecting the time constant in the control system model) it is possible for the action forces to depend on higher order derivatives of MBS position. We must then create a special state model of the controlled MBS, of which the equations of motion are a part.

Finally, the effect of the control system may be described as a relation of MBS coordinates using a differential equation without the actual force (physical) realization of the control action (e.g. the dependence of drive revolutions per time unit on the position of MBS elements). Such differential equations may then be included into the LEMT equations of motion as nonholonomic higher order constraints.

9.12 Applications

As an example of a MBS dynamic solution, we outline briefly the assembly of the equations of motion of the car front wheel trapezoidal suspension mechanism, a car front axle mechanism passing a surface unevenness and a new car wheel mechanism.

9.12.1 Car Front Wheel Trapezoidal Suspension

Figure 9.15 shows the unsimplified spatial model of the mechanism from the example in Section 9.3. We are going to investigate again its gross oscillations about the equilibrium position. We choose angles $\varphi_{12}, \varphi_{14}, \varphi_{23}$ and ψ_{23} as coordinates. We express the kinetic energy of the system, according

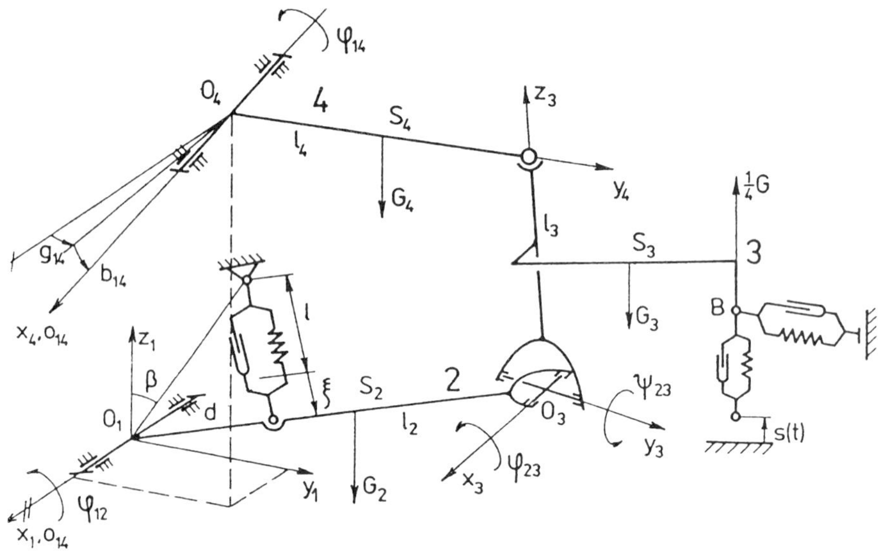

Figure 9.15. Spatial model of front wheel suspension.

to (9.34b), as

$$T = \frac{1}{2}(I_2\dot{\varphi}_{12}^2 + I_4\dot{\varphi}_{14}^2 + m_3(\dot{x}_{1S_3}^2 + \dot{y}_{1S_3}^2 + \dot{z}_{1S_3}^2) + I_{3x_3}\omega_{13x_3}^2 + \\ + I_{3y_3}\omega_{13y_3}^2 + I_{3z_3}\omega_{13z_3}^2 - 2I_{3x_3y_3}\omega_{13x_3}\omega_{13y_3} - \\ - 2I_{3x_3z_3}\omega_{13x_3}\omega_{13z_3} - 2I_{3y_3z_3}\omega_{13y_3}\omega_{13z_3}) \qquad \text{(a)}$$

We then express

$$\begin{aligned}
\mathbf{T}_{13} &= \mathbf{T}_{Z4}(\varphi_{12})\mathbf{T}_{Z2}(l_2)\mathbf{T}_{Z4}(\varphi_{23})\mathbf{T}_{Z5}(\psi_{23})\mathbf{T}_{Z3}(z_{3S_3})\mathbf{T}_{Z1}(x_{3S_3})\mathbf{T}_{Z2}(y_{3S_3}) \\
x_{1S_3} &= x_{3S_3}\mathrm{c}\psi_{23} + z_{3S_3}\mathrm{s}\psi_{23} \\
y_{1S_3} &= x_{3S_3}\mathrm{s}(\varphi_{12}+\varphi_{23})\mathrm{s}\psi_{23} + y_{3S_3}\mathrm{c}(\varphi_{12}+\varphi_{23}) - \\
&\quad - z_{3S_3}\mathrm{s}(\varphi_{12}+\varphi_{23})\mathrm{c}\psi_{23} + l_2\mathrm{c}\varphi_{12} \\
z_{1S_3} &= -x_{3S_3}\mathrm{c}(\varphi_{12}+\varphi_{23})\mathrm{s}\psi_{23} + y_{3S_3}\mathrm{s}(\varphi_{12}+\varphi_{23}) + \\
&\quad + z_{3S_3}\mathrm{c}(\varphi_{12}+\varphi_{23})\mathrm{c}\psi_{23} + l_2\mathrm{s}\varphi_{12}
\end{aligned}$$

$$\dot{x}_{1S_3}^2 + \dot{y}_{1S_3}^2 + \dot{z}_{1S_3}^2 = ((x_{3S_3}\mathrm{s}\psi_{23} - z_{3S_3}\mathrm{c}\psi_{23})^2 + y_{3S_3}^2)(\dot{\varphi}_{12} + \dot{\varphi}_{23})^2 +$$

$$
\begin{aligned}
+ & \quad l_2^2\dot{\varphi}_{12}^2 + (x_{3S_3}^2 + y_{3S_3}^2)\dot{\psi}_{23}^2 - 2l_2 x_{3S_3}(\mathrm{c}\varphi_{23}\mathrm{c}\psi_{23}\dot{\psi}_{23} - \\
- & \quad \mathrm{s}\varphi_{23}\mathrm{s}\psi_{23}(\dot{\varphi}_{12} + \dot{\varphi}_{23}))\dot{\varphi}_{12} - 2y_{3S_3}(x_{3S_3}\mathrm{c}\psi_{23} + z_{3S_3}\mathrm{s}\psi_{23}) \cdot \\
\cdot & \quad \dot{\psi}_{23}(\dot{\varphi}_{12} + \dot{\varphi}_{23}) + 2l_2 y_{3S_3}\mathrm{c}\varphi_{23}(\dot{\varphi}_{12} + \dot{\varphi}_{23})\dot{\varphi}_{12} + \\
- & \quad 2l_2 z_{3S_3}\mathrm{s}(\varphi_{23} + \psi_{23})\dot{\varphi}_{12}\dot{\psi}_{23}
\end{aligned}
$$

$$
\begin{aligned}
\boldsymbol{\Omega}_{13} &= \mathbf{S}_{Z5}(-\psi_{23})\mathbf{S}_{Z4}(-\varphi_{23})\mathbf{D}_{Z4}(\dot{\varphi}_{12})\mathbf{S}_{Z4}(\varphi_{23})\mathbf{S}_{Z5}(\psi_{23}) + \\
&\quad + \mathbf{S}_{Z5}(-\psi_{23})\mathbf{D}_{Z4}(\dot{\varphi}_{23})\mathbf{S}_{Z5}(\psi_{23}) + \mathbf{D}_{Z5}(\dot{\psi}_{23}) \\
\omega_{13x_3} &= (\dot{\varphi}_{12} + \dot{\varphi}_{23})\mathrm{c}\psi_{23} \\
\omega_{13y_3} &= \dot{\psi}_{23} \\
\omega_{13z_3} &= (\dot{\varphi}_{12} + \dot{\varphi}_{23})\mathrm{s}\psi_{23} \qquad \text{(b)}
\end{aligned}
$$

The coordinates are related by constraint equations which we obtain by disconnecting a mechanism loop in the spherical pair C

$$
\mathbf{T}_{Z4}(\varphi_{12})\mathbf{T}_{Z2}(l_2)\mathbf{T}_{Z4}(\varphi_{23})\mathbf{T}_{Z5}(\psi_{23})[0, 0, l_3, 1] =
$$

$$
= \mathbf{T}_{Z1}(x_{14})\mathbf{T}_{Z2}(y_{14})\mathbf{T}_{Z3}(z_{14})\mathbf{T}_{Z6}(\gamma_{14})\mathbf{T}_{Z5}(\beta_{14})\mathbf{T}_{Z4}(\varphi_{14})[0, l_4, 0, 1]
$$

$$
\begin{aligned}
l_3\mathrm{s}\psi_{23} - l_4(-\mathrm{s}\gamma_{14}\mathrm{c}\varphi_{14} + \mathrm{c}\gamma_{14}\mathrm{s}\beta_{14}\mathrm{s}\varphi_{14}) - x_{14} &= 0 \\
-l_3 s(\varphi_{12} + \varphi_{23})\mathrm{c}\psi_{23} + l_2\mathrm{c}\varphi_{12} - l_4(\mathrm{c}\gamma_{14}\mathrm{c}\varphi_{14} + & \\
+\mathrm{s}\gamma_{14}\mathrm{s}\beta_{14}\mathrm{s}\varphi_{14}) - y_{14} &= 0 \\
l_3 c(\varphi_{12} + \varphi_{23})\mathrm{c}\psi_{23} + l_2\mathrm{s}\varphi_{12} - l_4\mathrm{c}\beta_{14}\mathrm{s}\varphi_{14} - z_{14} &= 0 \qquad \text{(c)}
\end{aligned}
$$

The forces acting on the mechanism are the force of gravity G_2 on $-z_{1S_2}$, G_3 on $-z_{1S_3}$, G_4 on $-z_{1S_4}$, the force of the suspension spring $-F(\xi, \dot{\xi})$ on ξ, one quarter of the automobile weight $1/4G$ on z_{1B_3} and the forces of springs substituting the tyre F_{3z} on $-z_{1B_3}$, F_{3y} on $-y_{1B_3}$, where

$$
\begin{aligned}
z_{1S_2} &= y_{2S_2}\mathrm{s}\varphi_{12} \\
z_{1S_4} &= y_{4S_4}\mathrm{c}\beta_{14}\mathrm{s}\varphi_{14} + z_{14} \\
\xi &= \sqrt{a - 2bs(\varphi_{12} + \beta)} - l \\
y_{1B_3} &= x_{3B_3}\mathrm{s}(\varphi_{12} + \varphi_{23})\mathrm{s}\psi_{23} + y_{3B_3}\mathrm{c}(\varphi_{12} + \varphi_{23}) - \\
&\quad - z_{3B_3}\mathrm{s}(\varphi_{12} + \varphi_{23})\mathrm{c}\psi_{23} + l_2\mathrm{c}\varphi_{12} \\
z_{1B_3} &= -x_{3B_3}c(\varphi_{12} + \varphi_{23})\mathrm{s}\psi_{23} + y_{3B_3}\mathrm{s}(\varphi_{12} + \varphi_{23}) + \\
&\quad + z_{3B_3}\mathrm{c}(\varphi_{12} + \varphi_{23})\mathrm{c}\psi_{23} + l_2\mathrm{s}\varphi_{12}
\end{aligned}
$$

We now assemble the equations of motion applying LEMT from Section 9.3 and we may examine the MBS properties (see Section 10.4).

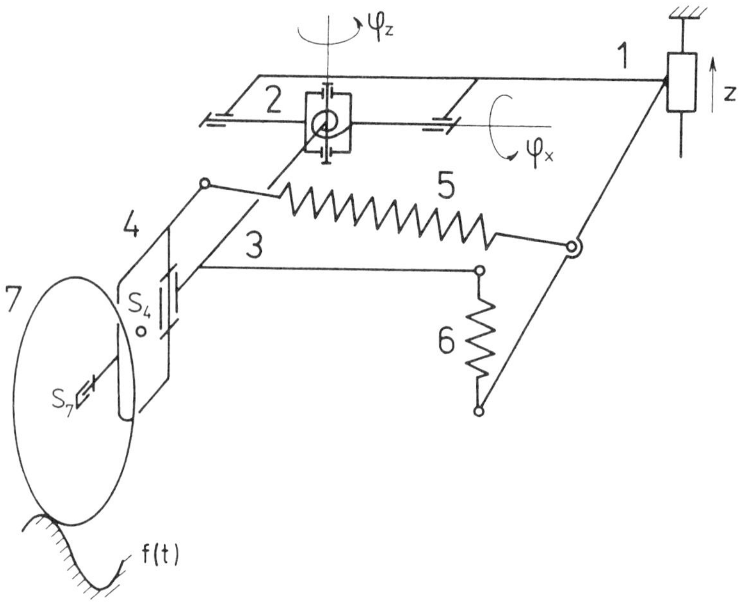

Figure 9.16. Substitute dynamic MBS model of car axle.

9.12.2 Car Axle on Passing a Surface Unevenness

When a car passes over a rectangular unevenness, oscillations appear in the steering and the front axle suspension. We wish to determine the influence of a large number of structure parameters on these oscillations. It is a complex dynamical process affected by nonlinear behaviour of the tyre, by flexibility of several mechanism elements, etc. For mastering the simulation and identification of the process, a substitute dynamic MBS model according to Figure 9.16 has been used. The tyre deformation has been substituted by an experimentally measured input-output relation of the wheel. The flexibility of elements has been substituted by springs. In spite of all these alterations, a relatively complex spatial MBS has resulted. Its position is described by 10 coordinates: φ_x, φ_y, z, coordinates x_4, y_4, z_4 of centre of element 4 in the system of coordinates of the frame, Cardan's angles $\varphi_{x4}, \varphi_{y4}, \varphi_{z4}$ of this element, and the angle φ_7 of wheel rotation. Using Cardan's angles here is very convenient; singular cases do not occur, as we are concerned with oscillations about an equilibrium position and the description is simpler than with Euler's parameters. The coordinates are related by constraint equations given by five conditions for the revolute pair 3–4 and input-output function $z_7 = f(t)$.

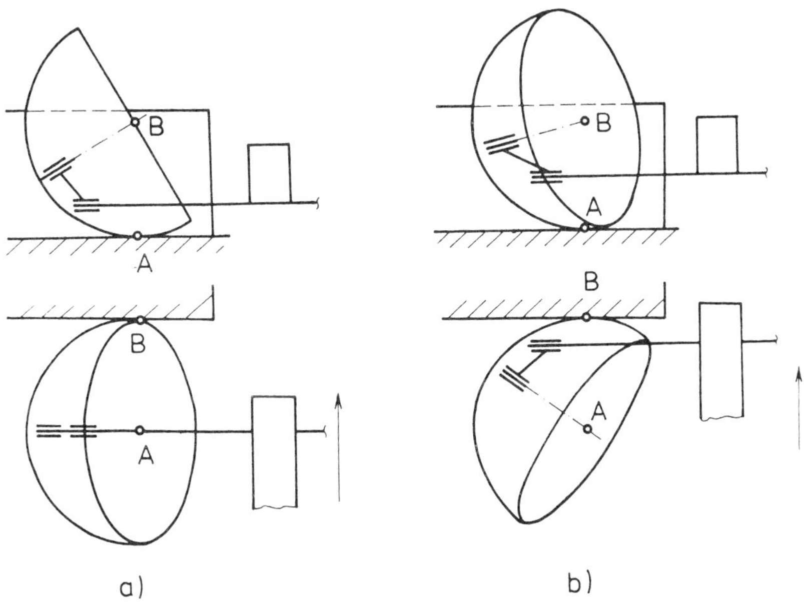

Figure 9.17. A new wheel suspension (Kovář and Suchý, 1982).

Although the reality has been simplified into the given MBS model, the simulation results correspond with the experimental values. The greatest influence on the oscillation amplitude arises from the rigidity of the axle mounting (seating support). Before performing the simulation experiments, it had not been possible to forecast this effect using only the designer's intuition. This is an example of the way in which CAD methods for MBS can improve the quality of a product (Kovanda and Valášek, 1987).

9.12.3 New Car Wheel Mechanism

Figure 9.17 shows a new car wheel suspension mechanism (Kovář and Suchý, 1982) which has advantageous properties in overcoming barriers or unevennesses. To explain the properties and to calculate the necessary forces (quasistatically or dynamically) is an interesting example of MBS with change of structure. Before reaching the unevenness, the wheel is moving as other traditional wheels. After contacting the unevenness, rotations occur around both axes; the wheel deviates up to the point where the friction force at point B is capable of overcoming the weight of the car, and the contact in point A is broken.

Chapter 10

General Computer Systems for MBS Analysis Automation

As it is evident from the preceding chapters, the study of spatial MBS behaviour is a problem for which the successful solution requires a good knowledge of mechanics, numerical mathematics and programming; without using a computer obtaining an error-free solution in acceptable time is practically impossible. For the creation of an operational model of the system being studied a whole series of approaches presented in Sections 9.2, 9.5 may be used. For the actual solution several methods of numerical mathematics are available. The variety of solved problems, the composition of real systems and the extent of their idealization mean that it is impossible to find a method which is in all aspects significantly superior to other methods. A consequence of this is that there are more than twenty programs or program systems described in literature, ranging from the simplest single purpose programs, e.g. for the calculation of centric crank mechanism reaction forces, to large systems capable of the analysis of dynamic properties of whole vehicles, robots or general space devices including impact, backlash, elasticity of elements and friction. Some of these programs have served as an experiment or as a verification of a certain idea and their development has not continued, while others have gained widespread acknowledgement and commercial success. Because of the vast variety of purpose, perfection, extent and function, comparison of these programs is not easy. It is even difficult to formulate the criteria for such a comparison. The range of MBS analysis activities carried out by computer could be one of the criteria. The use of a computer as a "solver" of a manually assembled model is nowadays already an exception. The main question is then whether the operational model (e.g. kinematic or dynamic equations) is to be assembled numerically or symbolically.

Another criterion is the formalism used for the assembly of the operational model. The non-recursive formalisms used most frequently may be divided into two main groups: the Newton-Euler formalism described in Sections 9.5.1.1 and 9.5.1.2, and the Lagrange formalism presented in Section 9.3 (see also (Chace, 1967), (Paul, 1975)) and in Section 9.5.1.4–9.5.1.5. Less frequent is the use of other formalisms: Hamilton equations (Section 9.2.2.6 (Shet and Uicker, 1972)), Kane equations (Section 9.5.1.3), compartments and vector networks (Section 9.7), natural coordinates (Section 9.5.1.6), etc. The latest and the most promising approaches are the recursive formalism presented in Section 9.5.2 and the residual algorithm in Section 9.5.1.9.

The type and number of coordinates used to describe the MBS position may be another criterion. For a system formed from N bodies, either the maximum number of $6N$ ($7N$) coordinates may be used or the minimum number of coordinates corresponding to the degrees of freedom, or a convenient number of coordinates between these two extremes. As to the type of coordinates used, the largest variety is in the description of the orientation of bodies. Euler angles, Cardan angles, Euler parameters (normed quaternions), screws, etc. are used. For the purpose of computation it is possible to distinguish systems solving large motions of MBS or small oscillations around a equilibrium position or both. Other systems are designed for solving the kinematics, or the operational model assembly only, whereas the actual solution is completely separated. Another important point is the capability of calculating reaction forces or possibly solving systems with friction, with elastic elements, etc. Some computer programs are even capable of solving controlled systems. Several of them offer the optimization of MBS according to prescribed criterion.

For the actual solution we may distinguish between procedures where the system of equations obtained, usually consisting of differential and algebraic equations, is solved without modification and procedures where, before actual solution, auxiliary variables are eliminated and the integration is carried out only for the minimum necessary number of coordinates. The numerical method used may be another criterion.

These criteria and some others were observed by assembling Table 10.1, which gives a summarized survey of data on the best-known programs for MBS dynamics solution. For further details see also (Schiehlen, 1990b, Kortüm and Sharp, 1993). We treat the properties of computer programs created in the scope of our research in more detail in the following text.

Table 10.1 Program systems

MULTIBODY SOFTWARE	ADAMS	COMPAMM	DADS	MADYMO	MEDYNA	MESA VERDE	MOTION	NEWEUL	OKAMODAM	SIMPACK	SDFAST
FORMALISM: Newton-Euler (NE) Lagrange (L), recursive (R)	L	L	L	R	NE	NE	R	NE	L	R	L
COORDINATES: absolute (A), relative (R)	AR	AR	AR	R	A	AR	R	AR	L	L	R
TOPOLOGY: open chains (O), closed chains (C)	OC	OC	OC	O	OC	OC	OC	OC	OC	OC	OC
ANALYSIS: statics (S), kinematics (K), dynamics (D)	SKD	KD	SKD	KD	SD	SKD	SKD	SKD	KD	SKD	SKD
FORMAT of eq. numeric (N), symbolic (S)	N	N	N	N	N	S	S	S	S	N	S
TYPE of eq. ODE (O), DAE (D)	OD	O	OD	O	O	O	D	OD	D	OD	OD
ELASTIC bodies (E)	E		E		E					E	
CONTROL (C)	C		C			C				C	
REACTION forces (R)	R	R	R	R	R	R	R	R	R	R	R
INVERSE dynamics (I)	I	I	I	I			I	I		I	I

10.1 System KIN

System KIN (Brát and Stejskal, 1975) can be classified as consistently using the numerical approach. Its theoretical basis is the method of basic matrices in kinematics. Its purpose is the solution of the position problem of all elements and significant points of a MBS with arbitrary structure containing only lower kinematic pairs. The system is realized as a set of sub-programs which are started by a control program (monitor) according to user commands.

The program system KIN was originally written in FORTRAN IV-PLUS. Nowadays its PC version in FORTRAN 77 is used. The updated version of KIN called MECH is also able to perform static analysis (Stejskal et al., 1988).

10.2 System OKAM2

System OKAM2 (Stejskal and Valášek, 1986) represents programs using symbolic algebraic manipulations for the assembly of a MBS operational model. Its theoretical basis, in common with system KIN, is the method of basic matrices. For the purpose of computer processing, two symbolic calculi have been created using special internal symbolic computer representations of expressions. The first calculus operates with symbolic matrix expressions of type (5.94) and (5.97), which serve to specify the problem. The purpose of the second calculus is the actual symbolic calculation of elements of the resultant matrices and vectors on the basis of symbolic matrix equations which are either given or which are specified by the first calculus. In addition, there is an algorithm for creating programs in FORTRAN for the numerical solution of the symbolically created operational model of the MBS kinematic analysis.

System OKAM2 may be characterized as a generator of FORTRAN programs for the complete kinematic analysis of spatial MBS (position, velocity and acceleration of points and bodies). It is written in LISP 1.5. A special problem-oriented language has been created for communication with the user. Its basic commands are:

CONSTRAINTS:	specifying matrix equations of type (5.94) and (5.97) describing the solved MBS,
MOTIONS:	specifying matrix equations describing the calculated motion of points (5.87) or bodies (5.93a),
INDEPENDENT:	specifying a list of identifiers used to designate independent variables,

DEPENDENT:	specifying a list of identifiers used to designate dependent variables,
TIME:	specifying an identifier designating time,
MODE:	specifying whether only position or velocity and acceleration are to be computed,
SUBSTITUTION:	specifying position vectors, non-standard matrices and expressions to be substituted,
DIMENSION:	specifying the dimensions of the solved MBS,
INPUT:	specifying a list of constant identifiers to be read in as data,
DRIVES:	specifying the time behaviour of independent variables,
ESTIMATION:	specifying estimates of initial values of dependent variables, which are the start values of the iteration process,
SOLUTION:	specifying the time interval in which the solution is to be carried out,
PRINT:	specifying the step and the interval from which the computed values are to be printed,
PLOT:	specifying a list of variables for output into a file to be plotted later on,
NAME:	specifying the problem designation.

Character $ ends the specification.

All commands may be written in any order and may be used more than once. The latter does hold for commands TIME: and MODE:. Default values are TIME: C, MODE: ACCELERATION, NAME: OKAM, ESTIMATION: $z_1 = 0.0, z_2 = 0.0, \ldots$ Algebraic expressions in the specification are allowed in the form of FORTRAN arithmetic expressions with a few limitations.

The numerical computation uses the same procedures as system KIN; however velocities and accelerations are also solved. The processing of computational results in the form of graphs is not a part of the system.

System OKAM2 was originally used on an ICL 4–72 computer under the operation system MULTIJOB. Version OKAM2PC for computers of the IBM PC class is also available.

The following example (see also Section 5.7.1), shows the use of system OKAM2 for the kinematic analysis of a car front wheel suspension where toe-in $\varrho = \text{atan}((x_N - x_M)/(y_N - y_M))$ with respect to spring length, is to be determined, Figure 5.16. The system OKAM2 input commands are of the form:

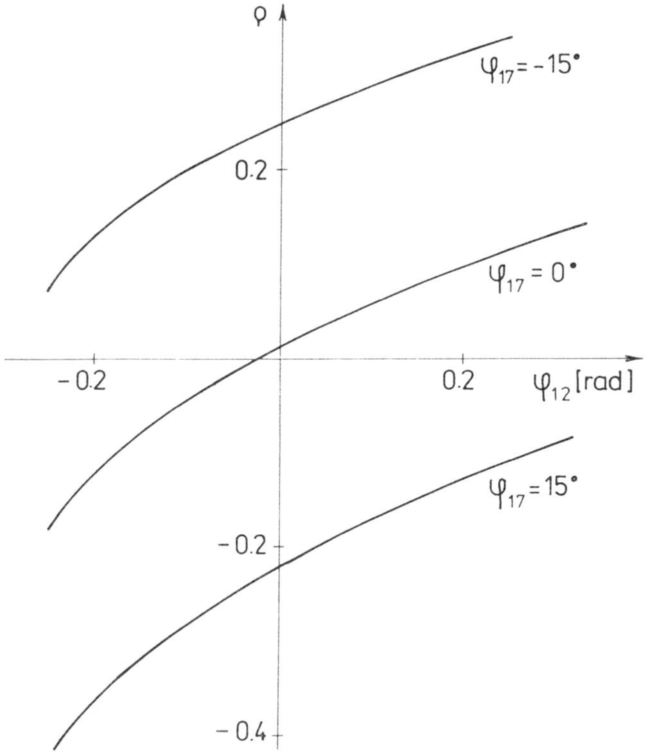

Figure 10.1. Toe-in $\varrho = \varrho(\varphi_{12}, \varphi_{17})$.

```
    NAME: TOE-IN
CONSTRAINS: (T12*T23*R3L=T14*R4L)
    (T12*T23*T35*R5K=T17*T76*R6K)
MOTIONS: R1M=(T12*T23*T35*R5M)
    R1N=(T12*T23*T35*R5N)
SUBSTITUTIONS: T23=(TZ2(D2)*TZ4(FIX3)*TZ5(FIY3))
    T12=(TZ1(XIA)*TZ2(ETA)*TZ3(DZA)*TZ5(BA)*
    TZ6(GA)*TZ4(FI12))
    T14=(TZ1(XIB)*TZ2(ETB)*TZ3(DZB)*TZ5(BB)*
    TZ6(GB)*TZ4(FI14))
    T35=(TZ3(D31)*TZ6(FI35))
    T17=(TZ1(XIR)*TZ2(ETR)*TZ3(DZR)*
```

```
    TZ4(AR)*TZ5(BR)*TZ6(FI17))
    T76=(TZ1(-D7)*TZ4(FIX6)*TZ6(FIZ6))
    R3L=((0.0 (0.0) (D3) (1.0)) R5M=((XM (YM) (ZM) (1.0))
    R4L=((0.0) (D4) (0.0) (1.0)) R5N=((XN) (YN) (ZN) (1.0))
    R5K=((-D5) (0.0) (0.0) (1.0)) R6K=((0.0) (D6) (0.0) (1.0))
INDEPENDENT: FI12 FI17
DEPENDENT: FI14 FIX3 FIY3 FIX6 FIZ6 FI35
TIME: C
DRIVES: FI12=(RYF12*C) FI17=(RYF17*C)
ESTIMATION: FI14=FI140 FIX3=FIX30 FIY3=FIY30
    FIX6=FIX60 FIZ6=FIZ60 FI35=FI350
INPUT: XIA ETA DZA BA XIB ETB DZB BB GB XIR ETR DZR
    AR BR D2 D3 D31 D4 D5 D6 D7 XM YM ZM XN YN ZN RYF12
    RYF17 FI140 FIX30 FIY30 FIX60 FIZ60 FI350 C0 CM CK CT
SOLUTION: C=(C0 CM CK)
PRINT: C=(C0 CM CT)
PLOT: R1M R1N
MODE: POSITION
$
```

The identifiers used in input commands are self-explanatory. For the numerical calculation, additional numerical values of dimensions and estimated values of dependent variables must be entered and the range of calculations must be specified. The numerical data are $\xi_A = \eta_A = \gamma_A = \gamma_B = \xi_R = \alpha_A = 0.0, \zeta_A = -14.0, \xi_B = -1.33, \eta_B = 15.0, \zeta_B = 10.66, \beta_B = 0.175, \eta_R = 11.6, \zeta_R = -1.5, \beta_R = 0.175, d_2 = 41.3, d_3 = 27.85, d_{31} = 12.66, d_4 = 21.65, d_5 = 12.45, d_6 = 27.66, d_7 = 11.0, x_M = -1.0, y_M = z_M = 0.0, x_N = 1.0, y_N = 8.5, z_N = -1.6, \dot{\varphi}_{12} = 0.25, \varphi_{140} = \varphi_{x30} = \varphi_{y30} = \varphi_{x60} = \varphi_{z60} = \varphi_{350} = 0.0, c_0 = 0.0, c_M = 1.0, c_K = 0.02, c_T = 1.0$.

The calculated toe-in behaviour is shown in Figure 10.1.

10.3 System DYN

System DYN (Valášek, 1986c) represents a system for MBS dynamical analysis based on a strictly numerical approach. It uses the compartment method described in Section 9.7. We present a specific application in the example of a compartment of an element of planar MBS connected to neighbouring elements by a revolute and a prismatic pairs, Figure 10.2, observed in a fixed system of coordinates Oxy connected to the frame. In the space of the i-body a local system $O_i x_i y_i$ with the origin in the centre of mass is chosen. The position of the body in the space of the frame is determined by coordinates x_i, y_i, φ_i. The revolute kinematic pair on the i-the element is

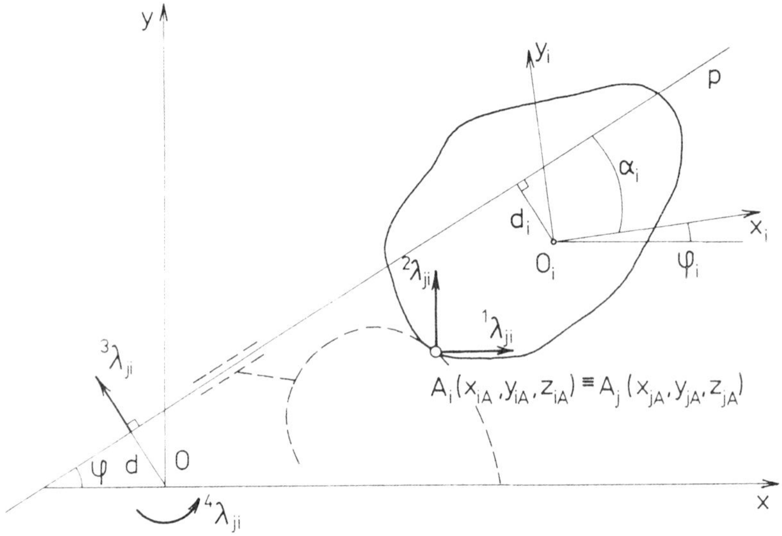

Figure 10.2. The compartment BRP.

determined by centre $A_i[x_{iA}, y_{iA}]$. Line p determined by oriented distance d_i from point O_i and angle α_i to axis x_i is the axis of the prismatic pair. Using the designation according to Figure 10.2, the constraint conditions may be written as follows

$$x_i + x_{iA}c\varphi_i - y_{iA}s\varphi_i = x_j + x_{jA}c\varphi_j - y_{jA}s\varphi_j \quad (10.1a)$$
$$y_i + x_{iA}s\varphi_i + y_{iA}c\varphi_i = y_j + x_{jA}s\varphi_j + y_{jA}c\varphi_j \quad (10.1b)$$
$$d_i + y_i c(\alpha_i + \varphi_i) - x_i s(\alpha_i + \varphi_i) =$$
$$= d_j + y_j c(\alpha_j + \varphi_j) - x_j s(\alpha_j + \varphi_j) \quad (10.2a)$$
$$\alpha_i + \varphi_i = \alpha_j + \varphi_j \quad (10.2b)$$

The former two equations express the identity of absolute coordinates of point A expressed through elements i and j connected by this pair. In a similar way the latter two equations express the identity of oriented distance of line p from origin O and the identity of its angle to axis x, both expressed through bodies i and j.

The equations of body motion assembled using Lagrange equations of mixed type are

$$m_i\ddot{x}_i = \sum_A {}^1\lambda_{ji} - \sum_p {}^3\lambda_{ji}s(\alpha_i + \varphi_i) \quad (10.3a)$$

$$m_i \ddot{y}_i = \sum_A {}^2\lambda_{ji} + \sum_p {}^3\lambda_{ji} \mathrm{c}(\alpha_i + \varphi_i) \tag{10.3b}$$

$$I_i \ddot{\varphi}_i = \sum_A (-{}^1\lambda_{ji}(x_{iA} \mathrm{s}\varphi_i + y_{iA} \mathrm{c}\varphi_i) +{}^2 \lambda_{ji}(x_{iA} \mathrm{c}\varphi_i - y_{iA} \mathrm{s}\varphi_i)) +$$
$$+ \sum_p ({}^3\lambda_{ji}(-y_i \mathrm{s}(\alpha_i + \varphi_i) - x_i \mathrm{c}(\alpha_i + \varphi_i)) +{}^4 \lambda_{ji}) \tag{10.3c}$$

where m_i, I_i are the mass and moment of inertia of body i with respect to the centre of mass, and ${}^k\lambda_{ji}$ corresponds to the k-th constraint in equations (10.1)–(10.2). For reaction forces at the contact of elements i, j

$$ {}^k\lambda_{ji} +{}^k \lambda_{ij} = 0, \qquad k = 1, 2, 3, 4. \tag{10.4}$$

Using the above equations, which comply with the condition given in Section 9.7, it is possible to create a so-called macromodel according to Figure 9.10 which can be used in system SADYS initially created for the analysis of electric circuits (Mann, 1986) and which complies with the same conditions.

The description of the macromodel called BRP (i.e. Body with Revolute and Prismatic pair) is of the following form:

```
BRP UAX UAY UBD UBF/M=1, JJ=1, XA=0, YA=0, AL=0;
:BINARY LINK WITH REVOLUTE AND PRISMATIC JOINTS
:DEFINITION OF BOUNDARY NODES AND PARAMETERS
:M — MASS OF THE LINK
:JJ — MOMENT OF INERTIA OF THE LINK
:XA YA — LOCAL COORDINATES OF THE CENTRE OF THE
    REVOLUTE JOINT A
:AL    ANGLE BETWEEN THE AXIS OF THE PRISMATIC JOINT
    B AND THE LOCAL COORDINATE AXIS X
SYSVAR FI, XI, YI, 1, 2, 3;
EAX UAX=VD.XI–(XA*SIN(FI)+YA*COS(FI))*VD.FI;
:CONSTRAINTS
EAY UAY=VD.YI–(XA*COS(FI) -YA*SIN(FI))*VD.FI;
EBD UBD=VD.YI*COS(AL+FI)–VD.XI*SIN(AL+FI)–YI
    *SIN(AL+FI)*VD.FI–XI*COS(AL+FI)*VD.FI;
EBF UBF=VD.FI;
:EQUATION OF MOTION IN THE X COORDINATE
J 0,1=I.EAX–I.EBD*SIN(AL+FI);
CX 1=M;
:EQUATION OF MOTION IN THE Y COORDINATE
J 0,2=I.EAY–I.EBD*COS(AL+FI);
CY 2=M
```

```
:EQUATION OF MOTION IN THE FI COORDINATE
J 0,3=-I.EAX*(XA*SIN(FI)+YA*COS(FI))+
   I.EAY*(XA*COS(FI)-YA*SIN(FI))+
   I.EBD*(-YI*SIN(AL+FI)-XI*SIN(AL+FI))+I.EBF;
CF 3=JJ;
:INTEGRATION OF DISPLACEMENT FROM VELOCITIES
BI XI=V.1;
BI YI=V.2;
BI FI=V.3;
EO@;
```

In the description, identifiers with initial character E correspond to source of voltage, J to source of current, C to capacitor, while BI designates the block of integration. The first line defines the macromodel (compartment) BRP with clamps UAX, UAY, UBD, UBF and parameters M, JJ, XA, YA, AL with default values. The comments start with colon. The command SYSVAR defines the inner variables of the macromodel. Each of them correspond to a node. Each node has its identifier (alfanumeric or natural number). An implicit node 0 for the ground is always introduced. The command E[*name*] *node1, node2 = expression* has the meaning that a source E-*name* of voltage between the nodes *node1* and *node2* with the value *expression* is introduced; if *node2* =0 it can be omitted. Similarly for source of current J. For a capacitor C the expression means its value, for a block of integration BI its input. The expression I.*name* means the current going through the element *name*. The expression V.*name* means the voltage between the node *name* and ground and VD.*name* means its time derivative. EO@ ends the macromodel definition.

Similar macromodels (BRR, BPP, BRRR, BRRP, etc.) may be created for other types of MBS planar elements with varying numbers of various types of kinematic pairs. Based on the definition of these macromodels, system SADYS automatically assembles and solves the necessary equations.

System DYN enables the modeling of all planar MBSs with lower kinematic pairs to be carried out by simple connection of the respective macromodels according to the construction diagram. The user does not have to know anything about analogy with electric circuits and may formulate his problems using the language and methods of his own discipline. An example of this approach, in the form of a diagram presenting the connection of macromodels by solving the motion of a car front wheel suspension from Figure 9.2, is shown in Figure 10.3. The description of connection of the macromodels follows

```
L1@BR3 AX AY BX BY EX EY/M=4, JJ=0.04, XA=-0.15, YA=0,
```

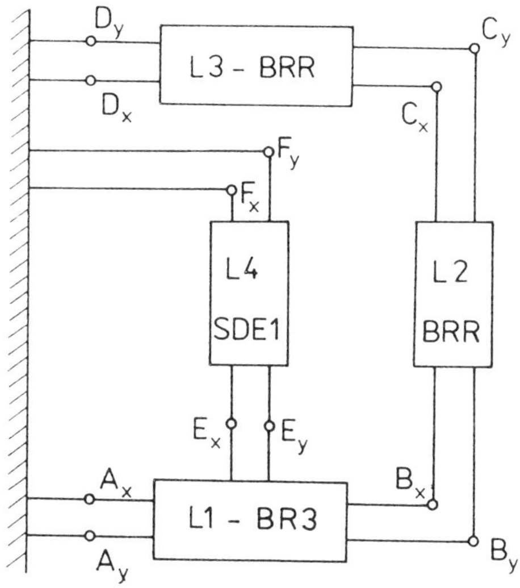

Figure 10.3. The compartment model of front wheel suspension.

```
        XB=0.23, YB=0, XC=0.07, YC=0;
L2@BRR BX BY CX CY/M=40, JJ=1.15, XA=-0.1, YA=0,
        XB=0.12, YB=0;
L3@BRR CX CY DX DY/M=0.5, JJ=0.00355, XA=-0.144, YA=0,
        XB=0.13, YB=0;
L4@SDE1 EX EY FX FY/L=0.29, K=4000, B=200;
```

It could also be very advantageous to create similar macromodels within system SADYS on the basis of the formalism of natural coordinates (Section 9.5.1.5).

10.4 System ODAM

System ODAM (Brát et al., 1981) has been created for spatial MBS dynamical analysis. Algebraic manipulations are used to assemble equations. It is written in LISP 1.5 and is based on similar principles as system OKAM2. Its theoretical bases are Lagrange equations of the mixed type. A substantial part of the derivation of the equations of motion consists of differentiation. We have introduced an expression for the i-th derivative of y with respect to x in the form

$$(Y \sqcup \underbrace{X \sqcup \ldots \sqcup X}_{i})$$

This derivative $\mathrm{d}^i y/\mathrm{d}x^i$ is called the implicit one to distinguish it from an explicit one $(\mathrm{d}(y^2)/\mathrm{d}y = 2y)$ the value of which can be directly computed in a symbolic way.

For the purpose of differentiating, two functions DERT and DERP have been created using the methods of language LISP. The former carries out the operation of total differentiation

$$\frac{\mathrm{d}}{\mathrm{dx}} R\left(\frac{\mathrm{d}y_1}{\mathrm{d}x}, \frac{\mathrm{d}y_2}{\mathrm{d}x}, \ldots, \frac{\mathrm{d}y_n}{\mathrm{d}x}, y_1, y_2, \ldots, y_n, x\right) =$$

$$= \sum_{i=1}^{n} \frac{\partial R}{\partial\left(\frac{\mathrm{d}y_i}{\mathrm{d}x}\right)} \frac{\mathrm{d}^2 y_i}{\mathrm{d}x^2} + \sum_{i=1}^{n} \frac{\partial R}{\partial y_i} \frac{dy_i}{dx} + \frac{\partial R}{\partial x}$$

while the latter computes the partial derivative

$$\frac{\partial}{\partial y_k} R(y_1, y_2, \ldots, y_n, x) = \frac{\partial R}{\partial y_k}$$

These functions are capable of differentiating symbolically algebraic expressions containing standard functions and operations of FORTRAN, i.e. addition, substraction, multiplication, division, raising to power, EXP, LN, SIN, COS, TG, ATAN, SQRT, implicit derivatives and the operations DERT and DERP. Expressions produced by functions DERT and DERP are modified to a certain extent.

The user of system ODAM specifies the problem using a problem-oriented language. The commands of the language are:

VARIABLES:	specifying a list of variables used,
TIME:	specifying the identifier for time,
KINERGIE:	specifying the expression for MBS kinetic energy,
DISSIPATION:	specifying the dissipation function for solving systems with damping,
CONSTRAINTS:	specifying constraint conditions,
FORCES:	specifying a list of forces and coordinates in which they act,
SUBSTITUTIONS:	specifying substitutions used in expressions for kinetic energy and in constraint equations,
INPUT:	specifying a list of constant identifiers to be read in as data.

After syntactic checks on problem correctness, the ODAM system uses symbolic manipulations to generate matrix elements contained in equation (9.68) in the form of full FORTRAN source code, which is compiled and executed numerically later on. For the integration of equations of motion, subroutine DIFSUB (Gear, 1971) or DE (Shampine and Gordon, 1975) with Baumgarte's stabilization or coordinate partitioning is used.

A run of system ODAM with additional data specified yields as output files of values of computed quantities which may be printed or plotted.

An example of the use of the ODAM system is the dynamic analysis of the spatial mechanism of a car front wheel suspension according to Figure 9.15 (see Section 9.12.1). The input commands of ODAM system are

```
VARIABLES: FI12 FI23 PSI23 FI14
TIME: C
KINERGIE: (0.5*(I2*(FI12 C)**2+I4*(FI14 C)* *2+M3*(((X3S3*
    SIN(PSI23)-Z3S3*COS(PSI23))**2 +Y3S3* *2)*((FI12 C)+(FI23 C))
    **2+L2**2*(FI12 C)**2+ (X3S3**2+Y3S3**2)*(PSI23 C)**2-2. *L2
    *X3S3*(COS(FI23)*COS(PSI23)*(PSI23 C) -SIN(FI23)*SIN(PSI23)
    *((FI12 C)+(FI23 C)))*(FI12 C)-2.*Y3S3 *(X3S3*COS(PSI23)+
    Z3S3*SIN(PSI23))*(PSI23 C)*((FI12 C)+(FI23 C))+ 2.*L2*Y3S3*
    COS(FI23)*((FI12 C)+(FI23 C))*(FI12 C)-2.* L2*Z3S3*SIN(FI23+
    PSI23)*(FI12 C)*(PSI23 C))+I3X3*OM13X3* *2+I3Y3*OM13Y3**2+
    I3Z3*OM13Z3**2-2.*D3X3Y3*OM13X3 *OM13Y3-2.*D3X3Z3*
    OM13X3*OM13Z3-2.*D3Y3Z3*OM13Y3*OM13Z3))
CONSTRAINTS: (L3*SIN(PSI23)-L4*(COS(G14)*SIN(B14) *SIN(FI14)-
    SIN(G14)*COS(FI14))-X14)
    (-L3*SIN(FI12+FI23) *COS(PSI23)+L2*COS(FI12)-L4*(COS(G14)
    *COS(FI14)+SIN(G14)*SIN(B14)*SIN(FI14))-Y14)
    (L3*COS(FI12+FI23)*COS(PSI23)+L2*SIN(FI12)
    -L4*COS(B14)*SIN(FI14)-Z14)
FORCES: ((-G2) Z1S2) ((-G3) Z1S3) ((-G4) Z1S4) ((-F) KSI)
     ((-F3Z) Z1B3) ((-F3Y) Y1B3) ((G/4.) Z1B3)
SUBSTITUTIONS: KSI=(SQRT(A-2.*B*SIN(FI12+BETA))-L)
    Z1S2=(Y2S2*SIN(FI12))
    X1S3=(X3S3*COS(PSI23)+ Z3S3*SIN(PSI23))
    Y1S3=(X3S3*SIN(FI12+FI23)*SIN(PSI23)+Y3S3 *COS(FI12+
       FI23)-Z3S3*SIN(FI12+FI23)*COS(PSI23)+L2*COS(FI12))
    OM13Z3=(((FI12 C)+(FI23 C))*SIN(PSI23))
    Z1S3=(-X3S3*COS(FI12+FI23)* SIN(PSI23)+Y3S3*SIN(FI12+
       FI23)+Z3S3*COS(FI12+FI23)* COS(PSI23)+L2*SIN(FI12))
    Z1S4=(Y4S4*COS(B14)*SIN(FI14)+Z14)
    Y1B3=(X3B3*SIN(FI12+FI23)*SIN(PSI23) +Y3B3*COS(FI12+
```

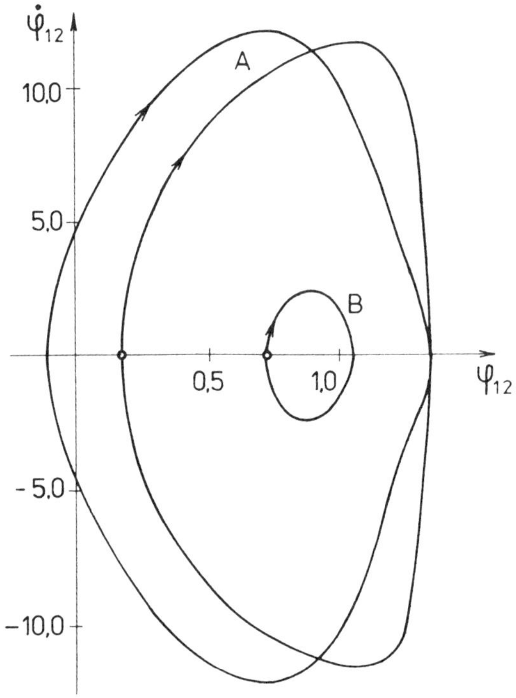

Figure 10.4a. Phase trajectory $\dot{\varphi}_{21} = f(\varphi_{21})$.

```
        FI23)-Z3B3*SIN(FI12+FI23)* COS(PSI23)+L2*COS(FI12))
     Z1B3=(-X3B3*COS(FI12+FI23) *SIN(PSI23)+Y3B3*SIN(FI12+
        FI23)+Z3B3*COS(FI12+FI23)* COS(PSI23)+L2*SIN(FI12))
     OM13X3=(((FI12 C)+(FI23 C))*COS(PSI23))
     OM13Y3=((PSI23 C))
     OM13Z3=(((FI12 C)+(FI23 C))*SIN(PSI23))
     F=(C1*KSI+C3*KSI**3+K1*DERT(KSI,C,(FI12))+
        K3*DERT(KSI,C,(FI12))**3)
     F3Z=(-ZK*Z1B3-ZB*DERT(Z1B3,C,(FI12,FI23,PSI23)))
     F3Y=(-YK*Y1B3-YB*DERT(Y1B3,C,(FI12,FI23,PSI23)))
INPUT: I2 I4 M3 I3X3 I3Y3 I3Z3 D3X3Y3 D3X3Z3 D3Y3Z3
     L2 L3 L4 D L X14 Y14 Z14 B14 G14 BETA A B
     X3S3 Y3S3 Z3S3 Y2S2 Y4S4 X3B3 Y3B3 Z3B3
     G2 G3 G4 G C1 C3 K1 K3 ZK ZB YK YB
$
```

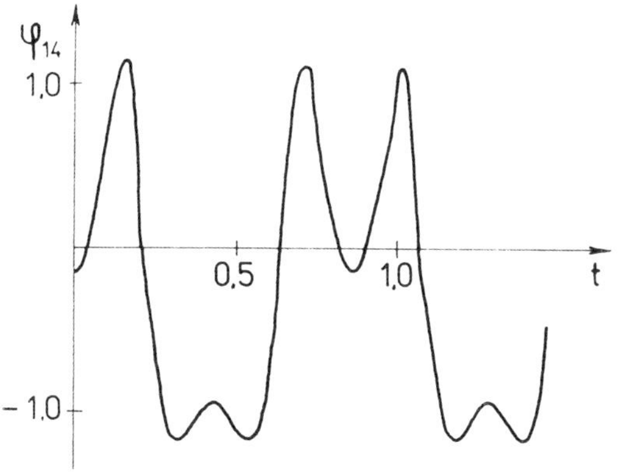

Figure 10.4b. Oscillations of the body 4.

The example has been solved for the following numerical values: $d = 0.22$ m, $l = 0.29$ m, $\beta = 17^o 11'$, $l_2 = 0.38$ m, $l_3 = 0.22$ m, $l_4 = 0.274$ m, $x_{14} = 0.05$ m, $y_{14} = 0.05$ m, $z_{14} = 0.32$ m, $b_{14} = 0.087\ 266\ 4$, $g_{14} = b_{14}$, $y_{4S4} = 0.17$ m, $y_{2S2} = 0.24$ m, $z_{3S3} = 0.1$ m, $y_{3S3} = 0.1$ m, $x_{3S3} = 0.02$ m, $x_{3B3} = 0$, $y_{3B3} = 0.12$ m, $z_{3B3} = -0.1$ m, $I_2 = 0.13$ kg m^2, $I_4 = 0.012$ kg m^2, $m_3 = 40$ kg, $I_{3x3} = 1.55$ kg m^2, $I_{3y3} = 1.55$ kg m^2, $I_{3z3} = 0.5$ kg m^2, $D_{3x3y3} = D_{3x3z3} = D_{3y3z3} = 0$, $a = 0.180\ 9$ m^2, $b = 0.08$ m^2, $c_1 = 4\ 000$ N m^{-1}, $c_3 = 20$ N m^{-3}, $k_1 = 10\ 000$ N m^{-1} s, $k_3 = 0$, $z_k = 40$ N m^{-1}, $z_b = 0$, $y_k = 4$ N m^{-1}, $y_b = 0$, $G_2 = 20$ N, $G_3 = 400$ N, $G= 10\ 000$ N, $G_4 = 4$ N, $F = c_1\xi + c_3\xi^3 + k_1\dot{\xi} + k_3\dot{\xi}^3$, $F_{3z} = -z_k(z_{1B3} - s) - z_b(\dot{z}_{1B3} - \dot{s})$, $F_{3y} = -y_k y_{1B3} - y_b \dot{y}_{1B3}$, $s = 0, \dot{s} = 0$.

The mechanism can oscillate in a very complicated manner and with large displacements. Its motion in the phase plane is shown in Figure 10.4a for initial values $\varphi_{12}(0) = 0.17453, \dot{\varphi}_{12}(0) = 0, \varphi_{23}(0) = 0.07892$ with damping neglected (line A). The time behaviour of φ_{14} in such a case can be seen from Figure 10.4b.

Small vibrations around the equilibrium position ($\varphi_{12e} = 1.1055, \varphi_{23e} = -3.86957$) are much simpler. They are shown in Figure 10.4a by line B for initial values $\varphi_{12}(0) = 0.71, \dot{\varphi}_{12}(0) = 0, \varphi_{23}(0) = -4$.

The transition between the two kinds of vibration is illustrated in Figure 10.4c. The initial values are the same as for line A but there is a damping ($k_1 = 430$ Nm^{-1}s).

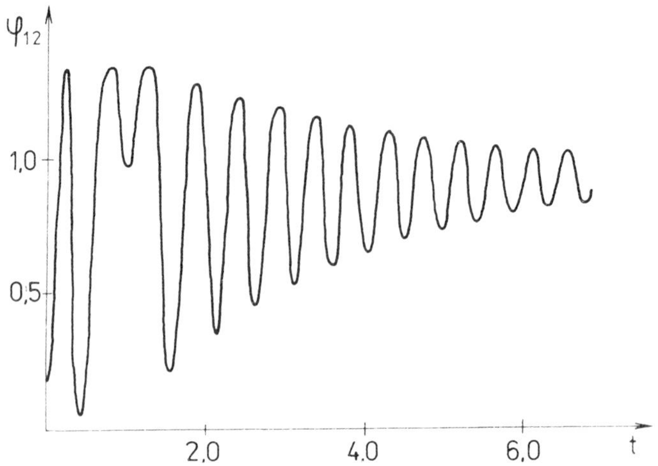

Figure 10.4c. Oscilations of the body 2.

Chapter 11

Accuracy and Sensitivity of MBSs

When modelling mechanisms, we suppose that their links are manufactured exactly according to the documentation, that kinematic pairs are ideal, that no temperature changes occur, etc. However, these assumptions are not fulfilled in practice. The result is erroneous function of mechanisms. Problem of incorrect function of mechanisms is especially important in robotics.

The basic purpose of robot manipulators is to manipulate objects, for example, to grip an object in a specified place and to move it to another place within its working space. In order to preserve the fidelity of motion, the correct position and orientation of the end-effector is needed in starting, in later points in the trajectory and in many cases during the whole trajectory. The main function of a spatial mechanism is reproduction (repetition) of the required motion. The reasons for the differences between theoretical and real motion, e.g. the errors in position and orientation (and also in velocity and acceleration) of the end-effector of the real manipulator include the dimensional errors of the manipulator's components, the effects of flexibility of its members, backlash in kinematic pairs, the changes of temperature and errors due to position accuracy of the joint variables (velocities and accelerations). This is, in general, the cause of misfunction of real mechanisms.

This chapter describes a method which enables the analysis to be made of the influence of errors on the reproduction of motion, velocity and acceleration of spatial mechanisms or robot manipulators. It also enables us to automate this analysis because the "manual" analysis is, in complicated cases, very laborious, unreliable and even unreal. We distinguish kinematic and dynamic accuracy of MBSs.

11.1 Kinematic Accuracy

In the present analysis we will take into account only the errors due to position accuracy of the joint variables and dimensional errors of the manipulator components. The other effects (flexibility of members, backlash, wearing off, changes of temperature) can be taken into account after they have been expressed in terms of the errors mentioned above. In this chapter, the exact model of errors will be developed first; the linear model of errors will then be developed, the latter being well suited for a statistical approach.

The closed or open kinematic chains which are mechanical models of spatial mechanisms and robot manipulators will be called ideal if their members possess nominal dimensions and the joint variables have their nominal values. The differences between the ideal and real positions, velocities and accelerations are errors. The errors can be grouped into two categories. The first category consists of the dimensional errors of members and of the errors in values of driving joint variables. They are mutually independent. The second category is formed by the errors in values of driven joint variables and the errors in position, velocity and acceleration of points and members of a mechanism. Errors in the second category are dependent on those in the first category.

A complete theory of kinematic accuracy of MBS dealing with all position, velocity and acceleration characteristics will be derived using a basic matrix approach. The relationship between the kinematic accuracy and the kinematic solution of MBS will also be determined; finally, the automated analysis of kinematic accuracy using the program OKAM2 for kinematic solution of MBS will be demonstrated.

11.1.1 Kinematic Accuracy of a Simple Open Chain

A motion of a simple open chain which consists of n members can be described by means of the transformation matrix $\mathbf{T}_{ab}$ (5.71). The motion of a point M of member b can according to (4.50), (5.70) be expressed as

$$\mathbf{r}_{aM} = \mathbf{T}_{ab}\mathbf{r}_{bM} \tag{11.1}$$

We assume the radius vector $\mathbf{r}_{bM}$ to be accurate. If not, we coincide the point M with the origin of space b and the corresponding error will be expressed by $\mathbf{T}_{ab}$ itself. The transformation matrix $\mathbf{T}_{ab}$ is a function of parameters p_j of the chain, i.e. the function of dimensions and values of the joint variables

$$\mathbf{T}_{ab} = \mathbf{T}_{ab}(p_j) \tag{11.2}$$

The expression (11.2) relates to ideal motion. During real motion the last member of the chain cannot reach the ideal position defined by space b because of differences Δp_j of its parameters p_j, but it will reach the real position that is determined by another space $\bar{b}$. This real motion is described by the transformation matrix $\mathbf{T}_{a\bar{b}}$. On the basis of (11.2) we can write

$$\mathbf{T}_{a\bar{b}} = \mathbf{T}_{ab}(p_j + \Delta p_j) \tag{11.3}$$

From the known model (11.2) of the kinematic chain and from the known errors Δp_j of its parameters we are able to express the real motion of the chain from (11.3) and also the exact error of motion as the difference

$$\Delta\mathbf{T}_{ab} = \mathbf{T}_{a\bar{b}} - \mathbf{T}_{ab} = \mathbf{T}_{ab}(p_j + \Delta p_j) - \mathbf{T}_{ab}(p_j) \tag{11.4}$$

or as a virtual motion from the ideal position b to the real one $\bar{b}$ by the relation

$$\mathbf{T}_{a\bar{b}} = \mathbf{T}_{ab}\mathbf{T}_{b\bar{b}} \tag{11.5}$$

$$\mathbf{T}_{b\bar{b}} = \mathbf{T}_{ab}^{-1}\mathbf{T}_{a\bar{b}} = \mathbf{T}_{ab}^{-1}(p_j)\mathbf{T}_{ab}(p_j + \Delta p_j) \tag{11.6}$$

The elements of matrix (11.4) express the differences of the respective elements of matrix $\mathbf{T}_{ab}$. The matrix $\mathbf{T}_{b\bar{b}}$ expresses the difference in composition of motions, i.e. the motion in which the ideal and the real one differ. Analogous expressions for velocities and accelerations of motions $a\bar{b}$, $b\bar{b}$ or the differences Δ can be derived (Valášek and Stejskal, 1987b).

These expressions constitute the exact model of errors (or exact accuracy model) of a simple open chain. They are valid even for large differences of chain parameters in which they are non-linear.

We will now derive the linear model of errors (or linear accuracy model) for a simple open chain that will be valid for small differences of chain parameters with suitable results.

Let the differential changes δp_j of kinematic parameters of the chain cause the differential changes of its motion, i.e. the differential changes in position and orientation of the space b. These changes will be expressed by the differential change $\delta\mathbf{T}_{ab}$ of the transformation matrix $\mathbf{T}_{ab}$. We can write

$$\delta p_j = \dot{p}_j \delta t \tag{11.7}$$

$$\delta\mathbf{T}_{ab} = \sum_j \frac{\partial \mathbf{T}_{ab}}{\partial p_j}\delta p_j = \sum_j \frac{\partial \mathbf{T}_{ab}}{\partial p_j}\dot{p}_j \delta t = \dot{\mathbf{T}}_{ab}\delta t \tag{11.8}$$

Using (4.61) we have for error motion

$$\delta\mathbf{T}_{ab} = \dot{\mathbf{T}}_{ab}\delta t = \mathbf{T}_{ab}\mathbf{V}_{b,ab}\delta t = \mathbf{T}_{ab}\mathbf{V}_{ab}\delta t \tag{11.9}$$

For $\mathbf{V}_{ab}\delta t$ it follows from (4.70)

$$\mathbf{V}_{ab}\delta t = \begin{bmatrix} 0 & -\omega_z\delta t & \omega_y\delta t & v_x\delta t \\ \omega_z\delta t & 0 & -\omega_x\delta t & v_y\delta t \\ -\omega_y\delta t & \omega_x\delta t & 0 & v_z\delta t \\ 0 & 0 & 0 & 0 \end{bmatrix} = \\ = \begin{bmatrix} 0 & -\delta\varphi_z & \delta\varphi_y & \delta x \\ \delta\varphi_z & 0 & -\delta\varphi_x & \delta y \\ -\delta\varphi_y & \delta\varphi_x & 0 & \delta z \\ 0 & 0 & 0 & 0 \end{bmatrix} \tag{11.10}$$

as the motion with angular or translational velocity during the time interval δt raises differential rotation $\delta\varphi$ around the corresponding axis or differential translation along the same axis. These motions (and differential changes of the position and orientation of the origin of space b) are expressed in space b.

In order to express these differential changes in space a, we use the other definition of the velocity matrix $\mathbf{V}_{a,ab}$ (4.77) which has the same elements as $\mathbf{V}_{b,ab}$ but expresses the velocities of the motion of the space a according to the space b with the opposite sign. Hence

$$\delta\mathbf{T}_{ab} = \dot{\mathbf{T}}_{ab}\delta t = \mathbf{V}_{a,ab}\mathbf{T}_{ab}\delta t = \mathbf{V}_{a,ab}\delta t\mathbf{T}_{ab} \tag{11.11}$$

Matrix

$$\mathbf{V}_{a,ab}\delta t = \mathbf{T}_{ab}\mathbf{V}_{ab}\delta t\mathbf{T}_{ab}^{-1} \tag{11.12}$$

expresses the differential changes of the position and orientation of the origin of the space a according to the space b in the space a with the opposite sign. The matrix $\delta\mathbf{T}_{ab}$ expresses the differential changes of the transformation matrix of the motion of space b according to space a. We thus obtain the differential changes of the position of the origin of space b from (11.10) expressed in space b and from (11.9) expressed in space a. We obtain the differential changes of the orientation of the origin of space b from eq. (11.10) expressed in space b and from (11.12) in space a, whereas from (11.9) we only have the differential changes of the direction cosines.

Assuming that differences Δp_j of chain parameters p_j are small, we neglect high order terms in (11.8) to (11.10) and use these expressions for the evaluation of finite differences. From (11.8) and (11.9) it follows that

$$\Delta\mathbf{T}_{ab} = \sum_j \frac{\partial\mathbf{T}_{ab}}{\partial p_j}\Delta p_j \tag{11.13}$$

$$\Delta\mathbf{T}_{ab} = \mathbf{T}_{ab}\mathbf{H}_{ab} \tag{11.14}$$

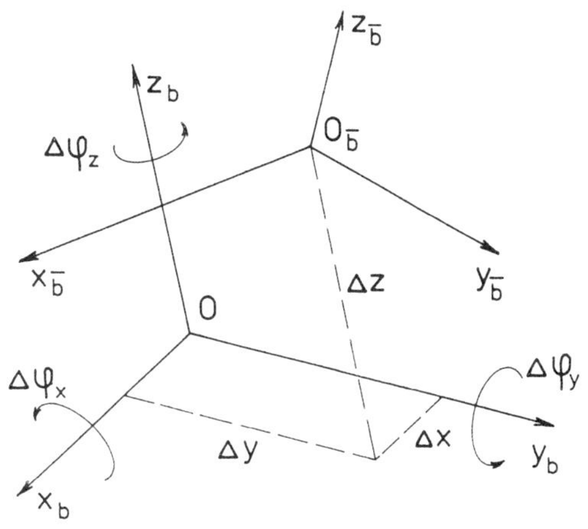

Figure 11.1. Position and orientation errors.

where

$$\mathbf{H}_{ab} = \mathbf{H}_{b,ab} = \mathbf{T}_{ab}^{-1}\Delta\mathbf{T}_{ab} = \mathbf{T}_{ab}^{-1}\sum_j \frac{\partial \mathbf{T}_{ab}}{\partial p_j}\Delta p_j =$$

$$= \begin{bmatrix} 0 & -\Delta\varphi_z & \Delta\varphi_y & \Delta x \\ \Delta\varphi_z & 0 & -\Delta\varphi_x & \Delta y \\ -\Delta\varphi_y & \Delta\varphi_x & 0 & \Delta z \\ 0 & 0 & 0 & 0 \end{bmatrix} \tag{11.15a}$$

is the error matrix expressed in space b. In this matrix, $\Delta x, \Delta y, \Delta z$ are the small differences in position of origins of spaces b and $\bar{b}$ and $\Delta\varphi_x, \Delta\varphi_y, \Delta\varphi_z$ stand for the small rotations around the x_b, y_b, z_b axes needed for coincidence of their directions with those of $x_{\bar{b}}, y_{\bar{b}}, z_{\bar{b}}$ axes (all expressed in space b), Figure 11.1. For error matrix $H_{a,ab}$ expressed in space a

$$\mathbf{H}_{a,ab} = \mathbf{T}_{ab}\mathbf{H}_{b,ab}\mathbf{T}_{ab}^{-1} \tag{11.15b}$$

For the transformation matrix of real motion in first order terms in case of the linear theory of errors we use the symbol $\mathbf{T}^*_{ab}$. Then

$$\mathbf{T}_{a\bar{b}} \doteq \mathbf{T}^*_{ab} = \mathbf{T}_{ab} + \Delta\mathbf{T}_{ab} = \mathbf{T}_{ab} + \mathbf{T}_{ab}\mathbf{H}_{ab} = \mathbf{T}_{ab}(\mathbf{E}(1) + \mathbf{H}_{ab}) \tag{11.16}$$

where $\mathbf{E}(1)$ is a unit matrix. We further introduce the correction matrix

$\mathbf{C}_{ab}$ for the motion of chain

$$\mathbf{C}_{ab} = \mathbf{C}_{b,ab} = \mathbf{E}(1) + \mathbf{H}_{ab} = \begin{bmatrix} 1 & -\Delta\varphi_z & \Delta\varphi_y & \Delta x \\ \Delta\varphi_z & 1 & -\Delta\varphi_x & \Delta y \\ -\Delta\varphi_y & \Delta\varphi_x & 1 & \Delta z \\ 0 & 0 & 0 & 1 \end{bmatrix} \tag{11.17}$$

From (11.16) it follows that

$$\mathbf{T}_{a\bar{b}} \doteq \mathbf{T}^*_{ab} = \mathbf{T}_{ab}\mathbf{C}_{ab} \tag{11.18}$$

Comparing (11.18) and (11.15), we learn that in linear theory of errors the matrix $\mathbf{C}_{ab}$ is the same as the transformation matrix of small motion from space b to space $\bar{b}$

$$\mathbf{T}_{b\bar{b}} \doteq \mathbf{T}^*_{bb} = \mathbf{C}_{ab} \tag{11.19}$$

It is very useful to define an error matrix and a correction matrix for basic motions. From (11.14) and (11.17) we have

$$\mathbf{H}_{Zk}(\Delta) = \mathbf{D}_{Zk}(\Delta) \tag{11.20}$$

$$\mathbf{C}_{Zk}(\Delta) = \mathbf{E}(1) + \mathbf{D}_{Zk}(\Delta) \tag{11.21}$$

The motion (11.2) will be expressed as a product of basic matrices and we have

$$\mathbf{T}_{ab} = \prod_{i=1}^{N} \mathbf{T}_{Zk_i}(p_{k_i}), \qquad k_i \in \{1,2,3,4,5,6\} \tag{11.22}$$

where p_{k_i} refers to the i-th parameter of basic motion (all parameters p_j are among them) and N is the number of all basic motions.
It follows from (11.13) that

$$\Delta\mathbf{T}_{ab} = \sum_{j=1}^{N} \mathbf{F}_j\mathbf{E}(\Delta p_{k_j}) \tag{11.23}$$

where $\mathbf{F}_j$ is defined by (5.90). The error matrix $\mathbf{H}_{ab}$ is then defined by the expression

$$\mathbf{H}_{ab} = \mathbf{T}^{-1}_{ab}\Delta\mathbf{T}_{ab} = \mathbf{T}^{-1}_{ab}\sum_{j=1}^{N} \mathbf{F}_j\mathbf{E}(\Delta p_{k_j}) = \sum_{j=1}^{N} \mathbf{G}_j\mathbf{E}(\Delta p_{k_j}) \tag{11.24}$$

where $\mathbf{G}_j$ is defined by (5.93e). The identity of the expression (11.24) for $\mathbf{H}_{ab}$ and equations for $\mathbf{V}_{ab}$ proves in another way the equation (11.15).

The correction matrix is

$$\mathbf{C}_{ab} = \mathbf{H}_{ab} + \mathbf{E}(1) \tag{11.25}$$

The expressions (11.16), (11.18), (11.19) and (11.23) to (11.25) derived above constitute the linear error theory of position accuracy for a simple open kinematic chain. They describe the real motion, errors of this motion and the small virtual motion $b \to \bar{b}$ by means of known parameters of a kinematic chain and their errors with higher order terms neglected. Matrix $\mathbf{G}_j$ shows the influence of the j-th parameter error (the sensitivity) on the motion error $\Delta\mathbf{T}_{ab}$ expressed in space b, and matrix $\mathbf{F}_j$ shows the influence of the same error in space a. So far we have been interested only in position.

We derive the expressions from (11.18) by differentiating with respect to time

$$\dot{\mathbf{T}}^*_{ab} = \dot{\mathbf{T}}_{ab}\mathbf{C}_{ab} + \mathbf{T}_{ab}\dot{\mathbf{C}}_{ab} \tag{11.26}$$

and according to the definition we have

$$\begin{aligned} \mathbf{V}^*_{ab} &= \mathbf{C}^{-1}_{ab}\mathbf{V}_{ab}\mathbf{C}_{ab} + \mathbf{C}^{-1}_{ab}\dot{\mathbf{C}}_{ab} && (11.27) \\ \mathbf{V}^*_{bb} &= \mathbf{C}^{-1}_{ab}\dot{\mathbf{C}}_{ab} && (11.28) \end{aligned}$$

where

$$\begin{aligned} \dot{\mathbf{C}}_{ab} &= \dot{\mathbf{H}}_{ab} = \sum_{j=1}^{N}(\dot{\mathbf{G}}_j\mathbf{E}(\Delta p_{k_j}) + \mathbf{G}_j\mathbf{E}(\Delta\dot{p}_{k_j})) = \\ &= \begin{bmatrix} 0 & -\Delta\dot{\varphi}_z & \Delta\dot{\varphi}_y & \Delta\dot{x} \\ \Delta\dot{\varphi}_z & 0 & -\Delta\dot{\varphi}_x & \Delta\dot{y} \\ -\Delta\dot{\varphi}_y & \Delta\dot{\varphi}_x & 0 & \Delta\dot{z} \\ 0 & 0 & 0 & 0 \end{bmatrix} && (11.29) \end{aligned}$$

Neglecting higher order terms,

$$\mathbf{C}^{-1}_{ab} = \mathbf{E}(1) - \mathbf{H}_{ab} \tag{11.30}$$

After rearranging (11.26) and (11.27), and neglecting quadratic terms, we obtain

$$\begin{aligned} \mathbf{V}^*_{ab} &= \mathbf{V}_{ab} + \Delta\mathbf{V}_{ab} = \mathbf{V}_{ab} - \mathbf{H}_{ab}\mathbf{V}_{ab} + \mathbf{V}_{ab}\mathbf{H}_{ab} + \dot{\mathbf{C}}_{ab} && (11.31) \\ \Delta\mathbf{V}_{ab} &= -\mathbf{H}_{ab}\mathbf{V}_{ab} + \mathbf{V}_{ab}\mathbf{H}_{ab} + \dot{\mathbf{C}}_{ab} && (11.32) \\ \mathbf{V}^*_{bb} &= \dot{\mathbf{C}}_{ab} && (11.33) \end{aligned}$$

The expressions (11.29), (11.31) and (11.32) constitute the linear theory of velocities for a simple opened kinematic chain.

Acceleration relations are to be obtained by differentiating (11.26) with respect to time

$$\ddot{\mathbf{T}}^*_{ab} = \ddot{\mathbf{T}}_{ab}\mathbf{C}_{ab} + 2\dot{\mathbf{T}}_{ab}\dot{\mathbf{C}}_{ab} + \mathbf{T}_{ab}\ddot{\mathbf{C}}_{ab} \tag{11.34}$$

After rearranging (11.34) according to the definitions, we have

$$\begin{aligned}
\mathbf{A}^*_{ab} &= -\mathbf{C}^{-1}_{ab}\dot{\mathbf{C}}_{ab}\mathbf{C}^{-1}_{ab}\mathbf{V}_{ab}\mathbf{C}_{ab} + \mathbf{C}^{-1}_{ab}\mathbf{V}_{ab}\dot{\mathbf{C}}_{ab} + \\
&\quad + \mathbf{C}^{-1}_{ab}\mathbf{A}_{ab}\mathbf{C}_{ab} - \mathbf{C}^{-1}_{ab}\dot{\mathbf{C}}_{ab}\mathbf{C}^{-1}_{ab}\dot{\mathbf{C}}_{ab} + \mathbf{C}^{-1}_{ab}\ddot{\mathbf{C}}_{ab} &(11.35)\\
\mathbf{B}^*_{ab} &= \mathbf{C}^{-1}_{ab}\mathbf{B}_{ab}\mathbf{C}_{ab} + 2\mathbf{C}^{-1}_{ab}\mathbf{V}_{ab}\dot{\mathbf{C}}_{ab} + \mathbf{C}^{-1}_{ab}\ddot{\mathbf{C}}_{ab} &(11.36)\\
\mathbf{A}^*_{bb} &= -\mathbf{C}^{-1}_{ab}\dot{\mathbf{C}}_{ab}\mathbf{C}^{-1}_{ab}\dot{\mathbf{C}}_{ab} + \mathbf{C}^{-1}_{ab}\ddot{\mathbf{C}}_{ab} &(11.37)\\
\mathbf{B}^*_{bb} &= \mathbf{C}^{-1}_{ab}\ddot{\mathbf{C}}_{ab} &(11.38)
\end{aligned}$$

where

$$\begin{aligned}
\ddot{\mathbf{C}}_{ab} &= \ddot{\mathbf{H}}_{ab} = \begin{bmatrix} 0 & -\Delta\ddot{\varphi}_z & \Delta\ddot{\varphi}_y & \Delta\ddot{x} \\ \Delta\ddot{\varphi}_z & 0 & -\Delta\ddot{\varphi}_x & \Delta\ddot{y} \\ -\Delta\ddot{\varphi}_y & \Delta\ddot{\varphi}_x & 0 & \Delta\ddot{z} \\ 0 & 0 & 0 & 0 \end{bmatrix} = \\
&= \sum_{j=1}^{N}(\ddot{\mathbf{G}}\mathbf{E}(\Delta p_{k_j}) + 2\dot{\mathbf{G}}_j\mathbf{E}(\Delta\dot{p}_{k_j}) + \mathbf{G}_j\mathbf{E}(\Delta\ddot{p}_{k_j})) \qquad (11.39)
\end{aligned}$$

Neglecting quadratic terms in (11.35) to (11.38), we obtain

$$\begin{aligned}
\mathbf{A}^*_{ab} &= \mathbf{A}_{ab} - \dot{\mathbf{C}}_{ab}\mathbf{V}_{ab} + \mathbf{V}_{ab}\dot{\mathbf{C}}_{ab} - \mathbf{H}_{ab}\mathbf{A}_{ab} + \mathbf{A}_{ab}\mathbf{H}_{ab} + \ddot{\mathbf{C}}_{ab} &(11.40)\\
\Delta\mathbf{A}_{ab} &= -\dot{\mathbf{C}}_{ab}\mathbf{V}_{ab} + \mathbf{V}_{ab}\dot{\mathbf{C}}_{ab} - \mathbf{H}_{ab}\mathbf{A}_{ab} + \mathbf{A}_{ab}\mathbf{H}_{ab} + \ddot{\mathbf{C}}_{ab} &(11.41)\\
\mathbf{B}^*_{ab} &= \mathbf{B}_{ab} - \mathbf{H}_{ab}\mathbf{B}_{ab} + \mathbf{B}_{ab}\mathbf{H}_{ab} + 2\mathbf{V}_{ab}\dot{\mathbf{C}}_{ab} + \ddot{\mathbf{C}}_{ab} &(11.42)\\
\Delta\mathbf{B}_{ab} &= -\mathbf{H}_{ab}\mathbf{B}_{ab} + \mathbf{B}_{ab}\mathbf{H}_{ab} + 2\mathbf{V}_{ab}\dot{\mathbf{C}}_{ab} + \ddot{\mathbf{C}}_{ab} &(11.43)\\
\mathbf{A}^*_{bb} &= \mathbf{B}^*_{bb} = \ddot{\mathbf{C}}_{ab} &(11.44)
\end{aligned}$$

The expressions (11.40) to (11.44) describe the linear error theory of acceleration for a simple open kinematic chain.

Although the expressions derived above as well as those derived below are not simple, the solution of kinematic accuracy based on these expressions is quite simply programmable and achievable by means of a computer. It is even possible to make use of the standard program system OKAM2 up to solution of the kinematic accuracy of velocities (see section 10.2).

11.1.2 Kinematic Accuracy of a Simple Closed Chain

The simple closed chain (mechanism) in Figure 5.13 consists of n members and has i degrees of freedom. The loop of the mechanism is assumed to be divided by a hypothetical cut into two open chains. The kinematic equation for an ideal mechanism is (5.75) and for a real one

$$\mathbf{T}_{1\overline{2}}\mathbf{T}_{\overline{23}}\ldots\mathbf{T}_{\overline{k-1,k}} = \mathbf{T}_{1\overline{n}}\mathbf{T}_{\overline{n},\overline{n-1}}\ldots\mathbf{T}_{\overline{k+1,k}} \qquad (11.45)$$

Contrary to the previous sections, we must now distinguish among the parameters p_j of the kinematic chain the independent parameters q_j and the dependent ones z_j. In the previous part all parameters were independent.

Substituting (11.2) into (5.75) and ordering the elements of matrix equation in the chosen fixed order, we obtain a system of nonlinear equations of the type (5.74).

Substitution of (11.3) into (11.45) leads after an analogous operation to

$$\mathbf{f}(\mathbf{z} + \Delta\mathbf{z}, \mathbf{q} + \Delta\mathbf{q}) = \mathbf{0} \tag{11.46}$$

The systems of equations (5.74) and (11.46) have the same form,

$$\Delta\mathbf{q} = \lceil \Delta q_1, \ldots, \Delta q_{n-i} \rceil \quad , \quad \Delta\mathbf{z} = \lceil \Delta z_1, \ldots, \Delta z_i \rceil$$

The equation (11.46) constitutes the exact model of position errors for a simple closed chain. Its numerical solution is analogous to the numerical solution of equation (5.74), and it enables us to obtain $\mathbf{z} + \Delta\mathbf{z}$ as a function of $\mathbf{q} + \Delta\mathbf{q}$, i.e. to find $\Delta\mathbf{z}$ expressed as a function $\mathbf{q}, \mathbf{z}, \Delta\mathbf{q}$.

Analogously, we obtain the exact model of velocity and acceleration errors from the solution of velocities and accelerations of the MBS (Section 5.4.2 and 5.8).

We shall now turn our attention to a linear model of errors for a simple closed chain. Substituting (11.18) into (11.45), we have

$$\mathbf{T}^*_{12}\mathbf{T}^*_{23}\ldots\mathbf{T}^*_{k-1,k} = \mathbf{T}^*_{1n}\mathbf{T}^*_{n,n-1}\cdots\mathbf{T}^*_{k+1,k} \tag{11.47}$$

$$\mathbf{T}_{12}\mathbf{C}_{12}\mathbf{T}_{23}\mathbf{C}_{23}\ldots\mathbf{T}_{k-1,k}\mathbf{C}_{k-1,k} = \mathbf{T}_{1n}\mathbf{C}_{1n}\mathbf{T}_{n,n-1}\mathbf{C}_{n,n-1}\ldots\mathbf{T}_{k+1,k}\mathbf{C}_{k+1,k} \tag{11.48}$$

These last equations can be simplified by ignoring the higher order terms. It will be better to use the basic matrix method instead of the general one. Using expressions (11.18) and (11.17) in (11.47) leads to equations

$$\mathbf{T}^*_{1k} = \mathbf{T}^*_{1nk} \tag{11.49a}$$

$$\mathbf{T}_{1k}\mathbf{C}_{1k} = \mathbf{T}_{1nk}\mathbf{C}_{1nk} \tag{11.49b}$$

$$\mathbf{T}_{1k}(\mathbf{E}(1) + \mathbf{H}_{1k}) = \mathbf{T}_{1nk}(\mathbf{E}(1) + \mathbf{H}_{1nk}) \tag{11.49c}$$

In our case, $\mathbf{T}_{1k} = \mathbf{T}_{1nk}$ and it means that we are interested in the error model rather than in a mechanism synthesis problem. Therefore it is important that

$$\mathbf{T}_{1k} = \prod_{j=1}^{k-1} \mathbf{T}_{j,j+1} \tag{11.50a}$$

$$\mathbf{T}^*_{1k} = \prod_{j=1}^{k-1} \mathbf{T}^*_{j,j+1} \tag{11.50b}$$

We therefore have

$$\mathbf{C}_{1k} = \mathbf{C}_{1nk} \tag{11.51a}$$
$$\mathbf{H}_{1k} = \mathbf{H}_{1nk} \tag{11.51b}$$

and, after substituting from (11.24),

$$\sum_{j=1}^{k} \mathbf{G}_j \mathbf{E}(\Delta p_{k_j}) = \sum_{j=k+1}^{n} \mathbf{G}_j \mathbf{E}(\Delta p_{k_j}) \tag{11.52}$$

These last equations have the same form as the velocity equation of mechanism because from

$$\mathbf{T}_{1k}(p_{k_j}) = \mathbf{T}_{1nk}(p_{k_j}) \tag{11.53a}$$
$$\mathbf{V}_{1k}(p_{k_j}, \dot{p}_{k_j}) = \mathbf{V}_{1nk}(p_{k_j}, \dot{p}_{k_j}) \tag{11.53b}$$

we obtain for errors

$$\mathbf{V}_{1k}(p_{k_j}, \Delta p_{k_j}) = \mathbf{V}_{1nk}(p_{k_j}, \Delta p_{k_j}) \tag{11.54}$$

This is the same equation as (11.52). We can therefore write in matrix form

$$\boldsymbol{\Phi}_z(\mathbf{z}, \mathbf{q})\Delta\mathbf{z} = -\boldsymbol{\Phi}_q(\mathbf{z}, \mathbf{q})\Delta\mathbf{q} \tag{11.55}$$

Application of equations (11.51) is advantageous because, according to the structure of (11.15), (11.17) or the velocity matrix, we can choose a smaller number of independent equations in regular cases, i.e. elements (3,2), (1,3), (2,1), (1,4), (2,4), (3,4) of the matrix equation.

From (11.55), we obtain for velocities and accelerations

$$\begin{aligned}\boldsymbol{\Phi}_z(\mathbf{z}, \mathbf{q})\Delta\dot{\mathbf{z}} &= -\boldsymbol{\Phi}_q(\mathbf{z}, \mathbf{q})\Delta\dot{\mathbf{q}} - \dot{\boldsymbol{\Phi}}_q(\mathbf{z}, \mathbf{q}, \dot{\mathbf{z}}, \dot{\mathbf{q}})\Delta\mathbf{q} - \\ &\quad -\dot{\boldsymbol{\Phi}}_z(\mathbf{z}, \mathbf{q}, \dot{\mathbf{z}}, \dot{\mathbf{q}})\Delta\mathbf{z} \end{aligned} \tag{11.56}$$

$$\begin{aligned}\boldsymbol{\Phi}_z(\mathbf{z}, \mathbf{q})\Delta\ddot{\mathbf{z}} &= -\boldsymbol{\Phi}_q(\mathbf{z}, \mathbf{q})\Delta\ddot{\mathbf{q}} - 2\dot{\boldsymbol{\Phi}}_q(\mathbf{z}, \mathbf{q}, \dot{\mathbf{z}}, \dot{\mathbf{q}})\Delta\dot{\mathbf{q}} - \\ &\quad -2\dot{\boldsymbol{\Phi}}_z(\mathbf{z}, \mathbf{q}, \dot{\mathbf{z}}, \dot{\mathbf{q}})\Delta\dot{\mathbf{z}} - \ddot{\boldsymbol{\Phi}}_q(\mathbf{z}, \mathbf{q}, \dot{\mathbf{z}}, \dot{\mathbf{q}}, \ddot{\mathbf{z}}, \ddot{\mathbf{q}})\Delta\mathbf{q} - \\ &\quad -\ddot{\boldsymbol{\Phi}}_z(\mathbf{z}, \mathbf{q}, \dot{\mathbf{z}}, \dot{\mathbf{q}}, \ddot{\mathbf{z}}, \ddot{\mathbf{q}})\Delta\mathbf{z} \end{aligned} \tag{11.57}$$

where the desired matrices can be derived from (11.55).

Another possibility is to make use of (11.49)

$$\dot{\mathbf{T}}^*_{1k} = \dot{\mathbf{T}}^*_{1nk} \tag{11.58a}$$
$$\mathbf{T}^*_{1k}\mathbf{V}^*_{1k} = \mathbf{T}^*_{1nk}\mathbf{V}^*_{1nk} \tag{11.58b}$$
$$\mathbf{V}^*_{1k} = \mathbf{V}^*_{1nk} \tag{11.58c}$$

Substituting from (11.31)

$$\mathbf{V}_{1k} - \mathbf{H}_{1k}\mathbf{V}_{1k} + \mathbf{V}_{1k}\mathbf{H}_{1k} + \dot{\mathbf{C}}_{1k} = \mathbf{V}_{1nk} - \mathbf{H}_{1nk}\mathbf{V}_{1nk} + \mathbf{V}_{1nk}\mathbf{H}_{1nk} + \dot{\mathbf{C}}_{1nk} \tag{11.59}$$

and, according to (11.51) and (11.58), we have

$$\dot{\mathbf{C}}_{1k} = \dot{\mathbf{C}}_{1nk} \tag{11.60}$$

Equation (11.60) is a matrix form of vector equation (11.56) after substitution from (11.24) and (11.29). Analogously from (11.58), it follows that

$$\ddot{\mathbf{C}}_{1k} = \ddot{\mathbf{C}}_{1nk} \tag{11.61}$$

Again, equation (11.61) is a matrix form of vector equation (11.57).

11.1.3 Kinematic Accuracy of Mixed Chains

The most general type of a mechanical system is that of several closed loops with an open chain attached. Such a model can arise in an accuracy study of a composed mechanism itself or in an accuracy study of an open chain whose motion is caused by a mechanism with several loops. The solution of such a problem can easily be obtained using the results derived in previous sections. For any loop we write equations (5.75) and (11.45), and we collect all elements of matrices in equations (5.74) and (11.46) in the specified order. For any open chain (or for a member in whose motion we are interested) we write equations (11.2) and (11.3). The combination of these equations is the desired error model. In this process it is important to distinguish between dependent $\mathbf{z}$ and independent $\mathbf{q}$ kinematic parameters.

The equation (5.74) describes the position of an ideal mechanism, and for the position of an ideal chain we have

$$\mathbf{T} = \mathbf{T}(\mathbf{z}, \mathbf{q}) \tag{11.62}$$

For the exact error model of the mechanism, we have equation (11.46) and for the exact model of a chain

$$\mathbf{T} + \Delta\mathbf{T} = \mathbf{T}(\mathbf{z} + \Delta\mathbf{z}, \mathbf{q} + \Delta\mathbf{q}) \tag{11.63}$$

Solving (11.46) for $\mathbf{z} + \Delta\mathbf{z}$ and substituting into (11.63) we have

$$\mathbf{z} + \Delta\mathbf{z} = \mathbf{f}^{-1}(\mathbf{q} + \Delta\mathbf{q}) \tag{11.64}$$

$$\mathbf{T} + \Delta\mathbf{T} = \mathbf{T}(\mathbf{f}^{-1}(\mathbf{q} + \Delta\mathbf{q}), \mathbf{q} + \Delta\mathbf{q}) \tag{11.65}$$

This is the exact accuracy model for the position of the mixed chain. For velocities and accelerations we obtain analogous formulae.

The linear accuracy model may be derived in a similar way to the exact model. Using the basic equation (5.74), we obtain for differences $\Delta\mathbf{z}, \Delta\mathbf{q}$ and their derivatives the equations (11.55) to (11.57). We obtain similarly from (11.62)

$$\Delta\mathbf{T} = \frac{\partial\mathbf{T}(\mathbf{z},\mathbf{q})}{\partial\mathbf{z}}\Delta\mathbf{z} + \frac{\partial\mathbf{T}(\mathbf{z},\mathbf{q})}{\partial\mathbf{q}}\Delta\mathbf{q} \tag{11.66}$$

Differentiating the last equation with respect to time we obtain $\Delta\dot{\mathbf{T}}$ and $\Delta\ddot{\mathbf{T}}$. Substituting $\Delta\mathbf{z}, \Delta\dot{\mathbf{z}}$ and $\Delta\ddot{\mathbf{z}}$ from (11.55) to (11.57) into the expressions for $\Delta\mathbf{T}, \Delta\dot{\mathbf{T}}$ and $\Delta\ddot{\mathbf{T}}$ yields equations of the type

$$\Delta\mathbf{T} = \mathbf{S}_1(\mathbf{z},\mathbf{q})\Delta\mathbf{q} \tag{11.67a}$$

$$\Delta\dot{\mathbf{T}} = \mathbf{S}_1(\mathbf{z},\mathbf{q})\Delta\dot{\mathbf{q}} + \mathbf{S}_2(\mathbf{z},\mathbf{q},\dot{\mathbf{z}},\dot{\mathbf{q}})\Delta\mathbf{q} \tag{11.67b}$$

$$\Delta\ddot{\mathbf{T}} = \mathbf{S}_1(\mathbf{z},\mathbf{q})\Delta\ddot{\mathbf{q}} + 2\mathbf{S}_2(\mathbf{z},\mathbf{q},\dot{\mathbf{z}},\dot{\mathbf{q}})\Delta\dot{\mathbf{q}} + \mathbf{S}_3(\mathbf{z},\mathbf{q},\dot{\mathbf{z}},\dot{\mathbf{q}},\ddot{\mathbf{z}},\ddot{\mathbf{q}})\Delta\mathbf{q} \tag{11.67c}$$

or — in matrix form —

$$\begin{bmatrix} \Delta\mathbf{T} \\ \Delta\dot{\mathbf{T}} \\ \Delta\ddot{\mathbf{T}} \end{bmatrix} = \begin{bmatrix} \mathbf{S}_1 & 0 & 0 \\ \mathbf{S}_1 & \mathbf{S}_2 & 0 \\ \mathbf{S}_1 & 2\mathbf{S}_2 & \mathbf{S}_3 \end{bmatrix} \begin{bmatrix} \Delta\mathbf{q} \\ \Delta\dot{\mathbf{q}} \\ \Delta\ddot{\mathbf{q}} \end{bmatrix} = \mathbf{S} \begin{bmatrix} \Delta\mathbf{q} \\ \Delta\dot{\mathbf{q}} \\ \Delta\ddot{\mathbf{q}} \end{bmatrix} \tag{11.68a}$$

$$\Delta\mathbf{W} = \mathbf{S}\Delta\mathbf{s} \tag{11.68b}$$

where

$$\mathbf{W} = [\mathbf{T}, \dot{\mathbf{T}}, \ddot{\mathbf{T}}] \quad , \quad \mathbf{s} = [\mathbf{q}, \dot{\mathbf{q}}, \ddot{\mathbf{q}}] \tag{11.69}$$

The detailed form of equations (11.67) is rather complicated (Valášek and Stejskal, 1987b).

11.1.4 Evaluation of Kinematic Accuracy and Tolerance Spaces

Until now we have determined the influence of the differences of independent kinematic parameters (errors of the first category) on the differences of dependent kinematic parameters (errors of the second category). The kinematic parameters consist of the geometric dimensions of the mechanism $\boldsymbol{\varrho}$ with errors $\Delta\boldsymbol{\varrho}$, of the driving joint variables $\mathbf{s}$ with errors $\Delta\mathbf{s}$ and of the driven joint variables $\mathbf{u}$ with errors $\Delta\mathbf{u}$. We can distinguish several problems

– the errors of the driving joint variables $\Delta\mathbf{s}$ and the errors of the dimensions $\Delta\boldsymbol{\varrho}$ are given and we are looking for the errors of the driven joint variables $\Delta\mathbf{u}$ and the errors of motion of points and bodies in which we are interested (the problem of kinematic accuracy analysis),

– the problem of kinematic accuracy analysis in tolerance spaces,

– the tolerance space of the driven joint variables or of the motion of points and bodies in which we are interested are given and we are looking for the tolerance space of the driving joint variables and dimensions (the problem of kinematic accuracy synthesis),

– the dependence between driven joint variables and driving, where joint variables are given (on the completed machine) and we are looking for the unknown errors in the dimensions (the problem of kinematic errors calibration),

– the dependence between driven and driving joint variables are given (on the proposed machine) and we are looking for such changes in the dimensions $\Delta\varrho$ that the prescribed dependence holds (the problem of a mechanism's synthesis).

The following are the most important and the most general among these many problems. Firstly there is the basic problem of kinematic accuracy analysis, i.e. the determination of the tolerance space of output motion from the tolerance spaces of dimensions and driving joint variables of the mechanism. Secondly there is the basic problem of kinematic accuracy synthesis, i.e. the determination of such tolerance spaces of dimensions and driving joint variables that the spatial mechanism realizes the output motion within the given tolerance space. For the solution of these two problems we use mapping from the space of independent parameters of the spatial mechanism into the space of dependent (output motion) variables given generally by the equations (11.67) – (11.69). In this section we use only the linear model of kinematic accuracy of spatial mechanisms because the results for the tolerance spaces within the exact error model involve many more numerical difficulties and, in any case, they are not necessary for practical problems. The basic difficulty is that the mapping has no unique inverse solution because the number of independent parameters is usually much greater than the number of dependent parameters. If the differences in position are small, we can limit our considerations to the nominal trajectory and to the linear model of accuracy in its neighbourhood. If this is not the case, we must make use of the exact model of accuracy and solve the problem according to the kinematic analysis of the whole motion (i.e. large differences as large motions). In the following text we have in mind the linear model of accuracy. Let us express the tolerance space of independent parameters as

$$|\bar{s}_i - s_i| = |\Delta s_i| \leq \Delta_i \tag{11.70}$$

We obtain the tolerance space of the output parameters being studied from equation (11.67). We will suppose that the general abstract space of these output parameters is divided into subspaces of dimension 3 (this is often the case for the practical interpretation of output motion). The generalization of further results on the spaces of different dimension is obvious and simple. In the following imaginary case, we suppose that the resulting motion can be characterized by the radius vector $\mathbf{r}_{ab}$ of the real position of the origin O_b of the Cartesian system of some body in which we are interested. For the differences of the position of the point O_b, it follows from equation (11.67) that

$$\begin{aligned}\Delta\mathbf{r}_{ab} &= \sum_{j=1}^{N}(\Delta x_j\mathbf{i}_a + \Delta y_j\mathbf{j}_a + \Delta z_j\mathbf{k}_a) = \\ &= \sum_{j=1}^{N}(S_{rx,j}\Delta s_j\mathbf{i}_a + S_{ry,j}\Delta s_j\mathbf{j}_a + S_{rz,j}\Delta s_j\mathbf{k}_a) = \\ &= \sum_{j=1}^{N}\mathbf{v}_j\pi_j \qquad (11.71)\end{aligned}$$

where

$$-1 \leq \pi_j = \frac{\Delta s_j}{\Delta_j} \leq 1 \qquad (11.72)$$

$$\mathbf{v}_j = S_{rx,j}\Delta_j\mathbf{i}_a + S_{ry,j}\Delta_j\mathbf{j}_a + S_{rz,j}\Delta_j\mathbf{k}_a \qquad (11.73)$$

and $\mathbf{i}_a, \mathbf{j}_a, \mathbf{k}_a$ are unit vectors of space a (frame), $\Delta x_j, \Delta y_j, \Delta z_j$ are the position errors of O_b in space a and, corresponding to the j-th parameter, $s_j, S_{rx,j} S_{ry,j} S_{rz,j}$ are particular elements of the matrix $\mathbf{S}$ in (11.67) corresponding to the position $\mathbf{r}_{ab}$, to the coordinates $\mathbf{i}_a, \mathbf{j}_a, \mathbf{k}_a$ and to the parameters s_j. The vectors $\mathbf{v}_j$ express the maximum possible error determined by the parameter s_j; parameters π_j reduce this maximum possible error to the real error. The tolerance space of the real position of the point O_b is given by (11.72) and (11.73); this creates the convex polyhedron with the edges determined by vectors $\mathbf{v}_j$. (The tolerance space on the basis of the exact error model is a general convex solid which approximates to a convex polyhedron.) A very important approximation of this tolerance polyhedron is the approximation by the tolerance rectangle with the edges parallel to the coordinate axes of space a which involves the all-tolerance polyhedron. Its centre of is the point O_b and the lengths of its edges are

$$K_x = 2\sum_j |S_{rx,j}|\Delta_j \qquad (11.74a)$$

$$K_y = 2\sum_j |S_{ry,j}|\Delta_j \qquad (11.74b)$$

$$K_z = 2\sum_j |S_{rz,j}|\Delta_j \qquad (11.74c)$$

We can estimate the maximum error by the diagonal of this tolerance rectangle.

Conversely, let us assume the tolerance space of the position of the point O_b by the tolerance rectangle described above with dimensions K_x, K_y, K_z. We obtain the tolerance space of kinematic parameters as the tolerance polyhedron determined by the conditions

$$2\,|\sum_j S_{rx,j}\Delta s_j| \;\leq\; K_x \tag{11.75a}$$

$$2\,|\sum_j S_{ry,j}\Delta s_j| \;\leq\; K_y \tag{11.75b}$$

$$2\,|\sum_j S_{rz,j}\Delta s_j| \;\leq\; K_z \tag{11.75c}$$

(In the exact error model, there is a convex solid in the space of parameters.) A very important approximation of this tolerance polyhedron in the space of parameters is the approximation by the tolerance rectangle in the N-dimensional space of parameters whose edges have the lengths proportional to the sensitivity of the particular parameter

$$|\Delta s_j| \leq \Delta_j = \min\left(\frac{K_x}{2N|S_{rx,j}|}, \frac{K_y}{2N|S_{ry,j}|}, \frac{K_z}{2N|S_{rz,j}|}\right) \tag{11.76}$$

This tolerance rectangle lies completely within the tolerance polyhedron of parameters and guarantees that the real position of the point O_b lies completely within the tolerance rectangle. For the whole motion the tolerance polyhedron or rectangle of parameters is determined by the intersection of all the tolerance polyhedrons (or rectangles) corresponding to the particular positions.

This is the solution of the basic problem of kinematic accuracy analysis and synthesis. The use of these results for more abstract vectors $\mathbf{r}$ is obvious.

For the problems of kinematic accuracy (and errors) of spatial mechanisms, there are important statistical considerations. Within the linear model of kinematic accuracy, the principle of superposition for the influences of parameter differences holds. From equation (11.67), it follows that, if the parameters are normal distributed with zero mean, the variables $\mathbf{W}$ are also normal distributed with mean

$$\mathcal{E}[\mathbf{W}] = \mathbf{S}\mathcal{E}[\mathrm{s}] = \mathbf{0} \tag{11.77}$$

and variance

$$\mathcal{D}[\mathbf{W}] = \mathcal{E}\big[(\mathbf{W} - \mathcal{E}[\mathbf{W}])(\mathbf{W} - \mathcal{E}[\mathbf{W}])^T\big] = \mathbf{S}\mathcal{D}[\mathrm{s}]\mathbf{S}^T \tag{11.78}$$

11.1.5 Calibration for Kinematic Errors

Another important application of the linear theory of kinematic accuracy is the determination of manufacturing errors in the kinematic parameters (dimensions) so that the kinematic equations more accurately represent the actual robot geometric model. We must at the same time distinguish a robot's repeatability and its absolute positioning accuracy. A robot's repeatability is the robot's positioning accuracy when one positions the robot by teaching. The robot is positioned in the desired location and the joint angles are recorded. It is influenced by the accuracy of the driving joint positioning. A robot's absolute positioning accuracy is its positioning accuracy when one positions the robot by the driving joint values calculated from its kinematic equations. From the desired end-effector location based on the Cartesian coordinates one calculates the driving joint values by the inverse kinematic solution. The robot's absolute positioning accuracy is influenced by the accuracy of the driving joint positioning as well as by the accuracy of the robot geometric model (e.g. its dimensions). This is necessary to enable robots to be used in an advanced CAD/CAM environment. The majority of industrial robots have good repeatability (for example (Hayati and Roston, 1986), about 0.1 millimeter for PUMA 560) but their absolute positioning accuracy is an order of magnitude less (for example, about 2–5 millimeters for PUMA 560). The technique of calibration can substantially increase the absolute positioning accuracy of robots (for example, to 0.5 millimeter for PUMA 560).

We use for the calibration the expression (11.67a) in the form

$$\Delta\mathbf{T} = \mathbf{S}_1 \Delta\mathbf{q} + \boldsymbol{\varepsilon} \tag{11.79}$$

where $\boldsymbol{\varepsilon}$ represents errors of measurement. The calibration procedure consists of placing the end-effector in various positions and comparing the measured results with the theoretical (Wu, 1984; Hayati and Roston, 1986). Using equation (11.79) we then determine for each measurement the value $\Delta\mathbf{q}$ representing the required errors of kinematic parameters using, for example, the least squares method. The matrix $\mathbf{S}_1$ is computed based on the nominal kinematic parameters. For equation (11.79) to hold, it is necessary that small variations in the position and orientation can be modelled by small variations of the kinematic parameters. This assumption is violated if two consecutive joints have parallel or near parallel joint axes and Denavit-Hartenberg's parameters from Section 5.11.1 are used. For parallel joint axes these are $s_i \neq 0, a_i \neq 0$; for small skew misalignment of these axes they intersect and the parameters are $s_i \rightarrow \infty, a_i \rightarrow 0$. Therefore in such cases a new set of link parameters would be defined or matrix $\mathbf{A}_i$ from Section 5.11.1 would be replaced by the transformation matrix

$\mathbf{D}_i = \mathbf{A}_i \mathbf{T}_{Z5}(\gamma_i)$ with skew rotation γ_i (Toogood and Warkentin, 1986). The greatest influence on the resultant accuracy comes from errors of twist α and skew γ. Considering these only, a straightforward simple method of measurement (Toogood and Warkentin, 1986) may be used to determine them.

Before calibration, we have been able to determine the inverse kinematic solution analytically, using a particular geometric robot configuration (see Section 5.13 the case of so-called simple robots). However, we find that, after correcting the kinematic parameters by calibration, such an analytical solution no longer exists; this is because a particular geometric robot configuration is no longer valid bacause of kinematic errors. In such a case, the solution can be found by iterations from the initial estimate; this is determined by analytical inversion for a simple robot corresponding to the given robot after neglecting some geometric parameter so that an analytical inverse solution can be found (Hayati and Roston, 1986). A robot with intersecting axes of partial motions of the wrist is an example of a simple robot, whereas the respective wrist axes of a non-simple robot do not intersect.

Let the kinematic equations of a given non-simple robot be $\mathbf{T} = \mathbf{f}(\mathbf{q})$, where $\mathbf{q}$ is the vector of six joint variables and $\mathbf{T}$ the transformation matrix of the robot end-effector. We are aiming at obtaining $\mathbf{f}^{-1}(\mathbf{T})$. We assign the given robot a simple robot for which $\mathbf{T} = \mathbf{f}_s(\mathbf{q}_s)$ and $\mathbf{f}_s^{-1}(\mathbf{T})$ can be determined analytically. The computation procedure is as follows. For the given position $\mathbf{T}$ we determine $\mathbf{q}_s$ for a simple robot

$$\mathbf{q}_s = \mathbf{f}_s^{-1}(\mathbf{T}) \tag{11.80}$$

We determine the position $\mathbf{T}_s$ which the given non-simple robot would have for these joint variables

$$\mathbf{T}_s = \mathbf{f}(\mathbf{q}_s) \tag{11.81}$$

Assuming the validity of linear theory of accuracy, we determine the correction matrix $\mathbf{C}$ between position $\mathbf{T}$ and $\mathbf{T}_s$ according to (11.15) and (11.17)

$$\begin{aligned} \mathbf{T}_s &= \mathbf{TC} = \mathbf{T}(\mathbf{E}(1) + \mathbf{H}) \\ \mathbf{C} &= \mathbf{T}^{-1}\mathbf{T}_s = \mathbf{T}^{-1}\mathbf{f}(\mathbf{f}_s^{-1}(\mathbf{T})) \\ \mathbf{H} &= \mathbf{T}^{-1}\mathbf{T}_s - \mathbf{E}(1) = \mathbf{T}^{-1}\mathbf{f}(\mathbf{f}_s^{-1}(\mathbf{T})) - \mathbf{E}(1) \end{aligned} \tag{11.82}$$

According to (11.17) matrix $\mathbf{C}$ contains six independent quantities. Expressing matrix $\mathbf{C}$ as a function of nominal quantities q_j and their corrections $\Delta q_j, (j = 1, \ldots, 6)$, it follows from (11.23) that

$$\mathbf{C} = \mathbf{T}^{-1} \sum_{j=1}^{6} \mathbf{F}_j(\mathbf{q})\mathbf{E}(\Delta q_j)$$

$$\mathbf{H} = \mathbf{T}^{-1} \sum_{j=1}^{6} \mathbf{F}_j(\mathbf{q})\mathbf{E}(\Delta q_j) - \mathbf{E}(1) \tag{11.83}$$

for non-singular cases, we obtain from (11.82), (11.83) and (11.15) a system of equations

$$\mathbf{J}(\mathbf{q})\Delta\mathbf{q} = \lceil \Delta x, \Delta y, \Delta z, \Delta\varphi_x, \Delta\varphi_y, \Delta\varphi_z \rceil \tag{11.84}$$

where $\mathbf{J}$ is the robot Jacobian.

It would be advantageous to assemble the inverse Jacobian symbolically. This can easily be done by differentiating the analytical solution of the inverse kinematic problem (Section 5.13) with respect to time. If, however, as in this case, we do not know an analytical solution, we are not able to assemble the inverse Jacobian symbolically. It is then more computationally efficient to solve the equation (11.84) numerically. After solving it we get

$$\mathbf{q} = \mathbf{q}_s - \Delta\mathbf{q} \tag{11.85}$$

This is the first iteration of the required solution $\mathbf{f}^{-1}(\mathbf{T})$. This is usually good enough for calibrated robots.

The above procedure may also be used for the inverse kinematic solution of general non-simple robots. To the given non-simple robot we attach a simple robot by neglecting some geometric parameters. The procedure described above can then be applied. However, more iterations are then necessary. We find them by repeating the computation starting with (11.81) for $\mathbf{q}_s = \mathbf{q}$ until we reach the required precision measured by $\mathbf{T}_s - \mathbf{T}, \mathbf{C}, \mathbf{H}$ or $\Delta\mathbf{q}$. It this way, an inverse kinematic solution for non-simple robots is generally achievable in real time. However, a new promissing algorithm (Valášek and Kalný, 1992) is $\mathbf{q}_{i+1} = \mathbf{f}_s^{-1}(\mathbf{T} - \mathbf{f}_r(\mathbf{q}_i))$ where $\mathbf{f}_r(\mathbf{q}) = \mathbf{f}(\mathbf{q}) - \mathbf{f}_s(\mathbf{q})$.

11.2 Dynamic Accuracy

An MBS should be designed using a certain model which contains a series of simplifications in relation to reality. These simplifications are made either on purpose in order to enable the synthesis or analysis of the given MBS to be made through its simplified model or as a consequence of production and operational effects which have not been considered. Using the method by which we studied the effect of such simplifications on the MBS properties from the view point of kinematics, we now study their effect from the veiwpoint of dynamics.

The MBS properties studied from the viewpoint of dynamics are kinematic quantities (position, velocity, acceleration) and force quantities, e.g.

reaction forces. The MBS properties (model simplification, errors) which affect its dynamic properties may be divided into parametric and structural. The parametric effects are given by deviation of the MBS model parameters from the real ones. The structural effects are given by deviations of the MBS model structure from the real MBS structure, for example by neglecting the MBS element elasticity, the backlash in a kinematic pair or friction. Parametric effects on an assembled dynamic MBS model may be studied in general using sensitivity analysis (Tomovic and Vukobratovic, 1972). The computation of sensitivity in dynamics is however more difficult than in kinematics. The reason for this is the higher complexity of the description of dynamic MBS behaviour expressed by differential equations.

The term "dynamic sensitivity" designates in general the sensitivity of the goal function ψ with respect to changes in the design parameters $\mathbf{p} = [p_1, p_2, \ldots, p_m]^T$. In most practical cases, the goal function may be used in the form

$$\psi = G(t^1, \mathbf{s}(t^1), \dot{\mathbf{s}}(t^1), \mathbf{p}) + \int_0^{t^1} F(t, \mathbf{s}, \dot{\mathbf{s}}, \ddot{\mathbf{s}}, \lambda, \mathbf{p}) dt \tag{11.86}$$

for the motion of the MBS in the time interval $[0, t^1]$ where the final time t^1 is generally described by the final condition

$$H^1(t^1, \mathbf{s}^1, \dot{\mathbf{s}}^1, \mathbf{p}) = 0 \tag{11.87}$$

where superscripts 1 or 0 denote that the particular expression is evaluated by the final values $t^1, \mathbf{s}^1 = \mathbf{s}(t^1), \ldots$ or by the initial values $t^0 = 0, \mathbf{s}^0 = \mathbf{s}(0), \ldots$ of variables $t, \mathbf{s} = [s_1, s_2, \ldots, s_n]^T$. The dynamic sensitivity is therefore the expression $\psi_p = \delta\psi/\delta\mathbf{p}^T$ which is a matrix of dimension $(1, m)$. The change $\delta\psi$ in the goal function also depends of course on changes $\delta\mathbf{s}, \delta\dot{\mathbf{s}}, \delta\ddot{\mathbf{s}}$ in the coordinates, velocities and accelerations and $\delta\boldsymbol{\lambda}$ of the Lagrange multipliers, which must be excluded. This may be done by the adjoint differential equation method (Haug and Arora, 1979; Haug et al., 1981; Haug et al., 1984; Bestle, 1989, 1994).

We now describe the basic principle of this method. Let the MBS dynamics be described by LEMT (9.68)

$$\mathbf{M}\ddot{\mathbf{s}} = \mathbf{g} + \boldsymbol{\Phi}_s^T \boldsymbol{\lambda}$$

or in state description

$$\begin{aligned} \dot{\mathbf{s}} &= \mathbf{w} \\ \mathbf{M}\dot{\mathbf{w}} &= \mathbf{g} + \boldsymbol{\Phi}_s^T \boldsymbol{\lambda} \end{aligned} \tag{11.88}$$

where $\mathbf{M}(\mathbf{s}, \mathbf{p})$ is the generalized symmetric mass matrix, $\mathbf{g}(t, \mathbf{s}, \dot{\mathbf{s}}, \mathbf{p})$ is the vector of generalized forces and other remaining terms. The coordinates $\mathbf{s}$

are subjected to r holonomic constraints

$$\mathbf{f}(\mathbf{s},\mathbf{p},t)=\mathbf{0} \tag{11.89}$$

with second time derivatives

$$\boldsymbol{\Phi}_s\dot{\mathbf{w}}+(\boldsymbol{\Phi}_s\mathbf{w})_s\mathbf{w}+\boldsymbol{\Phi}_{ts}\mathbf{w}+\boldsymbol{\Phi}_{tt}=\mathbf{0} \tag{11.90}$$

where subscripts denote a particular partial derivative; for example $\boldsymbol{\Phi}_s = \partial\mathbf{f}/\partial\mathbf{s}^T$ is a matrix of dimension (r,n), $\mathbf{t}_p=\partial t/\partial\mathbf{p}^T$ is a matrix of dimension $(1,m)$.

The initial conditions on position and velocity may be specified by functions of design parameters

$$\begin{aligned}\mathbf{C}^0(\mathbf{s}^0,\mathbf{p}) &= \mathbf{0}\\ \dot{\mathbf{C}}^0(\mathbf{s}^0,\mathbf{w}^0,\mathbf{p}) &= \mathbf{0}\end{aligned} \tag{11.91}$$

with non-singular Jacobian matrices. The partial derivative of the goal function (11.86) with respect to design parameters $\mathbf{p}$ is

$$\begin{aligned}\psi_p &= G^1_{t^1}\mathbf{t}^1_p+\mathbf{G}^1_s\mathbf{s}^1_p+\mathbf{G}^1_w\mathbf{w}^1_p+\mathbf{G}^1_p+\\ &+\int_0^{t^1}(\mathbf{F}_s\mathbf{s}_p+\mathbf{F}_w\mathbf{w}_p+\mathbf{F}_{\dot{w}}\dot{\mathbf{w}}_p+\mathbf{F}_\lambda\boldsymbol{\lambda}_p+\mathbf{F}_p)\mathrm{d}t+F^1\mathbf{t}^1_p\end{aligned} \tag{11.92}$$

For the final derivatives of the state variables,

$$\begin{aligned}\frac{\partial\mathbf{s}^1}{\partial\mathbf{p}^T} &= \mathbf{w}^1\mathbf{t}^1_p+\mathbf{s}^1_p\\ \frac{\partial\mathbf{w}^1}{\partial\mathbf{p}^T} &= \dot{\mathbf{w}}^1\mathbf{t}^1_p+\mathbf{w}^1_p\end{aligned} \tag{11.93}$$

where the first terms describe the changes due to the change of the final time and the second terms due to the change of the state trajectory. Hence

$$\begin{aligned}\psi_p &= (G^1_{t^1}+\mathbf{G}^1_s\mathbf{w}^1+\mathbf{G}^1_w\dot{\mathbf{w}}^1+F^1)\mathbf{t}^1_p+\mathbf{G}^1_s\mathbf{s}^1_p+\mathbf{G}^1_w\mathbf{w}^1_p+\mathbf{G}^1_p+\\ &+\int_0^{t^1}(\mathbf{F}_s\mathbf{s}_p+\mathbf{F}_w\mathbf{w}_p+\mathbf{F}_{\dot{w}}\dot{\mathbf{w}}_p+\mathbf{F}_\lambda\boldsymbol{\lambda}_p+\mathbf{F}_p)\mathrm{d}t\end{aligned} \tag{11.94}$$

Similarly the partial derivative of equations of motion (11.88) and constraints (11.87), (11.90), (11.91) with respect to design parameters $\mathbf{p}$ is

$$\dot{\mathbf{s}}_p-\mathbf{w}_p=\mathbf{0} \tag{11.95}$$

$$\begin{aligned}\mathbf{M}\dot{\mathbf{w}}_p &+ (\mathbf{M}\dot{\mathbf{w}})_s\mathbf{s}_p + \mathbf{M}_p\dot{\mathbf{w}} - \mathbf{g}_s\mathbf{s}_p - \mathbf{g}_w\mathbf{w}_p - \mathbf{g}_p - \mathbf{\Phi}_s^T\boldsymbol{\lambda}_p - \\ &- (\mathbf{\Phi}_s^T\boldsymbol{\lambda})_s\mathbf{s}_p - \mathbf{\Phi}_{sp}^T\boldsymbol{\lambda} = \mathbf{0}\end{aligned} \tag{11.96}$$

$$\begin{aligned}\mathbf{H}_p^1 = H_t^1\mathbf{t}_p^1 + \mathbf{H}_s^1\mathbf{s}_p^1 + \mathbf{H}_w^1\mathbf{w}_p + \mathbf{H}_p^1 &= (H_t^1 + \mathbf{H}_s^1\mathbf{w}^1 + \mathbf{H}_w^1\dot{\mathbf{w}}^1)\mathbf{t}_p^1 + \\ &+ \mathbf{H}_s^1\mathbf{s}_p^1 + \mathbf{H}_w^1\mathbf{w}_p^1 + \mathbf{H}_p^1 = \mathbf{0}\end{aligned} \tag{11.97}$$

$$\begin{aligned}\mathbf{\Phi}_s\dot{\mathbf{w}}_p &+ (\mathbf{\Phi}_s\dot{\mathbf{w}})_s\mathbf{s}_p + (\mathbf{\Phi}_s\mathbf{w})_s\mathbf{w}_p + (\mathbf{\Phi}_s\mathbf{w}_p)_s\mathbf{w} + (\mathbf{\Phi}_{sp}\mathbf{w})_s\mathbf{w} + \\ &+ ((\mathbf{\Phi}_s\mathbf{w})_s\mathbf{w})_s\mathbf{s}_p + \mathbf{\Phi}_{sp}\dot{\mathbf{w}} + (\mathbf{\Phi}_{ts}\mathbf{w})_s\mathbf{s}_p + \mathbf{\Phi}_{tsp}\mathbf{w} + \\ &+ \mathbf{\Phi}_{ts}\mathbf{w}_p + \mathbf{\Phi}_{tts}\mathbf{s}_p + \mathbf{\Phi}_{ttp} = \mathbf{0}\end{aligned} \tag{11.98}$$

$$\mathbf{C}_s^0\mathbf{s}_p^0 + \mathbf{C}_p^0 = \mathbf{0} \tag{11.99}$$

$$\dot{\mathbf{C}}_s^0\mathbf{s}_p^0 + \dot{\mathbf{C}}_w^0\mathbf{w}_p^0 + \dot{\mathbf{C}}_p^0 = \mathbf{0} \tag{11.100}$$

We introduce adjoint variables $\boldsymbol{\mu}, \boldsymbol{\nu}, \boldsymbol{\xi}$ in the number of coordinates $\mathbf{s}$, τ^1 in the number of constraint H^1, $\boldsymbol{\gamma}$ in the number of constraints $\mathbf{f}$ and $\boldsymbol{\zeta}^0, \boldsymbol{\eta}^0$ in the number of constraints $\mathbf{C}^0, \dot{\mathbf{C}}^0$. We multiply the equation (11.95) by $\boldsymbol{\mu}$, the equation (11.96) by $\boldsymbol{\nu}$ and by $\boldsymbol{\xi}$ and the equation (11.98) by $\boldsymbol{\gamma}$, and then integrate them from 0 to t^1. The equations multiplied by $\boldsymbol{\mu}$ and $\boldsymbol{\nu}$ are modified by integration per partes. Hence

$$\boldsymbol{\mu}^{1T}\mathbf{s}_p^1 - \boldsymbol{\mu}^{0T}\mathbf{s}_p^0 - \int_0^{t^1}(\dot{\boldsymbol{\mu}}^T\mathbf{s}_p + \boldsymbol{\mu}^T\mathbf{w}_p)\mathrm{d}t = 0$$

$$\begin{aligned}&\boldsymbol{\nu}^{1T}\mathbf{M}^1\mathbf{w}_p^1 - \boldsymbol{\nu}^{0T}\mathbf{M}^0\mathbf{w}_p^0 - \int_0^{t^1}\left(\frac{\mathrm{d}}{\mathrm{d}t}(\mathbf{M}\boldsymbol{\nu})^T\mathbf{w}_p - \boldsymbol{\nu}^T(\mathbf{M}\dot{\mathbf{w}})_s\mathbf{s}_p - \right.\\ &-\boldsymbol{\nu}^T\mathbf{M}_p\dot{\mathbf{w}} + \boldsymbol{\nu}^T\mathbf{g}_s\mathbf{s}_p + \boldsymbol{\nu}^T\mathbf{g}_w\mathbf{w}_p + \boldsymbol{\nu}^T\mathbf{g}_p + \boldsymbol{\nu}^T\mathbf{\Phi}_s^T\boldsymbol{\lambda}_p + \\ &\left.+\boldsymbol{\nu}^T(\mathbf{\Phi}_s^T\boldsymbol{\lambda})_s\mathbf{s}_p + \boldsymbol{\nu}^T\mathbf{\Phi}_{sp}^T\boldsymbol{\lambda}\right)\mathrm{d}t = 0\end{aligned}$$

$$\begin{aligned}&\int_0^{t^1}(\boldsymbol{\xi}^T\mathbf{M}\dot{\mathbf{w}}_p + \boldsymbol{\xi}^T(\mathbf{M}\dot{\mathbf{w}})_s\mathbf{s}_p + \boldsymbol{\xi}^T\mathbf{M}_p\dot{\mathbf{w}} - \boldsymbol{\xi}^T\mathbf{g}_s\mathbf{s}_p - \boldsymbol{\xi}^T\mathbf{g}_w\mathbf{w}_p - \\ &-\boldsymbol{\xi}^T\mathbf{g}_p - \boldsymbol{\xi}^T\mathbf{\Phi}_s^T\boldsymbol{\lambda}_p - \boldsymbol{\xi}^T(\mathbf{\Phi}_s^T\boldsymbol{\lambda})_s\mathbf{s}_p - \boldsymbol{\xi}^T\mathbf{\Phi}_{sp}^T\boldsymbol{\lambda})\mathrm{d}t = 0\end{aligned}$$

$$\begin{aligned}&\int_0^{t^1}(\boldsymbol{\gamma}^T\mathbf{\Phi}_s\dot{\mathbf{w}}_p + \boldsymbol{\gamma}^T(\mathbf{\Phi}_s\dot{\mathbf{w}})_s\mathbf{s}_p + \boldsymbol{\gamma}^T(\mathbf{\Phi}_s\mathbf{w})_s\mathbf{w}_p + \\ &+\boldsymbol{\gamma}^T(\mathbf{\Phi}_s\mathbf{w}_p)_s\mathbf{w} + \boldsymbol{\gamma}^T(\mathbf{\Phi}_{sp}\mathbf{w})_s\mathbf{w} + \boldsymbol{\gamma}^T((\mathbf{\Phi}_s\mathbf{w})_s\mathbf{w})_s\mathbf{s}_p + \\ &+\boldsymbol{\gamma}^T\mathbf{\Phi}_{sp}\dot{\mathbf{w}} + \boldsymbol{\gamma}^T(\mathbf{\Phi}_{ts}\mathbf{w})_s\mathbf{s}_p + \boldsymbol{\gamma}^T\mathbf{\Phi}_{tsp}\mathbf{w} + \boldsymbol{\gamma}^T\mathbf{\Phi}_{ts}\mathbf{w}_p + \\ &+\boldsymbol{\gamma}^T\mathbf{\Phi}_{tts}\mathbf{s}_p + \boldsymbol{\gamma}^T\mathbf{\Phi}_{ttp})\mathrm{d}t = \mathbf{0}\end{aligned} \tag{11.101}$$

The idea of using the state description (11.88) of equations of motion instead of (9.68) and of using the partial derivative (11.96) of equations of motion twice with two different adjoint variables $\boldsymbol{\nu}$ and $\boldsymbol{\xi}$ and with two different modifications originates in (Bestle, 1989) and enables us to solve the serious problem of tractable formulation of the adjoint variable method. Further, we multiply (11.97) by τ^1, (11.99) by $\boldsymbol{\zeta}^0$ and (11.100) by $\boldsymbol{\eta}^0$

$$\tau^1(H_t^1 + \mathbf{H}_s^1\mathbf{w}^1 + \mathbf{H}_w^1\dot{\mathbf{w}}^1)\mathbf{t}_p^1 + \tau^1\mathbf{H}_s^1\mathbf{s}_p + \tau^1\mathbf{H}_w^1\mathbf{w}_p + \tau^1\mathbf{H}_p^1 = \mathbf{0}$$

$$\begin{aligned} \boldsymbol{\zeta}^{0T}\mathbf{C}_s^0\mathbf{s}_p^0 + \boldsymbol{\zeta}^{0T}\mathbf{C}_p^0 &= \mathbf{0} \\ \boldsymbol{\eta}^{0T}\dot{\mathbf{C}}_s^0\mathbf{s}_p^0 + \boldsymbol{\eta}^{0T}\dot{\mathbf{C}}_w^0\mathbf{w}_p^0 + \boldsymbol{\eta}^{0T}\dot{\mathbf{C}}_p^0 &= \mathbf{0} \end{aligned} \tag{11.102}$$

All of these identities hold for arbitrary adjoint variables ($\boldsymbol{\mu}(t), \boldsymbol{\nu}(t), \boldsymbol{\xi}(t)$, $\boldsymbol{\gamma}(t)$ as a function of time, $\tau^1, \boldsymbol{\zeta}^0, \boldsymbol{\eta}^0$ as constants) and arbitrary parameters. The principle of adjoint variable methods is to select the arbitrary adjoint variables in such a way that terms involving derivatives of coordinates and Lagrange multipliers with respect to design parameters in (11.94) may be explicitly written in terms of quantities which are computed together with the MBS dynamic analysis.

The identities (11.101) to (11.102) are summed and substracted from the six right-hand sides of (11.94). Because the identities (11.101) to (11.102) are equal to zero, the value of ψ_p is not changed. If the coefficients of particular derivatives of coordinates and Lagrange multipliers in the sum of (11.101) to (11.102) are equated to corresponding coefficients in (11.94) by the suitable choice of adjoint variables, then ψ_p is exactly equal to the rest of terms in (11.101), (11.102) and (11.94), which do not explicitly depend on the particular derivatives of coordinates and Lagrange multipliers. Hence we obtain for the coefficient by t_p^1

$$\tau^1(H_t^1 + \mathbf{H}_s^1\mathbf{w}^1 + \mathbf{H}_w^1\dot{\mathbf{w}}^1) = G_{t^1}^1 + \mathbf{G}_s^1\mathbf{w}^1 + \mathbf{G}_w^1\dot{\mathbf{w}}^1 + F^1 \tag{11.103}$$

Similarly for the coefficient by $\mathbf{s}_p^1, \mathbf{w}_p^1$

$$\tau^1\mathbf{H}_s^1 + \boldsymbol{\mu}^{1T} = \mathbf{G}_s^1 \tag{11.104}$$

$$\tau^1\mathbf{H}_w^1 + \boldsymbol{\nu}^{1T}\mathbf{M}^1 = \mathbf{G}_w^1 \tag{11.105}$$

for the coefficient by $\mathbf{s}_p^0, \mathbf{w}_p^0$

$$-\boldsymbol{\mu}^{0T} + \boldsymbol{\zeta}^{0T}\mathbf{C}_s^0 + \boldsymbol{\eta}^{0T}\dot{\mathbf{C}}_s^0 = \mathbf{0} \tag{11.106}$$

$$-\boldsymbol{\nu}^{0T}\mathbf{M}^0 + \boldsymbol{\eta}^{0T}\dot{\mathbf{C}}_w^0 = \mathbf{0} \tag{11.107}$$

for the coefficient by $\mathbf{s}_p, \mathbf{w}_p, \dot{\mathbf{w}}_p$

$$\begin{aligned} &-\dot{\boldsymbol{\mu}}^T - \boldsymbol{\nu}^T\left(-(\mathbf{M}\dot{\mathbf{w}})_s + \mathbf{g}_s + (\boldsymbol{\Phi}_s^T\boldsymbol{\lambda})_s\right) - \boldsymbol{\xi}^T\left(-(\mathbf{M}\dot{\mathbf{w}})_s + \mathbf{g}_s + (\boldsymbol{\Phi}_s^T\boldsymbol{\lambda})_s\right) + \\ &+ \boldsymbol{\gamma}^T\left((\boldsymbol{\Phi}_s\dot{\mathbf{w}})_s + ((\boldsymbol{\Phi}_s\mathbf{w})_s\mathbf{w})_s + (\boldsymbol{\Phi}_{ts}\mathbf{w})_s + \boldsymbol{\Phi}_{tts}\right) = \mathbf{F}_s \end{aligned} \tag{11.108}$$

$$-\boldsymbol{\mu}^T - \frac{\mathrm{d}}{\mathrm{dt}}(\mathbf{M}\boldsymbol{\nu})^T - \boldsymbol{\nu}^T \mathrm{g}_w - \boldsymbol{\xi}^T \mathrm{g}_w + \gamma^T((\boldsymbol{\Phi}_s \mathbf{w})_s + \boldsymbol{\Phi}_{ss}^T \mathbf{w} + \boldsymbol{\Phi}_{ts}) = \mathbf{F}_w \tag{11.109}$$

$$\boldsymbol{\xi}^T \mathbf{M} + \gamma^T \boldsymbol{\Phi}_s = \mathbf{F}_{\dot{w}} \tag{11.110}$$

and for coefficient by $\boldsymbol{\lambda}_p$

$$-\boldsymbol{\nu}^T \boldsymbol{\Phi}_s^T - \boldsymbol{\xi}^T \boldsymbol{\Phi}_s^T = \mathbf{F}_\lambda \tag{11.111}$$

The method of computation is as follows. The motion of the MBS is solved from (11.88), (11.89), (11.91) and (11.87) as in Chapter 9. We store the behaviour of all the variables. From equation (11.103) we calculate $\tau^1 = \tau(t^1)$

$$\tau^1 = \frac{G_{t^1}^1 + \mathbf{G}_s^1 \mathbf{w}^1 + \mathbf{G}_w^1 \dot{\mathbf{w}}^1 + F^1}{H_t^1 + \mathbf{H}_s^1 \mathbf{w}^1 + \mathbf{H}_w^1 \dot{\mathbf{w}}^1} \tag{11.112}$$

from equation (11.104) and (11.112) we calculate $\boldsymbol{\mu}^1 = \boldsymbol{\mu}(t^1)$

$$\boldsymbol{\mu}^1 = \mathbf{G}_s^{1T} - \tau^1 \mathbf{H}_s^{1T} \tag{11.113}$$

from equation (11.105) and (11.112) we calculate $\boldsymbol{\nu}^1 = \boldsymbol{\nu}(t^1)$

$$\mathbf{M}^1 \boldsymbol{\nu}^1 = \mathbf{G}_w^{1T} - \tau^1 \mathbf{H}_w^{1T} \tag{11.114}$$

and from equation (11.108) to (11.111) we assemble the equations

$$\begin{bmatrix} \mathbf{M} & \boldsymbol{\Phi}_s^T \\ \boldsymbol{\Phi}_s & \mathbf{0} \end{bmatrix} \begin{bmatrix} \boldsymbol{\xi} \\ \gamma \end{bmatrix} = \begin{bmatrix} \mathbf{F}_{\dot{w}}^T \\ -\mathbf{F}_\lambda^T - \boldsymbol{\Phi}_s \boldsymbol{\nu} \end{bmatrix} \tag{11.115}$$

$$\begin{aligned} \dot{\boldsymbol{\mu}} &= -\left(-(\mathbf{M}\dot{\mathbf{w}})_s + \mathrm{g}_s + (\boldsymbol{\Phi}_s^T \boldsymbol{\lambda})_s\right)^T (\boldsymbol{\xi} + \boldsymbol{\nu}) + ((\boldsymbol{\Phi}_s \dot{\mathbf{w}})_s + \\ &+ \left((\boldsymbol{\Phi}_s \mathbf{w})_s \mathbf{w})_s + (\boldsymbol{\Phi}_{ts} \mathbf{w})_s + \boldsymbol{\Phi}_{tts}\right)^T \gamma - \mathbf{F}_s^T \end{aligned} \tag{11.116}$$

$$\begin{aligned} \mathbf{M}\dot{\boldsymbol{\nu}} &= -\boldsymbol{\mu} - (\mathbf{M}_s \mathbf{w} + \mathbf{M}_t)^T \boldsymbol{\nu} - \mathrm{g}_w^T (\boldsymbol{\xi} + \boldsymbol{\nu}) + \\ &+ \left((\boldsymbol{\Phi}_s \mathbf{w})_s + \boldsymbol{\Phi}_{ss}^T \mathbf{w} + \boldsymbol{\Phi}_{ts}\right)^T \gamma - \mathbf{F}_w^T \end{aligned} \tag{11.117}$$

The equations (11.116) and (11.117) are backward integrated from t^1 to 0 and the linear algebraic equations (11.115) are simultaneously solved. We obtain $\boldsymbol{\mu}(t), \boldsymbol{\nu}(t), \boldsymbol{\xi}(t)$ and $\gamma(t)$. Then, from (11.106) a (11.107), we determine $\boldsymbol{\eta}^0, \boldsymbol{\zeta}^0$

$$\dot{\mathbf{C}}_w^{0T} \boldsymbol{\eta}^0 = \mathbf{M}^0 \boldsymbol{\nu}^0 \tag{11.118}$$

$$\mathbf{C}_s^{0T} \boldsymbol{\zeta}^0 = \boldsymbol{\mu}^0 - \dot{\mathbf{C}}_s^{0T} \boldsymbol{\eta}^0 \tag{11.119}$$

ψ_p is then determined from (11.94) using (11.101) to (11.111)

$$\begin{aligned}\psi_p = \; & \mathbf{G}_p^1 - \tau^1 \mathbf{H}_p^1 - \boldsymbol{\zeta}^{0T} \mathbf{C}_p^0 - \boldsymbol{\eta}^{0T} \dot{\mathbf{C}}_p^0 + \\ + & \int_0^{t^1} \Big((\boldsymbol{\nu}^T + \boldsymbol{\xi}^T)(-\mathbf{M}_p \dot{\mathbf{w}} + \mathbf{g}_p + \boldsymbol{\Phi}_{sp}^T \boldsymbol{\lambda}) + \mathbf{F}_p - \\ - & \gamma^T ((\boldsymbol{\Phi}_{sp} \mathbf{w})_s \mathbf{w} + \boldsymbol{\Phi}_{sp} \dot{\mathbf{w}} + \boldsymbol{\Phi}_{tsp} \mathbf{w} + \boldsymbol{\Phi}_{ttp}) \Big) \mathrm{d}t \qquad (11.120)\end{aligned}$$

The advantage of this procedure is that adjoint equations (11.112) to (11.119) do not depend on design parameter derivatives. Therefore the adjoint variables are determined for all derivatives (11.120) only once. This is advantageous in the case of many parameters. However, even using this method, we must compute the partial derivatives of equations of motion with respect to both coordinates and parameters. It is especially easy only if the equations of motion are formulated by parts of the MBS, i.e. using the compartment or free body method. The proper equations of motion are created by large but simple expressions. This approach has been realized by the DADS system for which further generalization has been derived (Haug et al., 1984).

The structural effects may however be studied only by comparing the properties of a model containing a certain MBS structural property with a model not containing it. Because of the difficulty of assembling the MBS dynamic model for real problems, it is usually possible to study both kinds of effects only by comparing MBS ideal model properties with MBS real model properties. Thus, in principle, the problem of dynamic accuracy is converted to the problem of MBS dynamic analysis.

11.3 Applications

11.3.1 Kinematic Accuracy of a Robot

Let us solve the analytical and synthetic problems of the kinematic accuracy for the robot manipulator for which the kinematic schema is shown in Figure 11.2 (Valášek and Stejskal, 1987b). This is a hydraulically driven robot containing two loops. Its kinematic accuracy will be studied for the motion of the gripper along the straight line AB with the same gripper orientation as that of the basic system axis Oxyz during the whole motion. We will use the linear model of the kinematic accuracy, and we will make use of the relation between the analysis of the kinematic accuracy and the kinematic analysis which has been shown in the previous sections. If we can solve the problem of the kinematic analysis of a spatial mechanism up to velocities, we can also easily solve the analysis of kinematic accuracy of

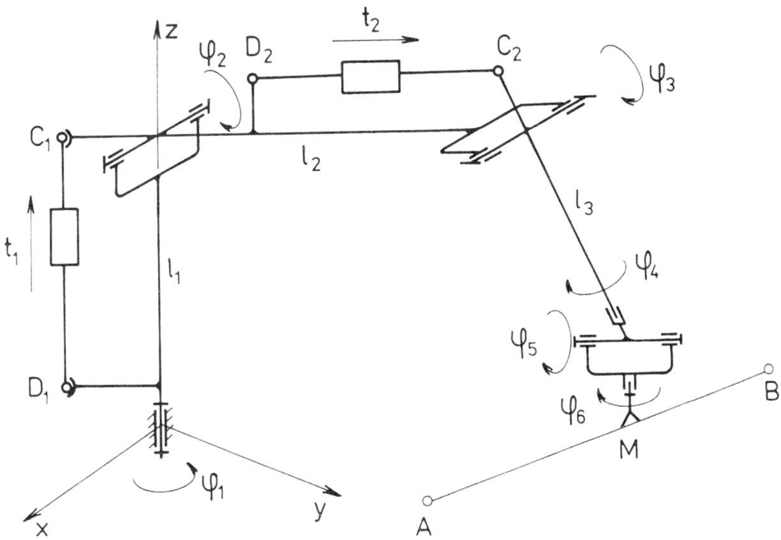

Figure 11.2. Kinematic accuracy of hydraulically driven robot.

position according to the linear theory. Further, if we can solve the problem of the kinematic analysis up to accelerations, we are able to obtain within this solution all the expressions needed for analysis of the kinematic accuracy of velocities. (The kinematic analysis is at the same time the exact model of accuracy, but we cannot make use of its results for the estimation of the tolerance spaces according to Section 11.1.4).

Let us start with the analysis of the kinematic accuracy of position. Let us assume that the tolerance spaces of the coordinates of the driven kinematic pairs

$$\begin{array}{lll} |\bar{\varphi}_1 - \varphi_1| \leq \Delta_1, & |\bar{t}_1 - t_1| \leq \Delta_2, & |\bar{t}_2 - t_2| \leq \Delta_3, \\ |\bar{\varphi}_4 - \varphi_4| \leq \Delta_4, & |\bar{\varphi}_5 - \varphi_5| \leq \Delta_5, & |\bar{\varphi}_6 - \varphi_6| \leq \Delta_6 \end{array}$$

are known. Our task is then to study the tolerance spaces of the position and orientation of the gripper. These tolerance spaces follow from (a) and a matrix **S** in (11.68). However, we must first know the values of all coordinates for the position in which we analyze the kinematic accuracy. System OKAM2 (see Section 10.2) can accomplish this task. We again find the elements of the **S** matrix which are necessary for the solution of the kinematic accuracy of position according to the Section 11.1.4 by means of system OKAM2. We alternate the input information for system OKAM2 in such a manner that, in all positions of interest on the trajectory AB, we successively substitute the unit velocity for independent parameter s_j (this means that in (11.67a) we successively substitute for $\Delta q_j = 1$).

```
NAME:           KINEMATIC ACCURACY OF A ROBOT
CONSTRAINTS:    (TZ3(XL1)*TZ5(FI2)*R3C1=TZ1(XD1)
                *TZ2(YD1)*TZ3(ZD1)*TZ4(FI7)*TZ5(FI8)
                *TZ3(T1)*N)
                (TZ3(ZD2)*TZ1(XD2)*TZ2(YD2)*TZ4(FI9)
                *TZ5(FI10)*TZ3(T2)*N=TZ3(XL3)*
                TZ5(FI3)*R4C2)
MOTIONS:        R1M=(TZ6(FI1)*TZ3(XL1)*TZ5(FI2)
                *TZ3(XL2)*TZ5(FI3)*TZ3(XL3)*TZ6(FI4)
                *TZ5(FI5)*TZ6(FI6)*R6M)
SUBSTITUTION:   R3C1=((XC1) (YC1) (ZC1) (1.0))
                R4C2=((XC2) (YC2) (ZC2) (1.0)
                R6M=((XM) (YM) (ZM) (1.0)
                N=((0.0) (0.0) (0.0) (1.0))
INDEPENDENT:    FI1 T1 T2 FI4 FI5 FI6
DEPENDENT:      FI2 FI3 FI7 FI8 FI9 FI10
DRIVES:         FI1=(FI11+V1*C)
                T1=(T10+V2*C)
                T2=(T20+V3*C)
                FI4=(FI40+V4*C)
                FI5=(FI50+V5*C)
                FI6=(FI60+V6*C)
SOLUTION:       C=(C0 CF DC)
PRINT:          C=(C0 CF TC)
MODE:           VELOCITY
INPUT:          XL1 XL2 XL3 XM YM ZM XD1 YD1 ZD1
                XC1 YC1 ZC1 XD2 YD2 ZD2 XC2 YC2 ZC2 T10
                T20 FI11 FI20 FI30 FI40 FI50 FI60 FI70 FI80
                FI90 FI100 V1 V2 V3 V4 V5 V6 C0 CF DC TC
ESTIMATION:     FI2=FI20 FI3=FI30 FI7=FI70 FI8=FI80
                FI9=FI90 FI10=FI100
$
```

The program input data statement MOTIONS: specifies the resulting motion of the gripper M, and identifiers Vi define the velocities of independent variables. In the position in which we are interested, we substitute unit vectors in V1, V2 V3 V4 V5 V6 and we obtain the sensitivities (elements of matrix $\mathbf{S}$) instead of the velocities of the point M. Our problem was solved for the following numerical data: XL1 = 727 mm, XL2 = 835 mm, XL3 = 825 mm, XM = 50 mm, YM = 50 mm, ZM = 220 mm, XD1 = −100 mm,

YD1 = 0, ZD1 = 0, XC1 = 0, YC1 = 0, ZC1 = −100 mm, ZD2 = 50 mm, XD2 = −100 mm, YD2 = 0, XC2 = 0, YC2 = 0, ZC2 = −100 mm,

FI11, T10, T20, FI40, FI50, FI60 and the estimations FI20, FI30, FI70, FI80, FI90, FI100 are computed according to the position in which we are interested on the trajectory AB, where $A = [730, 250, 500]$ and $B = [-70, 1250, 600]$, $C0 = CF = DC = 0, TC = 1$. In the initial position the matrices $\mathbf{S}_j$ have the following element values

$$\begin{aligned} \mathbf{S}_{\varphi_1} = [S_{rx,\varphi_1}, S_{ry,\varphi_1}, S_{rz,\varphi_1}] &= [-250.1, 730.4, 0] \\ \mathbf{S}_{t_1} &= [-2.460, -0.7238, -8.717] \\ \mathbf{S}_{t_2} &= [-6.593, -1.939, -0.3002] \\ \mathbf{S}_{\varphi_4} &= [47.23, -40.83, -1.462] \\ \mathbf{S}_{\varphi_5} &= [-211.1, -62.0, 62.05] \\ \mathbf{S}_{\varphi_6} &= [-49.96, 50.04, -0.006455] \end{aligned}$$

Let the tolerance space (11.70) of independent parameters be

$$\Delta_{\varphi_1} = \Delta_{\varphi_4} = \Delta_{\varphi_5} = \Delta_{\varphi_6} = \frac{2\pi}{10000} = 0.000628$$

$$\Delta_{t_1} = \Delta_{t_2} = 0.001$$

According to (11.74) we obtain the lengths of the edges of the tolerance rectangular parallelepiped as

$$\frac{1}{2}K_x = 0.35972 \quad , \quad \frac{1}{2}K_y = 0.557356 \quad , \quad \frac{1}{2}K_z = 0.039886$$

The length of its diagonal as the maximum existing error of position of the gripper is

$$\frac{1}{2}K = \sqrt{(\frac{1}{2}K_x)^2 + (\frac{1}{2}K_y)^2 + (\frac{1}{2}K_z)^2} = 0.66440 \text{ mm}$$

Similar solutions have been obtained in other points of the trajectory AB and the maximum value of this error has been found to be

$$\frac{1}{2}K = 0.96146 \text{ mm}$$

This is the result of the analytical problem of the position accuracy.

Alternatively, let us assume that the maximum allowed accuracy of the position

$$\frac{1}{2}K = 0.01 \text{ mm}$$

From this,

$$\frac{1}{2}K_x = \frac{1}{2}K_y = \frac{1}{2}K_z = \frac{1}{2}K/\sqrt{3} = 0.0057735$$

For the above values of $\mathbf{S}$ we can obtain from (11.76) for the initial position

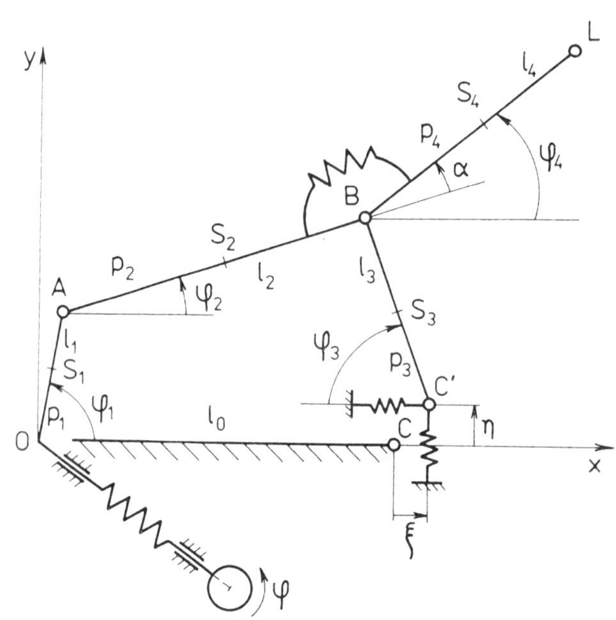

Figure 11.3. Dynamic accuracy of feeding machine.

$$\Delta_{\varphi_1} = 1.32 \times 10^{-6} \,, \qquad \Delta_{t_1} = 1.10 \times 10^{-4} \,, \qquad \Delta_{t_2} = 1.46 \times 10^{-4}$$
$$\Delta_{\varphi_4} = 2.04 \times 10^{-5} \,, \qquad \Delta_{\varphi_5} = 4.56 \times 10^{-6} \,, \qquad \Delta_{\varphi_6} = 1.92 \times 10^{-5}$$

The corresponding solution for the other points of the trajectory AB derives the tolerance space of independent variables. The minimum values of these on the whole trajectory are

$$\Delta_{\varphi_1} = 0.7 \times 10^{-6} \,, \qquad \Delta_{t_1} = 7 \times 10^{-5} \,, \qquad \Delta_{t_2} = 1.5 \times 10^{-4}$$
$$\Delta_{\varphi_4} = 6 \times 10^{-6} \,, \qquad \Delta_{\varphi_5} = 4 \times 10^{-6} \,, \qquad \Delta_{\varphi_6} = 1 \times 10^{-5}$$

This is the result of the synthetic problem of the position accuracy.

11.3.2 Feeding Machine Dynamic Accuracy

In the following problem (Brát and Stejskal, 1979b) we compare the path of the working point L of a feeder according to Figure 11.3 with an ideal path assuming a rigid mechanism, assuming constant angular velocity of the driving electromotor, a driving shaft which is compliant in torsion, and a connecting and bearing C which are compliant in bending.

We choose angles $\varphi_1, \varphi_2, \varphi_3, \varphi_4$ and displacements ξ, η of the bearing C for coordinates. Two constraint equations hold, expressing loop

$OABC'CO$ closure. We thus have a system with four degrees of freedom. We assemble the equations of motion using LEMT according to Section 9.3 in the form

$$\left[\begin{array}{cccccc|cc} A, & Cc\varphi_{12}, & 0, & Dc\varphi_{14}, & 0, & 0 & -l_1 s\varphi_1, & l_1 c\varphi_1 \\ & B, & 0, & Ec\varphi_{42}, & 0, & 0 & -l_2 s\varphi_2, & l_2 c\varphi_2 \\ & & I_3, & 0, & Fs\varphi_3, & Fc\varphi_3 & -l_3 s\varphi_3, & -l_3 c\varphi_3 \\ & & & I_4, & 0, & 0 & 0, & 0 \\ & & & & m_3, & 0 & -1, & 0 \\ & & & & & m_3 & 0, & -1 \\ \hline & & & & & & 0, & 0 \\ & \text{sym.} & & & & & & 0 \end{array}\right] \left[\begin{array}{c} \ddot{\varphi}_1 \\ \ddot{\varphi}_2 \\ \ddot{\varphi}_3 \\ \ddot{\varphi}_4 \\ \ddot{\xi} \\ \ddot{\eta} \\ \hline -\lambda_1 \\ -\lambda_2 \end{array}\right] = \left[\begin{array}{c} b_1 \\ b_2 \\ b_3 \\ b_4 \\ b_5 \\ b_6 \\ \hline b_7 \\ b_8 \end{array}\right]$$

where $\varphi 12 = \varphi_1 - \varphi_2, \varphi_{14} = \varphi_1 - \varphi_4, \varphi_{42} = \varphi_4 - \varphi_2$ and

$$\begin{array}{ll} A = I_1 + (m_2 + m_4) l_1^2, & B = I_{2S} + m_2 p_2^2 + m_4 l_2^2 \\ C = (m_2 p_2 + m_4 l_2) l_1, & D = m_4 l_1 p_4 \\ E = m_4 l_2 p_4, & F = m_3 p_3 \\ I_3 = I_{3S} + m_3 p_3^2, & I_4 = I_{4S} + m_4 p_4^2 \end{array}$$

$$\begin{aligned} b_1 &= C\dot{\varphi}_2(\dot{\varphi}_1 - \dot{\varphi}_2)s(\varphi_1 - \varphi_2) + D\dot{\varphi}_4(\dot{\varphi}_1 - \dot{\varphi}_4)s(\varphi_1 - \varphi_4) \\ &\quad C\dot{\varphi}_1\dot{\varphi}_2 s(\varphi_1 - \varphi_2) - D\dot{\varphi}_1\dot{\varphi}_4 s(\varphi_1 - \varphi_4) + Q_1 \\ b_2 &= C\dot{\varphi}_1(\dot{\varphi}_1 - \dot{\varphi}_2)s(\varphi_1 - \varphi_2) + E\dot{\varphi}_4(\dot{\varphi}_4 - \dot{\varphi}_2)s(\varphi_4 - \varphi_2) + \\ &\quad + C\dot{\varphi}_1\dot{\varphi}_2 s(\varphi_1 - \varphi_2) + E\dot{\varphi}_2\dot{\varphi}_4 s(\varphi_4 - \varphi_2) + Q_2 \\ b_3 &= 0 \\ b_4 &= D\dot{\varphi}_1^2 s(\varphi_1 - \varphi_4) - E\dot{\varphi}_2^2 s(\varphi_4 - \varphi_2) + Q_4 \\ b_5 &= -F\dot{\varphi}_3^2 c\varphi_3 + Q_5 \\ b_6 &= F\dot{\varphi}_3^2 s\varphi_3 + Q_6 \\ b_7 &= l_1\dot{\varphi}_1^2 c\varphi_1 + l_2\dot{\varphi}_2^2 c\varphi_2 + l_3\dot{\varphi}_3^2 c\varphi_3 \\ b_8 &= l_1\dot{\varphi}_1^2 s\varphi_1 + l_2\dot{\varphi}_2^2 s\varphi_2 - l_3\dot{\varphi}_3^2 s\varphi_3 \end{aligned}$$

The generalized forces have been assumed in the form

$$Q_1 = -(k_{11}(\varphi_1 - \omega t) + k_{13}(\varphi_1 - \omega t)^3 + b_{11}(\dot{\varphi}_1 - \omega) + b_{13}(\dot{\varphi}_1 - \omega)^3)$$

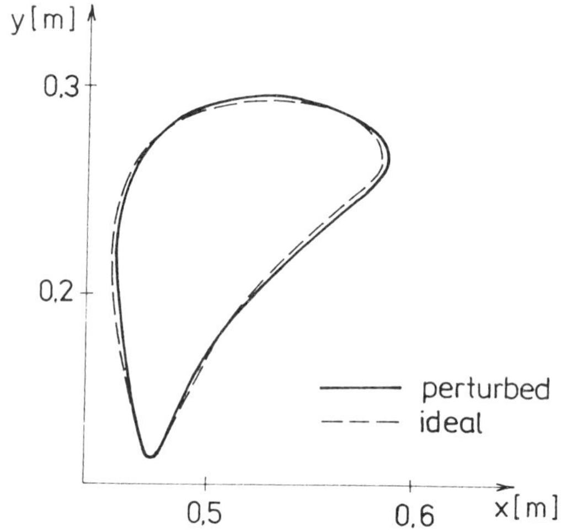

Figure 11.4. Ideal and perturbed trajectory.

$$
\begin{aligned}
Q_2 &= k_{21}(\varphi_4 - \varphi_2 - \alpha) + k_{23}(\varphi_4 - \varphi_2 - \alpha)^3 + b_{21}(\dot{\varphi}_4 - \dot{\varphi}_2) + \\
&\quad + b_{23}(\dot{\varphi}_4 - \dot{\varphi}_2)^3 \\
Q_4 &= -Q_2 \\
Q_5 &= -(k_{31}\xi + b_{31}\dot{\xi}) \\
Q_6 &= -(k_{41}\eta + b_{41}\dot{\eta})
\end{aligned}
$$

The numerical solution has been carried out for values $l_0 = 0.44$ m, $l_1 = 0.08$ m, $l_2 = 0.37$ m, $l_3 = 0.16$ m, $l_4 = 0.2$ m, $\alpha = 15°$, $p_1 = 0.0$ m, $p_2 = 0.18$ m, $p_3 = 0.08$ m, $p_4 = 0.1$ m, $m_1 = 0.15$ kg, $m_2 = 0.6$ kg, $m_3 = 0.25$ kg, $m_4 = 0.3$ kg, $I_1 = 0.008$ kg m^2, $I_{2S} = 0.007$ kg m^2, $I_{3S} = 0.006$ kg m^2, $I_{4S} = 0.001$ kg m^2, $\omega = 100$ rad s^{-1}, $k_{11} = 4000$ N m rad^{-1}, $k_{21} = 50000$ N m rad^{-1}, $k_{41} = k_{31} = 10^7$ N m^{-1}, $k_{13} = k_{23} = 0$, $b_{11} = 5$ N m s rad^{-1}, $b_{21} = 10$ N m s rad^{-1}, $b_{13} = b_{23} = 0$, $b_{31} = b_{41} = 200$ N m^{-1}s and initial conditions $t = 0$, $\varphi_1 = 0$, $\xi = \eta = 0$, $\varphi_4 = \varphi_2 + \alpha$, $\dot{\varphi}_1 = 100$ rad s^{-1}, $\dot{\xi} = \dot{\eta} = 0$, $\dot{\varphi}_4 = \dot{\varphi}_2$.

The computed k_L and ideal k_{Li} paths of point L are presented in Figure 11.4.

Chapter 12

MBS Synthesis and Optimization

The purpose of MBS synthesis is the study of design methods of mechanism and other MBSs which comply with certain functional and technological requirements. These requirements may concern only the position of certain MBS points or elements (e.g. a specific point should move along a line, or a specific body should perform a translational motion); in this case we speak of geometric synthesis. In other cases we have to deal with requirements on velocity and acceleration. The methods of fulfilling these are the topic of kinematic synthesis. Dynamic requirements (balancing, reduction of reaction forces, reduction of run irregularities, required motion behaviour) are solved by dynamic synthesis. In most cases the methods of synthesis have their starting point in idealized MBSs operating with rigid elements and ideal kinematic pairs without friction and backlash. If necessary, it is possible to take the above influences into account, but this results in increased computational complexity. Many multibody synthesis problems are transformed into optimization problems because they cannot be solved at all (i.e. the requirements cannot be fulfilled simultaneously) or because of their easier numerical solvability. On the other hand, any optimization problem can be always treated as a synthesis problem. Therefore we shall not further separately distinguish optimization problems and we shall study only synthesis problems.

Before the designer starts his work on the synthesis, he must have clearly formulated the principal requirements on the designed MBS (see Section 2.1). These usually concern its function. In addition, a series of subsidiary requirements must be taken into account as they are just as important for the future user or operator of the designed device. These might be, for example, requirements on safety, reliability, lifetime, ease of maintenance, etc. The design must also comply with the requirements of the manufacturer, who usually asks for an easy and inexpensive production method from easily accessible materials. From this incomplete outline of criteria, it is obvious that the designer's task is not an easy one and that the result of his work cannot satisfy all requirements. The solution is almost always a compromise.

In the remainder of this chapter we treat the topic of MBS design completely generally. We shall limit ourselves to the criteria which may be judged and satisfied using the laws of mechanics. We shall therefore treat only the kinematic and dynamic synthesis which have — especially at the beginning — much in common.

The first step at each synthesis is the choice of MBS type, where the designer must decide whether he would use for the given purpose a planar or spatial mechanism, a joint or cam mechanism, a strict MBS or an MBS with additional electric, hydraulic or pneumatic elements, a controlled or uncontrolled system; if controlled, would control be achieved in a traditional manner or using microprocessors, etc. The choice of structure, usually expressed by the structure diagram of the designed equipment, then follows. This is the topic of the so-called structure synthesis (Figure 2.1). For the structure analysis, as with the choice of type, the role of the designer is indispensable (see Section 2.3). For the decision making, apart from intuition and one's own experience, structures of a similar type described in literature and other sources are available; expert systems (see Section 2.4.5), databases and special programs generating various structures (e.g. satisfying prescribed mobility and containing various numbers of bodies and kinematic pairs of a specified type) may also be used.

Once the structure has been chosen, the process of kinematic synthesis involves the MBS dimensions being determined (dimension synthesis) in such a way as to satisfy the given requirements. It is possible to automate this activity to a great extent (see Section 2.5). It is exactly the same as the kinematic dimension synthesis of joint mechanisms (linkages), understood in the narrower sense of the term *synthesis of mechanisms*.

In the case of dynamic synthesis, not only the dimensions must be determined but also the mass parameters; these must be distributed in such a way that, for example, the mechanism is balanced, significant elements carry out a uniform motion, the input power is at its minimum, run irregularities are kept as low as possible, etc. Controlled MBS's form a topic in themselves, where the satisfaction of required properties results from active measures taken to influence the system.

In this chapter we shall deal mainly with the dimension kinematic synthesis of a MBS whose structure has been chosen. We distinguish between

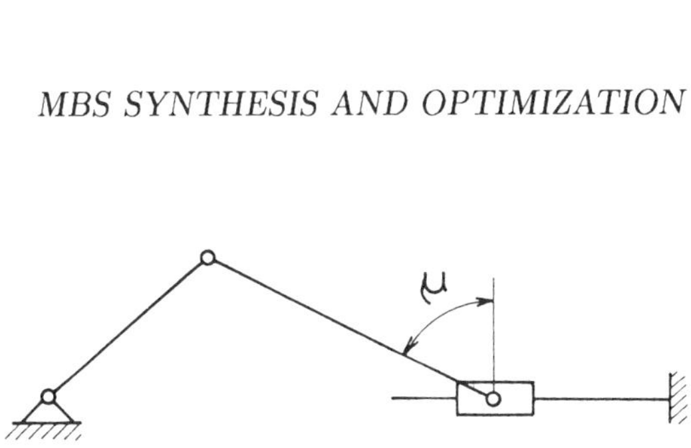

Figure 12.1. Transmission angle.

the synthesis of transmission mechanisms (function generation) and rigid body guidance mechanisms (path generation). We shall not treat here the synthesis of cam mechanisms which forms a separate part of kinematics and dynamics.

Let us stress once more that each synthesis is based on analysis (see Sections 2.1, 2.5), which enables required kinematic quantities to be calculated; for this reason, the largest part of this book is devoted to this subject. From the analysis methods described in Chapter 4, we use the matrix method.

From the introduction to this chapter, it is evident that the designer usually chooses and selects rather than uniquely computes most of the structure parameters. Only some of them are determined from uniquely given requirements. Because choices are possible, a number of constructions may satisfy the basic goal. Some of them, however, have an operational behaviour substantially different from expectations based on a theoretically perfect model. The reason is that universal criteria for quality of transmission of motion and forces are not known (and therefore not taken into account in design). However, partial criteria are known. For example, with planar mechanisms the criterion might be the transmission angle μ defined as the angle of the normals or tangents of the path of the common point of input and transmission elements of the mechanism by its absolute motion with respect to the frame and by its relative motion with respect to the driving element, Figure 12.1. Angle μ should maintain the range 40° to 140°.

An analogy to the transmission angle for spatial mechanisms is the transmission index TI (Sutherland and Roth, 1973), defined by the relation

$$TI = \left| \frac{(p+p')\cos\Theta - r\sin\Theta}{[(p+p')^2 + \varrho'^2]^{\frac{1}{2}}} \right| \tag{12.1}$$

where (Figure 12.2) p is the pitch of the output element instantaneous velocity screw axis ISA_f, p' is the pitch of the instantaneous transmission wrench screw TWS, Θ is the dual angle and ϱ the distance from ISA_f to

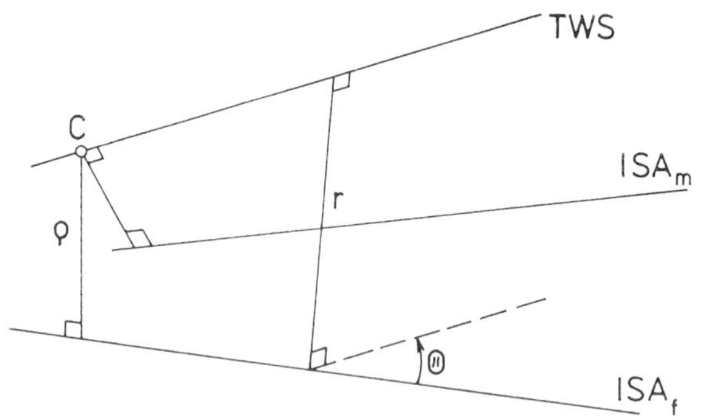

Figure 12.2. Transmission index.

the mechanism instantaneous characteristic point C which is the point of intersection of the shortest line segment connecting TWS and the instantaneous velocity screw axis ISA_m for the moving joint to the output link. The TI has values in the range between 0 (force and motion transmission impossible) and 1 (force and motion transmission at their optimum).

Another quantity characterizing the overall properties of the mechanism is the mechanical error expression denominator (Hartenberg and Denavit, 1964) defined by the relation

$$MED = \frac{\partial f}{\partial \psi} \tag{12.2}$$

where, in the case of a simple mechanism with one degree of freedom, $f(p, \varphi, \psi) = 0$ is the linkage closure equation containing structure parameters p, input coordinate φ and output coordinate ψ. Maximizing the value of MED, we can reduce the error in the position of the output element caused by manufacturing imprecision, dilatational changes of mechanism element lengths, or their plasticity and backlash.

For compound mechanisms, $\det \boldsymbol{\Phi}$ may also be used as a quality criterion, where $\boldsymbol{\Phi}$ is the Jacobian from equation (5.169) for velocity calculation. For the value of $\det \boldsymbol{\Phi}$ tending towards zero the position tends to a singular position, where the solution does not have to exist (see Section 7.2).

The common disadvantage of most criteria is the fact that they are not easily compared for mechanisms with the same purpose but a different structure. Although there exists no universally valid criterion for the quality of motion and force transmission in mechanisms, it may be stated that there is a relationship between the above criteria, namely improving one of them usually improves the rest of them. It is therefore reasonable to include these criteria among the requirements for the synthesised MBS (see Section 2.2).

12.1 Synthesis of Transmission Mechanisms

The problem of synthesis of transmission mechanisms may be formulated as follows:

The dimensions of a mechanism with a given structure are to be determined so that the input-output relation corresponds to the function

$$y = F(x), \qquad x \in < x_b, x_e > \tag{12.3}$$

Assigning independent variables x to the coordinate q of the input element and dependent variables y to coordinate s of the output element of the mechanism with one degree of freedom, and choosing limiting values q_b, q_e, s_b, s_e, we can transform the original function (12.3) using linear relations

$$\frac{s - s_b}{s_e - s_b} = \frac{y - y_b}{y_e - y_b}, \qquad \frac{q - q_b}{q_e - q_b} = \frac{x - x_b}{x_e - x_b} \tag{12.4}$$

into the form

$$s = f(q), \qquad q \in \langle q_b, q_e \rangle \tag{12.5}$$

This describes the required input-output relation.

Choosing the limiting values we choose scales for both variables. More generally, it is possible to substitute the function (12.3) by the function

$$\mathbf{y} = \mathbf{F}(\mathbf{x}), \qquad \mathbf{x} \in \langle \mathbf{x}_b, \mathbf{x}_e \rangle \tag{12.6}$$

with more variables; we can then assign to that function the input-output relation

$$\mathbf{s} = \mathbf{f}(\mathbf{q}), \qquad \mathbf{q} \in \langle \mathbf{q}_b, \mathbf{q}_e \rangle \tag{12.7}$$

This refers to a MBS with more degrees of freedom, the input (driving) elements of which have coordinates $\mathbf{q}$ and the output elements the coordinates $\mathbf{s}$.

The actual synthesis may be performed in two steps. In the first step, the input-output relation for the given structure must be derived in general in the form

$$\mathbf{g}(\mathbf{s}, \mathbf{q}, \boldsymbol{\varrho}) = \mathbf{0} \tag{12.8}$$

where ϱ_i, $i = 1, 2, \ldots, k$ are longitudinal and angular dimensions. By deriving relation (12.8), the matrix method of MBS analysis described in Chapter 5 demonstrates its advantages.

In the second step, dimensions ϱ must be computed. In general it cannot be expected that, for some ϱ, (12.5) could be precisely satisfied on the whole interval $\langle \mathbf{q}_b, \mathbf{q}_e \rangle$. We rely therefore on the original formulation (12.5) and shall be content with the solution of the following problem instead.

A function $s = f(q)$ is given on interval $q \in \langle q_b, q_e \rangle$ by its discrete values (s_i, q_i), $i = 1, 2, \ldots, m$. The dimensions ϱ of a MBS with a given structure are to be determined so that its input-output function f_g yields the same values as function f in the given discrete points. These discrete points are called precision points. It is often not necessary to assume that all MBS dimensions are unknown. We choose some of them and we take the remaining $n < k$ dimensions for synthesised (design) parameters $p_j, j = 1, 2, \ldots, n$.

The problem formulated above corresponds to the geometric synthesis. In kinematic synthesis, derivatives of coordinates and possibly time as a variable would appear in the arguments of required and generated functions.

The problem of geometric synthesis can be solved by three principal methods:

- the interpolation method (colocation method)
- the correction method
- optimizing methods.

The interpolation method in its basic form can be used in cases where we choose the same number of precision points and design parameters ($m = n$). The procedure is as follows.

We rewrite the input-output relation (12.8) in the form

$$\mathbf{g}(\mathbf{s}, \mathbf{q}, p_1, p_2, ..., p_n) = \mathbf{0} \tag{12.9}$$

After substituting the precision points, we obtain n strongly non-linear equations of the type

$$g_i(s_i, q_i, \mathbf{p}) = 0, \qquad i = 1, 2, \ldots, n \tag{12.10}$$

for the calculation of n design parameters $\mathbf{p}$. The direct numerical solution of equations (12.10) usually requires a good initial estimate $\mathbf{p}^0$ which, in more complex cases, can be determined only with difficulty.

More efficient than the direct solution is an approach based on the fact that in most cases it is possible to modify the function (12.9) into the form

$$\mathbf{g}(\mathbf{s}, \mathbf{q}, \mathbf{p}) = \sum_{j=0}^{n} c_j(\mathbf{p}) h_j(\mathbf{s}, \mathbf{q}) \tag{12.11}$$

where functions $c_j(\mathbf{p})$ containing all non-linearity aspects of the problem depend only on unknown parameters $\mathbf{p}$, and h_j are linearly independent functions of coordinates $\mathbf{s}, \mathbf{q}$. We choose $c_0 = 1$. Substituting precision points into (12.11), we obtain a system of n linear equations for functions $c_j, j = 1, \ldots, n$. As a precondition for a successful solution, we must convert the function (12.8) into the form(12.10), and non-linear functions c must be explicitly solvable for $\mathbf{p}$. In several cases this approach is efficient (Rao et al., 1973).

Another possibility is the correction method (Brát, 1981). We set out using a mechanism whose dimensions are chosen to generate a function similar to the required relation (12.5). In the process of calculation, the values of structure parameters $\mathbf{p}$ are corrected by iteration aiming at a better satisfaction of (12.7). The precision points are again the criteria. To present the method, we use the results of Chapter 11 and we limit our presentation to closed chains. Using correction matrix $\mathbf{C}$, a chain may be described by equation (11.48). Of course, only the corrections of design parameters which we express by basic motion correction matrices according to (11.21) are relevant here. After multiplying the matrices, neglecting terms containing products of corrections and choosing independent equations, we obtain from (11.48) the matrix equation

$$\mathbf{P}(\mathbf{z}, \mathbf{q}, \mathbf{p})\Delta\mathbf{p} = \mathbf{h}(\mathbf{z}, \mathbf{q}, \mathbf{p}) \tag{12.12}$$

where $\mathbf{P}$ is a $(6, n)$-matrix, $\Delta\mathbf{p}$ is an n-element correction vector and vector $\mathbf{h}$ has dimension 6.

The computation of corrections $\Delta\mathbf{p}$ is an iterative process. It is performed, for example, by solving the position problem in the chosen m precision points with initial mechanism dimensions with known coordinates $q_i, z_i,\ i = 1, 2, \ldots, m$ into (12.12) and getting a system of $6m$ linear equations for corrections $\Delta\mathbf{p}^0$. We solve it by the methods described in Chapter 7. The new dimensions in the point iteration are

$$\mathbf{p}^1 = \mathbf{p}^0 + \Delta\mathbf{p}^0 \tag{12.13}$$

Due to the linearization of equation (11.48) and in the usual cases where $m \neq n$, the required function is not even going to be satisfied at the precision points; we have to carry on iterating until the condition $|\Delta\mathbf{p}| < \varepsilon$ or $|\Delta\mathbf{g}| < \varepsilon, \varepsilon > 0$ is satisfied.

Using the optimization method for the solution of dimension synthesis, we usually set out from the requirements on the function of deviations $\delta(q)$ between the generated function $f_g(q)$ and the required behaviour (12.7). Thus

$$\delta(q) = f_g(q) - f(q) \tag{12.14}$$

These requirements are expressed by the optimization criterion. This is usually the Gauss criterion of the least square of deviations expressed by the requirement

$$\int_{q_b}^{q_e} \delta^2(q)\mathrm{d}q \longrightarrow \min \tag{12.15}$$

which by numerical calculation converts into the form

$$\sum_{i=1}^{m} \delta^2(q_i) \longrightarrow \min \tag{12.16}$$

The same criterion is used for solution by the generalized interpolation correction method when there are more precision points than are allowed by the number of design parameters. The row dimension of matrix $\mathbf{P}$ from (12.12) is then greater than the column dimension, and we solve the equations after having converted them into the form (see also (7.2))

$$\mathbf{P}^T\mathbf{P}\Delta\mathbf{p} = \mathbf{P}^T\mathbf{h} \tag{12.17}$$

The best approximation (minimax) might be another criterion, requiring

$$\max|\delta(q)| \longrightarrow \min, \qquad q \in \langle q_b, q_e\rangle \tag{12.18}$$

It is possible to use several other cost functions (Angeles, 1982). The satisfaction of the basic optimization requirement expressed by the cost function is of course not the only requirement in problems in actual practice. The dimensions sought must for constructional reasons usually lie in a certain interval or must possibly fulfil other requirements. We then have the case of optimizing with subsidiary conditions, whether in the form of equalities or inequalities.

The above methods of solution of simple MBS dimension synthesis can also be applied for compound MBS and MBS with a greater number of degrees of freedom. However, the computational complexity increases (Sohoni and Haug, 1982).

In principle the above methods are also applicable in the case of kinematic synthesis, where requirements are set not only on the input-output function but also on the behaviour of the velocity or even the acceleration of some element. In that case the basic equations also contain derivatives. We obtain these by differentiating equation (12.12) with respect to time. Relations derived in Chapter 11 can be used for this purpose.

When solving a specific problem, the question of distribution of precision points in interval $< q_b, q_e >$ is important. In the attempt to obtain the best possible approximation of the required relation (12.5) in the whole interval, we choose the allocation of precision points according to the so-called Chebyshev polynomials. A Chebyshev polynomial of degree k has the property of having in a closed interval k zero points and $k+1$ local extremes

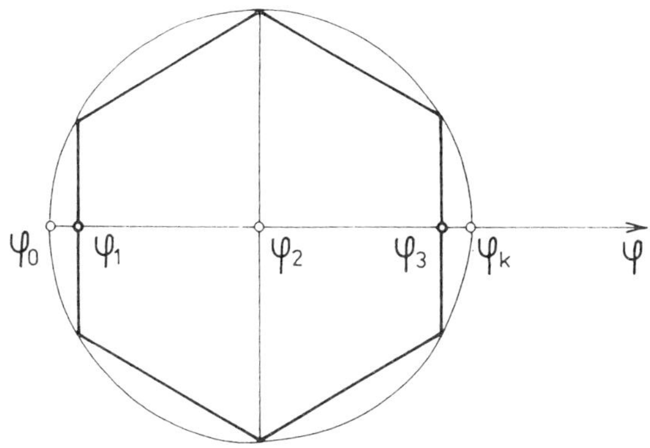

Figure 12.3. Precision points by Chebyshev's polynomial.

with the same absolute value and alternating signs. For the allocation of precision points,

$$q_i = \frac{q_b + q_e}{2} - \frac{q_e - q_b}{2} \cos\frac{p(2i-1)}{2i}, \qquad i = 1, 2, \ldots, n \tag{12.19}$$

The graphical presentation using a regular $2n$-polygon (see Figure 12.3) is more intuitive than the formula. Should the deviation $\delta(q)$ according to (12.14) indeed be a polynomial, the above choice of precision points would yield a uniform approximation of prescribed and actual input-output function. As this is not the case, we get local extremes of varying absolute value. Reallocating a precision point placed between extreme deviations of different absolute values in the direction of the greater deviation is expected to lead to balancing of the two absolute values. An iterative process can be used according to the relation

$$q_i^1 = q_i - \varrho(|d_i| - |d_{i+1}|)(\gamma_{i+1} - \gamma_i) \tag{12.20}$$

where q_i^1 is the new position and q_i the initial position of i-th precision point, γ_i is the position of extreme value δ_i and ϱ is a constant dependent on the absolute value of deviations δ_1 to δ_{n+1}.

Let us once more consider the problem of scales introduced in (12.4). The limiting values used, q_b, q_e, s_b, s_e, are a matter of choice. Obviously this choice could to a great extent affect the resultant solution. It is therefore possible for the synthesis solution to recommend methods where, for example, the start point of interval is one of the synthetized parameters.

We have been considering synthesis with discrete and different precision points and we have required the identity of the required and generated input-output relations in those points. The identity of the input-output relations and of their derivatives in just one precision point, e.g.

$$f = f_g, \quad \frac{\partial f}{\partial q} = \frac{\partial f_g}{\partial q}, \ldots, \frac{\partial f^{(n-1)}}{\partial q^{(n-1)}} = \frac{\partial f_g^{(n-1)}}{\partial q^{n-1)}} \tag{12.21}$$

might be another requirement. We then speak of interpolation using n infinitesimally separated positions or of colocation in n infinitesimally separated positions. For the solution of this problem, the methods presented can be used.

12.2 Synthesis of Guiding Mechanisms

For the geometric synthesis of a guidance MBS we set out from the requirement for a certain point to move along a given path or for a certain body to maintain prescribed positions. For kinematic synthesis, requirements on velocity or even acceleration are added. In the same way as with the synthesis of transmission MBS, the given requirements are usually specified for several precision points. Expressing, for example, the required velocity through the function $f(q, \dot{q}, \varrho, t)$, the function generated by the designed mechanism $f_g(q, \dot{q}, \varrho, t)$, the deviation function δ through a relation similar to (12.14), it is evident that from the viewpoint of mathematics there is no difference between the synthesis of transmission and guidance mechanisms. In both cases, the solution methods described in Section 12.1 can be used.

For the sake of completeness we note that simple problems of guidance mechanism synthesis may be solved using graphical methods (Hartenberg and Denavit, 1964).

12.3 Applications

12.3.1 A Planar Guiding Mechanism

Let us show as an illustration the procedure for planar guiding mechanism synthesis. A mechanism is to transfer a container with bottles from one conveyor to another with a prescribed change of orientation, Figure 12.4a. From the mechanics point of view we are dealing with motion of a body from one position to another including prescribed orientation in these positions. The positions are described in the basic system of coordinates Oxy by points H_1, H_2 and angles Θ_1, Θ_2.

We choose a six-member planar mechanism according to Figure 12.4b for fulfillment of these requirements. This choice is based on experience (it is a synthesis of structure according to Chapter 2). The synthesized

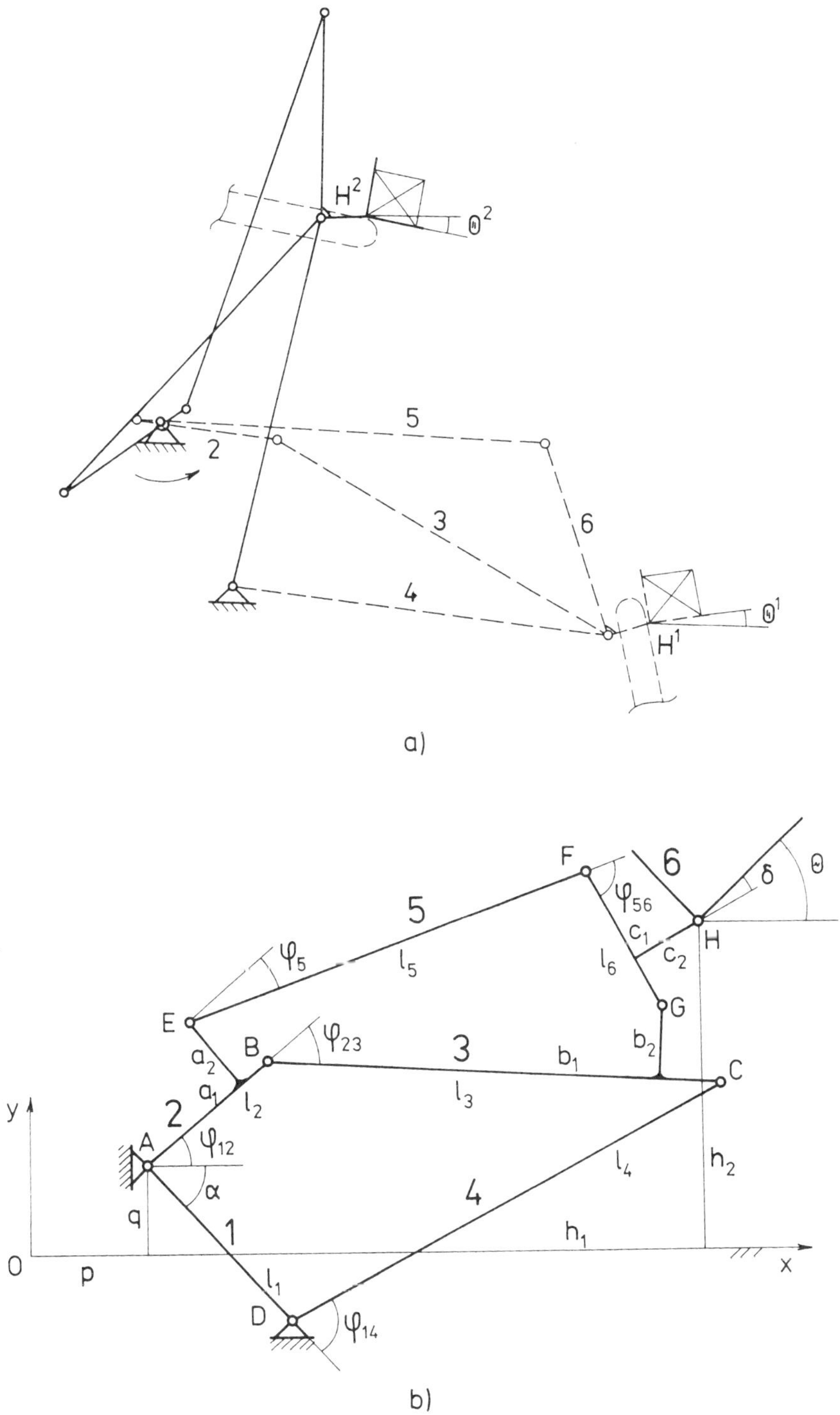

Figure 12.4. Guiding mechanism for a container transfer.

mechanism is described by 18 parameters: $l_1, l_2, l_3, l_4, l_5, l_6, p, q, \alpha, a_1, a_2, b_1, b_2, c_1, c_2, \delta, \varphi_{12,1}, \varphi_{12,2}$. Among them there are also angles $\varphi_{12,1}, \varphi_{12,2}$ defining the angle position of crank 2 (and through it all the rest coordinates of the mechanism) in two working positions. The aim of the synthesis is to determine the parameters mentioned.

Let us specify the required positions of the container by transformation matrices

$$\mathbf{T}^r_{1bi} = \begin{bmatrix} c\Theta_i, & -s\Theta_i, & h_{1,i} \\ s\Theta_i, & c\Theta_i, & h_{2,i} \\ 0, & 0, & 1 \end{bmatrix}, \qquad i = 1,2 \tag{a}$$

The superscript r expresses the fact that we are using reduced vectors the dimensions of which are (3,3) and (3,1) respectively (see Section 4.7.2). We drop the superscript r in the following.

The main aim of synthesis is characterized by equation

$$\mathbf{T}_{1Hi} = \mathbf{T}_{Z1}(p)\mathbf{T}_{Z2}(q)\mathbf{T}_{Z6}(\varphi_{12,i})\mathbf{T}_{Z1}(a_1)\mathbf{T}_{Z2}(a_2)\mathbf{T}_{Z6}(\varphi_{25,i})\cdot$$

$$\cdot\mathbf{T}_{Z1}(l_5)\mathbf{T}_{Z6}(\varphi_{56,i})\mathbf{T}_{Z1}(c_1)\mathbf{T}_{Z2}(c_2)\mathbf{T}_{Z6}(\delta) = \mathbf{T}_{1bi}, \tag{b}$$

We have two closed loop conditions for determination of the rest of the MBS coordinates $\varphi_{23,i}, \varphi_{14,i}, \varphi_{25,i}, \varphi_{56,i}$. For the loop $ABCDA$ disconnected in point C

$$\mathbf{T}_{Z6}(\varphi_{12,i})\mathbf{T}_{Z1}(l_2)\mathbf{T}_{Z6}(\varphi_{23,i})\mathbf{r}_{3C} = \mathbf{T}_{Z6}(\alpha)\mathbf{T}_{Z1}(l_1)\mathbf{T}_{Z6}(\varphi_{14,i})\mathbf{r}_{4C}, \quad i = 1,2$$

$$\mathbf{r}_{3C} = [l_3, 0, 1], \qquad \mathbf{r}_{4C} = [l_4, 0, 1] \tag{c}$$

Similarly, for the loop $BEFGB$, we have

$$\mathbf{T}_{Z1}(a_1 - l_2)\mathbf{T}_{Z2}(a_2)\mathbf{T}_{Z6}(\varphi_{25,i})\mathbf{T}_{Z1}(l_5)\mathbf{T}_{Z6}(\varphi_{56,i})\mathbf{r}_{6G} =$$

$$= \mathbf{T}_{Z6}(\varphi_{23,i})\mathbf{r}_{3G}, \qquad i = 1,2$$

$$\mathbf{r}_{6G} = [l_4, 0, 1], \qquad \mathbf{r}_{3G} = [b_1, b_2, 1] \tag{d}$$

Equations (c), (d) give 4 scalar conditions for calculation of $\varphi_{23,i}, \varphi_{14,i}, \varphi_{25,i}, \varphi_{56,i}$ in each position.

The whole problem of synthesis can be solved using 20 nonlinear equations (b) to (d) (14 of which are independent) for 18 parameters and 8 unknown coordinates $\varphi_{23,i}, \varphi_{14,i}, \varphi_{25,i}, \varphi_{56,i}$. For this purpose we can use the modified Newton method with selection of regular core according to Section 5.12. It is necessary to use the modified Newton method, as the starting estimate of parameters can generally be very far from the solution.

A similar procedure for solution can be suggested on the basis of Chapter 11 and the results of the foregoing Section 12.1. We proceed in such a way that we estimate the starting values of parameters and collect them into the parameters vector $\mathbf{p}^0$ of null approximation. We select two positions of crank 2 and we find the rest of coordinates $\varphi_{23,i}, \varphi_{14,i}, \varphi_{25,i}, \varphi_{56,i}$ from equations (c), (d) using the procedures of kinematic analysis according to Chapter 5. However, these values of parameters $\mathbf{p}^0$ and coordinates found do not meet the synthesis requirements (b). For the estimated parameters, the mechanism reaches actual positions $\mathbf{T}_{1bi}^{ac}$

$$\mathbf{T}_{1Hi} = \mathbf{T}_{1bi}^{ac} \tag{e}$$

We now find a change of parameters $\Delta\mathbf{p}^0$ for the modified mechanism to reach the required position

$$\mathbf{T}_{1Hi} = \mathbf{T}_{1bi} \tag{f}$$

According to the linear theory of accuracy (11.18)

$$\mathbf{T}_{1Hi} \doteq \mathbf{T}_{1Hi}^{*} = \mathbf{T}_{1bi} \tag{g}$$

Using correction matrices $\mathbf{C}$ from (b) according to (11.50b) and (11.18), we obtain

$$\begin{aligned}
&\mathbf{T}_{Z1}(p)\mathbf{C}_{Z1}(\Delta p)\mathbf{T}_{Z2}(q)\mathbf{C}_{Z2}(\Delta q)\mathbf{T}_{Z6}(\varphi_{12,i})\mathbf{C}_{Z6}(\Delta\varphi_{12,i})\mathbf{T}_{Z1}(a_1).\\
&.\mathbf{C}_{Z1}(\Delta a_1)\mathbf{T}_{Z2}(a_2)\mathbf{C}_{Z2}(\Delta a_2)\mathbf{T}_{Z6}(\varphi_{25,i})\mathbf{C}_{Z6}(\Delta\varphi_{25,i}).\\
&.\mathbf{T}_{Z1}(l_5)\mathbf{C}_{Z1}(\Delta l_5)\mathbf{T}_{Z6}(\varphi_{56,i})\mathbf{C}_{Z6}(\Delta\varphi_{56,i})\mathbf{T}_{Z1}(c_1)\mathbf{C}_{Z1}(Dc_1).\\
&.\mathbf{T}_{Z2}(c_2)\mathbf{C}_{Z2}(\Delta c_2)\mathbf{T}_{Z6}(\delta)\mathbf{C}_{Z6}(\Delta\delta) = \mathbf{T}_{1bi}^{ac}\mathbf{C}_{1bi} = \mathbf{T}_{1bi}
\end{aligned} \tag{h1}$$

where

$$\mathbf{C}_{1bi} = (\mathbf{T}_{1bi}^{ac})^{-1}\mathbf{T}_{1bi}$$

We obtain linearized conditions by deleting the higher order terms in (h1). This operation is rather laborious. We can avoid it by using matrices $\mathbf{F}$ and $\mathbf{G}$ from Section 11.1.1.

According to (11.23)

$$\Delta\mathbf{T}_{1Hi} = \mathbf{T}_{1bi} - \mathbf{T}_{1bi}^{ac}$$

$$\begin{aligned}
&\mathbf{F}_1\mathbf{E}(\Delta p) + \mathbf{F}_2\mathbf{E}(\Delta q) + \mathbf{F}_3\mathbf{E}(\Delta\varphi_{12,i}) + \mathbf{F}_4\mathbf{E}(\Delta a_1) + \mathbf{F}_5\mathbf{E}(\Delta a_2)+\\
&+\mathbf{F}_6\mathbf{E}(\Delta\varphi_{25,i}) + \mathbf{F}_7\mathbf{E}(\Delta l_5) + \mathbf{F}_8\mathbf{E}(\Delta\varphi_{56,i}) + \mathbf{F}_9\mathbf{E}(\Delta c_1) +\\
&+\mathbf{F}_{10}\mathbf{E}(\Delta c_2) + \mathbf{F}_{11}\mathbf{E}(\Delta\delta) = \mathbf{T}_{1bi} - \mathbf{T}_{1bi}^{ac}
\end{aligned} \tag{h2}$$

where, from (5.90), we have, for example

$$\mathbf{F}_8 = \mathbf{T}_{Z1}(p)\mathbf{T}_{Z2}(q)\mathbf{T}_{Z6}(\varphi_{12,i})\mathbf{T}_{Z1}(a_1)\mathbf{T}_{Z2}(a_2)\mathbf{T}_{Z6}(\varphi_{25,i}).$$

$$\cdot\mathbf{T}_{Z1}(l_5)\mathbf{T}_{Z6}(\varphi_{56,i})\mathbf{D}_{Z6}(1)\mathbf{T}_{Z1}(c_1)\mathbf{T}_{Z2}(c_2)\mathbf{T}_{Z6}(\delta)$$

Condition (h2) expressed using matrix $\mathbf{G}$ is from (11.24)

$$\Delta\mathbf{T}_{1Hi} = \mathbf{T}_{1Hi}\mathbf{H}_{1Hi} = \mathbf{T}_{1bi} - \mathbf{T}_{1bi}^{ac}$$

$$\mathbf{H}_{1Hi} = (\mathbf{T}_{1bi}^{ac})^{-1}(\mathbf{T}_{1bi} - \mathbf{T}_{1bi}^{ac}) = (\mathbf{T}_{1bi}^{ac})^{-1}\mathbf{T}_{1bi} - \mathbf{E}(1)$$

$$\mathbf{H}_{1Hi} = \mathbf{G}_1\mathbf{E}(\Delta p) + \mathbf{G}_2\mathbf{E}(\Delta q) + \mathbf{G}_3\mathbf{E}(\Delta\varphi_{12,i}) + \mathbf{G}_4\mathbf{E}(\Delta a_1) + \mathbf{G}_5\mathbf{E}(\Delta a_2) +$$
$$+\mathbf{G}_6\mathbf{E}(\Delta\varphi_{25,i}) + \mathbf{G}_7\mathbf{E}(\Delta l_5) + \mathbf{G}_8\mathbf{E}(\Delta\varphi_{56,i}) + \mathbf{G}_9\mathbf{E}(\Delta c_1) +$$
$$+ \mathbf{G}_{10}\mathbf{E}(\Delta c_2) + \mathbf{G}_{11}\mathbf{E}(\Delta\delta) = (\mathbf{T}_{1bi}^{ac})^{-1}\mathbf{T}_{1bi} - \mathbf{E}(1) \qquad \text{(h3)}$$

where from (5.93e) we have, for example

$$\mathbf{G}_8 = \mathbf{T}_{Z6}(-\delta)\mathbf{T}_{Z2}(-c_2)\mathbf{T}_{Z1}(-c_1)\mathbf{D}_{Z6}(1)\mathbf{T}_{Z1}(c_1)\mathbf{T}_{Z2}(c_2)\mathbf{T}_{Z6}(\delta)$$

The difference between expressions (h2) and (h3) is that in the former the total errors are expressed in basic space, while in the latter they are expressed in the space of container H.

The parameters of the modified mechanism must however meet the closed loop conditions (c) and (d). For example, for the loop $ABCDA$, analogously to (h1, 2, 3),

$$\mathbf{T}_{Z6}(\varphi_{12,i})\mathbf{C}_{Z6}(\Delta\varphi_{12,i})\mathbf{T}_{Z1}(l_2)\mathbf{C}_{Z1}(\Delta l_2)\mathbf{T}_{Z6}(\varphi_{23,i})\mathbf{C}_{Z6}(\Delta\varphi_{23,i})\mathbf{r}_{3C}^{*} =$$
$$= \mathbf{T}_{Z6}(\alpha)\mathbf{C}_{Z6}(\Delta\alpha)\mathbf{T}_{Z1}(l_1)\mathbf{C}_{Z1}(\Delta l_1)\mathbf{T}_{Z6}(\varphi_{14,i})\mathbf{C}_{Z6}(\Delta\varphi_{14,i})\mathbf{r}_{4C}^{*} \qquad \text{(i1)}$$

where

$$\mathbf{r}_{3C}^{*} = [l_3 + \Delta l_3, 0, 1], \qquad \mathbf{r}_{4C}^{*} = [l_4 + \Delta l_4, 0, 1]$$

Further

$$(\mathbf{F}_1^L\mathbf{E}(\Delta\varphi_{12,i}) + \mathbf{F}_2^L\mathbf{E}(\Delta l_2) + \mathbf{F}_3^L\mathbf{E}(\Delta\varphi_{23,i}))\mathbf{r}_{3C} +$$
$$+\mathbf{T}_{Z6}(\varphi_{12,i})\mathbf{T}_{Z1}(l_2)\mathbf{T}_{Z6}(\varphi_{23,i})\Delta\mathbf{r}_{3C} = (\mathbf{F}_1^R\mathbf{E}(\Delta\alpha) + \mathbf{F}_2^R\mathbf{E}(\Delta l_1) +$$
$$+ \mathbf{F}_3^R\mathbf{E}(\Delta\varphi_{14,i}))\mathbf{r}_{4C} + \mathbf{T}_{Z6}(\alpha)\mathbf{T}_{Z1}(l_1)\mathbf{T}_{Z6}(\varphi_{14,i})\Delta\mathbf{r}_{4C} \qquad \text{(i2)}$$

where, according to (5.90),

$$\mathbf{F}_3^L = \mathbf{T}_{Z6}(\varphi_{12,i})\mathbf{T}_{Z1}(l_2)\mathbf{T}_{Z6}(\varphi_{23,i})\mathbf{D}_{Z6}(1) =$$
$$= \begin{bmatrix} -s(\varphi_{12}+\varphi_{23}), & -c(\varphi_{12}+\varphi_{23}), & 0 \\ c(\varphi_{12}+\varphi_{23}), & -s(\varphi_{12}+\varphi_{23}), & 0 \\ 0, & 0, & 0 \end{bmatrix}$$

$$\Delta\mathbf{r}_{3C} = [\Delta l_3, 0, 0], \qquad \Delta\mathbf{r}_{4C} = [\Delta l_4, 0, 0]$$

Also

$$(\mathbf{G}_1^L\mathbf{E}(\Delta\varphi_{12,i}) + \mathbf{G}_2^L\mathbf{E}(\Delta l_2) + \mathbf{G}_3^L\mathbf{E}(\Delta\varphi_{23,i}))\mathbf{r}_{3C} + \Delta\mathbf{r}_{3C} =$$

$$= (\mathbf{G}_1^R\mathbf{E}(\Delta\alpha) + \mathbf{G}_2^R\mathbf{E}(\Delta l_1) + \mathbf{G}_3^R\mathbf{E}(\Delta\varphi_{14,i}))\mathbf{r}_{4C} + \Delta\mathbf{r}_{4C} \qquad \text{(i3)}$$

where, according to (5.93e),

$$\mathbf{G}_3^L = \mathbf{D}_{Z6}(1) = \begin{bmatrix} 0 & -1 & 0 \\ 1 & 0 & 0 \\ 0 & 0 & 0 \end{bmatrix}$$

For the second loop $BEFGB$ we have analogously

$$(\mathbf{G}_1^L\mathbf{E}(\Delta a_1 - \Delta l_2) + \mathbf{G}_2^L\mathbf{E}(\Delta a_2) + \mathbf{G}_3^L\mathbf{E}(\Delta\varphi_{25,i}) + \mathbf{G}_4^L\mathbf{E}(\Delta l_5) +$$

$$+ \mathbf{G}_5^L\mathbf{E}(\Delta\varphi_{25,i}))\mathbf{r}_{6G} + \Delta\mathbf{r}_{6G} = \mathbf{G}_1^R\mathbf{E}(\Delta\varphi_{23,i})\mathbf{r}_{3G} + \Delta\mathbf{r}_{3G} \qquad \text{(j)}$$

Equations (h), (i), (j) give altogether 20 scalar conditions (14 of which are independent) for 18 changes of parameters and 8 changes of coordinates in two positions $i = 1, 2$. The number of dependent equations may be reduced according to (11.51) and the note after (11.55) in Section 11.1.2. Again we have more unknowns than equations. We can solve the problem according Section 5.10 by selection of regular core with superfluous parameters being constant (the differences between them equalling zero). We change, for example, only $a_2, c_1, c_2, \delta, \varphi_{12,1}$ and $\varphi_{12,2}$. Parameters which we keep constant cannot be excluded from the synthesis as a matter of principle, because a solution exists for values of these parameters belonging only to certain intervals.

Another possibility for the solution of synthesis equations is pseudo-inversion according to (12.17). This means solution in the sense of least sum of squared deviations.

Having starting values of parameters $\mathbf{p}^0$ and their corrections $\Delta\mathbf{p}^0$ in zero approximation, we are able to find values of first approximation

$$\mathbf{p}^1 = \mathbf{p}^0 + \varkappa\Delta\mathbf{p}^0 \qquad \text{(k)}$$

with the coefficient $\varkappa \in [0, 1]$, the meaning of which is explained in Section 5.10.

We find the new accurate values of coordinates $\varphi_{23,i}, \varphi_{14,i}, \varphi_{56,i}$ for new parameters from (c), and (d). The differences $\Delta\varphi_{23,i}, \Delta\varphi_{14,i}, \Delta\varphi_{25,i}, \Delta\varphi_{56,i}$ were among the corrections of parameters for which we have searched. We then proceed in this manner until $|\Delta\mathbf{p}^k| < \varepsilon$ or $|\mathbf{C}_{1bi}| < \varepsilon$.

The existence of free parameters in the above formulation of the task offers the possibility of meeting other requirements concerning, for example, velocities. In this way, from position synthesis, we have a more general problem of kinematic synthesis.

The natural requirement for a synthetized mechanism is to pick up a container smoothly from the first conveyor and to put it on the second conveyor in the same manner. So we want the vector of the point H velocity to lie on such a line that moves point H in direction away from the first conveyor in the first position and towards the second conveyor in the second position. Further we want the ratio of components of point H velocities in the direction of the conveyor movement to be the prescribed value equal to the ratio of the velocities of the two conveyors. We also want the ratio of angular velocities and translation velocities to be the prescribed value.

These requirements we formulate using 6 constants $\beta_i, t_i, r_i,\ i = 1,2$. From these, $\beta_i,\ i = 1,2$ specify the position of the velocity vector at an angle to the x axis, and r_i and t_i express the ratio of the angular and translational velocity in the i-th position with respect to the driving velocity $\dot{\varphi}_{12,i}$. The velocity requirements can be expressed by the velocity matrix

$$\mathbf{V}_{1Hi} = \mathbf{V}_{1bi} \tag{l}$$

where

$$\mathbf{V}_{1bi} = \begin{bmatrix} 0, & -r_i\dot{\varphi}_{12,i}, & \dot{\varphi}_{12,i}t_i c\beta_i \\ r_i\dot{\varphi}_{12,i}, & 0, & \dot{\varphi}_{12,i}t_i s\beta_i \\ 0, & 0, & 0 \end{bmatrix}$$

represents three equations constituted by elements (1,2), (1,3) and (2,3).

The requirements just formulated can be expressed from (f) by division of the equations for translational and angular velocity components in and between the both positions. Therefore we only need to know the ratios $t_1/t_2, r_1/t_1, r_2/t_2$ and the values of β_1, β_2 i.e. only five equations. For conciseness we represent the conditons (f) with 6 equations in the following way.

It is evident that the velocity conditions (c), (d) must be fulfilled. According to (5.98)

$$\begin{aligned} \dot{\mathbf{T}}_{13}\mathbf{r}_{3C} &= \dot{\mathbf{T}}_{14}\mathbf{r}_{4C} \\ \dot{\mathbf{T}}_{16}\mathbf{r}_{6G} &= \dot{\mathbf{T}}_{13}\mathbf{r}_{3G} \end{aligned} \tag{m}$$

Velocities $\dot{\varphi}_{23,i}, \dot{\varphi}_{14,i}, \dot{\varphi}_{25,i}, \dot{\varphi}_{56,i}$ can be found from (m) and substituted into (l). After that all elements of the two matrices are proportional to $\dot{\varphi}_{12,i}$ and $\dot{\varphi}_{12,i}$ can be cancelled. We can therefore take $\dot{\varphi}_{12,i} = 1$ without loss of generality. Altogether we have 34 equations (28 of which are independent) for 18 parameters and 16 coordinates. These equations can be solved using the modified Newton method or by the method of corrections. Let us write the equation of corrections only for the velocity condition (1). For the estimated parameters

$$\mathbf{V}_{1Hi} = \mathbf{V}_{1bi}^{ac}$$

We find such changes of parameters $\Delta\mathbf{p}^0$ to satisfy the equality

$$\mathbf{V}_{1\bar{H}i} = \mathbf{V}_{1bi}$$

According to linear accuracy theory (11.18) we have

$$\mathbf{V}_{1\bar{H}i} \doteq \mathbf{V}_{1Hi}^{*} = \mathbf{V}_{1bi}$$

The analogy to relation (h1) can be found using (11.27) and the analogy of relations (h2) and (h3) follows from (11.31). Hence

$$\begin{aligned} \mathbf{V}_{1Hi} - \mathbf{H}_{1Hi}\mathbf{V}_{1Hi} + \mathbf{V}_{1Hi}\mathbf{H}_{1Hi} + \dot{\mathbf{H}}_{1Hi} &= \mathbf{V}_{1bi} \\ -\mathbf{H}_{1Hi}\mathbf{V}_{1Hi} + \mathbf{V}_{1Hi}\mathbf{H}_{1Hi} + \dot{\mathbf{H}}_{1Hi} &= \mathbf{V}_{1bi} - \mathbf{V}_{1bi}^{ac} \end{aligned}$$

where $\mathbf{H}_{1Hi}$ from (h3) and according to (11.29) is given by

$$\dot{\mathbf{H}}_{1Hi} = \mathbf{G}_3\mathbf{E}(\Delta\dot{\varphi}_{12,i}) + \mathbf{G}_6\mathbf{E}(\Delta\dot{\varphi}_{25,i}) + \mathbf{G}_8\mathbf{E}(\Delta\dot{\varphi}_{56,i}) + \dot{\mathbf{G}}_1\mathbf{E}(\Delta p)+$$

$$+\dot{\mathbf{G}}_2\mathbf{E}(\Delta q) + \dot{\mathbf{G}}_3\mathbf{E}(\Delta\varphi_{12,i}) + \dot{\mathbf{G}}_4\mathbf{E}(\Delta a_1) + \dot{\mathbf{G}}_5\mathbf{E}(\Delta a_2)+$$

$$+\dot{\mathbf{G}}_6\mathbf{E}(\Delta\varphi_{25,i}) + \dot{\mathbf{G}}_7\mathbf{E}(\Delta l_5)$$

Also, for example,

$$\dot{\mathbf{G}}_8 = \dot{\mathbf{G}}_9 = \dot{\mathbf{G}}_{10} = \dot{\mathbf{G}}_{11} = \mathbf{0}$$

Matrix $\dot{\mathbf{G}}_7$, for example, is

$$\dot{\mathbf{G}}_7 = \mathbf{T}_{Z6}(-\delta)\mathbf{T}_{Z2}(-c_2)\mathbf{T}_{Z1}(-c_1)\mathbf{T}_{Z6}(-\varphi_{56,i})\mathbf{D}_{Z6}(-\dot{\varphi}_{56,i}).$$

$$.\mathbf{D}_{Z1}(1)\mathbf{T}_{Z6}(\varphi_{56,i})\mathbf{T}_{Z1}(c_1)\mathbf{T}_{Z2}(c_2)\mathbf{T}_{Z6}(\delta)+$$

$$+\mathbf{T}_{Z6}(-\delta)\mathbf{T}_{Z2}(-c_2)\mathbf{T}_{Z1}(-c_1)\mathbf{T}_{Z6}(-\varphi_{56,i})\mathbf{D}_{Z1}(1).$$

$$.\mathbf{T}_{Z6}(\varphi_{56,i})\mathbf{D}_{Z6}(\dot{\varphi}_{56,i})\mathbf{T}_{Z1}(c_1)\mathbf{T}_{Z2}(c_2)\mathbf{T}_{Z6}(\delta)$$

For the loop ABCDA we have according to (11.60) similarly

$$\mathbf{T}_{13}^{*}\mathbf{r}_{3C} + \mathbf{T}_{13}\Delta\mathbf{r}_{3C} = \mathbf{T}_{14}^{*}\mathbf{r}_{4C} + \mathbf{T}_{14}\Delta\mathbf{r}_{4C}$$

$$\mathbf{H}_{13}\mathbf{r}_{3C} + \Delta\mathbf{r}_{3C} = \mathbf{H}_{14}\mathbf{r}_{4C} + \Delta\mathbf{r}_{4C}$$

$$\mathbf{H}_{13}\dot{\mathbf{r}}_{3C} + \dot{\mathbf{H}}_{13}\mathbf{r}_{3C} + \Delta\dot{\mathbf{r}}_{3C} = \mathbf{H}_{14}\dot{\mathbf{r}}_{4C} + \dot{\mathbf{H}}_{14}\mathbf{r}_{4C} + \Delta\dot{\mathbf{r}}_{4C}$$

$$\dot{\mathbf{r}}_{3C} = \dot{\mathbf{r}}_{4C} = \Delta\dot{\mathbf{r}}_{3C} = \Delta\dot{\mathbf{r}}_{4C} = \mathbf{0}$$

$$(\mathbf{G}_1^L\mathbf{E}(\Delta\dot{\varphi}_{12,i}) + \mathbf{G}_3^L\mathbf{E}(\Delta\dot{\varphi}_{23,i}) + \dot{\mathbf{G}}_1^L\mathbf{E}(\Delta\varphi_{12,i}) + \dot{\mathbf{G}}_2^L\mathbf{E}(\Delta l_2))\mathbf{r}_{3C} =$$
$$= (\mathbf{G}_3^R\mathbf{E}(\Delta\dot{\varphi}_{14,i}) + \dot{\mathbf{G}}_1^R\mathbf{E}(\Delta\alpha) + \dot{\mathbf{G}}_2^R\mathbf{E}(\Delta l_1))\mathbf{r}_{4C}$$

The solution of the kinematic synthesis problem leads via the method of corrections to the large system of nonlinear or linearized equations. Equality in the number of equations and the number of parameters is very rare and if it is the case it does not mean that a solution exists and is applicable in practice. It depends especially on the proper choice of mechanism structure (it must make it possible to realize the required function) and also on suitable estimation of the initial dimensions. It is common for the solutions of real problems (if they exist) to be impractical because the lengths of some members are too large (or on the contrary too small). Usually there are also some requirements that are not described analytically and fulfillment of which is not so easy; for example, in our illustration the manipulation task is to be achieved by the motion of φ_{12} in the whole interval $< 0, 2\pi >$. It is advisable to use fewer synthetized parameters selected from a large number of feasible ones.

12.3.2 Car Front Wheel Suspension

We are to design a front wheel suspension mechanism. We use the mechanism from Section 10.2. We require the mechanism to keep the change of toe-in as a function of length to be minimal or to fulfil the spatial analogy of Ackermann's conditions.

We formulate the former requirement as the minimum tangent slope of the toe-in behaviour as a function of spring length. We solve this problem using methods of optimization.

We formulate the latter requirement as the minimum distance of projections of intersection points of the front wheel axes with the vertical plane containing the rear axle on to the road plane. Again we treat this problem as an optimization problem.

In both cases we formulate the synthesis problems as optimization problems because we cannot guarantee exactly to reach the required conditions. The problems may be solved using well known optimization procedures or by the procedure according to Figure 2.4 from Section 2.5.

12.4 Dynamic Synthesis

The problems of MBS dynamic synthesis can be divided into two groups. The first concerns the MBS function as such and includes problems of precision of realized motion, values of permissible forces in kinematic pairs (joints), permissible loading of mechanism elements, small-run irregularities, etc.

The second group of problems concerns the influence of a MBS on its neighborhood and includes requirements on the size and frequency of forces transmitted from the mechanism on to the frame (balancing), noise reduction, power input minimization, etc.

These requirements can be satisfied by passive influences (change of dimensions, mass and force parameters), by active influences (traditional control or microprocessor control) or by both.

Compared with kinematic synthesis, the set of design parameters to be chosen from is larger here. In addition to the MBS element dimensions, masses, moments of inertia, section areas, spring stiffnesses, damping factors, coordinates of centers of mass, controller setting, etc. can be design parameters.

With regard to the normal disproportion between the number of design parameters and the number of requirements set on the mechanism, the dynamic synthesis is in most cases solved as an optimizing problem. The main difficulty is the choice of a suitable cost funtion and a suitable numerical method (Polak, 1971; Fleury and Braibant, 1984; Gill et al., 1984 and references cited by them). In particular, the gradient projection method is very efficient for systems with constraints, e.g. MBS (Haug and Arora, 1979).

MBS dynamic synthesis problems may be divided into two groups. One concerns the MBS dynamic synthesis primarily specified as finite-dimensional. We consider a certain finite number of MBS parameters which we treat in order to satisfy the required function and MBS properties. We speak therefore of parametric dynamic synthesis.

The other group of problems in MBS dynamic synthesis consists of problems primarily specified as infinite-dimensional. Their basis is the synthesis of a certain functional behaviour, and the determination of a general function would require the determination of an infinite number of values. To achieve a proper numerical solution these problems would be converted into discrete ones, i.e. finite-dimensional. The methods of their solution depend on their initial specification. Problems such as the time behaviour synthesis of a control quantity realized by cam form or by a microprocessor control system would mostly be considered in this group of MBS synthesis problems. There are problems of other types in this group, such as the synthesis of spring stiffness behaviour, MBS element elasticity, continuous distribution of element mass, etc.

There is in principle no difference between the solution of problems of parametric dynamic synthesis and kinematic synthesis. The criterion expressing the required MBS function is specified as the required behaviour of a certain characteristic or as the required value of an integral criterion of the form (11.86). For the purpose of parametric synthesis represented by seeking a finite number of parameters, the behaviour requirements are either discontinued or substituted by a finite number of integral characteristics (e.g. expressing the extent to which the approximation is optimal). The problem is thus transformed into finding a finite number of parameters to satisfy a finite number of conditions. If the number of parameters and conditions is the same, the problem can be solved by seeking the parameters as a solution of a system of nonlinear equations formed by the required conditions.

The number of parameters and conditions is often not the same. The problem is then to be solved as an optimization problem, where we have to determine the desired parameters in the way which best satisfies the required conditions. The satisfaction of the required conditions is usually converted into an optimality criterion (12.15)–(12.18). The solution method of kinematic synthesis (interpolation method, correction method, optimization methods) presented in more detail can be completely utilized for the parametric dynamic synthesis. Only the form of the criteria expressing the required MBS function has a great variety in terms of dynamic behaviour.

As an example we solve the problem from Section 9.12.2 as a parametric dynamic synthesis problem. We need to decrease the value of the vibration forces in the suspension mechanism due to unevenness of the road below the specified value. The model of the suspension mechanism involves many parameters. Firstly we evaluate the sensitivities of particular parameters to the forces. We choose those with the greatest sensitivity for which the changes can be realized. Then we solve the synthesis as an optimization problem.

The solution of infinite-dimensional dynamic systems is more difficult. These problems, with some exceptions to be discussed further on, are optimization problems. They may be specified in general as follows (Haug and Arora, 1979).

A function $\mathbf{u}(\mathbf{x}, t)$ is to be determined describing the desired MBS (e.g. spatial form or control with respect to time), which is a function of spatial coordinates $\mathbf{x}$ from the set $\Omega^k \in E^k$ of the k-dimensional real numbers space. The MBS behaviour is described by quantities of state $\mathbf{z}(\mathbf{x}, t)$ using a general operator equation

$$\mathcal{S}(\mathbf{u})\mathbf{z} + \mathcal{T}(\mathbf{u})\mathbf{z} = \mathbf{Q}(\mathbf{x}, t) \tag{12.22}$$

in the set $\Omega \times < 0, \tau >$, where $< 0, \tau >$ is the time interval of MBS motion being considered with boundary conditions

$$\mathcal{B}\mathbf{z} = \mathbf{q}(\mathbf{x}, t) \tag{12.23}$$

on the boundary $\partial\Omega\times <0,\tau>$ and initial conditions

$$\mathcal{C}\mathbf{z} = \mathbf{r}(\mathbf{x}) \tag{12.24}$$

in the set Ω at time $t=0$. The differential operator $\mathcal{S}$ is a space operator and $\mathcal{T}$ is a time operator. We want to determine the function $\mathbf{u}$ minimizing the criterion

$$\psi_0 = \int_0^\tau \int\int_\Omega \mathbf{F}_0(\mathbf{z},\mathbf{u})\ \mathrm{d}\mathbf{x}\ \mathrm{d}t \tag{12.25}$$

with limiting constraints

$$\psi_\alpha = \int_0^\tau \int\int_\Omega \mathbf{F}_\alpha(\mathbf{z},\mathbf{u})\ \mathrm{d}\mathbf{x}\ \mathrm{d}t = 0 \left\{ \begin{array}{l} =0 \\ \leq 0 \end{array} \right. \tag{12.26a}$$

$$\varphi_\beta(\mathbf{x},\mathbf{u}) = \left\{ \begin{array}{l} =0 \\ \leq 0 \end{array} \right. \tag{12.26b}$$

Operators $\mathcal{S},\mathcal{T},\mathcal{B},\mathcal{C}$ and other functions in (12.22)–(12.26) can be functions of $\mathbf{x}$ and t. The solution of such a general problem is very demanding. We are not going to describe the solution here. A general algorithm for the numerical solution of such problems is given in (Haug and Arora, 1979) where partial examples are presented and solved.

From the above problem of dynamic synthesis it can be seen that the problem of finding the control is very typical for MBS with feedback control. In general we have the requirement on the MBS input-output relation or on the path of its point of body motion to have a certain behaviour. We have here an obvious analogy with problems of kinematic synthesis. We attempt to satisfy the required behaviour either precisely or approximately. The required behaviour is given either by a prescribed time relation or by an optimizing criterion

The problems of control are solved either by including the discrete controller according to the principle of feedback into the MBS structure and synthesizing its parameters to satisfy the required behaviour or by synthetitizing the behaviour of the control quantity, which is then realized nowadays mostly by microprocessor control or alternatively by a mechanical cam. The synthesis of discrete controller parameters is a problem of parametric dynamic synthesis and in general allows only approximate satisfaction of the required behaviour, even with minimum of overcontrol, optimum damping, etc.

The synthesis of the control quantity behaviour is usually even more difficult as we do not require a once-for-all computation unlike when computing the cam form where the problem is solved by one inverse dynamic solution. We require the MBS control system to be capable of realizing any required path in real time. We thus reach the most difficult problems of the theory of control. We solve it nowadays in general by seeking feedbacks realizable by microprocessors (e.g. algebraic) in order to obtain a decoupled system (i.e. MBS and control system with hydraulic or electric drives). Controlled MBSs usually have more degrees of freedom. In a decoupled system each degree of freedom is controlled by one control quantity without being influenced by others. Each degree of freedom is thus controlled on its own.

The constraints for decoupling of a controlled MBS are determined either by constructing a transformation of the given system into a decoupled system (Freund, 1982; Valášek, 1982, 1987a) or at least by constructing a transformation of the given non-linear system into a linear one (Hunt et al., 1983).

The decoupling problem is simpler for linear systems and has been solved (Wonham and Morse, 1970, 1971). For many non-trivial non-linear MBSs the control decoupling problem is insoluble. An example of this is the control of crane motion. Independent motion of the load and the crane crab is not possible during the whole time interval. An independent (arbitrary) dynamic state of the load and the crab is possible only for discontinuous time instants (Valášek, 1986a). The control problem is then even more difficult.

The finding of optimum control represent another group of problems. Their formulation forms a part of the general problem (12.22)–(12.26); solution methods are analogous but they are more elaborate (Sage and White, 1977, Fedorenko, 1978).

In the case of linear MBS dynamic models, many control problems have a special and simpler solution (Sage and White, 1977). Research into complex system control, and hence MBS, is still an open and developing problem.

Let us show the formulation and solution of dynamic synthesis problems in the case of a typical problem of MBS control — the control of an industrial robot. Let us consider an industrial robot with n degrees of freedom, whose kinematic structure is formed by an open simple chain and for which each kinematic pair (revolute or prismatic) is driven by an electric DC servo. We shall model the robot as a MBS with rigid elements. We assemble the equations of motion by LEMT (9.48) obtaining (Valášek, 1986b)

$$\mathbf{I}(\mathbf{s})\ddot{\mathbf{s}} + \dot{\mathbf{s}}^T\mathbf{C}(\mathbf{s})\dot{\mathbf{s}} = \mathbf{n} - \mathbf{V}(\mathbf{s})\dot{\mathbf{s}} + \mathbf{g}(\mathbf{s}) \tag{12.27}$$

where $\mathbf{s} = \lceil s_1, s_2, \ldots, s_n \rceil$ is the n-dimensional vector of the robot's joint coordinates, $\mathbf{I}(\mathbf{s})$ is the $n \times n$ generalized inertia tensor, $\mathbf{C}(\mathbf{s})$ is the $n \times n \times n$ generalized tensor in the formulation of Coriolis and centrifugal forces, $\mathbf{V}(\mathbf{s})$ is the $n \times n$ generalized tensor in the formulation of viscous friction forces, $\mathbf{g}(\mathbf{s})$ is the n-dimensional vector of gravity forces and $\mathbf{n} = \lceil n_1, n_2, \ldots, n_n \rceil$

is the n-dimensional vector of joint torques. These torques are limited. For DC electric motors with permanent magnets the torque restrictions have the form

$$\mathbf{n}^- - \mathbf{F}(\mathbf{s})\dot{\mathbf{s}} \leq \mathbf{n} \leq \mathbf{n}^+ - \mathbf{F}(\mathbf{s})\dot{\mathbf{s}} \tag{12.28}$$

where $\mathbf{n}^+ = \lceil n_1^+, \ldots, n_n^+ \rceil$ and $\mathbf{n}^- = \lceil n_1^-, \ldots, n_n^- \rceil$ are maximum and minimum constant torque limits. The robot manipulates an object from point A to point B in time T. The path from A to B and the velocity profile along the path are given. We have to determine whether the path is dynamically realizable, i.e. whether the torque (12.27) required for the motion along the prescribed path with prescribed velocities and accelerations would satisfy the limiting conditions (12.28). If not, we have to determine a new time parameterization of this geometric path in order to allow dynamic realizability (Hollerbach, 1984; Valášek, 1987a). Another problem is to find the time parameterization so as to minimize the energy dissipation in robot actuators (Valášek, 1986b). Finally we seek the whole trajectory from A to B to satisfy the dynamic realizability and minimize the energy dissipation in robot actuators (Valášek, 1986b). The problems presented are problems of dynamic synthesis of control quantity behaviour. When the computation has been completed, we need a control system securing the robot motion along the computed path.

Here is an outline of the solution of the last two problems. First of all we transform the MBS-robot dynamic model (12.27)–(12.28) into a state space description by introducing state variables

$$\begin{aligned} x_{1i} &= s_i \\ x_{2i} &= \dot{s}_i, \qquad i = 1, \ldots, n \end{aligned} \tag{12.29}$$

and control variables

$$u_i = \frac{n_i - \frac{n_i^+ + n_i^-}{2}}{\frac{n_i^+ - n_i^-}{2}}, \qquad i = 1, \ldots, n \tag{12.30}$$

We then obtain

$$\dot{x}_{1i} = x_{2i}, \qquad k = 1, 2, \qquad j = 1, \ldots, n \tag{12.31}$$

$$\dot{x}_{2i} = f_i(u_m, x_{kj}) = \sum_{m=1}^{n} H_{im}(x_{kj}) u_m + L_i(x_{kj}) \tag{12.32}$$

with limits on the control variable

$$|u_i| \leq 1$$

We now use the method of the change of dynamic properties of a system by the transformation of input variables (Valášek, 1982) for the derivation of the new robot control structure with dynamic properties

$$\begin{aligned} \dot{y}_{1i} &= y_{2i} \\ \dot{y}_{2i} &= \sum_{m=1}^{n} K_{im} w_m \end{aligned} \tag{12.33}$$

with inputs w_m and outputs y_{ki}. We require the agreement of outputs of both systems $y_{ki} = x_{ki}$. From this condition we derive the transformation equations between input variables u_i and w_m

$$u_i = Q_i(w_m, x_{kj}) = \left([H_{im}(x_{kj})]^{-1} \left[\sum_m K_{im} w_m - L_i(x_{kj}) \right] \right)_i =$$

$$= \sum_{m=1}^{n} M_{im}(x_{kj}) w_m - N_i(x_{kj}) \tag{12.34}$$

If these equations (12.34) are realized in real time by the control system, the original dynamic system would have new dynamic properties given by equations

$$\begin{aligned} \dot{x}_{1i} &= x_{2i} \\ \dot{x}_{2i} &= \sum_{m=1}^{n} K_{im} w_m \end{aligned} \tag{12.35}$$

with new inputs w_m. If we calculate the control variable according to (12.35) for any prescribed trajectory and apply it to the robot control using (12.34), the robot would move along the prescribed trajectory only if the model (12.31) were absolutely complete. It cannot be practically satisfied. We must therefore use the feedback principle for the compensation of incomplete identification. The suggested control structure is shown in Figure 12.5 for one degree of freedom. The control system generates the control variable w_i and the corresponding required position x_{1ir} and velocity x_{2ir}. These required values are compared with the real values of the robot, and the differences are transferred by the controllers R_1, R_2 to the correction of the control variable w_i which is transformed in the compensator Q_i according to (12.34) to the real control variable u_i. The properties of this control structure are better compared with traditional ones even by when (12.34) is incomplete. This control structure is able to realize the robot's motion along the dynamically acceptable trajectory quite precisely

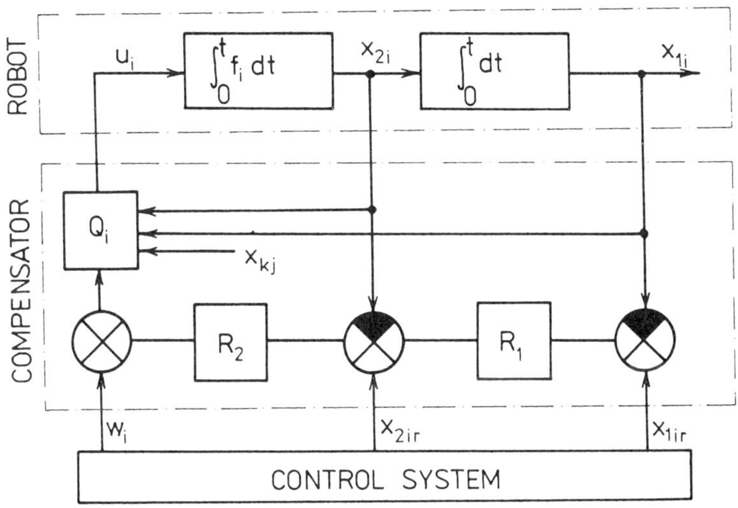

Figure 12.5. Robot control system with dynamic compensation.

(Valášek, 1987b). The basis for the dynamical compensation is the solution of the robot inverse dynamic problem, which is why we paid so much attention to it — see Section 8.1 and 9.5.1.8.

This solution of the problem of realization of robot control can be easily generalized for the majority of MBS's. The dynamic model of MBS (9.68) is transformed into a form equivalent to (12.27) by excluding the reaction forces (Lagrange's multipliers). If the generalized inertia tensor $\mathbf{I}(s)$ has full rank we can obtain the state-space model (12.31). If the matrix $\mathbf{H}(x_{kj})$ has full rank then the exact realization of control is possible. These assumptions are valid for a great variety of MBSs. This solution fails for the MBS in which the ratio of control to state variables may be zero, e.g. a vehicle suspension controlled by damping ratio, a crane, an elastic robot, etc.

The criterion expressing energy dissipation in robot actuators by robot motion is given by the relation

$$\psi_0 = \int_0^T (\sum_{i=1}^{n} e_i u_i^2) \mathrm{d}t \tag{12.36}$$

with weight coefficients e_i expressing the contribution of individual actuators. For the synthesis of the optimum path from A to B, the robot description (12.35) is used after transformation (12.34). We use the necessary optimization condition according to Pontriagin and Krotov (Krotov and Gurman, 1973). We assemble the Hamiltonian with adjoint variables ψ_{kj}

$$\mathcal{H} = -\sum_{i=1}^{n} e_i Q_i^2(w_m, x_{kj}) + \sum_{i=1}^{n} \left(\psi_{1i} x_{2i} + \psi_{2i} (\sum_{m=1}^{n} K_{im} w_m) \right) \tag{12.37}$$

and the corresponding system of adjoint differential equations

$$\begin{aligned} \dot{\psi}_{1j} &= -\frac{\partial \mathcal{H}}{\partial x_{1j}} = \sum_{i=1}^{n} 2e_i \left(\sum_{m=1}^{n} M_{im}(x_{kj}) w_m - N_i(x_{kj}) \right) \left(\sum_{m=1}^{n} \frac{\partial M_{im}}{\partial x_{1j}} w_m - \right. \\ & \left. - \frac{\partial N_i}{\partial x_{1j}} \right), \qquad j = 1, \ldots, n \\ \dot{\psi}_{2j} &= -\frac{\partial \mathcal{H}}{\partial x_{2j}} = \psi_{1j} + \sum_{i=1}^{n} 2e_i \left(\sum_{m=1}^{n} M_{im}(x_{kj}) w_m - N_i(x_{kj}) \right) \cdot \\ & \left(\sum_{m=1}^{n} \frac{\partial M_{im}}{\partial x_{2j}} w_m - \frac{\partial N_i}{\partial x_{2j}} \right) \end{aligned} \tag{12.38}$$

together with the Hamiltonian maximum condition

$$\frac{\partial \mathcal{H}}{\partial w_m} = 0, \qquad m = 1, \ldots, n$$

$$\begin{aligned} \sum_{i=1}^{n} e_i \left(\sum_{m=1}^{n} M_{im}(x_{kj}) M_{il}(x_{kj}) w_m \right) &= \frac{1}{2} \sum_{i=1}^{n} \psi_{2i} K_{il} + \\ + \sum_{i=1}^{n} e_i N_i(x_{kj}) M_{il}(x_{kj}), & \qquad l = 1, \ldots, n \end{aligned} \tag{12.39}$$

We use the special form of these equations for an efficient iterative calculation of optimum control. Knowing the J-th approximation $w_m^J(t)$, $x_{1i}^J(t)$, $x_{2i}^J(t)$, satisfying all boundary conditions (passes A and B) and limitations, we substitute into the right-hand sides of (12.38) for the calculation of control w_m from (12.39) and motion behaviour from (12.35). This is linearly dependent only in the initial values $\psi_{kj}(0)$; these we determine from the boundary conditions for the robot motion to pass through points A and B. We thus obtain the (J^*)-th approximation satisfying all boundary conditions. We determine the next $(J+1)$-th approximation as a linear combination

$$\begin{aligned} w_m^{J+1} &= (1-\alpha) w_m^J + \alpha w_m^{J^*} \\ x_{2i}^{J+1} &= (1-\alpha) x_{2i}^J + \alpha x_{2i}^{J^*} \\ x_{1i}^{J+1} &= (1-\alpha) x_{1i}^J + \alpha x_{1i}^{J^*} \end{aligned}$$

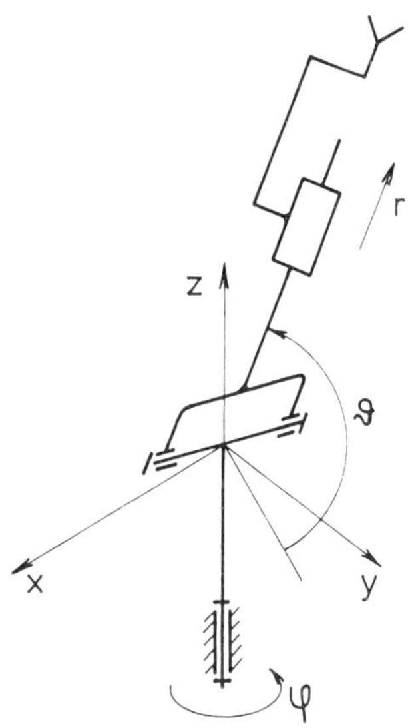

Figure 12.6. Diagram of a robot (UNIMATE).

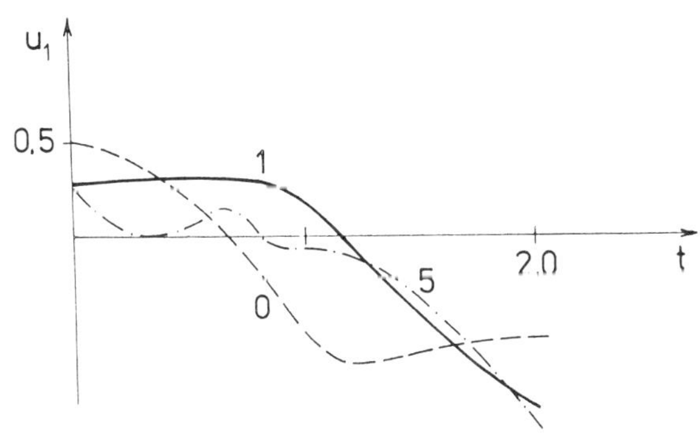

Figure 12.7. Optimal control of a robot.

which, due to linearity of (12.35), also satisfies the boundary conditions. We determine the constant α from a one-dimensional minimization of energy dissipation $\psi_0^{J+1} = \psi_0^{J+1}(\alpha)$. The convergence of this method was proved (Valášek, 1986b). A great advantage of this method is that each approximation represents a usable path satisfying boundary conditions. In practical examples the method converges quickly. Figure 12.6 shows a kine-

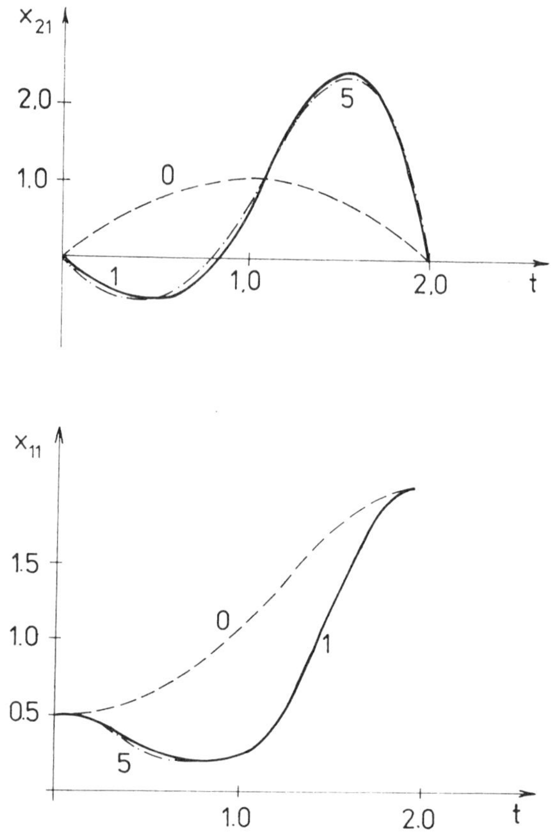

Figure 12.8. Optimal trajectory of a robot.

matic diagram of a robot for which optimum control has been determined using this method. The number of approximations necessary has been only 5, and the reduction in energy consumption has been enormous (Figure 12.7 and 12.8).

Chapter 13

Actual Research Problems and Further Developments

Continuing the principal aim of this book, which is to present the reader with a survey of the most important methods of mechanics of spatial mechanisms, we now present our subjective survey of promising lines of research and further development of the methods described.

The ability to investigate MBSs, even with their flexibility described by characteristics from FEM models, enables us to investigate very accurate models of a great variety of mechanisms and machines, with exactly the required detailed description. Finally we can construct dynamic models of machines which correspond to reality. Hence the applicability of the methods described should in future be even greater.

Let us first summarize the most important trends in research in MBS. An ideal MBS model should correspond closely to reality. It should be computationally efficient and should be easily connected with CAD tools. The most actual research problems in this respect are

– connection between FEM and MBS. It is not clear so far whether the FEM model should be integrated into MBS (see Section 9.10) or whether MBS should be an integrated part of FEM (Geradin, 1993). What could contribute the p-method FEM formulation (Babuska, 1988) to the integration of the FEM and MBS modeling? In the case of the common present approach of FEM integrated into MBS there is the actual problem of suitable selection of modes (see Section 9.10) and the usage of nonlinear elastic models, for example, for beams.

– connection between MBS and mechatronics. The extension of modelling capability of MBS towards mechatronical systems consisting of combined physical subsystems is very promising. The integration of modelling approaches of MBS, hydraulic systems and electrical networks has not yet been completed, especially from the point of computational efficiency or distributed integration.

– design of control of MBS. Only the restricted subset of MBS models is described by ODE and can be represented by state space description. The absolute majority of control design procedures – with the exception of Müller's approach (Schüpphaus and Müller, 1993) – is based on state space description. For MBS with loops a control design for systems described by DAE is needed.

– computational efficiency. We need further improvement of MBS solution procedures from the point of view of computational complexity despite the continuous rapid increase of computer power. There are at least three reasons for it: Many repetitive simulations of MBS motion are required for iterative design or optimization. The new technique in CAD called virtual prototyping is based on real time solution of MBS dynamics (Haug, 1993).

The improvement of computational complexity of MBS solution can be expected from new MBS formalism like residual algorithm (see Section 9.5.1.9) or Borri's approach, from the new integration procedures like DAE solvers and then especially from parallel computation. Massive parallelization of MBS solution is not easy and is still an open problem.

– DAE solvers. Large developments are being done for DAE solutions. This can make easier the MBS formulation and can improve the computational compexity.

– connection between CAD and MBS. The standardized data formats for MBS and mechatronic systems are needed. Some solutions for MBS are in (Otter et al., 1993). These should be extended towards mechatronics, where they could enable the direct automated connection of MBS with CAD geometric models (descriptive models) both for modelling and for visualization. The vizualization of MBS solution is becoming important for the creation of virtual reality within virtual prototyping.

Now we shall inspect and compare the techniques described in particular chapters of this book from the point of view of further development.

Among the methods for kinematic solution of MBSs there are three important approaches. The first approach is the basic matrix method from Section 5.8 but with several modifications. For the disconnecting of a loop we use either the second variant from Section 5.11.3, i.e. the method of removing a body with kinematic pairs, or the equality of transformation matrices in the cut together with the reduction of resultant equations by the transformation to the equality of corresponding quaternions from Section 5.12. In both cases we solve the equations with respect to unknowns sin(arg) and cos(arg) together with the constraint $\sin^2(\text{arg})+\cos^2(\text{arg}) = 1$ (see end of Section 5.11.2). The second approach is the method of natural coordinates from Section 5.11.4. The comparison of computational complexity of both these approaches ought to be realized. A very promising approach is the third one, the geometric diagram and parametric method from Section 5.11.6. However, it is necessary to prepare the methods of solution of singular cases within this approach. Another very promising method comes from robotics. The need to solve the inverse kinematics analytically has necessitated the development of methods of analytical solution of kinematics from Section 5.13. Analytical solution of MBS kinematics should be extended for MBS with many loops. Then the method of structural approximation from Section 11.1.5 enables us to find a very good estimation for an analytically unsolvable loop. Future development of this method of structural approximation for MBSs with several loops is desirable. An advantageous future method of kinematic solution of MBS could be the combination of structural approximation, which gives accurate estimation of the solution, and the geometric diagram and parametric method, which enables us to solve differential kinematic analysis with minimum computational complexity. The differential kinematic analysis is also the basis of the numerical solution of positional kinematic analysis for analytically unsolvable MBSs.

Among the methods for kinetostatic solution of MBSs (determination of reaction and driven forces) there are also three different approaches. The first approach is the method of free body described in Section 8.1. The second approach is the application of LEMT to kinetostatics from the end of Section 9.6. This second approach has substantially smaller computational complexity than the first because we need to solve a system of linear algebraic equations with only the necessary number of reactions in particular loops in order to be able to solve the rest of bodies in each loop from the equilibrium of every body of the loop. However, the most promising one is the third approach — the parametric method and the transformation of the MBS into the equivalent truss from Section 8.2, which is the method most likely to be advantageous in the future. The advantage of the equivalent truss is that, in its structure, the truss automatically fulfils many conditions on reaction forces which, in other approaches, can be computed only with difficulty in arriving at the solution.

The kinetostatic solution of MBSs with friction always leads to the solution of nonlinear equations by iterations. There is an advantageous way of using such iterations which have direct physical interpretation at least at the beginning of the iteration process (see end of Section 8.1). The first iteration is just the kinetostatic solution of MBS without friction.

Every further iteration is also the kinetostatic solution of MBS, where the friction forces are computed from the previous iteration of reaction forces and are considered as applied forces.

Among the methods for static solution of MBSs (determination of equilibrium position) there are also three different approaches. We either solve the equilibrium equations directly as a system of nonlinear equations, or we solve the equilibrium equations subsequently during the quasistatic loading (see end of Section 8.3), or we determine the equilibrium position of MBS as the solution of a special dynamic problem. We formulate the simulation problem of a particular MBS under the action of the potential energy or equivalent applied forces, and we then add large additional damping forces which depend on velocities. We start the simulation from some real initial position and we continue the integration until the still position, which is the equilibrium positition. The first approach has great problems because many different equilibrium positions often exist which can never or can only very rarely be reached because they are often unstable. In such circumstances, it is extremely difficult to find the real solution of a system of highly nonlinear equations. The other two approaches solve these problems by different simulation of the way in which the real equilibrium position of the MBS is achieved in reality. The application of all these methods depends further on the kinematic description (coordinates) used for the position of the MBS. It is necessary to investigate further the most suitable choice of coordinates, for example, to compare the computational complexity of different methods with relative (Lagrangian, minimum) coordinates, with cartesian and with natural coordinates, etc.

For the solution of direct dynamic problems there are many different methods (multibody formalisms) described and compared in Section 9.5. Surprisingly the most efficient procedure is the residual algorithm from Section 9.5.1.9. For smaller MBSs (for example for simple open kinematic chains up to 8 quite general revolute joints) the composite rigid-body formalism from Section 9.5.1.6 is very efficient. For larger MBSs the recursive formalism (articulated body-inertia) from Section 9.5.2 is the more efficient. A major research problem is still the most efficient way of solving MBSs with kinematic loops. There are two approaches. Either we use the method of disconnecting kinematic loops and Lagrange multipliers or we use the method of help variables (see Section 9.5.1.1). The first approach together with non-recursive formalisms leads to the system of differential-algebraic (DAE) equations or with recursive formalisms to differential equations and systems of linear algebraic equations for Lagrange multipliers. The second approach together with recursive formalism, based on the combination of the geometric diagram and the parametric method in kinematics and recursive formalism for MBSs with kinematic loops from Section 9.5.2.2, could be particularly efficient. It leads only to differential equations. However, further computational comparisons are necessary. Another research problem is the application of efficient recursive formalism to the MBSs with flexible elements possibly described by FEM method.

Another major research problem is to find a method for the integration of equations of motion which is efficient and sufficiently accurate and stable. From the point of view of computational complexity the choice of particular multibody formalism is also connected with the choice of the method of integration of equations of motion described in Section 9.8. There are still two groups of approaches; either we use redundant coordinates and we must solve DAE or we use minimum coordinates and we solve only differential equations. The use of DAE is very convenient because the equations of motion can be formulated simply by adding their parts, for example by the compartment method. The numerical solution of DAE has many problems described in Section 9.8, and many new approaches are still being developed. The solution of differential equations of equations of motion formulated by minimum coordinates is straightforward. The greatest problem is the choice of independent coordinates with respect to possible singular cases. Generally there exists no one choice of independent coordinates which can be suitable for all motions of MBS. Therefore the choice of independent coordinates must generally be repeated several times during the motion of MBS. The problems which must be solved are the re-start of integration after the new choice of coordinates and a reliable test of the need to change the chosen coordinates. An advantageous future method of direct dynamic solution of MBS could be the recursive formalism for MBSs with kinematic loops from Section 9.5.2.2 in combination with help variables and the geometric diagram and the parametric method in kinematics.

Another important investigation is the application of the compartment method in dynamics which would facilitate the investigation of systems combined from parts with a different physical nature such as mechanical systems driven by hydraulic actuators and controlled by electronic components, i.e. mechatronical systems. It is a great problem to know whether the recursive formalism can be generalized to other physical systems or whether such systems described by different formalisms must be connected by the compartment method.

Robotics needs the solution of many different problems of mechanics. The fundamental difficulty is that these problems must be solved in real time because they are the basis of robotic advanced control systems. The structure of robots will be even more complex in the future due to redundancy, flexibility and parallel construction. The solution of inverse kinematics (see Section 5.13, 5.11.6 and 11.1.5) and the solution of inverse dynamics (see Section 8.1 and 9.5.1.6.1) are the most important. An advantageous future method of inverse kinematic solution of robots could be the combination of structural approximation, which gives an accurate estimation of the solution, and the geometric diagram and parametric method, which enables us to solve differential kinematic analysis for the computation of the next iteration with minimum computation complexity. An advantageous future method of inverse dynamic solution of robots (which can be very complex MBSs) could always be derived from the corresponding efficient method of solution of the direct dynamic problem as in Section 9.5.1.8.

The kinematic and dynamic sensitivity analysis is based on the particular method used for kinematic and dynamic analysis. Hence the conclusions about kinematic and dynamic methods determine the methods of kinematic and dynamic sensitivity analysis. The most advantageous and promising method of dynamic sensitivity analysis is the method of adjoint variables. It is necessary to develop the efficient combination of recursive multibody formalism and the method of adjoint variables.

The methods of kinematic and dynamic synthesis are determined by the methods of kinematic and dynamic analysis. Their great problem is the lack of qualitative criteria of a machine's behaviour from the point of view of kinematics and dynamics. A great part of the dynamic synthesis of MBSs is the synthesis of their control systems. Future investigation would be necessary in the development of control systems for MBSs which cannot be directly decoupled. There are many such systems, for example, a crane, a walking robot, a flexible robot and general flexible mechanical structures.

All the methods described must ultimately be applied within a CAD system. The methods described must be connected with traditional CAD systems such as graphical systems and databases. The data from traditional CAD systems (e.g. structure of MBS, dimensions, mass and inertia properties, modes of vibrations from FEM systems) must be applicable in analyses of MBSs and the results of these analyses must be capable of further use within traditional CAD systems (e.g. for animation, numerical parameters). Particular interfaces must therefore be developed according to the principles of modularity from Section 2.4.4.

References

ALTSHULLER, G. S. (1984). *Creativity as an Exact Science (The Theory of the Solution of Inventive Problems)*. New York: Gordon and Breach Science Publishers

ANDREWS, G. C. (1977). *Dynamics using vector network techniques*. Waterloo: Department of Mechanical Engineering, University of Waterloo, Ontario, Canada

ANDREWS, G. C. and KESAVAN, H. K. (1978). Simulation of multibody systems using the vector network model. In *Dynamics of Multibody Systems*, edited by K. Magnus, pp. 1–13. Berlin: Springer-Verlag

ANDRZEJEWSKI, Th., BOCK, H. G., EICH, E. and VON SCHWERIN, R. (1993). Recent Advances in the Numerical Integration of Multibody Systems. In *Advanced Multibody System Dynamics, Simulation and Software Tools*, edited by W. Schiehlen, pp. 127–152. Dordrecht: Kluwer Academic Publishers

ANGELES, J. (1982). *Spatial Kinematic Chains*. Berlin: Springer-Verlag

ARGAWAL, V. P. and RAO, J. S. (1987). Structural classification of kinematic chains and mechanisms. *Mechanism and Machine Theory*, **22**, 489–496

ARMSTRONG, W. W. (1979). Recursive solution to the equations of motion of an n link manipulator. In *Proceedings of the 5th World Congress on Theory of Machines and Mechanisms*, Vol. 2, pp. 1343–1346. Montreal: ASME

ARTOBOLEVSKIJ, I. I. (1975). *Theory of Mechanisms and Machines*. Moscow: Nauka (in Russian)

ASSUR, L. V. (1952). *Investigation of Planar Mechanisms Containing Lower Kinematic Pairs from the Point of View of Their Structure and Classification.* Moscow: Izdatelstvo Akademii nauk (in Russian)

BABUŠKA, L. (1988). The p- and hp-version of the finite element method, the state of the art. In *Finite Elements: Theory and Applications*, edited by D. L. Dwoyer, M. Y. Hussaini and R. G. Voigt. New York: Springer Verlag

BAE, D. S. and HAUG, E. J. (1987). A recursive formulation for constrained mechanical system dynamics: Part I: Open loop systems, Part II: Closed loop systems. *Mechanical Structures and Machines*, **15**, 3, 4, 359–382, 481–506

BAE, D. S., HWANG, R. S. and HAUG, E. J. (1988). A recursive formulation for real time dynamic simulation. *Advances in Design Automation*, ASME, DE–14, 499–508

BAE, D. S. and YANG, S. M. (1991). A stabilization method for kinematic and kinetic constraint equations. In *Real-Time Integration Methods for Mechanical System Simulation*, edited by E. J. Haug and R. C. Deyo, pp. 209–232. Berlin: Springer-Verlag

BATHE, K. J. (1982). *Finite Element Procedures in Engineering Analysis.* Englewood Cliffs: Prentice-Hall

BAUMGARTE, J. (1972). Stabilization of constraints and integrals of motion in dynamical systems. *Computer Methods in Applied Mechanics and Engineering*, **1**, 1–16

BAZERGHI, A., GOLDENBERG, A. A. and APKARIAN, J. (1984). An exact kinematic model of PUMA 600 manipulator. *IEEE Transactions on Systems, Man, and Cybernetics*, **14**, 483–487

BEGGS, J. S. (1983). *Kinematics.* Washington: Hemisphere Publishing Corporation

BENEDICT, C. E. and TESAR, D. (1978). Model formulation of complex mechanisms with multiple inputs. Part 1 – Geometry, Part 2 – The dynamic model. *Transactions of ASME, Journal of Mechanical Design*, **100**, 747–754, 755–761

BERDICHEVSKIJ, V. L. (1983). *Variational Principles of Structural Mechanics.* Moscow: Nauka (in Russian)

BESTLE, D. (1989). Zur Optimierung von Mehrkörpersystemen. *Institutsbericht IB–14.* Stuttgart: Universität in Stuttgart, Institut B für Mechanik

BESTLE, D. (1994). Analyse und Optimierung von Mehrkörpersystemen. Berlin: Springer-Verlag

BLAJER, W. (1993). Dynamic analysis of constrained multibody systems in orthonormalized tangent space. In *Advanced Multibody System Dynamics, Simulation and Software Tools*, edited by W. Schiehlen, pp. 415–420. Dordrecht: Kluwer Academic Publishers

BORRI, M., MELLO, F. and ATLURU, S.N. (1990). Variational approaches for dynamics: Numerical studies. In *Dynamical Problems of Rigid-Elastic Systems and Structures*, edited by N. V. Banichuk, D. M. Klimov and W. Schiehlen, pp. 27–35. Berlin: Springer-Verlag

BORRI, M., BOTTASSO, C. and MANTEGAZZA, P. (1991). A new form of the Lagrangian multipliers in constrained multibody dynamics. In *Abstracts of EUROMECH 1st European Solid Mechanics Conference*, pp. 37–39. Munich: Technical University

BOTTEMA, O. and ROTH, B. (1979). *Theoretical Kinematics.* Amsterdam: North Holland

BRADY, M., HOLLERBACH, J. M., JOHNSON, T. L., LOZANO PEREZ, T. and MASON M. T. (1982). *Robot Motion: Planning and Control.* Cambridge, Massachusetts: MIT Press

BRANDL, H., JOHANNI, R. and OTTER, M. (1988). A very efficient algorithm for the simulation of robots and similar multibody systems without inversion of the mass matrix. In *Theory of Robots*, edited by P. Kopacek, I. Troch and K. Desoyer. Oxford: Pergamon Press

BRANDL, H., JOHANNI, R. and OTTER, M. (1987). An algorithm for the simulation of multibody systems with kinematic loops. In *Proceedings of the 7th World Congress of IFToMM* (Sevilla, 1987), edited by E. Bautista, J. Garcia Lomas and A. Navarro, pp. 407–411. Oxford: Pergamon Press

BRÁT, V. (1960). Parametric method and its use in mechanics of system of bodies. *Research report FSI ČVUT 60/6.* Prague: Czech Technical University (in Czech)

BRÁT, V. (1969). A six link spatial mechanism. *Journal of Mechanisms*, 4, 325–336

BRÁT, V. (1981). *The Matrix Methods in the Analysis and Synthesis of Spatial Constrained Mechanical Systems.* Prague: Academia (in Czech)

BRÁT, V. and LEDERER, P. (1973). KIDYAN – computer aided kinematic and dynamic analysis of planar mechanism. *Mechanism and Machine Theory*, 8, 457–467

BRÁT, V. and STEJSKAL, V. (1975). Kinematic solution of spatial mechanisms of wheel and axles suspensions. Theory. *Research Report Z-93/75.* Prague: The Motorcar Research Institute (in Czech)

BRÁT, V. and STEJSKAL, V. (1976). Kinematic solution of spatial mechanisms of wheel and axles suspensions. Algorithm. *Research Report Z-47/76.* Prague: The Motorcar Research Institute (in Czech)

BRÁT, V. and STEJSKAL, V. (1979a). Kinetostatic analysis of spatial constrained mechanial systems. *Strojírenství*, **29**, 323–329 (in Czech)

BRÁT, V. and STEJSKAL, V. (1979b). Dynamic accuracy of mechanisms. *Strojnícky časopis*, **30**, 559–570 (in Czech)

BRÁT, V., BÍLA, J. and VALÁŠEK, M. (1981). The automatic derivation of equation of motion of a mechanical system. *Acta Technica ČSAV*, 1981, 645–655

BREPTA, R. (1986). The relation between the model of mechanical system and the way of its loading. In *Proceedings of Conference on Mathematical Methods in Engineering*, pp. 563–566. Plzeň: Central Research Institute ŠKODA Enterprise

BUTENIN, N. V. (1971). *Introduction to Analytical Mechanics.* Moscow: Nauka (in Russian)

CELIGÜETA, J. T. and GARCIA de JALON, J. (1987). Analysis of mechanisms with flexible links. In *Proceedings of the 7th World Congress of*

IFToMM (Sevilla, 1987), edited by E. Bautista, J. Garcia Lomas and A. Navarro, pp. 505–509. Oxford: Pergamon Press

CHACE, M. A. (1967). Analysis of the time-dependence of multifreedom mechanical systems in relative coordinates. *Journal of Engineering for Industry*, 1967, pp. 119–125

CHEN WAI KAI (1971). *Applied Graph Theory.* Amsterdam: North Holland

CORBEN, H. C. and STEHLE, P. (1950). *Classical Mechanics.* New York: J. Wiley and Sons

DENAVIT, J. and HARTENBERG, R. S. (1955). A kinematic notation for lower pair mechanisms based on matrices. *Transactions of ASME, Journal of Applied Mechanics*, **22**, 215–221

DENNIS, J. E. Jr. and SCHNABEL, R. B. (1983). *Numerical Methods for Unconstrained Optimization and Nonlinear Equations.* Englewood Cliffs: Prentice-Hall

DIMENTBERG, F. M. (1971). *The Method of Screws in Applied Mechanics.* Moscow: Mashinostroenie (in Russian)

DIMENTBERG, F. M. (1978). *Theory of Screw and its Application.* Moscow: Nauka (in Russian)

DOBRONRAVOV, V. V. (1970). *Fundamentals of Mechanics of Nonholonomic Systems.* Moscow: Vysschaya Schkola (in Russian)

DOBRONRAVOV V. V. (1976). *Foundamentals of Analytical Mechanics.* Moscow: Vysschaya Schkola (in Russian)

DOBROVODSKÝ, K. (1986). Representation of geometric relations in robotic kinematic structures. In *Počítačová grafika CG'86.* Smolenice: JSMF (in Slovak)

DRÁTOVSKÝ, Z. (1978). Graph manipulation language GRM. *Research Report Z-87/78.* Prague: The Motorcar Research Institute (in Czech)

EICHBERGER, A., FÜHRER, C. and SCHWERTASSEK, R. (1993). The benefits of parallel multibody simulation and its application to vehicle dynamics. In *Advanced Multibody System Dynamics, Simulation and Software Tools*, edited by W. Schiehlen, pp. 107–126. Dordrecht: Kluwer Academic Publishers

ELMQVIST, H. (1994). Dymola – Dynamic Modeling Language, User's Manual. Lund: Dynasim AB

FEATHERSTONE, R. (1983). The calculation of robot dynamics using articulated body inertias. *The International Journal of Robotic Research*, **2**, 13–20

FEATHERSTONE, R. (1987). *Robot Dynamics Algorithms.* Dodrecht: Kluwer Academic Publishers

FEDORENKO, R. P. (1978). *Approximate Solutions of Optimal Control Problems.* Moscow: Nauka (in Russian)

FLEURY, C. and BRAIBANT, V. (1984). Optimization methods. In *Computer Aided Analysis and Optimization of Mechanical System Dynamics*, edited by E. J. Haug, pp. 637–662. Heidelberg: Springer-Verlag

FORSYTHE, G. E., MALCOLM, M. A. and MOLER, C. B. (1977). *Computer Methods for Mathematical Computations.* New York: Prentice-Hall

FREEMAN, R. A. and TESAR, D. (1982). The generalized coordinate selection for the dynamics of complex planar mechanical systems. *Transactions of ASME, Journal of Mechanical Design*, **104**, 206–217

FREUND, E. (1982). Fast nonlinear control with arbitrary pole placement for industrial robots and manipulators. *International Journal of Robotic Research*, **1**, 65–78

FÜHRER, C. (1988). *Differential-algebraische Gleichungssysteme in mechanischen Mehrkörpersystemen.* München: Technische Universität, Mathematisches Institut München

GARCIA de JALÓN, J., UNDA, J. and AVELLO A. (1986). Natural coordinates for the computer analysis of multibody systems. *Computer Methods in Applied Mechanics and Engineering*, **3**, 309–327

GARCIA de JALÓN, J., AVELLO, A. and JIMENEZ, J. M. (1987). Basis for the nullspace for the Jacobean matrix of constrained multibody systems.

In *Proceedings of the 7th World Congress of IFToMM* (Sevilla, 1987), edited by E. Bautista, J. Garcia Lomas and A. Navarro, pp. 501–504. Oxford: Pergamon Press

GARCIA de JALÓN, J. and BAYO E. (1994). *Kinematic and Dynamic Simulation of Multibody Systems. The Real-Time Challenge.* New York: Springer-Verlag

GEAR, C. W. (1971). *Numerical Initial Value Problem in Ordinary Differential Equations.* Englewood Cliffs: Prentice-Hall

GEAR, C. W. (1984). Differential algebraic equations. In *Computer Aided analysis and Optimization of Mechanical System Dynamics*, edited by E. J. Haug, pp. 323–334. Heidelberg: Springer-Verlag

GEAR, C. W. (1991). An Introduction to Numerical Methods for ODEs and DAEs. In *Real-Time Integration Methods for Mechanical System Simulation*, edited by E. J. Haug and R. C. Deyo, pp. 115–126. Berlin: Springer-Verlag

GEAR, C. W. and PETZOLD, L. (1984). ODE methods for solution of differential/algebraic systems. *SIAM Journal of Numerical Analysis*, **21**, 716–728

GEAR, C. W., LEIMKUHLER, F. and GUPTA, K. (1985). Automatic integration of Euler-Lagrange equations with constraints. *Journal of Computational and Applied Mathematics*, **12**, 77–99

GERADIN, M. (1993). Commputational aspects of the finite element approach to flexible multibody systems. In *Advanced Multibody Dynamics. Simulation and Software Tools*, edited by W. Schiehlen, pp. 337–354. Dodrecht: Kluwer Academic Publishers

GILL, P. E., MURRAY, W., SANNDARS, M. A. and WRIGHT, M. H. (1984). Sequential quadratic programming methods for nonlinear programming. In *Computer Aided Analysis and Optimization of Mechanical System Dynamics*, edited by E. J. Haug, pp. 679–700. Heidelberg: Springer-Verlag

GOLDSTEIN, H. (1950). *Classical Mechanics.* Reading: Addison-Wesley

GOLST, G., RELVIK, Ch. and SILDE O. (1979). *Fundamental Questions of Analytical Mechanics.* Talin: Valgus (in Russian)

GRIEPENTORG, E. and MÄRZ, R. (1986). *Differential Algebraic Equations and Their Numerical Treatment.* Leipzig: Teubner Verlag

HAIRER, E., LUBICH, Ch. and ROCHE, M. (1989). *The Numerical Solution of DA Systems by Runge Kutta Methods. LN in Mathematics 1409.* Berlin: Springer-Verlag

HALL, G., WATT, J. M. (1976) Editors. *Modern Numerical Methods for Ordinary Differential Equations.* Oxford: Clarendon Press

HARTENBERG, R. S. and DENAVIT, J. (1964). *Kinematic Synthesis of Linkages.* New York: McGraw-Hill

HAUG, E. J. (1989). *Computer-Aided Kinematics and Dynamics of Mechanical Systems, Volume I: Basic Methods.* Boston: Allyn and Bacon

HAUG, E. J. (1993). High speed multibody dynamic simulation and its impact on man-machine systems. In *Advanced Multibody System Dynamics, Simulation and Software Tools*, edited by W. Schiehlen, pp. 1–18. Dordrecht: Kluwer Academic Publishers

HAUG, E. J., NIKRAVESH, P. E., SOHONI, V. N. and WEHAGE, R. A. (1982). Computer aided analysis of large scale constrained mechanical systems. In *Proceedings of 4th International Symposium on Large Engineering Systems*, Calgary, Alberta

HAUG, E. J. and ARORA, J. S. (1979). *Applied Optimal Design.* New York: John Wiley

HAUG, E. J., WEHAGE, R. A. and BARMAN, N. C. (1981). Design sensitivity analysis of planar mechanism and machine dynamics. *Transactions of ASME, Journal of Mechanical Design*, **103**, 560–570

HAUG, E. J. and SOHONI, V. N. (1984). Design sensitivity analysis and optimization of kinematically driven systems. In *Computer Aided Analysis and Optimization of Mechanical System Dynamics*, edited by E. J. Haug, pp. 449–534. Heidelberg: Springer-Verlag

HAUG, E. J., MANI, N. K. and KRISHNASWANI, P. (1984). Design sensitivity analysis and optimization of dynamically driven systems. In *Computer Aided Analysis and Optimization of Mechanical System Dynamics*, edited by E. J. Haug, pp. 535–636. Heidelberg: Springer-Verlag

HAUG, E. J., WU, S. C. and YANG, S. M. (1986). Dynamics of mechanical systems with Coulomb friction, stiction, impact and constraint addition-deletion I. *Mechanism and Machine Theory,* **21**, 401–406

HAUG E. J. and DEYO, R. C. (1991) Editors. *Real-Time Integration Methods for Mechanical System Simulation.* Berlin: Springer-Verlag

HAYATI, S. A. and ROSTON, G. P. (1986). Inverse kinematic solution for near-simple robots and its application to robot calibration. In *Recent Trends in Robotics,* edited by M. Jamshidi, L. Y. S. Luh and M. Shakinpoor, pp. 41–49. Amsterdam: Elsevier

HEWITT, J.R. (1992). *Lecture notes for CISM Advance School of Mechatronic.* CISM, Udine

HOLLERBACH, J. M. (1980). A recursive formulation of Lagrangean manipulator dynamics and a comparative study of dynamics formulation complexity. *IEEE Transactions on System, Man, and Cybernetics,* **10**, 730–736

HOLLERBACH, J. M. (1984). Dynamics scaling and manipulators trajectories. *Transactions of ASME, Journal of Dynamic Systems, Measurement and Control,* **106**, 102–106

HUBKA, V. (1984). *Theorie Technischer Systeme.* Berlin: Springer-Verlag

HUNT, L. R., SU, R. and MEYER, G. (1983). Design of multi-input nonlinear systems. In *Differential Geometric Control Theory,* edited by R. Brockett. Birkhauser

HURTY, W. C. (1991). Dynamic analysis of structural systems using component modes. *AIAA Journal,* **32**, pp. 1637–1655

HUSTON R. L. (1990). *Multibody Dynamics.* Boston: Butterworth-Heinemann

IBM Application Program, System/360, Scientific Subroutine Package, Version III, Programmer's Manual, Subroutine MFGR

KANE, T. R. and LEVINSON, D. A. (1983). The use of Kane's dynamical equations in robotics. *The International Journal of Robotic Research,* **2**, 3–21

KANE, T. R., LIKINS, P. W. and LEVINSON, D. A. (1983). *Spacecraft Dynamics.* New York: McGraw-Hill

KARGER, A. and NOVÁK, J. (1978). *Spatial Kinematics and Lie's Groups.* Prague: SNTL (in Czech)

KIM, S. S. and HAUG, E. J. (1988). A Recursive Formulation for Flexible Multibody Dynamics, Part I: Open-Loop Systems. *Computer Methods in Applied Mechanics and Engineering,* **71**, pp. 293–314

KOENIG, H. E., TOKAD, Y., KESAVAN, H. K. and HEDGES, H. G. (1967). *Analysis of Discrete Physical Systems.* New York: McGraw-Hill

KORTÜM W. and SHARP R. S. (1993) Editors. *Multibody Computer Codes in Vehicle System Dynamics.* Amsterdam: Swets & Zeitlinger

KOVANDA, J. and VALÁŠEK, M. (1987). Dynamic optimization of axle suspension mechanism by means of CAD. In *Proceedings of Conference on Automation in Car Industry.* Bratislava: DT ČSVTS (in Czech)

KOVÁŘ, J. and SUCHÝ, M. (1982). Suspension of vehicles for rigid terrain. *U. S. Patent 4-353-428*

KREUZER, E. (1979). *Symbolische Berechnung von Bewegungsgleichungen von Mehrkörpersystemen.* Düsseldorf: VDI Verlag

KRON, G. (1963). *Diakoptics: The Piecewise Solution of Large Scale Systems.* London: McDonald

KROTOV, V. F. and GURMAN, V. I. (1973). *Methods and Problems of Optimal Control.* Moscow: Nauka (in Russian)

LEE, H. Y. and LIANG, C. G. (1988). Displacement analysis of the general spatial 7-link 7R mechanism. *Mechanism and Machine Theory,* **23**, 219–226

LEE, H. Y., WOERNLE, C. and HILLER, M. (1992). A complete solution for the inverse kinematic problems of the general 6R robot manipulator. *Transactions of ASME Journal of Mechanical Design,* **113**, 481–486

LEECH, J. W. (1958). *Classical Mechanics.* Frome, Butler and Tanner Ltd.

LEISTER, G. (1990). Wahl geeigneter Koordinaten zur Beschreibung von

Mehrkörpersysteme. *Zwischenbericht ZB-49.* Stuttgart: Universität in Stuttgart, Institut B für Mechanik

LIANG, C. G. and LANCE, G. M. (1987). A differentiable null space method for constrained dynamic analysis. *Journal of Mechanisms, Transmissions, and Automation in Design*, **109**, 405–411

LITVIN, F. L. (1970). *Determination of the position of spatial mechanism by condition disconnecting of the loop.* Mashinovedenie, 1970 (in Russian)

LIVERMORE, D. F. (1967). The determination of equilibrium configurations of springrestrained mechanisms using 4×4 matrix method. *Transactions of ASME, Journal of Engineering for Industry*, **78**, 87–93

LUBICH, Ch. (1989). h^2 Extrapolation method for differential-algebraic equations of index 2. *Technical Report.* Innsbruck: Universität in Innsbruck, Institut für Mathematik und Geometrie

LUH, J. S. Y., WALKER, M. W. and PAHL, R. D. C. (1980). On-line computational scheme for mechanical manipulators. *Transactions of ASME, Journal of Dynamic Systems, Measurement, and Control*, **102**, 69–76

MAISSER, P. (1988). Analytische Dynamik von Mehrkörpersystemen. *ZAMM*, **68**, 463–481

MAISSER, P. (1990). Analytical multibody dynamics. In *Dynamical Problems of Rigid-Elastic Systems and Structures*, edited by N. V. Banichuk, D. M. Klimov and W. Schiehlen, pp. 143–150. Berlin: Springer-Verlag

MANN, H. (1984). *The Use of Computer in Electric Design.* Prague: SNTL (in Czech)

MANN, H. (1986). *Automation in Dynamic System Design.* Prague: Dum techniky ČSVTS (in Czech)

MAYEDA, W. (1972). *Graph Theory.* New York: Wiley-Interscience

NEIMARK Iu. I. and FUFAEV N. A. (1972). *Dynamics of Nonholonomic Systems.* Providence: American Mathematical Society

NEUMAN, C. P. and MURRAY, J. J. (1985). Computational robot dynamics: Foundations and applications. *Journal of Robotic Systems*, **2(4)**, 425–452

NEUMAN, C. P. and MURRAY, J. J. (1987a). Customized computational robot dynamics. *Journal of Robotic Systems*, **4(4)**, 503–526

NEUMAN, C. P. and MURRAY, J. J. (1987b). Symbolically Efficient Formulations for Computational Robot Dynamics. *Journal of Robotic Systems*, **4(6)**, 743–769

NIELAN, P. E. (1986). *Efficient Computer Simulation of Motions of Multibody Systems.* Ph.D. Thesis, Stanford University, Stanford

NOVOTNÝ, J. (1983). *Theory of Chains of Rigid and Flexible Bodies.* DrSc. Thesis, VUT, Brno

NIKRAVESH, P. E. (1988). *Computer-Aided Analysis of Mechanical Systems.* Englewood Cliffs: Prentice Hall

ORLANDEA, N., CHACE, M. A. and CALAHAN, D. A. (1976). A sparsity-oriented approach to the dynamic analysis and design of mechanical systems – Parts 1 and 2. *ASME Journal of Engineering for Industry*, Paper No. 76-DET-19, ASME Mechanisms Conference, Montreal, Canada

OSTERMAYER, G. (1991). On Baumgarte stabilization for differential algebraic equations. In *Real-Time Integration Methods for Mechanical System Simulation*, edited by E. J. Haug and R. C. Deyo, pp. 193–208. Berlin: Springer Verlag

OTTER, M., HOCKE, M., DABERKOW, A. and LEISTER, G. (1993). An object-oriented data model for multibody systems. In *Advanced Multibody System Dynamics, Simulation and Software Tools*, edited by W. Schiehlen, pp. 19–48. Dordrecht: Kluwer Academic Publishers

PAUL, B. (1975). Analytical dynamics of mechanisms — a computer oriented overview. *Mechanism and Machine Theory*, **10**, 481–507

PAUL, R. P. (1981). *Robot Manipulators: Mathematics, Programming, and Control.* Cambridge, Massachussetts: MIT Press

PETZOLD L. R. (1991). Methods and software for differential-algebraic systems. In *Real-Time Integration Methods for Mechanical System Simulation*, edited by E. J. Haug and R. C. Deyo. Berlin: Springer-Verlag

POLAK, E. (1971). *Computational Methods in Optimization, A Unified Approach.* New York: Academic Press

RAO, A. V. M., SANDOR, G. N., KOHLI, D. and SONI, A. H. (1973). Closed form synthesis of spatial function generating mechanism for the maximum number of precision points. *Transactions of ASME, Journal of Engineering for Industry*, **84**, 725–736

REKTORYS, K. (1968). *Review of Applied Mathematics*. Prague: SNTL (in Czech)

ROBERSON, R. E. and SCHWERTASSEK, R. (1987). *Introduction to the Dynamics of Multibody Systems*. Berlin: Springer-Verlag

RONGVED, L. and FLETCHER, H. J. (1964). Rotational coordinates. *Journal of The Franklin Institute*, **277**, 414–421

ROONEY, J. (1977). A survey of representations of spatial rotation about a fixed point. *Environment and Planning B*, **4**, 180–210

ROONEY, J. (1978). A comparison of representations of general spatial screw displacement. *Environment and planning B*, 5, 45–88

RYAN R. R. (1990). ADAMS – Multibody system analysis software. In *Multibody Systems Handbook*, edited by W. Schiehlen. Berlin: Springer-Verlag

SADOVSKIJ, V. N. (1974). *Fundamentals of General System Theory*. Moscow: Nauka (in Russian)

SAGE, A. P. and WHITE, C. C. III. (1977). *Optimum System Control*. Englewood Cliffs: Prentice-Hall

SCHIEHLEN, W. (1984). Dynamics of complex multibody systems. *SM Archives 9*, 159–195

SCHIEHLEN, W. (1990a) Editor. *Multibody Systems Handbook*. Berlin: Springer-Verlag

SCHIEHLEN, W. (1990b). Simulation and animation of multibody systems. In *Dynamical Problems of Rigid-Elastic Systems and Structures*, edited by N. V. Banichuk, D. M. Klimov and W. Schiehlen, pp. 215–226. Berlin: Springer-Verlag

SCHIEHLEN, W., KREUZER, E. (1978). Symbolic computerized derivation of equations of motion. In *Dynamics of Multibody Systems*, edited by K. Magnus. Berlin: Springer-Verlag

SCHMOLL, K. P. (1988). Modularer Aufbau von Mehrkörpersystemen unter Verwendung der Relativkinematik. *Fortschr.-Ber. der VDI-Zeitschriften*, Reihe 18, Nr. 57. Düsseldorf: VDI-Verlag

SCHÜPPHAUS, R. and MÜLLER, P. C. (1993). Control analysis and synthesis of linear mechanical descriptor systems. In *Advanced Multibody System Dynamics, Simulation and Software Tools*, edited by W. Schiehlen, pp. 463–468. Dordrecht: Kluwer Academic Publishers

SCHWERTASSEK, R. and ROBERSON, R. E. (1985). A perspective on computer-oriented multibody dynamical formalisms and their implementation. In *Dynamics of Multibody Systems*, edited by G. Bianchi and W. Schiehlen, pp. 125–143. Berlin: Springer-Verlag

SHABANA, A. A. (1989). *Dynamics of Multibody Systems*. New York: John Wiley & Sons

SHAMPINE, L. F. and GORDON, M. K. (1975). *Computer Solution of Ordinary Differential Equations*. San Francisco: W. J. Freeman

SHET, P. N. and UICKER, J. J. Jr. (1972). A generalized notation for mechanism. *Transactions of ASME, Ser. B, Journal of Engineering for Industry*, **93(1)**, 18–33

SIMEON, B. and RENTROP, P. (1993). An extended descriptor form for the simulation of constrained mechanical systems. In *Advanced Multibody System Dynamics, Simulation and Software Tools*, edited by W. Schiehlen, pp. 469–474. Dordrecht: Kluwer Academic Publishers

SOHONI, V. and HAUG, E. J. (1982). A state space method for kinematic optimization of mechanisms and machines. *Transactions of ASME, Journal of Mechanical Design*, **104**, 101–107

SOMMERFELD, A. (1952). *Mechanics*. New York: Academic Press

SPRING, K. W. (1986). Euler parameters and the use of quaternion algebra in the manipulation of finite rotations: A review. *Mechanism and Machine Theory*, **21**, 365–373

ŠREJTR, J. (1954). *Technical Mechanics II, Kinematics 1, 2*. Prague: SNTL (in Czech)

STEHNO, J. (1973). *Motion in E3 and quaternions. Solution of spatial mechanisms.* Ph.D. Thesis. Prague: ČVUT, FSI (in Czech)

STEJSKAL, V. (1977a). Synthesis of mechanism for given force behaviour. *Strojnícky časopis*, **28**, 283–292 (in Czech)

STEJSKAL, V. (1977b). Graph theory and its application in technical mechanics. Research *Report FSI ČVUT 231/77/1*. Prague: Czech Technical University (in Czech)

STEJSKAL, V. (1978). Automatic assembly of equilibrium equations of spatial system of bodies. *Strojnícky časopis*, 29, 83–94 (in Czech)

STEJSKAL, V. and VALÁŠEK, M. (1986). Automated general kinematic analysis of spatial mechanisms program OKAM2. *Acta Technica ČSAV*, 1986, 137–148

STEJSKAL, V., KLOCPERK, J., KOVANDA, J., OPIČKA, F. and PROVAZNÍK, J. (1988). Program system MECH for kinematic analysis of spatial constrained mechanical systems. *Strojírenství*, 38, 730–734 (in Czech)

STEJSKAL, V., OPIČKA. F., VALÁŠEK, M. and JANOVSKÝ, V. (1990). Singular cases in rigid body mechanics. *Acta Technica ČSAV*, 1990, 65–82

STEWART, D. (1965). A platform with six degrees of freedom. *Proceedings Inst. Mech. Engrs.* **180**, Part 1, No. 15, 371

STUELPNAGEL, J. (1964). On the parameterization of the three dimensional rotation group. *SIAM Review*, **6**, 422–430

SUTHERLAND, G. H. and ROTH, B. (1973). A transmission index for spatial mechanisms. *Transaction of ASME, Journal of Engineering for Industry*, **95**, 589–597

TICHONOV, A. N. and ARSENIN, B. Ja. (1986). *Methods of the Solution of Incorrect Problems.* Moscow: Nauka (in Russian)

TOMOVIC, R. and VUKOBRATOVIC, M. (1972). *General Sensitivity Theory.* New York: Elsevier

TOOGOOD, R. W. (1987). Symbolic generation of robot dynamics equations. *Technical Report 87-04*. Alberta: Department of Mechanical Engineering, University of Alberta, Canada

TOOGOOD, R. W. (1988). Robot direct dynamics algorithms using symbolic generation. *Technical Report 88*. Alberta: Department of Mechanical Engineering, University of Alberta, Canada

TOOGOOD, R. W. and WARKENTIN, J. H. (1986). Robot calibration using an on-board vision system. *Technical report 86*. Alberta: Department of Mechanical Engineering, University of Alberta, Canada

UICKER, J. J. Jr. (1965). *On the dynamic analysis of spatial linkages using 4 by 4 matrices*. Ph.D. Thesis. Department of Mechanical Engineering and Astronautical Sciences, Northwestern University, Evanston, Illinois

UICKER, J. J. Jr., DENAVIT, J. and HARTENBERG, R. S. (1964). An iterative method for the displacement analysis of spatial mechanisms. *Transaction of ASME, Journal of Applied Mechanics*, **31**, 309–314

UICKER, J. J. Jr. and RAICU, A. (1975). A method for the identification of equivalence of kinematic chains. *Mechanism and Machine Theory*, 10, 375–383

UNDA, J., GARCIA de JALÓN, J., LOSANTOS, F. and ENPARANTZA, R. (1986). A comparative study on some different formulations of the dynamic equations of constrained mechanical systems. *ASME Paper 86-DET-138*

VALÁŠEK, M. (1982). The change of dynamic properties of a system by the transformation of its input variables. *Acta Polytechnica*, **10**, II, 4, 159–164 (in Czech)

VALÁŠEK, M. (1986a). Programme control of a crane in real time. *Automatizace*, **29**, 211–215 (in Czech)

VALÁŠEK, M. (1986b). Synthesis of optimal trajectory of industrial robots. *Kybernetika*, **22**, 409–424

VALÁŠEK, M. (1986c). Compartment modelling of planar constrained mechanical systems. In *Automation in Dynamic System Design*, edited by H. Mann. Prague: Dum techniky ČSVTS (in Czech)

VALÁŠEK, M. (1987a). Dynamic time parameterization of manipulator trajectories. *Kybernetika*, **23**, 154–174

VALÁŠEK, M. (1987b). Energetically optimal control of industrial robots. In *Proceedings of the 7th World Congress of IFToMM* (Sevilla, 1987), edited by E. Bautista, J. Garcia-Lomas and A. Navarro, pp. 1465–1468. Oxford: Pergamon Press

VALÁŠEK, M. (1987c). Simulation model assembly for any physical system by its compartment composition. In *Proceedings of European Congress on Simulation ECS'87*, Volume B, pp. 227–231. Prague: Academia

VALÁŠEK, M. (1988). Principles of computer-aided system operation. In *Cybernetics and Systems '88*, edited by R. Trappl, pp. 203–210. Dordrecht: Kluwer Academic Publishers

VALÁŠEK, M. (1990a). On the efficient implementation of multibody system formalism, *IB-17*. Stuttgart: Universität in Stuttgart, Institut B für Mechanik

VALÁŠEK, M. (1990b). Contributions to kinematics, statics and dynamics of multibody systems solvable in linear computational complexity, *IB-18*. Stuttgart: Universität in Stuttgart, Institut B für Mechanik

VALÁŠEK, M. (1990c). Differential kinematic analysis of general 6R robot manipulators without the inverse of Jacobian matrix. In *Proceedings of Advances in Robot Kinematics*. Linz: Johannes Kepler University

VALÁŠEK, M. and STEJSKAL, V. (1986). The complete equivalence of Newton-Euler equations of motion and Lagrange's equations of mixed type. *Acta Technica ČSAV*, 1986, 607–623

VALÁŠEK, M. and STEJSKAL, V. (1987a). The complete equivalence of Kane's equations of motion and Lagrange's equations of mixed type written in quasicoordinates. *Acta Technica ČSAV*, 1987, 30–35

VALÁŠEK, M. and STEJSKAL, V. (1987b). The kinematic accuracy of spatial mechanisms and robot manipulators. *Acta Technica ČSAV*, 1987, 453–478

VALÁŠEK, M. and STEJSKAL, V. (1988). The comparison of matrix and quaternion approaches towards kinematics. *Acta Technica ČSAV*, 1988, 118–136

VALÁŠEK, M. and STEJSKAL, V. (1990). Contributions to kinematics, statics and dynamics of multibody systems solvable in linear computational complexity. In *Proceedings of IUTAM Symposium on Dynamical Problems of Rigid-Elastic Systems and Structures*, edited by N.V. Banichuk, D.M. Klimov and W. Schiehlen. Berlin: Springer-Verlag

VALÁŠEK, M. and KALNÝ, R. (1992). Structural approximation algorithm for inverse kinematic tranformation of non-simple robots. In *Proceedings of 3rd International Workshop on Advances in Robot Kinematics*, pp. 75–79. Ferrara: University of Ferrara

VALÁŠEK, M., ZNAMENÁČEK, J. and VAMPOLA, T. (1995). Dynamics of flexible multibody systems. In *Proceedings of Ninth World IFToMM Congress on the Theory of Machines and Mechanisms*, pp. 1667–1671. Milano: Edizioni Unicopli

WALKER, M. W. and ORIN, D. E. (1982). Efficient dynamic computer simulation of robotic mechanisms. *Journal of Dynamic Systems, Measurment and Control*, **104**, 205–211

WANG, J. T. and HUSTON, R. L. (1987). Kane's equations with undetermined multipliers — application to constrained multibody systems. *Transactions of ASME, Journal of Applied Mechanics*, **54**, 424–429

WEHAGE, R. A. and HAUG, E. J. (1982). Generalized coordinate partitioning for dimension reduction in analysis of constrained dynamical systems. *Transactions of ASME, Journal of Mechanical Design*, **104**, 247–255

WEHAGE, R. A. (1984). Quaternions and Euler parameters — a brief exposition. In *Computer Aided Analysis and Optimization of Mechanical Systems Dynamics*, edited by E. J. Haug. Berlin: Springer-Verlag

WINSTON, P. H. (1977). *Artificial Intelligence*. Reading, Massachusetts: Addison-Wesley Publishing Company

WITTENBAUER, F. (1923). *Graphische Dynamik*. Berlin: Verlag von Julius Springer

WITTENBURG, J. (1977). *Dynamics of Systems of Rigid Bodies*. Stuttgart: Teubner B. G.

WITTENBURG, J. and WOLZ, U. (1985a). MESA VERDE: A symbolic program for nonlinear articulated-rigid-body dynamics. *ASME Paper 85-DET-151*, Presented at the ASME Design Engineering Division Conference and Exhibition on Mechanical Vibration and Noise, Cincinnati, Ohio

WITTENBURG, J. and WOLZ, U. (1985b). MESA VERDE: Ein Computerprogramm zur Simulation der nichtlinearen Dynamik von Vielkörpersystemen. *Robotersysteme*, **1**, 7–18

WOERNLE, C. (1988). Ein Systematisches Verfahren zur Aufstellung der geometrischen Schliessbedingungen in kinematische Schleifen mit Anwendung bei der Rückwartstransformation für Industrieroboter. *Fortschnittberichte VDI*, Reihe 18, Nr. 59. Düsseldorf: VDI Verlag

WONHAM, W. M. and MORSE, A. S. (1970). Decoupling and pole assignment in linear multivarible systems — A geometric approach. *SIAM Journal of Control*, **8**, 1–18

WONHAM, W. M. and MORSE, A. S. (1971). Status of noninteracting control. *IEEE Transactions on Automatic Control*, AC–16, 568–581

WU, Ch. (1984). A kinematic CAD tool for the design and control of a robot manipulator. *International Journal of Robotic Research*, **3**, 58–67

YANG, A. T. and FREUDENSTEIN, F. (1964). Application of dual-number quaternion algebra to the analysis of spatial mechanisms. *Transactions of ASME, Journal of Applied Mechanics*, **86**, 300–308

YOO, W. S. and HAUG, E. J. (1985). Dynamics of flexible mechanical systems using vibration and static correction modes. *ASME Paper* No. 85-Det-71

YOSHIKAWA, H. (1983). Automation of thinking in design. In *CAPE'83 Computer Application in Production and Engineering*, edited by E. A. Warman, pp. 405–417. Amsterdam: North-Holland

YUCHEN, T., and WENHAM, Q. (1986). Solvability of practical 6R manipulator. In *Recent Trends in Robotics: Modeling, Control, and Education*, edited by M. Jamshidi, J.Y.S. Luh and M. Shahinpoor. New York: North-Holland

ZINOVEV, V. A. (1952). *Spatial Mechanisms with Lower Kinematic Pairs*. Moscow: Gostechizdat (in Russian)

Index

Acceleration, 7, 46, 49, 66, 125, 136
 angular, 46, 49
 complete matrix of, 67
 Coriolis, 55
 error, 412
 extended, 68
 matrix, 66
 normal, 47
 reminder, 312
 Resal, 53
 tangential, 47
Accuracy, 405
 kinematic, 406, 428
 dynamic, 422, 432
ADAMS, 301, 391
Algebra modulo 2, 40
Algorithm,
 residual, 318
Analysis, 1, 5, 18
 dynamic, 257
 kinematic, 109
 kinetostatic, 231
 static, 242
 structure, 21
Angles,
 Cardan, 120, 126
 Euler, 115
Approach,
 ADAMS, 301, 391
 analytical, 260, 267
 articulated-body inertia, 260, 321
 Borri's, 358
 composite rigid-body, 260, 265
 DADS, 303, 391
 Geradin's, 259
 Kane, 298
 natural coordinate, 304
 NEWEUL, 290, 391
 MESA VERDE, 294, 391
 recursive, 260, 319
 vector, 260
Articulated-body inertia, 260, 321
Assur's classification, 25
Atan2, 118
Automated,
 analysis, 18
 synthesis, 18
Basic,
 decomposition, 47,67, 114
 matrix, 94
 motion, 94
 quaternion, 94, 103
 screw, 94, 104
Baumgarte stabilisation, 356
Branches,
 of a graph, 36
CAD, 1, 5
 structure of software, 17
Calibration, 420
Cardan angles, 120, 126
Centre of mass, 19, 261, 310
Chain, 22, 24, 129

classification, 22
Chasles theorem, 47
Chebyshev polynomial, 442
Chord,
of graph, 36
Circle, 34
basic, 37
Classification,
of kinematic pairs, 23
of mechanisms, 29
of singular cases, 214
Compartment, 30
method, 178, 260, 346
Composite rigid-body inertia, 311
Computational complexity, 105, 332
Computer programs,
for MBS dynamics, 389
Constrained mechanical systems, 21
Constraint, 21, 169, 340
condition, 165, 169
equations, 171
force, 36
holonomic, 272, 259
nonholonomic, 259, 273
violation stabilization, 356
Control, 455, 464
Coordinate, 22
absolute, 22
composite, 365
dependent, 29, 38, 258
combined, 272
generalized, 260
homogeneous, 60
independent, 22, 258, 361
modal, 381
natural, 167
Plücker, 90
physical, 356
quasi, 277
relative, 22
Coordinate partitioning method, 303
Coriolis,
acceleration, 55
force, 309
Cutset,
basic, 39
D'Alembert's principle, 265
DADS, 303, 391
DAE, 354, 467
solver, 360
Denavit-Hartenberg's notation, 159, 191, 311
Decomposition, 9, 61
basic, 47, 67, 114
Choleski, 333, 289
LU, 215, 358
QR, 358
singular-value (SVD), 217, 358
Degree of freedom, 22, 25
Description, 2, 45
matrix, 45
screw, 88
structural, 31
vector, 46
Design, 5, 10
of mechanism, 10
parameter, 423
process, 6
Diakoptics, 346
Differential-algebraic equations, 254, 467
Differential operator, 95
Differentiation,
of vector, 47
Direction cosine matrix, 56, 311
inverse, 58
Dual,
angle, 91
number, 84
Dymola, 352
Dynamics, 257
approaches, 260
multibody, 269
principles, 260

Edges,
  of a graph, 33
Energy,
  deformation, 379
  kinetic, 267, 378
  potential, 242, 379
Equations of motion, 257, 263, 271
Equilibrium, 235, 240
  position, 242, 466
Equivalence of vector and analytical approaches, 281
Euler,
  angles, 115
  dynamic equation, 233
  kinematic equation, 116
  parameters, 80, 124
  theorem, 47, 124
Feedback control, 458
Finite element method (FEM), 375, 381, 463
Flexible multibody systems, 375, 467
  displacements in, 377
Force, 231
  generalized, 273
  inertia, 233
  moment of, 232
  resultant, 233
Formalism,
  inverse multibody, 318
  matrix, 45
  quaternion, 45
  multibody, 289
  non-recursive, 290
  recursive, 260, 319
Formula,
  Grübler, 25
  MBS, 31
Frame, 22, 27, 29, 31, 133
Friction, 239, 369
Generalized,
  constraint force, 36
  coordinate, 260
  force, 273
  graph, 35
  inertia tenzor, 433
  inverse, 215
  momentum, 270
  vector method, 161
Graph,
  associated, 33, 248
  manipulation language, 42
  oriented, 41
  theory, 32
Grübler-Kutzbach formula, 25
Hamilton's,
  equations, 271
  principle, 269
Hinge point, 309
Holonomic constraint equations, 272
Impact, 372
Incidence matrix, 34
  reduced, 35
Inertia,
  couple, 233, 250
  matrix, 233
  transformations, 264
Inverse,
  generalized of matrix, 215
  matrix, 3, 58, 62
  motion, 68
Jacobian matrix, 177
Jourdain's principle, 268
Kane's method, 298
KIN, 392
Kinematic pair, 21, 23
  chain, 22
  cylindrical, 113
  flat, 128
  helical, 113
  higher, 197
  lower, 109
  planar, 128
  prismatic, 112
  revolute, 111

spherical, 114
general, 198
Kinetostatics, 231
Knowledge engineering, 17
Lagrange's equation, 270
of mixed type, 271
Lagrange multipliers, 335
LEMT, 271
Loop, 22, 32, 37
open, 320
closed, 323
Manipulator, 193
Matrix,
acceleration, 66
basic, 94
complete acceleration, 67
correction, 410
Denavit-Hartenberg's, 159
error, 409
incidence, 34
inverse, 3, 58, 62
Jacobian, 177
notation, 60
of angular acceleration, 67
of angular velocity, 65
of circles, 37
of cutsets, 39
of direction cosines, 57
of inertia, 233
of shape functions, 377
orthogonal, 58
permutation, 101
pseudoinverse, 215
rank of, 37
transformation, 61
transmission, 155
velocity, 63
MBS, 1
description, 270
formula, 31
dynamic analysis, 257
kinematic analysis, 109
with feedback control, 383
with flexible elements, 375
Mechanisms, 1, 5, 29
Altmann, 221
Bennett, 222
composed, 147
design, 5
guiding, 444
mobility, 21, 29, 220
overconstrained, 213
simple, 133
transmission, 438
Mechatronics, 10
MESA VERDE, 394
Method,
adjoint variables, 423, 426
basic matrices, 152
closed loop, 134
compartment, 170
disconnected loop, 135
generalized vector, 161
geometric diagram and parametric, 172, 240
natural coordinates, 167
of successive approximations, 10
MFGR subroutine, 215
Mobility, 21, 25, 220
Model, 14
accuracy, 406
computational, 14
descriptive, 14
hierarchical structure of, 14
mathematical, 19, 270
mechanical, 18
metric, 15
of design proces, 6
operational, 14
topological, 15
Moment,
of inertia, 262
of force, 232
of momentum, 262

Momentum, 261
Motion,
basic, 94
inverse, 68
complex, 46
planar, 107
revolute, 46
spatial, 46
spherical, 47, 114
simultaneous, 51, 70
translational, 46
Motor, 90, 232
Multibody system, 1, 21
computer programs, 391
NEWEUL, 290
Newton method, 177
modified, 179
Newton-Euler method, 260
Nodes of a graph, 33
Numerical methods, 177
nonlinear algebraic equations, 177
numerical integration, 353
ODE, 354
ODAM, 399
OKAM2, 392
Optimization, 435
Orientation,
of a body, 115
Palette of tools, 16
Parallel computation, 464
Path generation, 444
Permutation matrix, 101
Pitch, 123
Position,
equilibrium, 242
of a rigid body, 61
Principle,
d'Alembert's, 265
Gauss's, 269
Hamilton's, 269
Jourdain's, 268
of CAD, 14
of transference, 90
of virtual work, 268
Projection, 75
method, 363
Quasi-coordinate, 77
Quaternion, 78
basic, 103
conjugated, 78
dual, 82
multiplication, 78
norm, 79
shortened dual, 84
Radius-vector, 47
Reaction force, 233, 340
Recursive computation, 260, 319
Redundancy, 182
Reference,
frame, 22
point, 47
Reflection, 77
Relative,
coordinates, 22
motion, 22
Residual algorithm, 318
Robot, 2, 130, 149, 182, 186, 194, 211, 429
accuracy, 406, 428
analytical solution, 182
control, 455, 464
inverse kinematics, 182, 422
inverse dynamics, 351, 476
multibody analyses, 135, 156
non-simple, 422
parallel, 149
repeatibility, 420
simple, 422
Rodrigues,
formula, 93
parameter, 128
Roll, 123
Rotation,
about a point, 48

about an axis, 46
Screw, 88
affinor, 92
axis, 91
basic, 104
Sensitivity of MBS, 405, 423, 468
Simulation, 14
Singular case, 178
Singular-value decomposition, 217
Skelet,
of a graph, 36
Solution,
analytical, 182
numerical, 177
Space,
null, 366
tolerance, 416
Spring,
potential energy of, 242
Spherical,
motion, 107
pair, 114
State variables, 457
Stiction, 368
Structure,
analysis, 21
description, 31
change of, 368
of model, 14
Subroutine,
DE, 354
DIFSUB, 354
MFGR, 215
Synthesis, 435
dynamic, 452
kinematic, 439
parameter, 20
System,
groups, 22, 25, 28
topology, 32
Tensor of inertia, 233
Theorem,
Chasles, 47
Euler, 47
Hamilton, 94
Steiner, 265
Torsional springs and dampers, 254
Transformation, 75
matrix, 61
of models, 16
Translation, 46
Transmission,
angle, 437
index, 437
Tree configuration, 36
Triple vector product identities, 60
Turbula, 224
Universal joint, 138
Variables,
adjoint, 425
help, 291
Vector,
description, 46
product, 59
Velocity, 48
absolute, 54, 124
angular, 46
matrix, 63
partial, 298
relative, 54
translational, 48
Virtual
displacement, 268
velocity, 268
work, 260, 268
Work,
virtual, 260, 268
Yaw, 123